D1693715

T. Beth, G. Leuchs (Eds.)

Quantum Information Processing

2., revised and enlarged Edition

Quantum Information Processing

Second, revised and enlarged Edition

Edited by
Thomas Beth, Gerd Leuchs

WILEY-VCH Verlag GmbH & Co. KGaA

Editors

Prof. Dr.-Ing. Thomas Beth
Universität Karlsruhe
Fakultät für Informatik
eiss_office@ira.uka.de

Prof. Dr. Gerd Leuchs
Universität Erlangen
Institut für Optik, Information und Photonik
leuchs@physik.uni-erlangen.de

1st Edition 2003
2nd Edition 2005

Cover Picture
Main: Poincaré sphere showing quantum polarisation states
Courtesy of Christine Silberhorn and Joel Heersink
Lower right: Diagram for a controlled NOT-gate

Library of Congress Card No.: Applied for

British Library Cataloging-in-Publication Data:
A catalogue record for this book is available from the British Library

Bibliographic information published by Die Deutsche Bibliothek

Die Deutsche Bibliothek lists this publication in the Deutsche Nationalbibliografie; detailed bibliographic data is available in the Internet at <http://dnb.ddb.de>.

Printed in the Federal Republic of Germany

Printed on acid-free paper

Printing Strauss GmbH, Mörlenbach
Bookbinding Litges & Dopf Buchbinderei GmbH, Heppenheim

ISBN-13: 978-3-527-40541-1
ISBN-10: 3-527-40541-0

Contents

Preface to the First Edition

The Senate of the German Research Council (Deutsche Forschungsgemeinschaft) has set up a Focused Research Program "Quantum Information Processing" in July 1998. The Focused Program was jointly initiated by Th. Beth, G. Leuchs, W. Mathis, and W. Schleich. The present volume surveys the results of this work during the years 1999–2002.

The main thrive of this Focused Program is the research of the foundations of quantum information processing by means of controlled manipulation of entangled states. First experiments show that one can generate entangled states in a controlled way and manipulate them to a certain extent. This progress was made possible due to the recently developing unique worldwide synergy between theoretical and experimental physics, computer sciences, telecommunications and mathematics in the field of so called quantum information and physics of computation founded by Feynman.

The goal of this interdisciplinary Focused Program "Quantum Information Processing" is the systematic investigation of quantum systems aimed at their exact theoretical modelling and experimental manipulation, at tests of the foundations of quantum information and quantum physics and at the applications in the computer sciences, telecommunication, cryptography and high-precision measurement, quantum-error control and switching technologies. The main goal is meant to be not the examination of systems using quantum mechanical methods, as for example in the case of spectroscopy, but merely the controlled manipulation and exploitation of entangled states.

The development of some fields of modern physics leads to the possibility to isolate and to control quantum phenomena ever better. The transfer of the properties of a quantum state over a silica fibre, the high-precision backaction-evading, partially even interaction- free, measurements and the "computation" with quantum state are examples for that. All this is based on the principles of the state superposition and entanglement of various systems. The dream of a successful solution of exponentially difficult computational problems, like optimization or pattern identification could be thus fulfilled, despite of limited material hardware resources. According to general considerations these problems cannot be handled by means of classical computers with reasonable success chances.

Entangled states have played an important role in the attempt to understand quantum phenomena in Gedanken-experiments already at the beginning of quantum physics. This topic has surfaced again due to novel possibilities for the actual implementation of these Gedanken-experiments by means of modern physics. The new experiments on the foundations of physics provide more and more demanding tests of the quantum theory and the question of non-locality. In the last 7 years new surprising effects were discovered in this field. A spectacular example of such effects is teleportation making use of the non-locality of entangled states to

Quantum Information Processing, 2nd Edition. Edited by Thomas Beth and Gerd Leuchs

ISBN: 3-527-40541-0

transfer the properties of a quantum state or the possibility of a truly exponential speed-up of quantum algorithms over classical algorithms.

Apart from this, there are some recent suggestions to employ the specific features of quantum information for technological applications. Such as quantum key distribution and the use of multipartite entanglement in quantum communication protocols. Recent discoveries on the information-theoretical side suggest that new surprising applications be provided by such technologies still to be invented – exceeding the state of today's Quantum Key Exchange apparatus.

One main goal is to construct a highly parallel computer which works according to a completely new paradigm and which in all probability can solve problems, which are not efficiently solvable with a conventional computer. Further applications open up in optical communication and in cryptographic key distribution. The exploitation of the entanglement to be studied here is developing into an important concept. It is the basis for the effects of acceleration and precision predicted by theory. It is hoped that the results and perception gained from the studies of pure demonstration systems in the frame of this program will help to increase the know-how required towards building future industrial products such as a quantum processing machine.

Delimitation from other Quantum Mechanical Problems: The further optimization of the up-to-date CMOS-Technology is expected to be completed in fifteen years at the latest. All the alternatives to CMOS-Technology (e.g. single electron transistor (SET), resonant tunneling devices (RTD), rapid single flux quantum devices (RSFQ), molecular nanoelectronics (ME) and spin-devices/magneto-electronics (SD)) recognizable today are based on quantum phenomena. However they do not make use of the entanglement principle. This eliminates the advantage of the exponential gain via scalable entangled pure quantum states provided by quantum information processing.

In a similar way the goals of the quantum information processing differ from the methods of NMR- and laser spectroscopy. The main goal of spectroscopy is to examine an unknown system (for example a molecule or a solid) making use of quantum mechanical phenomena. Quantum information processing in contrast abstracts first from the system itself. The subject of this Focused Program is the research and controlled exploitation of the fundamental quantum mechanical effects in pure, experimentally realizable systems, keeping in mind the technical feasibility. All the applications have been judged accordingly. General experimental or theoretical investigations to e.g. the foundations of quantum mechanics without direct connection are not part of the focused program.

The contributed chapters in this volume are grouped according to the following list of topics:

- quantum algorithms and modelling and construction of elementary quantum logic elements and gates
- theoretical studies of the exploitation of quantum information and entanglement, as well as of their quantification
- analysis and control of the decoherence in quantum systems
- experimental realization of the controlled entangled quantum states in as pure as possible, well modelled systems
- quantum communication and cryptography

The most recent results of the Schwerpunktprogramm can be found by calling the home page http://www.quiv.de.

On behalf of all participants of the Schwerpunktprogramm we gratefully acknowledge the long term funding of this research project by the Deutsche Forschungsgemeinschaft (DFG). We are especially indebted to Dr. A. Szillinsky, Dr. A. Engelke, Dr. K. Wefelmeier, and Dr. S. Krückeberg for their guidance during the development and the pursuit of this research programme. For the first renewal of the research projects there was an unforeseen large number of applications. We are most grateful to the Bundesministerium für Bildung und Forschung for helping with additional funding through its VDI-office and we especially thank Dr. M. Böltau for his support.

We express our sincere thanks to Gerlinde Gardavsky, Dr. Markus Grassl, Priv. Doz. Dr. Natalia Korolkova, Christoph Marquardt, Dr. Martin Rötteler, Jessica Schneider, and Robert Zeier for their help during the preparation of this volume.

Thomas Beth and Gerd Leuchs

October 2002

Preface to the Second Edition

Since the publication of the first edition the results extensively described there have become established knowledge in Quantum Information Processing. Beyond this, new directions of research have evolved from amongst these areas. Thus in addition to revised chapters of the first edition, this second edition contains the additional new chapters 4, 12, 13, 24, 27–30 and 32.

The editors are pleased to note that the support for quantum information has been increasing during these last two years within Germany. The Deutsche Forschungsgemeinschaft is now funding a Graduiertenkolleg in Dortmund and a Sonderforschungsbereich in München. In addition several of the groups contributing to this volume received funding from the project "Quantum Information Highway A8" which is jointly sponsored by the Landesstiftung Baden-Würtemberg GmbH and the Bayerisches Staatsministerium für Forschung, Wissenschaft und Kunst. Our community is most grateful for this support and the present volume displays a cross section of the funded work.

Gerd Leuchs and Thomas Beth

November 2004

List of Contributors

- K. Abich, p. 251
  abich@physnet.uni-hamburg.de
- R. Ahlswede, p. 70
  ahlswede@mathematik.uni-bielefeld.de
- G. Alber, p. 14
  gernot.alber@physik.tu-darmstadt.de
- W. Alt, p. 265
  w.alt@iap.uni-bonn.de
- U. Andersen, p. 425
  andersen@kerr.physik.uni-erlangen.de
- Ch. Balzer, p. 237, p. 251
  balzerc@physnet.uni-hamburg.de
- H. Becker, p. 418
  e-mail: see H. Roskos
- Th. Beth, p. 1
  eiss_office@ira.uka.de
- G. Birkl, p. 287
  birkl@iqo.uni-hannover.de
- R.H. Blick, p. 338
  blick@eceserv0.ece.wisc.edu
- D. Born, p. 327
  e-mail: see W. Krech
- M. Bourennane, p. 393
  m.bourennane@mpq.mpg.de
- H.J. Briegel, p. 28
  hans.briegel@uibk.ac.at
- D. Bruß, p. 83
  bruss@thphy.uni-duesseldorf.de
- T.C. Bschorr, p. 113
  thorsten.bschorr@physik.uni-ulm.de
- F.B.J. Buchkremer, p. 287
  email: see G. Birkl
- N. Cai, p. 70
  cai@mathematik.uni-bielefeld.de
- G. Cennini, p. 275
  cennini@pit.physik.uni-tuebingen.de
- M. Curty, p. 44
  curty@kerr.physik.uni-erlangen.de
- A. Delgado, p. 14
  delgado@tangelo.phys.unm.edu
- R. de Vivie–Riedle, p. 312
  rdvpc@cup.uni-muenchen.de
- Ch. DiFidio, p. 198
  e-mail: see W. Vogel
- W. Dultz, p. 418
  e-mail: see H. Roskos
- R. Dumke, p. 287
  e-mail: see W. Ertmer
- K. Eckert, p. 83
  Kai.Eckert@itp.uni-hannover.de
- M. Eibl, p. 393
  m.eibl@mpq.mpg.de

Quantum Information Processing, 2nd Edition. Edited by Thomas Beth and Gerd Leuchs

ISBN: 3-527-40541-0

- W. Ertmer, p. 287
  wolfgang.ertmer@iqo.uni-hannover.de
- A.F. Fahmy, p. 58
  amr@deas.harvard.edu
- D.G. Fischer, p. 113
  dietmar.g.fischer@siemens.com
- R. Folman, p. 298
  folman@bgu.ac.il
- M. Freyberger, p. 113
  matthias.freyberger@physik.uni-ulm.de
- S. Gaertner, p. 393
  s.gaertner@mpq.mpg.de
- C. Geckeler, p. 275
  geckeler@pit.physik.uni-tuebingen.de
- S.J. Glaser, p. 58
  glaser@ch.tum.de
- O. Glöckl, p. 425
  gloeckl@kerr.physik.uni-erlangen.de
- V. Gomer, p. 265
  gomer@pbh.de
- M. Grajcar, p. 327
  grajcar@smbh.uniba.sk
- M. Grassl, p. 1
  grassl@ira.uka.de
- O. Gühne, p.44, 83
  otfried.guehne@uibk.ac.at
- A. Haase, p. 298
  haase@physi.uni-heidelberg.de
- P. Hänggi, p. 186
  Hanggi@Physik.Uni-Augsburg.DE
- Th. Hannemann, p. 237, p. 251
  hannemann@physnet.uni-hamburg.de
- K. Hayasaka, p. 209
  Hayasaka@crl.go.jp
- J. Heersink, p. 425
  heersink@kerr.physik.uni-erlangen.de
- M. Hennrich, p. 223
  markus.hennrich@icfo.es
- A.W. Holleitner, p. 338
  aholleitner@iquest.ucsb.edu
- T. Hübner, p. 327
  uwe.huebner@ipht-jena.de
- F. Hulpke, p. 83
  Florian.Hulpke@itp.uni-hannover.de
- A.K. Hüttel, p. 338
  andreas@akhuettel.de
- P. Hyllus, p. 83
  hyllus@itp.uni-hannover.de
- D. Janzing, p. 1
  janzing@ira.uka.de
- F. Jelezko, p. 150
  f.jelezko@physik.uni-stuttgart.de
- A. Kemp, p. 162
  kemp@physik.uni-erlangen.de
- M. Keller, p. 209
  Matthias.Keller@mpq.mpg.de
- N. Khaneja, p. 58
  navin@hrl.harvard.edu
- N. Kiesel, p. 393
  nikolai.kiesel@gmx.de
- S. Kohler, p. 186
  Sigmund.Kohler@Physik.Uni-Augsburg.DE
- J. Korbicz, p. 83
  Jarek.Korbicz@itp.uni-hannover.de
- N. Korolkova, p. 405, 425
  nvk@st-andrews.ac.uk

- W. Krech, p. 327
  wolfram.krech@uni-jena.de
- P. Krüger, p. 298
  peter.krueger@physi.uni-heidelberg.de
- A. Kuhn, p. 223
  Axel.Kuhn@mpq.mpg.de
- S. Kuhr, p. 265
  kuhr@iap.uni-bonn.de
- C. Kurtsiefer, p. 393
  Christian_kurtsiefer@nus.edu.sg
- B. Lange, p. 209
  e-mail: see M. Keller
- W. Lange, p. 209
  W.Lange@sussex.ac.uk
- G. Leuchs, p. 425
  leuchs@physik.uni-erlangen.de
- M. Lewenstein, p. 44, 83
  lewen@itp.uni-hannover.de
- H. Lorenz, p. 338
  bert.lorenz@physik.uni-muenchen.de
- S. Lorenz, p. 425
  lorenz@kerr.physik.uni-erlangen.de
- N. Lütkenhaus, p. 44
  luetkenhaus@kerr.physik.uni-erlangen.de
- H. Mack, p. 113
  holger.mack@physik.uni-ulm.de
- G. Mahler, p. 135
  mahler@theo1.physik.uni-stuttgart.de
- C.S. Maierle, p. 125
  chris@e3.physik.uni-dortmund.de
- C. Marquardt, p. 425
  marquardt@kerr.physik.uni-erlangen.de
- W. Martienssen, p. 418
  Martienssen@Physik.uni-frankfurt.de
- R. Marx, p. 58
  Raimund.Marx@ch.tum.de
- W. Mathis, p. 353
  mathis@tet.uni-hannover.de
- D. Meschede, p. 265
  meschede@iap.uni-bonn.de
- M. Michel, p. 135
  mathias@theo1.physik.uni-stuttgart.de
- M. Mihalik, p. 327
  e-mail: see W. Krech
- F. Mintert, p. 251
  e-mail: see P.E. Toschek
- J. Mompart, p. 83
  jordi.mompart@uab.es
- M. Mussinger, p. 14
  michael.mussinger@physik.uni-ulm.de
- T. Müther, p. 287
  tmuether@gmx.de
- J.M. Myers, p. 58
  myers@deas.harvard.edu
- W. Neuhauser, p. 237, p. 251
  neuhauser@physnet.uni-hamburg.de
- L. Pescini, p. 338
  laura.pescini@physik.uni-muenchen.de
- R. Raussendorf, p. 28
  raussen@cs.caltech.edu
- T. Reiss, p. 58
  Timo.Reiss@ch.tum.de
- D. Reiß, p. 237, p. 251
  e-mail: see P.E. Toschek
- G. Rempe, p. 223
  Gerhard.Rempe@mpq.mpg.de

- G. Ritt, p. 275
  ritt@pit.physik.uni-tuebingen.de
- H. Roskos, p. 418
  roskos@physik.uni-frankfurt.de
- M. Rötteler, p. 1
  mroetteler@uwaterloo.ca
- A. Sanpera, p. 83
  Anna.Sanpera@itp.uni-hannover.de
- S. Scheel, p. 100, 382
  s.scheel@imperial.ac.uk
- R. Scheunemann, p. 275
  e-mail: see M. Weitz
- W.P. Schleich, p. 113
  wolfgang.schleich@physik.uni-ulm.de
- H. Schmidt, p. 135
  harry@theo1.physik.uni-stuttgart.de
- K. Schmid, p. 418
  schmid@stud.uni-frankfurt.de
- J. Schmiedmayer, p. 298
  joerg.schmiedmayer@physi.uni-heidelberg.de
- D. Schrader, p. 265
  schrader@iap.uni-bonn.de
- T. Schulte–Herbrüggen, p. 58
  Tosh@ch.tum.de
- Ch. Silberhorn, p. 425
  csilberhorn@optik.uni-erlangen.de
- D. Suter, p. 125
  Dieter.Suter@physik.uni-dortmund.de
- C.M. Tesch, p. 312
  carmen.tesch@cup.uni-muenchen.de
- F. Tonner, p. 135
  tonner@theo1.physik.uni-stuttgart.de
- P.E. Toschek, p. 237, p. 251
  toschek@physnet.uni-hamburg.de
- U. Troppmann, p. 312
  ulrike.troppmann@cup.uni-muenchen.de
- A.V. Ustinov, p. 162
  ustinov@physik.uni-erlangen.de
- W. Vogel, p. 198
  werner.vogel@physik.uni-rostock.de
- M. Volk, p. 287
  volk@iqo.uni-hannover.de
- T. Wagner, p. 327
  thomas.wagner@ipht-jena.de
- A. Wallraff, p. 162
  andreas.wallraff@yale.edu
- H. Walther, p. 209
  Herbert.Walther@mpq.mpg.de
- H. Weinfurter, p. 393
  harald.weinfurter@physik.uni-muenchen.de
- M. Weitz, p. 275
  mweitz@pit.physik.uni-tuebingen.de
- D.-G. Welsch, p. 100
  D.-G.Welsch@tpi.uni-jena.de
- P. Wocjan, p. 1
  wocjan@ira.uka.de
- J. Wrachtrup, p. 150
  j.wrachtrup@physik.uni-stuttgart.de
- Ch. Wunderlich, p. 237, p. 251
  wunderlich@physnet.uni-hamburg.de
- R. Zeier, p. 1
  zeier@ira.uka.de
- X.-B. Zou, p. 353
  zou@tet.uni-hannover.de
- M. Zukowski, p. 393
  fizmz@univ.gda.pl

1 Algorithms for Quantum Systems — Quantum Algorithms

Th. Beth, M. Grassl, D. Janzing, M. Rötteler, P. Wocjan, and R. Zeier

Institut für Algorithmen und Kognitive Systeme
Universität Karlsruhe
Germany

1.1 Introduction

Since the presentation of polynomial time quantum algorithms for discrete log and factoring [Sho94] it is generally accepted that quantum computers may—at least for some problems—outperform classical ones. But the field of quantum information processing does not only have implications for computational problems. Using quantum mechanical systems for information processing naturally leads to the problem of finding efficient ways to control quantum mechanical systems.

Here we address several algorithmic aspects of quantum information processing in different areas. First, the state of the art of quantum signal transforms which are at the core of a huge class of quantum algorithms, namely hidden subgroup problems, is presented. Second, we discuss aspects of quantum error-correcting codes, in particular an interesting view on stabilizer codes which relates them to simple interaction Hamiltonians. The efficient implementation of unitary operations by given Hamiltonians is investigated next. Finally, results on the simulation of one quantum mechanical system by another are discussed.

1.2 Fast Quantum Signal Transforms

A basic task in classical signal processing is to find fast algorithms which compute the matrix-vector-product of a given transformation with an arbitrary input vector. Suppose that the input is a vector of length N with complex entries. A transformation is said to have a *fast* algorithm if the number of arithmetic operations needed to compute the matrix vector product—i. e., the number of additions and multiplications—is bounded by $O(N \log^c N)$, for some constant c. Amongst the most useful algorithms in computer science, physics, and engineering is the discrete Fourier transform DFT_N which is given by the unitary matrix

$$F_N := \frac{1}{\sqrt{N}} \cdot \left[\omega^{i \cdot j}\right]_{i,j=0,\ldots,N-1},$$

where $\omega = e^{2\pi i/N}$ denotes a primitive N-th root of unity. The computational complexity of computing the product $F_N \cdot x$ for an input vector $x \in \mathbb{C}^N$ is $O(N \log N)$.

In quantum computing the unitary transformation F_N is used in the following way. Suppose that a system consisting of n qubits holds a normalized state vector $|\psi\rangle = \sum_{i=1}^{N} x_i |i\rangle \in$

Quantum Information Processing, 2nd Edition. Edited by Thomas Beth and Gerd Leuchs

ISBN: 3-527-40541-0

$\mathbb{C}^N$, where $N := 2^n$. Note that here the information is encoded into the amplitudes of the basis states. To this state the unitary transformation F_N can be applied resulting in a state $F_N|\psi\rangle$. Unlike the situation in classical signal processing the components of $|\psi\rangle$ and $F_N|\psi\rangle$ are not directly accessible; they merely can be extracted by (POVM) measurements. However, for feature detecting purposes like the location of spectral peaks this approach is well-suited.

A quantum Fourier transform can be computed using $O(\log^2 N)$ elementary operations. This exponential speed-up compared to the classical complexity of the DFT, which surprisingly enough is obtained by a direct adaptation of the classic Cooley-Tukey algorithm to the quantum circuit model, is an essential indication for the power of quantum computing. This becomes manifest in the fact that the ability to compute a DFT in polylogarithmic time is the backbone of Shor's algorithms for factoring and computing the discrete logarithm [Sho94].

A natural question is whether other signal transforms which have desirable feature extraction properties in classical signal processing can be used for quantum algorithms. In a series of works [PRB99, RPB99, KR00, ABH+01, KR01, Röt02] it has been shown that many well-known signal transforms allow highly efficient realizations on a quantum computer. In particular the following classes of unitary transformations have been shown to be efficiently implementable on a quantum computer:

- Discrete Fourier transforms for finite abelian groups [ABH+01, Röt02].
- Generalized Fourier transforms for
  - 2-groups with maximal cyclic normal subgroup [PRB99, Röt02].
  - wreath products of the form $G = A \wr Z_2$, where A is abelian [RB98, Röt02].
  - Heisenberg groups over the finite fields $\mathbb{F}_{2^n}$ [Röt02].
- Discrete cosine and sine transforms of types I, II, III, and IV [RB99, KR01, Röt02].
- Discrete Hartley transforms [KR00, Röt02].

More precisely, we have shown that for discrete cosine transforms, discrete sine transforms, and discrete Hartley transforms $O(\log^2 N)$ elementary quantum gates are sufficient to implement any of those transforms for input sequences of length N. The Fourier transforms for finite groups which have been mentioned above have the same computational complexity, except for the Heisenberg group where an implementation using $O(\log^3 N)$ gates has been found.

The underlying theory which allows to find efficient factorizations in the above mentioned cases relies on two techniques which have been developed at the Institut für Algorithmen und Kognitive Systeme. The first technique is the so-called method of symmetry-based matrix factorizations. Here a matrix $M \in \mathbb{C}^{n \times n}$ is said to have symmetry (G, ϕ, ψ) if ϕ and ψ are representations of a finite group G and furthermore

$$\phi(g) \cdot M = M \cdot \psi(g), \quad \text{for all } g \in G.$$

The importance of symmetry in connection with generalized Fourier transforms was first recognized in the work of Th. Beth [Bet84]. An important feature of this approach is that classical fast algorithms can be explained—and automatically derived—in terms of symmetry of matrices [Min93, Egn97, Püs98].

In the thesis of M. Rötteler [Röt02] symmetry-based matrix factorizations have been taken as a starting point for further optimizations. This has led to efficient implementations of several generalized Fourier transforms, trigonometric transforms, and the Hartley transform.

A second technique has been developed by A. Klappenecker and M. Rötteler and is described in [Röt02]and [KR03]. The basic idea is to reuse previously found factorizations for the construction of higher level operations. A prime example is given by the problem of implementing functions of unitary transformations, i. e., operations of the form $f(U)$, where U is a unitary transformation of finite order and $f(U)$ is also unitary.

1.3 Quantum Error-correcting Codes

Owing to the high sensitivity of quantum mechanical systems to even small perturbations, means of error protection are essential for any computation or communication process based on quantum mechanics. A general theory of quantum error-correcting codes (QECC) has been developed (see, e. g., [KL97]), but many algorithmic aspects are still open. The main tasks are to find methods for constructing good QECC and efficient algorithms for encoding and decoding, including the correction of the errors.

A large family of QECC can be derived from cyclic codes (see [GB99, GGB99, GB00]). Those QECC admit various techniques for encoding and decoding, e. g., based on Fourier transformations over finite fields and quantum shift registers [BG00]. An interesting new class of QECC has been developed in cooperation with the group of G. Alber [ABC+01, AMD03]. Those so-called jump codes exploit side-information about the errors due to the emission of quanta (quantum jumps). The side-information is obtained by continuously monitoring the quantum system and recording which subsystem emitted, e. g., a photon. An interesting connection to design theory leads to various constructions of jump codes [BCG+03].

Another new concept of QECC linking various groups in the *Schwerpunktprogramm QIV* are so-called graph codes which have been introduced by D. Schlingemann and R. F. Werner [SW02]. Similar to the cluster states used in the one-way quantum computer of R. Raussendorf and H. J. Briegel [RB01], the basis states of a graph code are defined via an interaction graph corresponding to the spin–spin coupling Hamiltonian which can be used for encoding. Compared to the standard gate model of quantum computation, such a coupling Hamiltonian yields a very efficient—intrinsically parallel—algorithm.

We have shown that the concepts of graph codes and that of so-called stabilizer codes are equivalent, i. e., any graph code is a stabilizer code and vice versa [GKR02, Sch02, WSR+02]. In the following, we will present the main ideas used in the proof.

Starting from an α-dimensional Hilbert space $\mathcal{H}$, a graph code is an α^k-dimensional subspace Q of $\mathcal{H}^{\otimes n}$ which is spanned by the vectors

$$|\underline{x}\rangle = \frac{1}{\sqrt{\alpha^n}} \sum_{y \in \mathbb{F}_{p^m}^n} \Big(\prod_{j=1}^{k+n} \prod_{i=1}^{j-1} \chi(z_i, z_j)^{\Gamma_{ij}} \Big) |y\rangle. \tag{1.1}$$

Here the dimension α is a prime power, i. e., $\alpha = p^m$ for some prime number p. The basis states $|y\rangle$ of $\mathcal{H}^{\otimes n}$ are labeled by vectors $y \in \mathbb{F}_{p^m}^n$ over the finite field $\mathbb{F}_{p^m}$, and the encoded states $|\underline{x}\rangle$ are labeled by vectors $x \in \mathbb{F}_{p^m}^k$. The coefficients on the right hand side are given by the values of a non-degenerate symmetric bicharacter χ on $\mathbb{F}_{p^m} \times \mathbb{F}_{p^m}$, where $z = (x, y)$. Finally, the exponents Γ_{ij} are given by the adjacency matrix Γ of a weighted undirected graph with integral weights (corresponding to the coupling strength between the subsystems i and j). Equivalently, the states (1.1) of the graph code Q can be expressed in the form

$$|\underline{x}\rangle = \frac{1}{\sqrt{\alpha^n}} \sum_{y \in \mathbb{F}_{p^m}^n} \zeta(q(x+y))|y\rangle, \tag{1.2}$$

where ζ is a non-trivial additive character of $\mathbb{F}_p$ and q is a quadratic form on $\mathbb{F}_{p^m}^{k+n} \cong \mathbb{F}_p^{m(k+n)}$. This isomorphism shows that it is sufficient to consider graphical quantum codes which are defined over prime fields.

First we show that a graph code defined over the finite field $\mathbb{F}_p$ is a stabilizer code, i. e., a joint eigenspace of an abelian subgroup of the error group G generated by

$$X^a := \sum_{y \in \mathbb{F}_p^n} |y + a\rangle\langle y|, \quad \text{and } Z^d := \sum_{y \in \mathbb{F}_p^n} \exp(2\pi i/p)^{d^t y} |y\rangle\langle y|,$$

for $a, d \in \mathbb{F}_p^n$ (cf. [KR02a, KR02b]). Starting from (1.2) one can show that the stabilizer of Q is given by

$$S_Q = \{\exp(2\pi i/p)^{q(a)} X^a Z^{aM_y} \mid a \in \mathbb{F}_p^n \text{ such that } Ba = 0\}.$$

The matrices $M_y \in \mathbb{F}_p^{n \times n}$ and $B \in \mathbb{F}_p^{k \times n}$ are submatrices of the adjacency matrix Γ defining the quadratic form q which can be written as

$$\Gamma = \left(\begin{array}{c|c} M_x & B \\ B^t & M_y \end{array} \right).$$

Additionally, the orthogonal projection onto the joint eigenspace for the eigenvalue 1 of the operators in S_Q coincides with Q, showing that Q indeed is a stabilizer code.

In order to show that any stabilizer code defined over the finite field $\mathbb{F}_{p^m}$ can be realized as a graph code, we first note that those stabilizer codes can equivalently be considered as stabilizer codes over $\mathbb{F}_p$ (see [AK01]). To any stabilizer code corresponds a (classical) symplectic code $\mathcal{C}$ over $\mathbb{F}_{p^2}$. The generator matrix of $\mathcal{C}$ can be written as $(X|Z)$ where $X, Z \in \mathbb{F}_p^{(n-k) \times n}$ and $XZ^t - ZX^t = 0$. As the code $\mathcal{C}$ is self-orthogonal, i. e., contained in the orthogonal code $\mathcal{C}^\perp$ with respect to the symplectic inner product on $\mathbb{F}_{p^2}^n$, there exists a self-dual code $\mathcal{D}$ with $\mathcal{C} \subseteq \mathcal{D} = \mathcal{D}^\perp \subseteq \mathcal{C}^\perp$. We can choose a generator matrix for $\mathcal{D}$ of the form

$$G' := (X'|Z') = \left(\begin{array}{c|c} X & Z \\ \widetilde{X} & \widetilde{Z} \end{array} \right).$$

Similar to Gauß' algorithm, the matrix G' can be transformed into standard form $(\, I \mid C)$ where C is symmetric [Gra02a, Gra02b]. Quantum mechanically, the transformations correspond to local unitary operations which do not change the error-correcting properties of the QECC. Applying the same transformations to the subcode $\mathcal{C}$ of $\mathcal{D}$, we obtain an equivalent code whose generator matrix can be written as $(D|DC)$ where $D \in \mathbb{F}^{(n-k)\times n}$. Finally, we obtain the symmetric matrix

$$\Gamma := \left(\begin{array}{c|c} 0 & B \\ \hline B^t & C \end{array}\right),$$

where $DB^t = 0$. Using this matrix Γ as adjacency matrix in (1.1), we obtain a graph code which is equivalent to the stabilizer code corresponding to $\mathcal{C}$. Hence, for any stabilizer code over $\mathbb{F}_{p^m}$ there exists an equivalent graph code.

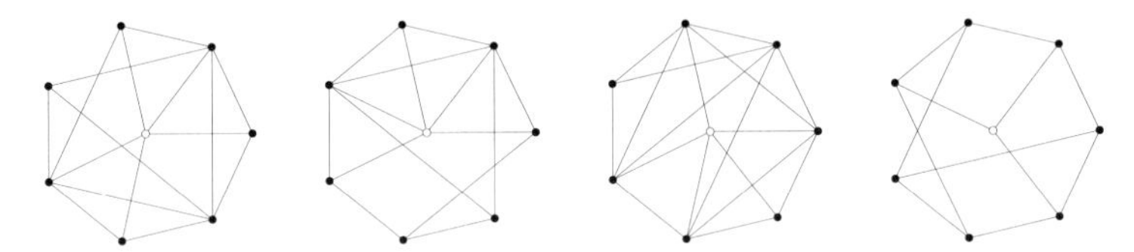

Figure 1.1: Non-isomorphic graphs which all yield graphical quantum codes that are equivalent to the CSS code $[\![7,1,3]\!]$.

In general, the corresponding graph is not uniquely determined, as illustrated in Fig. 1.1. All four graphs yield equivalent QECC, but the graphs are non-isomorphic. Therefore it is not straightforward to relate basic properties of the graph and the error-correcting properties of the resulting graph code. But the possibility to choose among different graphs results in different interaction Hamiltonians which may yield more efficient algorithms for encoding (cf. Section 1.5).

1.4 Efficient Decomposition of Quantum Operations into Given One-parameter Groups

The time evolution of all quantum mechanical systems is governed by the Schrödinger equation. To control the time evolution of a concrete physical implementation, we apply different Hamilton operators which are supposed to be time independent. This gives us the ability to implement computational operators on the state space. Equipped with m one-parameter groups

$$\{t \mapsto \exp(-i\mathrm{H}_1 t), t \mapsto \exp(-i\mathrm{H}_2 t), \ldots, t \mapsto \exp(-i\mathrm{H}_m t)\}$$

generated by the different Hamilton operators H_j, the goal is to build up each unitary U as a product of elements of the given one-parameter groups:

$$\mathrm{U} = \prod_{j=1}^{l} \exp(-i\mathrm{H}_{p_j} t_{p_j}), \quad \text{where } \mathrm{H}_{p_j} \in \{\mathrm{H}_1, \mathrm{H}_2, \ldots, \mathrm{H}_m\} \text{ and } t_{p_j} \in \mathbb{R}, t_{p_j} > 0.$$

As an efficiency measure we consider the minimal number of terms needed to express each unitary as a product of the given one-parameter groups. This efficiency measure is called the order of generation [Low71]. In a physical implementation of the given unitary operator this corresponds to the number of switches between different Hamilton operators.

As a first step we derive conditions on the set of Hamilton operators to be universal, i.e., to be able to generate each unitary. From the work of Chow [Cho39] follows the Lie algebra rank condition which states that each unitary can be implemented if the rank of the subalgebra generated by the given Hamilton operators equals the dimension of the Lie algebra belonging to the considered unitary operators. The work of Jurdjevic and Sussmann [JS72] states in addition that the order of generation is finite if the Lie algebra rank condition is fulfilled. To exclude—in the case of compact and connected Lie groups—the possibility of an infinite order of generation in the closure the work in [D'A02] can be used.

We emphasize that the order of generation depends heavily on the given one-parameter groups. The problem of determining bounds on the order of generation in the case of up to three Hamilton operators from the $\mathrm{su}(2)$ was considered in [ZB02]. Describing Hamilton operators as infinitesimal rotations of the Bloch sphere we can depict the application of an Hamilton operator in the complex plane after a stereographic projection from the Bloch sphere to the one-point compactification of the complex numbers. It is known [Gau19] that under stereographic projection the rotations of the sphere correspond to rotational Möbius transformations:

$$z \mapsto \frac{az - \bar{c}}{cz - \bar{a}}, \quad \text{where } a\bar{a} + c\bar{c} = 1.$$

The orbit of a rotational Möbius transformation is an Apollonian circle in the complex plane.

Beginning with the case of two Hamilton operators—we suppose that the Lie algebra rank condition is fulfilled—we follow [Low71] in order to derive bounds on the order of generation. Since Hamilton operators are identified with infinitesimal rotations of the Bloch sphere, we assume without loss of generality that one rotation axis is normalized to the z-axis and the other rotation axis lies in the x-z-plane. The orbits of an Hamilton operator correspond to a system of Apollonian circles in the complex plane. We track the alternating application of the two Hamilton operators on the state space starting from zero in the complex plane, which corresponds to the south pole of the Bloch sphere. Going from one circle to a tangent one we use a greedy-type algorithm to cover all of the complex plane. Figure 1.2 shows tangent circles corresponding to the alternating application of Hamilton operators to the state space visualized in the complex plane. One further step in the alternating application of Hamilton operators is necessary and sufficient to cover all of the complex plane in Figure 1.2.

By an analysis of figures similar to Figure 1.2 it is possible to get the minimal number of terms in a product of two one-parameter groups needed to transform each point on the Bloch sphere to the south pole, which corresponds in the complex plane to a transformation of each complex number into zero. We obtain a bound on the order of generation adding one to this minimal number. The reason for this is that each point of the Bloch sphere (and each point in the complex plane) can be identified pointwise with right cosets of rotations of the Bloch sphere representing all unitary operations. To distinguish the rotations in one coset in general this additional step is necessary.

In [Low71], figures similar to Figure 1.2 are analyzed. The analysis implies:

Theorem 1.1 *Let α denote the angle between two Hamilton operators in the Bloch vector representation and ξ the order of generation.*

1. $\alpha = \pi/2 \implies \xi = 3$ *(Euler decomposition),*
2. $\forall\ k \geq 2: \quad \pi/(k+1) \leq \alpha < \pi/k \implies \xi \leq k+2.$

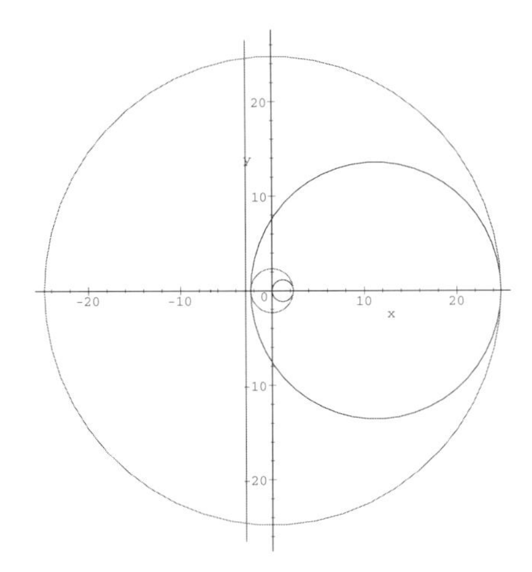

Figure 1.2: As described in the text, the figure shows tangent circles corresponding to the alternating application of two Hamilton operators to the state space visualized through stereographic projection to the complex plane. One further step in the alternating application of Hamilton operators is necessary and sufficient to cover all of the complex plane. Notice that for better orientation one Apollonian circle with infinite radius is pictured by a vertical line. The angle between the rotation axes in the Bloch sphere representation belonging to this figure is $(2 \cdot \pi/4 + \pi/5)/3$. In this case we get an order of generation less than or equal to 6.

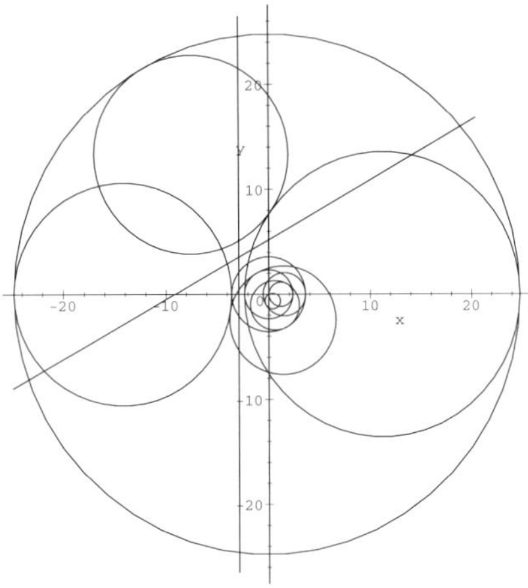

Figure 1.3: As in Figure 1.2 tangent circles corresponding to the application of three Hamilton operators to the stereographically projected state space are shown in the complex plane. To cover all of the complex plane no further step in the alternating application of Hamilton operators is needed. Notice that for better orientation two Apollonian circles with infinite radius are shown by two lines. In this example the angles between the rotation axes are $(2 \cdot \pi/4 + \pi/5)/3$, $\pi/5$, and $\pi/6$. Though all angles are less than or equal to the angle in example of Figure 1.2 we get a smaller upper bound of 5 steps.

The next step is to generalize the analysis from two to three Hamilton operators. Since we have a third Hamilton operator we have to consider a third axis of rotation in the representation of Hamilton operators as infinitesimal rotations of the Bloch sphere leading to a third system of Apollonian circles. The three axes of rotation can be characterized by specifying the three angles α_1, α_2, and α_3 between them. We get a valid triple of angles if

$$\alpha_3 \in [\min\{\pi - (\alpha_1 + \alpha_2), |\alpha_1 - \alpha_2|\}, \pi/2] .$$

As in the case of two Hamilton operators it is important to determine the minimal number of terms in a product of one-parameter groups necessary to transform each point in the complex plane to zero. In Figure 1.3, the tangent circles describing the application of the different Hamilton operators can be received from a similar greedy-type algorithm. As remarked in Figure 1.3 we can get with this type of analysis better bounds on the order of generation when using three Hamilton operators even though no greater angles between two rotation axes are present. Summarizing, this is a first step towards a systematic and efficient implementation of unitary operators.

1.5 Simulation of Hamiltonians

Historically, the first motivation for constructing a quantum computer was the efficient simulation of quantum dynamics. Feynman observed a profound difference in the nature of physical evolution governed by the laws of quantum physics as compared to the evolution under the laws of classical physics. It appears that the simulation of general quantum dynamics by any classical computers involves an unavoidable exponential slowdown in running time. Therefore it would be an interesting application of quantum computers to simulate Hamiltonian dynamics of many-particle systems. In our approach to this problem, we do not construct sequences of quantum gates to simulate the Hamiltonian evolution of a quantum system. The computational resource in our setting is the interaction of the qubits or qudits in the quantum register. The natural time evolution according to this interaction is interspersed with external control operations in such a way that the resulting dynamics is the evolution according to the simulated Hamiltonian evolution. We have restricted ourselves to the mutual simulation of Hamiltonians that are given by pair-interactions among n qudits.

The theoretical framework is as follows. The underlying Hilbert space of each subsystem is d-dimensional and the joint space of n qudits is $\mathcal{H} = \mathbb{C}^d \otimes \mathbb{C}^d \otimes \cdots \otimes \mathbb{C}^d$. The $(d^2 - 1)$-dimensional Lie algebra of traceless Hermitian matrices of size d is denoted by $su(d)$. We set in the following $m := d^2 - 1$. Let $B = \{\sigma_\alpha \mid \alpha := 1, \ldots, m\}$ be an orthogonal basis of $su(d)$ with respect to the trace inner product $\langle A|B\rangle := \mathrm{tr}(A^\dagger B)/d$. We denote by $\sigma_\alpha^{(k)}$ the operator that acts as σ_α on the kth qudit and as identity on the other qudits. We represent the interaction (which is assumed to contain only two-body terms) by a so-called coupling matrix $J = (J_{kl;\alpha\beta})$ of size $nm \times nm$ such that

$$H = \sum_{k<l} \sum_{\alpha\beta} J_{kl;\alpha\beta} \sigma_\alpha^{(k)} \sigma_\beta^{(l)} .$$

We chose the coupling matrix to be symmetric and set $J_{kk;\alpha\beta} = 0$ for all k. Now we use the so-called fast control limit and assume that all operations in $K = U(d) \otimes U(d) \otimes \cdots \otimes U(d)$ can be performed arbitrarily fast, where $U(d)$ denotes the group of unitary operations on a qudit. This assumption is a good approximation if the external control operations act on a considerably smaller time scale than the natural evolution of the considered Hamiltonians. For instance, this is the case for NMR-experiments.

A starting point for simulating an infinitesimal time evolution is that if a Hamiltonian $\tilde{H}$ can be written as the sum $\tilde{H} = \sum_{j=1}^{N} H_j$, then the concatenation of time evolutions according to each H_j for a small time approximates the evolution according to $\tilde{H}$ by the Trotter formula. This "average Hamiltonian approach" leads to the following definition [WJB02, WRJB02a, JWB03].

Definition 1.2 *Let $\tilde{H}$ be an arbitrary Hamiltonian. We say $\tilde{H}$ can be simulated by H with time overhead τ, written $\tilde{H} \leq \tau H$, and N time steps if there are N positive real numbers τ_j summing to τ and N control operations $U_j \in K$ such that*

$$\tilde{H} = \sum_{j=1}^{N} \tau_j U_j^{\dagger} H U_j \,, \tag{1.3}$$

The time overhead gives the slowdown of the simulated interaction with respect to the system Hamiltonian. Time optimal simulation leads therefore to a convex optimization problem.

To derive lower bounds on the time overhead and the number of time steps it is possible to work with the coupling matrices instead of the Hamiltonians themselves. This is because the condition in Eq. (1.3) translates into

$$\tilde{J} = \sum_{j=1}^{N} \tau_j O_j J O_j^T \,, \tag{1.4}$$

where O_j are orthogonal transformations corresponding to the local operations U_j for $j = 1, \ldots, N$ [WRJB02a]. The advantage is that the size of the coupling matrices grows only linearly with n, whereas the size of the Hamiltonians grows exponentially.

Let H and $\tilde{H}$ be arbitrary Hamiltonians with coupling matrices J and $\tilde{J}$, respectively. A necessary condition that H can simulate $\tilde{H}$ with time overhead τ follows from Eq. (1.4) and Uhlmann's theorem [WJB02, WRJB02a]: If $\tilde{H} \leq \tau H$ then we necessarily have

$$\sum_{i=1}^{k} \lambda_i(\tilde{J}) \leq \tau \sum_{i=1}^{k} \lambda_i(J) \,,$$

for $k = 1, \ldots, nm$, where $\lambda_1(\tilde{J}) \geq \lambda_2(\tilde{J}) \geq \cdots \geq \lambda_{nm}(\tilde{J})$ and $\lambda_1(J) \geq \lambda_2(J) \geq \cdots \geq \lambda_{nm}(J)$ denote the eigenvalues of $\tilde{J}$ and J, respectively. This so-called *spectral majorization criterion* has still to be satisfied after rescaling the strengths $J_{kl;\alpha\beta}$ and $\tilde{J}_{kl;\alpha\beta}$ of corresponding interactions in H and $\tilde{H}$ by the same real factor $s_{kl} \in \mathbb{R}$. The reason is that the same sequence of local operations may be used for the rescaled problem.

One of the nicest applications of spectral majorization theory is to derive lower bounds on time-reversal schemes [JWB02], i. e., to simulate $-H$ when H is the quantum computer's Hamiltonian. This is the usual problem of efficient refocusing techniques in NMR. For the case of coupled qubits, i. e., $d = 2$, we consider the Hamiltonian $H = \sum_{k<l} w_{kl} \sigma_z^{(k)} \sigma_z^{(l)}$ such

that all w_{kl} are non-zero. We denote by K the matrix whose all diagonal entries are 0 and whose all off-diagonal entries are 1. Then after rescaling the coupling between the qubits k and l by $-w_{kl}$ we have $J' := -K \otimes \mathrm{diag}(0,0,1)$ and $\tilde{J}' := K \otimes \mathrm{diag}(0,0,1)$. By considering the largest eigenvalues of J' and $\tilde{J}'$ we obtain $n-1$ as a lower bound on the time overhead for time-reversal. A related problem is to switch off some interactions if there is an interaction between all qudits, i. e., to reduce the interaction graph. Then the lower bounds on the time overhead depend on the spectrum of the adjacency matrices corresponding to the graphs that indicate the remaining interactions. The intuitive meaning of the time overhead is the factor by which the remaining interactions are weakened.

The time overhead is not the only complexity measure for simulation schemes. The number N of time steps is also of importance. We have presented a general method to obtain lower bounds on the number of time steps by comparing the rank of coupling matrices [JWB03]. Consider the problem to switch off an interaction with coupling matrix J, i. e., $\sum_j \tau_j O_j J O_j^T = 0$. By adding τA with an arbitrary matrix A to the equation we have

$$\tau A = \sum_{j=1}^{N} \tau_j O_j (J + O_j^T A O_j) O_j^T \,.$$

By suitably choosing the matrix A it is possible to give an upper bound on the rank of each term $J + O_j^T A O_j$. In this way we obtain lower bounds on N. Especially, the method yields a classification of spin–spin interactions with respect to the complexity of the required schemes for decoupling and time reversal [JWB02].

Upper bounds on the time overhead and the number of time steps for mutual simulation of Hamiltonians have been constructed using selective decoupling techniques based on the concepts of orthogonal arrays from combinatorics and nice error bases from quantum coding theory [WRJB02a, JWB03]. The assumption that all local operations are allowed can be weakened. Finite groups of control operations to enable universal simulation of Hamiltonians are characterized in [WRJB02b].

References

[ABC+01] G. Alber, Th. Beth, Ch. Charnes, A. Delgado, M. Grassl, and M. Mussinger. Stabilizing Distinguishable Qubits against Spontaneous Decay by Detected-Jump Correcting Quantum Codes. *Phys. Rev. Lett.*, 86(19):4402–4405, May 2001. See also LANL preprint quant-ph/0103042.

[ABH+01] G. Alber, Th. Beth, M. Horodecki, P. Horodecki, R. Horodecki, M. Rötteler, H. Weinfurter, R. Werner, and A. Zeilinger. *Quantum Information: An Introduction to Basic Theoretical Concepts and Experiments, Springer Texts in Modern Physics*, vol. 173. Springer, 2001.

[AK01] A. Ashikhmin and E. Knill. Nonbinary quantum stabilizer codes. *IEEE Trans. Inf. Theory*, 47(7):3065–3072, November 2001.

[AMD03] G. Alber, M. Mussinger, and A. Delgado. Quantum information processing and error correction with jump codes. In Th. Beth and G. Leuchs, eds., *Quantum Information Processing*. Wiley VCH, 2002, pp. 14-27.

[BCG+03] Th. Beth, Ch. Charnes, M. Grassl, G. Alber, A. Delgado, and M. Mussinger. A New Class of Designs which Protect against Quantum Jumps. *Designs, Codes and Cryptography*, 29(1-3):51-70, 2003.

[Bet84] Th. Beth. *Verfahren der Schnellen Fouriertransformation*. Teubner, 1984.

[BG00] Th. Beth and M. Grassl. Algorithmen zur spektralen Codierung und Decodierung von zyklischen Quantencodes. Vortrag beim DFG-Kolloquium in Bad Honnef, 10.–12. January 2000.

[Cho39] W.-L. Chow. Über Systeme von linearen partiellen Differentialgleichungen erster Ordnung. *Math. Ann.*, 117:98–105, 1939.

[D'A02] D. D'Alessandro. Uniform Finite Generation of Compact Lie Groups. *Systems & Control Lettes*, September 2002. See also LANL preprint quant-ph/0111133.

[Egn97] S. Egner. Zur Algorithmischen Zerlegungstheorie Linearer Transformationen mit Symmetrie. Dissertation, Universität Karlsruhe, Informatik, 1997.

[Gau19] C. F. Gauss. Die Kugel. *Werke*, 8:351–356, c. 1819.

[GB99] M. Grassl and Th. Beth. Quantum BCH Codes. In W. Mathis and T. Schindler, eds., *Proceedings X. International Symposium on Theoretical Electrical Engineering*, pp. 207–212, Magdeburg, September 1999. Universität Magdeburg.

[GB00] M. Grassl and Th. Beth. Cyclic quantum error-correcting codes and quantum shift registers. *Proceedings of the Royal Society London A*, 456(2003):2689–2706, November 2000. See also LANL preprint quant-ph/9910061.

[GGB99] M. Grassl, W. Geiselmann, and Th. Beth. Quantum Reed-Solomon Codes. In M. Fossorier, H. Imai, S. Lin, and A. Poli, eds., *Proceedings Applied Algebra, Algebraic Algorithms and Error-Correcting Codes (AAECC-13), Lecture Notes in Computer Science*, vol. 1719, pp. 231–244, Honolulu, Hawaii, November 1999. Springer. See also LANL preprint quant-ph/9910059.

[GKR02] M. Grassl, A. Klappenecker, and M. Rötteler. Graphs, Quadratic Forms, and Quantum Codes. In *Proceedings 2002 IEEE International Symposium on Information Theory*, Lausanne, 2002, p. 45.

[Gra02a] M. Grassl. *Fehlerkorrigierende Codes für Quantensysteme: Konstruktionen und Algorithmen*. Shaker, Aachen, 2002. Zugl.: Universität Karlsruhe, Dissertation, 2001.

[Gra02b] M. Grassl. Algorithmic aspects of quantum error-correcting codes. In R. K. Brylinski and G. Chen, eds., *Mathematics of Quantum Computation*, pp. 223–252. CRC-Press, 2002.

[JS72] V. Jurdjevic and H. J. Sussmann. Control systems on Lie groups. *J. Differ. Equations*, 12:313–329, 1972.

[JWB02] D. Janzing, P. Wocjan, and Th. Beth. Complexity of decoupling and time-reversal for n spins with pair-interactions: Arrow of time in quantum control. *Phys. Rev. A*, 66:042311, 2002. See also LANL-preprint quant-ph/0106085v2.

[JWB03] D. Janzing, P. Wocjan, and Th. Beth. On the Computational Power of Physical Interactions: Bounds on the Number of Time Steps for Simulating Arbitrary Interaction Graphs. *International Journal of Foundations of Computer Science*, 14(5):889-903, 2003. See also LANL-preprint quant-ph/0203061.

[KL97] Emanuel Knill and Raymond Laflamme. Theory of quantum error-correcting codes. *Phys. Rev. A*, 55(2):900–911, February 1997. See also LANL preprint quant-ph/9604034.

[KR00] A. Klappenecker and M. Rötteler. On the irresistible efficiency of signal processing methods in quantum computing. In R. Creutzburg and K. Egiazarian, eds., *Spectral Techniques and Logic Design for Future Digital Systems: Proceedings of the First International Workshop on Spectral Techniques and Logic Design for Future Digital Systems, Tampere, Finland, June 2–3*, 2000.

[KR01] A. Klappenecker and M. Rötteler. Discrete Cosine Transforms on Quantum Computers. In S. Loncaric and H. Babic, eds., *Proceedings of the 2nd International Symposium on Image and Signal Processing and Analysis (ISPA01)*, pp. 464–468, Pula, Croatia, 2001.

[KR02a] A. Klappenecker and M. Rötteler. Beyond Stabilizer Codes I: Nice Error Bases. *IEEE Trans. Inf. Theory*, 48(8):2392–2395, August 2002. See also LANL preprint quant-ph/0010082.

[KR02b] A. Klappenecker and M. Rötteler. Beyond Stabilizer Codes II: Clifford Codes. *IEEE Trans. Inf. Theory*, 48(8):2396–2399, August 2002. See also LANL preprint quant-ph/0010076.

[KR03] A. Klappenecker and M. Rötteler. Engineering Functional Quantum Algorithms. *Phys. Rev. A*, 67:010302(R), 2003. See also LANL preprint quant-ph/0109088.

[Low71] F. Lowenthal. Uniform finite generation of the rotation group. *Rocky Mountain J. Math.*, 1:575–586, 1971.

[Min93] T. Minkwitz. Algorithmensynthese für lineare Systeme mit Symmetrie. Dissertation, Universität Karlsruhe, Informatik, 1993.

[PRB99] M. Püschel, M. Rötteler, and Th. Beth. Fast Quantum Fourier Transforms for a Class of non-abelian Groups. In M. Fossorier, H. Imai, S. Lin, and A. Poli, eds., *Proceedings Applied Algebra, Algebraic Algorithms and Error-Correcting Codes (AAECC-13), Lecture Notes in Computer Science*, vol. 1719, pp. 148–159, Honolulu, Hawaii, November 1999. Springer. See also LANL preprint quant-ph/9910059.

[Püs98] M. Püschel. Konstruktive Darstellungstheorie und Algorithmengenerierung. Dissertation, Universität Karlsruhe, Informatik, 1998.

[RB98] M. Rötteler and Th. Beth. Polynomial-time solution to the hidden subgroup problem for a class of non-abelian groups. LANL preprint quant-ph/9812070, 1998.

[RB99] M. Rötteler and Th. Beth. Efficient Realisation of Discrete Cosine Transforms on a Quantum Computer. *Proceedings X. International Symposium on Theoretical Electrical Engineering*, pp. 85–89, Magdeburg, September 1999. Universität Magdeburg.

[RB01] R. Raussendorf and H. J. Briegel. A One-Way Quantum Computer. *Phys. Rev. Lett.*, 86(22):5188–5191, May 2001. See also LANL preprint quant-ph/0010033.

[Röt02] M. Rötteler. *Schnelle Signaltransformationen für Quantenrechner*. Shaker, Aachen, 2002. Zugl.: Universität Karlsruhe, Dissertation, 2001.

[RPB99] M. Rötteler, M. Püschel, and Th. Beth. Fast Signal Transforms for Quantum Computers. In W. Kluge, ed., *Proceedings of the Workshop on Physics and Computer Science, Heidelberg*, DPG–Frühjahrstagung, Heidelberg, 1999.

[Sch02] D. Schlingemann. Stabilizer codes can be realized as graph codes. *Quant. Inf. Comp.*, 2(4):307-323, 2002. See also LANL preprint quant-ph/0111080, 2001.

[Sho94] P. W. Shor. Algorithms for quantum computation: discrete logarithm and factoring. In *Proc. FOCS 94*, pp. 124–134. IEEE Computer Society Press, 1994.

[SW02] D. Schlingemann and R. F. Werner. Quantum error-correcting codes associated with graphs. *Phys. Rev. A*, 65(1):012308, 2002. See also LANL preprint quant-ph/0012111.

[WJB02] P. Wocjan, D. Janzing, and Th. Beth. Simulating arbitrary pair-interactions by a given Hamiltonian: Graph-theoretical bounds on the time complexity. *Quant. Inform. & Comp.*, 2(2):117–132, 2002. See also LANL-preprint quant-ph/0106077.

[WRJB02a] P. Wocjan, M. Rötteler, D. Janzing, and Th. Beth. Simulating Hamiltonians in quantum Networks: Efficient schemes and complexity bounds. *Phys. Rev. A*, 65(4):042309, April 2002. See also LANL preprint quant-ph/0109088.

[WRJB02b] P. Wocjan, M. Rötteler, D. Janzing, and Th. Beth. Universal Simulation of Hamiltonians using a finite set of control operations. *Quant. Inform. & Comp.*, 2(2):133–150, 2002. See also LANL-preprint quant-ph/0109063.

[WSR+02] R. F. Werner, D. Schlingemann, M. Reimpell, Th. Beth, M. Grassl, A. Klappenecker, and M. Rötteler. Quantum error correction: Graph codes and stabilizer codes. Poster at the DFG-Colloquium in Bad Honnef, 28.–30. January 2002.

[ZB02] R. Zeier and Th. Beth. Efficient decomposition of quantum operations into given one-parameter groups. Poster at the DFG-Colloquium in Bad Honnef, 28.–30. January 2002.

2 Quantum Information Processing and Error Correction with Jump Codes

G. Alber[1], *M. Mussinger*[1], *and A. Delgado*[2]

[1] Institut für Angewandte Physik
Technische Universität Darmstadt
Darmstadt
Germany

[2] Department of Physics and Astronomy
University of New Mexico
Albuquerque, New Mexiko
USA

2.1 Introduction

Current developments in quantum information processing demonstrate in an impressive way the practical potential of quantum physics [1–3]. In quantum computation, for example, characteristic quantum phenomena, such as interference and entanglement, are exploited for solving computational tasks more efficiently than it is possible by any known classical means [4–7]. However, interference and entanglement are fragile phenomena which can be destroyed easily by uncontrolled interactions with an environment. In order to protect quantum information against the resulting decoherence powerful methods of error correction [8–18] have been developed over the last few years.

The main aim of quantum error correction is to reverse the perturbing influence of an uncontrollable environment. Whether such an inversion is possible or not and how it can be achieved most efficiently depends on the physical interaction between the quantum system considered and its environment. In the subsequent sections we discuss main ideas underlying a recently developed new class of error correcting quantum codes which are capable of correcting a frequently occurring class of errors arising from spontaneous decay processes [19]. In quantum optical systems such spontaneous decay processes may arise from the spontaneous emission of photons and in solid state devices, for example, they may originate from the spontaneous emission of phonons. These jump codes exploit in an optimal way information about errors which is obtained from continuous observation of the environment. It will be demonstrated that on the basis of Heisenberg- and Ising-type Hamiltonians universal quantum gates can be constructed for these jump codes. They guarantee that any error can be corrected even if it occurred during the action of one of these gates. Thus, with the help of these quantum gates it is possible to stabilize quantum information processing units against spontaneous decay processes.

This contribution is organized as follows: In Sec. 2 we summarize basic facts about the inversion of general quantum operations or generalized measurements. One of the particularly useful results arising from the systematic analysis of this general problem is an algebraic criterion for the inversion of error operators. In Sec. 3 we discuss the theoretical description of spontaneous decay processes and of continuous measurement processes by master equations.

Quantum Information Processing, 2nd Edition. Edited by Thomas Beth and Gerd Leuchs

ISBN: 3-527-40541-0

The practical need of inverting events involving zero- and one-photon (or phonon) emission processes leads directly to one-error correcting jump codes. These quantum codes exploit in an optimal way information about error times and error positions by monitoring the environment continuously. In Sec. 4 we address the problem of stabilizing the coherent dynamics of a quantum system against spontaneous decay processes. An example of such a coherent dynamics is a quantum algorithm performed by a quantum information processing unit. In particular, we address two main problems which arise in this context. Firstly, we deal with the question how one can implement any unitary transformation entirely within the code space of a jump code without leaving it at any time. Secondly, we propose a universal entanglement gate which allows one to entangle two arbitrary basic quantum registers of a quantum information processing unit. This entanglement gate does not leave the error correcting code space of a jump code at any time. Together with the local unitary transformations which can be performed on any of the basic quantum registers it forms a universal set of quantum gates.

2.2 Invertible Quantum Operations and Error Correction

A typical quantum information processing unit is composed of a system of N two-level quantum systems, so called qubits, which can be addressed individually. According to the linear superposition principle of quantum mechanics an arbitrary pure quantum state of such a N-qubit quantum register is of the form

$$|\psi\rangle = \sum_{i_1,i_2,\ldots,i_N=0,1} a_{i_1 i_2 \cdots i_N} |i_N, \ldots, i_2, i_1\rangle \tag{2.1}$$

with $|0_\alpha\rangle$ and $|1_\alpha\rangle$ denoting two orthogonal basis states of qubit α. The corresponding orthonormal basis states of the N-qubit Hilbert space $\mathcal{H}$ are denoted $|i_N, i_{N-1}, \cdots, i_1\rangle \equiv |i_N\rangle \otimes |i_{N-1}\rangle \otimes \cdots \otimes |i_1\rangle$. The complex coefficients $a_{i_1 i_2 \cdots i_N}$ fulfill the normalization condition $\sum_{i_1,\cdots,i_N=0,1} \mid a_{i_1 i_2 \cdots i_N} \mid^2 = 1$. The power of quantum computation relies on the ability to preserve the quantum coherence of such a register-state. Any coupling to an external environment which involves uncontrollable degrees of freedom may destroy linear superpositions thus causing decoherence [20]. This phenomenon which is undesirable from the point of view of quantum information processing can be overcome by quantum mechanical error correction techniques. Shor [8] demonstrated that quantum error correcting codes are possible. By now many different classes of error correcting quantum codes have been developed [9–18].

A main aim of quantum error correction is to reverse the dynamical influence of an external environment on the states of a quantum register [21–23]. The most general dynamical influence of this kind can be represented by a unitary joint evolution of a quantum register with an environment followed by a von Neumann measurement performed on the environment. If initially the quantum register and the environment are not entangled and if the various possible measurement results are discarded, this way a trace-preserving or deterministic quantum operation $\mathcal{E}$ is obtained. Its action on an arbitrary register state with density operator ρ (and proper normalization $\mathrm{Tr}\rho = 1$) can be characterized by a set of Kraus-operators [24] $\{K_{lm}\}$. These Kraus- or error operators characterize all possible environmental influences which may occur and they satisfy the completeness relation $\sum_{lm} K_{lm}^\dagger K_{lm} = \mathbf{1}$. The quantum state resulting from a deterministic quantum operation is given by

$$\mathcal{E}: \ \rho \rightarrow \mathcal{E}(\rho) \ = \ \sum_l p_l \rho_l. \tag{2.2}$$

The labels l characterize all possible measurement results which occur with probabilities $p_l = \mathrm{Tr}(\sum_m K_{lm}^\dagger K_{lm}\rho)$. Observation of a particular measurement result, say l, implies that immediately afterwards the register is in the normalized state $\rho_l = \sum_m K_{lm}\rho K_{lm}^\dagger / p_l$.

Typically a set of Kraus-operators $\{K_{lm}\}$ which defines a quantum operation (or generalized measurement) is not unique. Any two sets of Kraus-operators, say $\{\overline{K}_{\lambda\mu}\}$ and $\{K_{lm}\}$, give rise to the same quantum operation if and only if they are related by a unitary matrix $\mathcal{U}_{\lambda\mu,lm}$, i.e. $\overline{K}_{\lambda\mu} = \sum_{\lambda\mu,lm} \mathcal{U}_{\lambda\mu,lm} K_{lm}$ [25]. An important special case of deterministic quantum operations are pure ones. They are characterized by the property that for each measurement result l the associated quantum state ρ_l involves one Kraus-operator $\{K_l\}$ only, i.e.

$$\mathcal{E}_p: \ \rho \rightarrow \mathcal{E}_p(\rho) \ = \ \sum_l p_l \rho_l \tag{2.3}$$

with $p_l = \mathrm{Tr}(K_l^\dagger K_l \rho)$, $\rho_l = K_l \rho K_l^\dagger / p_l$ and $\sum_l K_l^\dagger K_l = \mathbf{1}$. Pure quantum operations correspond to situations where a maximum amount of information about the register state is extracted from the quantum state of an environment [21–23]. As a result, an initially prepared pure register state, say $|\psi\rangle$, remains pure, i.e. $|\psi\rangle \rightarrow |\psi'\rangle = K_l|\psi\rangle / \sqrt{\langle\psi|K_l^\dagger K_l|\psi\rangle}$.

A quantum operation $\mathcal{E}$ is reversible, if one can construct a deterministic quantum operation $\mathcal{R}$ such that $\mathcal{R}(\mathcal{E}(\rho)) = \rho$ for any density operator ρ. The recovery operation $\mathcal{R}$ is required to be deterministic because we want the reversal definitely to occur not just with some probability. In general such an inverse quantum operation cannot be constructed over the whole state space of a quantum register. The main problem in quantum error correction is to find an appropriate, sufficiently high dimensional subspace $\mathcal{C} \subset \mathcal{H}$ over which such an inversion operation can be defined.

It has been shown by Knill and Laflamme [26] and by Bennett et al. [27] that a quantum operation is reversible on a subspace $\mathcal{C}$ if and only if there exists a non-negative matrix $\Lambda_{ll'}$ such that

$$P_{\mathcal{C}} K_l^\dagger K_{l'} P_{\mathcal{C}} \ = \ \Lambda_{ll'} P_{\mathcal{C}} \tag{2.4}$$

for all possible error (or Kraus-) operators K_l and $K_{l'}$. Thereby $P_{\mathcal{C}}$ denotes the projection operator onto the desired subspace $\mathcal{C}$ which is usually called a quantum error-correcting code space or code. Its code words which may be identified with classical bit-strings are formed by an orthonormal basis of states, say $\{|c_i\rangle, i = 1, \cdots, L\}$. The difference r between the dimension of the original Hilbert space $\mathcal{H}$ and the dimension of $\mathcal{C}$, i.e. $r = 2^N - L \geq 0$, is a measure of the redundancy which has to be introduced in order to guarantee successful error correction. For the actual reversal of a quantum operation one has to identify first of all the character of the error (i.e. its syndrome) by an appropriate measurement and subsequently one has to apply an appropriate unitary recovery operation which reverses this quantum operation [21–23]. The criterion of Eq. (2.4) guarantees the existence of such a measurement process and its associated unitary recovery operation. These two basic steps, namely determination of the character of an error and subsequent application of a (nontrivial) unitary recovery operation, constitute the basic elements of any kind of active quantum error correction.

A special situation arises, if one is able to identify a subspace $\mathcal{C}'$ which fulfills not only Eq. (2.4) but also the more stringent condition

$$K_l P_{\mathcal{C}'} = \lambda_l P_{\mathcal{C}'} \tag{2.5}$$

for all possible error operators K_l considered. In this case the quantity $\Lambda_{ll'}$ of Eq. (2.4) factorizes according to $\Lambda_{ll'} \equiv \lambda_l^* \lambda_{l'}$. It is apparent that in this case all the required unitary recovery operations are trivial as they are equal to the identity operation over the code space $\mathcal{C}'$. Thus, no recovery operation has to be performed at all. Such a passive error correction [16–18] is not only capable of correcting single but also multiple errors of arbitrary order. However, so far only very few physical situations are known in which sufficiently high dimensional decoherence-free subspaces (DFSs) $\mathcal{C}'$ can be constructed. In many cases the relevant DFSs are one-dimensional so that they are not of any practical interest for purposes of quantum information processing.

In practical applications one is interested in constructing error correcting methods which tend to decrease not only the number of recovery operations but which also minimize redundancy. For this purpose it may be advantageous to combine passive and active methods of quantum error correction. In the subsequent sections we discuss such a family of error correcting quantum codes which is capable of correcting spontaneous decay processes of the distinguishable qubits of a quantum information processor.

2.3 Quantum Error Correction by Jump Codes

2.3.1 Spontaneous Decay and Quantum Trajectories

Any interaction of a quantum system with an environment whose degrees of freedom are not accessible to observation leads to decoherence. An example of such an interaction is the coupling of a quantum register to the unoccupied vacuum modes of the electromagnetic field (compare with Fig. 2.1). As a result an excited qubit can decay spontaneously by emission of a photon. For the sake of quantum information processing situations are of particular interest in which no spontaneous decay process affects the distinguishability of the qubits involved. This is guaranteed whenever the wave lengths λ of the spontaneously emitted photons are much smaller than typical distances d between adjacent qubits and therefore the qubits decay into statistically independent environments. In this case the time evolution of the state of the quantum register $\rho(t)$ is given by a quantum master equation of the form [28]

$$\frac{d\rho}{dt}(t) = -\frac{i}{\hbar}\left[H, \rho(t)\right] + \sum_{\alpha}\left\{\left[L_\alpha, \rho(t) L_\alpha^\dagger\right] + \left[L_\alpha \rho(t), L_\alpha^\dagger\right]\right\}. \tag{2.6}$$

Thereby the Hamiltonian H describes the coherent dynamics of the quantum register in the absence of any coupling to its environment. This coherent dynamics might represent a quantum algorithm, for example. The Lindblad operators [29] $L_\alpha = \sqrt{\kappa_\alpha}|0_\alpha\rangle\langle 1_\alpha|$ with $\alpha = 1, \cdots, N$ characterize the influence of the environment on the quantum register. The spontaneous decay rate of qubit α is denoted by κ_α. It should be mentioned that the Born- and Markov approximations underlying the derivation of Eq. (2.6) are applicable whenever the interaction

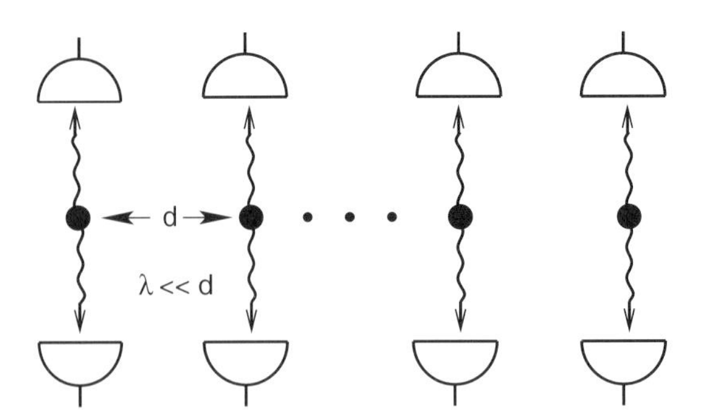

Figure 2.1: Schematic representation of an array of distinguishable qubits whose spontaneous emission times t_i and error positions α_i are monitored continuously by photodetectors.

between system and environment is weak and, in addition, the environmental correlation time is small. Typically these conditions are well fulfilled for quantum optical systems. Sometimes they are also fulfilled for other quantum systems, such as solid state devices with phononic decay processes, provided the environmental temperature is sufficiently high [30].

If the initial state of a quantum register is pure, a formal solution of the quantum master equation (2.6) is given by [31, 32]

$$\rho(t) = \sum_{n=0}^{\infty} \sum_{\alpha_1,\ldots,\alpha_n} \int_0^t dt_n \int_0^{t_n} dt_{n-1} \cdots \int_0^{t_2} dt_1 \, |t; t_n, \alpha_n; \ldots; t_1, \alpha_1\rangle \langle t; t_n, \alpha_n; \ldots; t_1, \alpha_1| \tag{2.7}$$

with the unnormalized pure states

$$|t; t_n, \alpha_n; \ldots; t_1, \alpha_1\rangle = e^{-\frac{i}{\hbar} H_{\text{eff}}(t-t_n)} L_{\alpha_n} \ldots L_{\alpha_2} e^{-\frac{i}{\hbar} H_{\text{eff}}(t_2-t_1)} L_{\alpha_1} e^{-\frac{i}{\hbar} H_{\text{eff}} t_1} |t = 0\rangle . \tag{2.8}$$

According to Eq. (2.7) the state of the register at time t is unravelled into a sum of contributions which are associated with all possible numbers n of spontaneously emitted photons. For a given number n of emitted photons the quantum state is unravelled into a sum of all contributions which describe all possible sequences of emission events taking place at emission times $t_1 \leq t_2 \leq \cdots \leq t_n$ and affecting qubits $\alpha_1, \cdots, \alpha_n$. The pure state $|t; t_n, \alpha_n; \ldots; t_1, \alpha_1\rangle$ of Eq. (2.8) describes the resulting quantum state of the register [31, 32]. The quantum jumps of the qubits from their excited to their ground states due to spontaneous decay processes are characterized by the Lindblad operators L_{α_i}. The time evolution between two successive quantum jumps with no photon emission in between is described by the effective Hamiltonian

$$H_{\text{eff}} = H - \frac{i\hbar}{2} \sum_{\alpha} L_{\alpha}^{\dagger} L_{\alpha}. \tag{2.9}$$

The norm of the quantum state of Eq. (2.8) yields the probability with which a particular measurement record characterized by a quantum trajectory [32–34] $(t_n, \alpha_n; \ldots; t_1, \alpha_1)$ contributes to the density operator $\rho(t)$. The formal solution of Eq. (2.7) describes the dynamics of the quantum register under the influence of the environment in cases in which the environment is monitored continuously by photodetectors [31, 32] but the measurement results are discarded. According to Sec. 2.2 the formal solution of Eq. (2.7) describes a deterministic quantum process where each quantum trajectory characterizes a particular measurement record.

2.3.2 Jump Codes

How can we stabilize a quantum system, such as the one depicted in Fig. 2.1, against spontaneous decay processes, if we are able to monitor the distinguishable qubits continuously by photodetectors? According to Eq. (2.7) we have to tackle two major tasks. Firstly, we have to correct the modifications taking place during successive photon emission events. These modifications are described by the effective (non-hermitian) Hamiltonian of Eq. (2.9). Secondly, we have to invert each quantum jump which is caused by the spontaneous emission of a photon. These quantum jumps are described by the Lindblad operators appearing in Eq. (2.8).

For the sake of simplicity let us concentrate in this section on the case of a quantum memory without any intrinsic coherent time evolution, i.e. $H \equiv 0$ in Eq. (2.6). If we want to correct the errors taking place during successive photon emission events, we must invert the pure quantum operation which is characterized by the one-parameter family of Kraus-operators

$$K_0(t) = e^{-\sum_\alpha L_\alpha^\dagger L_\alpha t/2}. \tag{2.10}$$

Specializing the criterion of Eq. (2.4) to the case of these hermitian error operators an inversion is possible over a subspace $\mathcal{C}$ if and only if

$$P_\mathcal{C} K_0(2t) P_\mathcal{C} = \Lambda_{00}(t) P_\mathcal{C} \tag{2.11}$$

with $\Lambda_{00}(t) \geq 0$. Stated differently, over the code space $\mathcal{C}$ the undesired modification appearing in the effective Hamiltonian of Eq. (2.9) has to act as a (non-negative) multiple of the unit operator. Thus, the code space we are looking for is a DFS of the effective Hamiltonian with $H \equiv 0$.

In the subsequent discussion we focus on the important special case in which the spontaneous decay rates of all qubits are equal, i.e. $\kappa_\alpha \equiv \kappa$. The corresponding DFSs can be found easily because the relevant operator, i.e. $\sum_\alpha L_\alpha^\dagger L_\alpha = \kappa \sum_\alpha |1_\alpha\rangle\langle_\alpha 1|$, just enumerates the number of excited qubits. Therefore, any set of orthonormal states which all involve the same number of excited qubits constitutes a passive error correcting code space for the Kraus-operators $K_0(t)$. The dimension D of a DFS involving N physical qubits k of which are excited, i.e. a DFS $-$ (N, k), is given by $D = \binom{N}{k} \equiv N!/[k!(N-k)!]$. For a given number of physical qubits N this dimension is maximal, if half of the qubits are excited, i.e. for $k = [N/2]$. ($[x]$ denotes the largest integer smaller or equal to x.) Such a DFS of maximal dimension involving four physical qubits, for example, is formed by the set of code words $\{|1100\rangle, |0011\rangle, |1010\rangle, |0101\rangle, |1001\rangle, |0110\rangle\}$.

In general, arbitrary linear superpositions of code words of such a DFS cannot be stabilized against quantum jumps arising from spontaneous decay processes. If we also want to invert each individual quantum jump, we have to find an appropriate subspace $\mathcal{C}' \subseteq \mathcal{C}$ over which any of the quantum operations appearing in Eq. (2.7) is reversible. For this purpose we note, that within any DFS $-(N,k)$ the time evolution between successive quantum jumps is proportional to the unit operator, i.e. $e^{-\sum_\alpha L_\alpha^\dagger L_\alpha t/2}|_{\mathcal{C}} \equiv e^{-k\kappa t/2} P_{\mathcal{C}}$. Therefore, we have to find appropriate subspaces $\mathcal{C}' \subseteq \mathcal{C}$ over which the Lindblad operators appearing in Eq. (2.8) are reversible. The details of the construction of an active error correcting quantum code capable of correcting one quantum jump at a time, for example, depends very much on whether the error position is known or not. In the case of an unknown error position one has to fulfill the criterion of Eq. (2.4) for all possible Lindblad operators L_α with $\alpha \in \{1, \cdots, N\}$. Plenio et al. [35] have been able to find such a code which requires at least eight physical qubits for the encoding of one logical qubit, i.e. for two orthonormal logical states. In contrast, if the error position α of a quantum jump characterized by Lindblad operator L_α is known, the redundancy of such an active one-error correcting quantum code which is embedded into a passive code can be lowered significantly.

The simplest example of such an embedded quantum code or jump code which is capable of correcting one error at a time can be constructed with the help of four physical qubits [19]. The (unnormalized) code words of this particular jump code represent a logical qutrit and are given by

$$|c_0\rangle = |0011\rangle + e^{i\varphi}|1100\rangle\,,\ |c_1\rangle = |0101\rangle + e^{i\varphi}|1010\rangle\,,\ |c_2\rangle = |0110\rangle + e^{i\varphi}|1001\rangle \quad (2.12)$$

with an arbitrary phase φ. Obviously, the code words of this jump code consist of four-qubit states in which half of the qubits are excited. The equal number of excited qubits involved in this code guarantees that the effective time evolution between successive quantum jumps is corrected passively. A characteristic feature of this quantum code is the complementary pairing of states with equal probabilities. This latter property guarantees the validity of the necessary and sufficient conditions of Eq. (2.4) provided the error position is known. This one-error correcting jump code involves three logical states and four physical qubits two of which are excited. Therefore, let us call it jump code $1-JC(4,2,3)$. This construction of a one-error correcting embedded quantum code can be generalized easily to any even number N of physical qubits. Thus, any jump code $1-JC(N, N/2, \binom{N-1}{N/2-1})$ can be constructed by an analogous complementary paring of N-qubit states half of which are excited. This way one obtains $\binom{N-1}{N/2-1}$ orthogonal code words which form a one-error correcting embedded quantum code for spontaneous decay processes. It can be shown that this family of one-error correcting quantum codes is optimal in the sense that their redundancy cannot be reduced any further [19]. Asymptotically, for large numbers of physical qubits the effective number of logical qubits L_q which can be encoded by the jump code $1-JC((N, N/2, \binom{N-1}{N/2-1})$ is given by $L_q \equiv \log_2\binom{N-1}{N/2-1} = N - \log_2\sqrt{N} + O(1)$. In addition, a far reaching link between these jump codes and fundamental structures of combinatorial design theory [36] can be established. Exploiting this link jump codes can be constructed which are even capable of correcting more than one error at a time [19, 37].

Provided the decay rates of all qubits are equal, error position and error time can be determined perfectly and recovery operations are applied immediately after the observation of a quantum jump, spontaneous decay processes can be corrected perfectly with these jump codes. But in reality, typically none of these conditions is fulfilled precisely. However, numerical simulations demonstrate that quantum states can be stabilized against various types of imperfections still to a high degree even if some of these conditions are not fulfilled perfectly [38].

2.4 Universal Quantum Gates in Code Spaces

The previously discussed error correcting jump codes allow one to stabilize a quantum memory against spontaneous decay processes. However, in order to be useful also for purposes of quantum information processing and quantum computation two major additional requirements have to be fulfilled. Firstly, one should be able to manipulate pure quantum states in such a way that a chosen error correcting code space is not left at any time during the performance of a quantum algorithm. This can be achieved by using a universal set of quantum gates which operates entirely within an error correcting code space and which is implemented by a set of Hamiltonians leaving this code space invariant. Such a Hamiltonian implementation of universal quantum gates guarantees that any quantum algorithm which is implemented with the help of these quantum gates does not leave this code space at any time even during the application of one of these quantum gates. Secondly, analogous to classical computer architecture, it is desirable to develop quantum information processors which are based on small quantum registers and, in addition, to design quantum gates in such a way that in each step at most two basic quantum registers are entangled. This ensures that the same set of quantum gates can be used for an arbitrarily large quantum information processing unit. For a recent proposal on implementing these ideas on suitable subspaces of our detected jump-error correcting codes see [39]. In the subsequent sections we present an example how a quantum information processing unit meeting these two major requirements can be constructed on the basis of elementary four-qubit registers each of which constitutes a local qutrit of the jump code $1 - JC(4, 2, 3)$.

2.4.1 Universal Sets of Quantum Gates for Qudit-Systems

Universal sets of quantum gates for qubit-systems were considered by D. DiVincenzo [40], A. Barenco et al. [41] and S. Lloyd [42]. These authors have shown that with a few Hamiltonians acting on single qubits and with one particular two-qubit Hamiltonian it is possible to generate any unitary transformation for a quantum register consisting of qubits. All possible one-qubit operations are members of the continuous group $SU(2)$ (suppressing a trivial $U(1)$ operation) and the two qubit operation entangles any two separable qubits. The lowest dimensional member of our previously discussed jump codes, namely the $1 - JC(4, 2, 3)$-code, provides a logical qutrit and therefore the most general unitary qutrit-operations needed for quantum information processing within this code space are members of the continuous group $SU(3)$ (again suppressing a trivial $U(1)$ operation). Thus the natural question arises which set of quantum gates is universal and thus capable of generating an arbitrary unitary transformation within the state space of a qutrit.

Jean-Luc and Ranee Brylinski [43] derived a generalization of the results of D. DiVincenzo, A. Barenco et al. and S. Lloyd. In particular, they demonstrated that for d-dimensional elementary data carriers, so called qudits, every N-qudit gate can be obtained by combinations of all one-qudit gates and a certain two-qudit entanglement gate. In particular, these authors call a collection $\mathcal{G}$ of one-qudit and two-qudit gates universal (exactly universal), if every $N-$qudit gate with $N \geq 2$ can be approximated with arbitrary accuracy (represented exactly) by a circuit made up of N-qudit gates of this collection $\mathcal{G}$. A (unitary) two-qudit gate V is called primitive, if it maps separable pure states again to separable pure states. Thus, if $|x\rangle$ and $|y\rangle$ are qudit-states, we can find qudit-states $|u\rangle$ and $|v\rangle$ such that $V|x\rangle|y\rangle = |u\rangle|v\rangle$. If V is not primitive, it is called imprimitive. Suppose we are given a two-qudit gate V. Then the collection of all one-qudit gates together with V is universal if and only if V is imprimitive. In particular, J.-L. and R. Brylinski [43] have proved the useful criterion that, if a (unitary) two-qudit gate V is diagonal in a computational basis, i.e. $V|j\rangle|k\rangle = \exp(i\theta_{jk})|j\rangle|k\rangle$, V is primitive if and only if we have

$$\theta_{jk} + \theta_{pq} \equiv \theta_{jq} + \theta_{pk} \pmod{2\pi} \tag{2.13}$$

for all possible values of j, k, p, q.

In general, the difficulty of finding an appropriate set of Hamiltonians by which one can generate a universal set of quantum gates operating entirely within an error correcting code space depends on the physical interactions available. Typical physical two-body interaction Hamiltonians which are expected to be realizable in laboratory are Heisenberg and Ising Hamiltonians H_{He} and H_{Is}, i.e.

$$\begin{aligned} H_{He} &= \sum_{\alpha\beta} C_{\alpha\beta}(t) \left(\sigma_\alpha^{(x)}\sigma_\beta^{(x)} + \sigma_\alpha^{(y)}\sigma_\beta^{(y)} + \sigma_\alpha^{(z)}\sigma_\beta^{(z)}\right), \\ H_{Is} &= \sum_{\alpha\beta} D_{\alpha\beta}(t)\sigma_\alpha^{(z)}\sigma_\beta^{(z)}. \end{aligned} \tag{2.14}$$

Thereby, $\sigma_\alpha^{(x)}, \sigma_\alpha^{(y)}, \sigma_\alpha^{(z)}$ denote the three Cartesian components of the Pauli spin operators of qubit α and the quantities $C_{\alpha\beta}(t)$ and $D_{\alpha\beta}(t)$ denote coupling coefficients of qubits α and β. These latter coefficients are assumed to be tunable arbitrarily. If it is not possible to realize particular linear combinations or commutators of these Hamiltonians by appropriate tunings of these coupling coefficients, one may use appropriate products, such as

$$\begin{aligned} e^{i(t_1 H_1 + t_2 H_2)} &= \left(e^{i\frac{t_1}{n}H_1} e^{i\frac{t_2}{n}H_2}\right)^n + O\left(\frac{1}{n}\right), \\ e^{i(i[t_1 H_1, t_2 H_2])} &= \left(e^{i\frac{t_1}{\sqrt{n}}H_1} e^{i\frac{t_2}{\sqrt{n}}H_2} e^{-i\frac{t_1}{\sqrt{n}}H_1} e^{-i\frac{t_2}{\sqrt{n}}H_2}\right)^n + O\left(\frac{1}{\sqrt{n}}\right). \end{aligned} \tag{2.15}$$

According to Eqs.(2.15) one needs infinite products for representing unitary transformations corresponding to sums or commutators of Hamiltonians exactly. However, it can be shown that in many cases exact representations can also be obtained which involve finite products only [44,45].

2.4.2 Universal One-Qutrit Gates

In this section we address the question how arbitrary unitary transformations can be implemented in the error correcting code spaces of jump codes with the help of Heisenberg-type and Ising-type Hamiltonians. Thereby the Hamiltonians considered are expected to leave these code spaces invariant so that during the application of an arbitrary sequence of unitary transformations the error correcting code space is not left at any time. This requirement guarantees that any error due to a spontaneous decay process occurring during the processing of a quantum state can be corrected. As an example we consider the implementation of arbitrary unitary transformations in the lowest dimensional one-error correcting jump code, i.e. the $1 - JC(4, 2, 3)$-code [46].

Two classes of two-particle Hamiltonians of the Heisenberg- and Ising-type acting on physical qubits will be needed for this construction, namely

$$\begin{aligned} E_{\alpha\beta} &= \frac{1}{2}\left(P_{\alpha,\beta} + \sigma_\alpha^{(x)}\sigma_\beta^{(x)} + \sigma_\alpha^{(y)}\sigma_\beta^{(y)} + \sigma_\alpha^{(z)}\sigma_\beta^{(z)}\right) \text{ and} \\ F_{\alpha\beta} &= \frac{1}{2}\left(P_{\alpha,\beta} + \sigma_\alpha^{(z)}\sigma_\beta^{(z)}\right) \end{aligned} \tag{2.16}$$

with $\alpha, \beta = 1, \ldots, N$. Any member of this family of two-particle Hamiltonians acts on the physical qubits α and β only leaving all other qubits unaffected. The terms of Eqs.(2.16) involving the projection operator $P_{\alpha,\beta}$ represent an energy shift of the two qubits. The residual interaction terms are of Heisenberg- and Ising-type. From these Hamiltonians we can select the six members E_{12}, E_{23}, E_{13}, F_{12}, F_{13}, F_{13}, for example. Their action on the code words of the jump code $1 - JC(4, 2, 3)$ with $\varphi = 0$ (compare with Eq. (2.12)) can be represented by the matrices

$$\begin{aligned} E_{12} &= \begin{pmatrix} 1 & 0 & 0 \\ 0 & 0 & 1 \\ 0 & 1 & 0 \end{pmatrix}, \quad E_{23} = \begin{pmatrix} 0 & 1 & 0 \\ 1 & 0 & 0 \\ 0 & 0 & 1 \end{pmatrix}, \quad E_{13} = \begin{pmatrix} 0 & 0 & 1 \\ 0 & 1 & 0 \\ 1 & 0 & 0 \end{pmatrix}, \\ F_{12} &= \begin{pmatrix} 1 & 0 & 0 \\ 0 & 0 & 0 \\ 0 & 0 & 0 \end{pmatrix}, \quad F_{13} = \begin{pmatrix} 0 & 0 & 0 \\ 0 & 1 & 0 \\ 0 & 0 & 0 \end{pmatrix}, \quad F_{23} = \begin{pmatrix} 0 & 0 & 0 \\ 0 & 0 & 0 \\ 0 & 0 & 1 \end{pmatrix}. \end{aligned} \tag{2.17}$$

Accordingly, the unitary transformations resulting from the Hamiltonians E_{12}, E_{13} and E_{23} swap two codewords and change the phase of the third code word. The unitary transformations resulting from the Hamiltonians F_{12}, F_{13} and F_{23} change the phase of exactly one of the three code words. It is straight forward to demonstrate that the six operators

$$\begin{aligned} C_{12}^+ &= E_{23} - F_{23}, \quad & C_{13}^+ &= E_{13} - F_{13}, \quad & C_{23}^+ &= E_{12} - F_{12}, \\ C_{12}^- &= i[C_{13}^+, C_{23}^+], \quad & C_{13}^- &= i[C_{12}^+, C_{23}^+], \quad & C_{23}^- &= i[C_{12}^+, C_{13}^+] \end{aligned} \tag{2.18}$$

and the two operators F_{12} and F_{13} form a basis for the Lie Algebra of the continuous group $SU(3)$. Thus by an appropriate linear combination of these eight generators any unitary transformation belonging to the continuous group $SU(3)$ can be represented on the code space of the jump code $1 - JC(4, 2, 3)$.

2.4.3 A Universal Entanglement Gate

In computer science it is common practice to use basic registers of a fixed size and to scale an information processing unit by using several of these basic registers. Consequently, on the one hand an algorithm consists of the manipulation of single basic registers, and on the other hand of the interaction between any two of these registers at a time. Such an architecture ensures

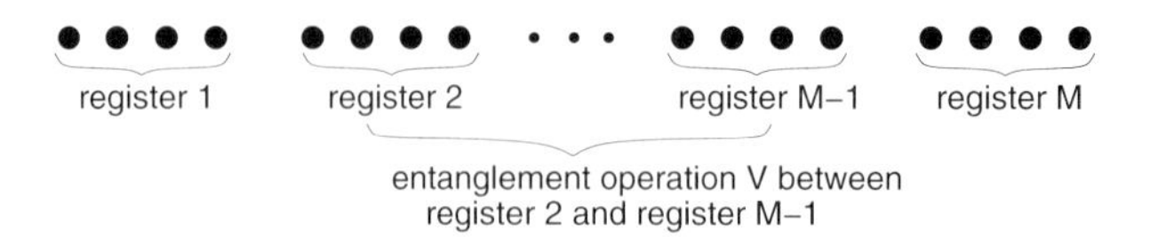

Figure 2.2: Schematic representation of an array of qubits consisting of M basic registers. Each basic register carries a logical qutrit of the jump code $1-JC(4,2,3)$. Any two basic registers, such as registers 2 and $(M-1)$, can be entangled by the entanglement gate V.

that the same set of gates can be used for an arbitrarily scaled device. In addition, new registers can be added to the information processing unit at any time even during a computation without necessitating a new encoding of all qubits involved. If one applies this idea to a quantum processing unit, the basic registers are formed by an appropriate number of qubits. If one wants to correct errors originating from spontaneous decay processes, the simplest basic register has to consist of four physical qubits which form a jump code $1-JC(4,2,3)$. Thus, an appropriate quantum information processing unit capable of stabilizing quantum algorithms against spontaneous decay processes would consist of an array of such four-qubit clusters (compare with Fig. 2.2). We have already demonstrated in the previous section that any unitary transformation within such a four-qubit basic quantum register can be implemented with the help of Heisenberg- and Ising-type Hamiltonians. Here we present a universal entanglement gate which is capable of entangling two arbitrary four-qubit basic registers and which is based on Ising-type Hamiltonians. Together with the unitary transformations discussed in the previous section this entanglement gate forms a universal set of quantum gates for a quantum information processing unit which is based on four-qubit registers. In addition, the presented entanglement gate ensures that all errors due to spontaneous decay processes can be corrected even if they take place during the application of a quantum gate. Let us consider first of all the nine tensor product states which are associated with two basic four-qubit registers. These states are constituted by the product states of two jump codes $1-JC(4,2,3)$, namely

$$\begin{aligned}
|00\rangle_L &= |00110011\rangle + |11001100\rangle + |00111100\rangle + |11000011\rangle, \\
|01\rangle_L &= |00110101\rangle + |11001010\rangle + |00111010\rangle + |11000101\rangle, \\
|02\rangle_L &= |00110110\rangle + |11001001\rangle + |00111001\rangle + |11000110\rangle, \\
|10\rangle_L &= |01010011\rangle + |10101100\rangle + |01011100\rangle + |10100011\rangle, \\
|11\rangle_L &= |01010101\rangle + |10101010\rangle + |01011010\rangle + |10100101\rangle, \\
|12\rangle_L &= |01010110\rangle + |10101001\rangle + |01011001\rangle + |10100110\rangle, \\
|20\rangle_L &= |01100011\rangle + |10011100\rangle + |01101100\rangle + |10010011\rangle, \\
|21\rangle_L &= |01100101\rangle + |10011010\rangle + |01101010\rangle + |10010101\rangle, \\
|22\rangle_L &= |01100110\rangle + |10011001\rangle + |01101001\rangle + |10010110\rangle.
\end{aligned}$$

The linear subspace spanned by these states is denoted by $\mathcal{C}_9$. It is apparent that these states are linear superpositions of code words of the one-error correcting jump code $1-JC(8,4,35)$. Let us assume that it is possible to implement the Ising-type Hamiltonian

$$H_{ent} = 1/2(F_{26} + F_{36} + F_{27} + F_{37}) \tag{2.19}$$

by an appropriate tuning of the coupling coefficients of Eq. (2.14). This Hamiltonian leaves the code space of the jump code $1 - JC(8, 4, 35)$ invariant so that any spontaneous decay process can be corrected. Let us denote the linear subspace spanned by the eight orthonormal states $\{|00\rangle_L, |01\rangle_L, |02\rangle_L, |10\rangle_L, |11\rangle_L, |12\rangle_L, |20\rangle_L, |21\rangle_L\}$ by A and the state subspace by the two orthonormal states $\{|22+\rangle_L = |01100110\rangle + |10011001\rangle$ and $|22-\rangle_L = |01101001\rangle + |10010110\rangle\}$ by B. With this notation the action of the Hamiltonian can be represented by $H_{ent} = P_A \oplus 2|22+\rangle_{LL}\langle 22+|$ with P_A denoting the projection operator onto subspace A. Therefore, the Hamiltonian H_{ent} acts in the subspaces A and B differently. Applying this Hamiltonian for the (dimensionless) time τ yields the unitary transformation

$$U(t) = e^{-iH\tau} = e^{-i\tau} P_A \oplus (e^{-i2\tau}|22+\rangle_L + |22-\rangle_L)_L\langle 22|. \tag{2.20}$$

Though states $|22+\rangle_L$ and $|22-\rangle_L$ are affected differently by this Hamiltonian the unitary transformation of Eq. (2.20) does not leave the one-error correcting code space $1 - JC(8, 4, 35)$ at any time. Therefore, any spontaneous emission event can be corrected. In order to implement an entanglement operation within the tensor product space of two basic four-qubit registers we choose the (dimensionless) interaction time so that $\tau = \pi$. This implies that all code words in subspace A are multiplied by a factor (-1) and states $|22+\rangle_L$ and $|22-\rangle_L$ are both multiplied by a factor $(+1)$. Applying an additional global factor of (-1) results in the conditional phase gate V

$$V = P_A - |22\rangle_{LL}\langle 22|. \tag{2.21}$$

This conditional phase gate is a universal entanglement gate because, consistent with the notation of Eq. (2.13), $\theta_{ij} = 0$ for all $(i, j) \neq (2, 2)$ and $\theta_{22} = \pi$. Therefore, $\theta_{12} + \theta_{21} = 0 \not\equiv \pi = \theta_{11} + \theta_{22} \pmod{2\pi}$ and according to the criterion of Eq. (2.13) V is a universal entanglement gate.

2.5 Summary and Outlook

We discussed main ideas underlying a recently introduced class of error correcting quantum codes, the so called jump codes, which are capable of correcting spontaneous decay processes originating from the coupling of distinguishable qubits to statistically independent environments. These quantum codes exploit information about error times and error positions in an optimal way by monitoring the environment continuously. We also addressed the practical question how these error correcting quantum codes can be used for stabilizing a quantum algorithm against these types of errors. For this purpose we presented a set of universal quantum gates which guarantees that any error due to a spontaneous decay process can be corrected even if it occurred during the application of one of these quantum gates. This is possible because these quantum gates are based on Heisenberg- and Ising-type Hamiltonians which leave the code space of a jump code invariant.

Though our discussion concentrated on one-error correcting quantum jump codes, the already mentioned connection with basic concepts of combinatorial design theory may offer interesting perspectives also for the construction of multiple-error correcting jump codes with minimal redundancy. Such optimal multiple-error correcting quantum codes are expected to be particularly useful for stabilizing the dynamics of quantum information processing units against environmental influences.

Acknowledgments

Discussions with T. Beth, I. Cirac, M. Grassl, R. Laflamme, D. Lidar and D. Shepelyansky are gratefully acknowledged.

References

[1] D. Bouwmeester, A. Ekert and A. Zeilinger (Eds.), *The physics of quantum information*, Springer Berlin (2000).

[2] M. A. Nielsen, I. L. Chuang, *Quantum Computation and Quantum Information,* Cambridge University Press (2000).

[3] G. Alber, Th. Beth, M. Horodecki, P. Horodecki, R. Horodecki, M. Rötteler, H. Weinfurter, R. Werner, and A. Zeilinger, *Quantum Information*, Springer, Berlin (2001).

[4] D. Deutsch, Proc. Roy. Soc. Lond. A **400**, 97 (1985).

[5] W. Shor, In *Proceedings of the 35th Annual Symposium on Foundations of Computer Science*, edited by S. Goldwasser (IEEE Computer Society, Los Alamitos, CA, 1994), p 124.

[6] D. Simon, In *Proceedings of the 35th Annual Symposium on the Foundations of Computer Science, 1994, Los Alamitos, California* (IEEE Computer Society Press, New York 1994) p. 116.

[7] L. K. Grover, In *Proceedings of the 28th Annual Symposium on the Theory of Computing, 1996, Philadelphia, Pennsylvania* (ACM Press, New York, 1996) p. 212; Phys. Rev. Lett. **79**, 325 (1997).

[8] P. W. Shor, Phys. Rev. A **52**, R2493 (1995).

[9] H. Mabuchi and P. Zoller, Phys. Rev. Lett. **76**, 3108 (1996).

[10] R. Laflamme, C. Miquel, J. P. Paz, W. Zurek, Phys. Rev. Lett. **77** 198 (1996).

[11] A. M. Steane, Phys. Rev. Lett. **77**, 793 (1996).

[12] A. Ekert and C. Macchiavello, Phys. Rev. Lett. **77**, 2585 (1996).

[13] A. R. Calderbank and P. W. Shor, Phys. Rev. A **54**, 1098 (1996).

[14] D. Gottesman, Phys. Rev. A **54**, 1862 (1996).

[15] T. Pellizzari, Th. Beth, M. Grassl, and J. Müller-Quade, Phys. Rev. A. **54**, 2698 (1996).

[16] L. M. Duan and G. C. Guo, Phys. Rev. Lett. **79**, 1953 (1997).

[17] P. Zanardi, Phys. Rev. Lett. **79**, 3306 (1997).

[18] D. A. Lidar, I. L. Chuang, and K. B. Whaley, Phys. Rev. Lett. **81**, 2594 (1998).

[19] G. Alber, Th. Beth, Ch. Charnes, A. Delgado, M. Grassl, M. Mussinger, Phys. Rev. Lett. **86**, 4402 (2001).

[20] D. Giulini, K. Kiefer, J. Kupsch, I. O. Stamatescu, and H. D. Zeh, *Decoherence and the Appearance of a Classical World in Quantum Theory* (Springer Verlag, Berlin, 1996).

[21] M. A. Nielsen, C. M. Caves, Phys. Rev. A **55**, 2547 (1997).

[22] M. A. Nielsen, C. M. Caves, B. Schumacher, H. Barnum, Proc. Roy. Soc. Lond. A **454**, 277 (1998), e-print: quant-ph/9706064.

[23] C. M. Caves, J. of Superconductivity **12**, 707 (1999) e-print: quant-ph/9811082.

[24] K. Kraus, *States, Effects and Operations: Fundamental Notions of Quantum Theory*, Springer Verlag Berlin (1983).

[25] M.-D. Choi, Linear Algebra and Its Applications **10**, 285 (1975).

[26] E. Knill and R. Laflamme, Phys. Rev. A **55**, 900 (1997).

[27] C. H. Bennett, D. P. DiVincenzo, J. A. Smolin, W. K. Wootters, Phys. Rev. A **54**, 3824 (1996).

[28] H. J. Carmichael, *Statistical Methods in Quantum Optics 1,* Springer Verlag, Berlin (1999).

[29] G. Lindblad, Commun. Math. Phys. **48**, 119 (1976).

[30] U. Weiss, *Quantum Dissipative Systems,* Series in Modern Condensed Matter Physics, Vol. 2, World Scientific, Singapore (1993).

[31] B. R. Mollow, Phys. Rev. A **12**, 1919 (1975).

[32] H. J. Carmichael, in *An Open Systems Approach to Quantum Optics*, Springer Verlag, Berlin (1993).

[33] J. Dalibard, Y. Castin, K. Mølmer, Phys. Rev. Lett. **68**, 580 (1992).

[34] R. Dum, A. S. Farkins, P. Zoller and C. W. Gardiner, Phys. Rev. A **46** , 4382 (1992).

[35] M. B. Plenio, V. Vedral and P. L. Knight, Phys. Rev. A **55**, 67 (1997).

[36] T. Beth, D. Jungnickel and H. Lenz, *Design Theory*, Vol. 1 and 2, Cambridge University Press (1999).

[37] T. Beth, Ch. Charnes, M. Grassl, G. Alber, A. Delgado, M. Mussinger, to be published.

[38] G. Alber, Th. Beth, Ch. Charnes, A. Delgado, M. Grassl, M. Mussinger, in preparation

[39] K. Khodjasteh and D. A. Lidar, Phys. Rev. Lett. **89**, 197904 (2002).

[40] D. P. DiVincenzo, Phys. Rev. A **51**, 1015 (1995).

[41] A. Barenco, C. H. Bennett, R. Cleve, D. P. DiVincenzo, N. Margolus, P. Shor, T. Sleator, J. A. Smolin, and H. Weinfurter, Phys. Rev. A **52**, 3457 (1995).

[42] S. Lloyd, Phys. Rev. Lett. **75** 346 (1995).

[43] J. L. Brylinski and R. Brylinski, e-print quant-ph/0108062, to appear in *Mathematics of Quantum Computation*, CRC Press (2002).

[44] D. D'Alessandro, e-print: quant-ph/0111133, to appear in System and Control Letters, Elsevier Science.

[45] Robert M. Zeier, Thomas Beth, private communication.

[46] G. Alber, M. Mussinger and A. Delgado, Fortschr. Phys. **49**, 901 (2001).

3 Computational Model for the One-Way Quantum Computer: Concepts and Summary

Robert Raussendorf[1] *and Hans J. Briegel*[2]

Sektion Physik
Ludwig-Maximilians-Universität München
Germany

The one-way quantum computer ($QC_{\mathcal{C}}$) is a universal scheme of quantum computation consisting only of one-qubit measurements on a particular entangled multi-qubit state, the cluster state. The computational model underlying the $QC_{\mathcal{C}}$ is different from the quantum logic network model and it is based on different constituents. It has no quantum register and does not consist of quantum gates. The $QC_{\mathcal{C}}$ is nevertheless quantum mechanical since it uses a highly entangled cluster state as the central physical resource. The scheme works by measuring quantum correlations of the universal cluster state.

3.1 Introduction

Quantum computational models play a twofold role in the development of quantum information science. On the theoretical side, they provide the framework in which mathematical concepts such as a "computation" or an "algorithmic procedure" become connected to the laws of physics. Basic notions of computer science such as "computational complexity" or 'logical depth" are usually derived with reference to such a model. On the practical side, computational models can have a strong influence on the design of actual experiments that try to realize a quantum computer in the laboratory.

The first model of a quantum computer, the quantum Turing machine (QTM) introduced by Deutsch [1] and further developed by Bernstein and Vazirani [2], connects a computation to a unitary transformation on the Hilbert space spanned by all possible "configurations" of the machine. Unlike its classical analog, it can be in a coherent superposition of many different configurations at the same time, which allows for the interference of different computational paths during a computation. This distinct feature of the QTM has opened the room for the invention of more efficient (quantum) algorithms that make use of interference effects.

On the other hand, most proposals for implementing a quantum computer in real physical systems do not follow the model of a quantum Turing machine. The design of most of todays experiments follow instead the model of a quantum logic network (QLN) [3], [4]. Although it was shown to be computationally equivalent to the QTM [3], [4], this model has been used

[1]Present address: California Institute of Technology, 1200 East California Boulevard, Pasadena, California
[2]Present address: Institute of Theoretical Physics, University of Innsbruck, Technikerstr. 25, Innsbruck, Austria

Quantum Information Processing, 2nd Edition. Edited by Thomas Beth and Gerd Leuchs

ISBN: 3-527-40541-0

more commonly in both theoretical and experimental investigations. The notion of quantum gates makes it much simpler to formulate quantum algorithms in the network language and most of the quantum algorithms that one knows of today – including Shor's celebrated factoring algorithm [5] – have been formulated within the network model. Furthermore, the fact that universal sets of quantum gates can be realized from only two-qubit interactions [6] has considerably simplified the problem of identifying specific physical systems that are suitable [7] for quantum computation.

In both the QTM and the QLN model of a quantum computer, unitary evolution plays a key role, even though the way how such a unitary evolution is generated is quite different. Recently, it has become clear that quantum gates (and thus general unitary transformations) need not be generated from a coherent Hamiltonian dynamics. Instead several schemes [8]-[13] have been proposed in which projective von Neumann measurements play a constitutive role.

Recently, we introduced the scheme of the one-way quantum computer [11]. This scheme uses a given entangled state, the so-called cluster state [14], as its central physical resource. The entire quantum computation consists only of a sequence one-qubit projective measurements on this entangled state. We called this scheme the "one-way quantum computer" ($QC_{\mathcal{C}}$) since the entanglement in the cluster state is destroyed by the one-qubit measurements and therefore it can only be used once. While it is possible to simulate any unitary evolution with the one-way quantum computer, the computational model of the $QC_{\mathcal{C}}$ makes no reference to the concept of unitary evolution. A quantum computation corresponds, instead, to a sequence of simple projections in the Hilbert space of the cluster state. The information that is processed is extracted from the measurement outcomes and is thus a purely classical quantity.

As we have shown in [11], any quantum logic network can be simulated efficiently on the one-way quantum computer. This shows that the one-way quantum computer is, in fact, universal. Surprisingly, it turns out that for many algorithms the simulation of a unitary network can be parallelized to a higher degree than the original network itself. As an example, circuits in the Clifford group – which is generated by the CNOT-gates, Hadamard-gates and $\pi/2$-phase shifts – can be performed by a $QC_{\mathcal{C}}$ in a single time step, i.e. all the measurements to implement such a circuit can be carried out at the same time. More generally, in a simulation of a quantum logic network by a one-way quantum computer, the temporal ordering of the gates of the network is transformed into a spatial pattern of measurement bases for the individual qubits on the resource cluster state. For the temporal ordering of the measurements there is, however, no counterpart in the network model. Therefore, the question of complexity of a quantum computation must be possibly revisited.

In the following we would like to give an introduction to the computational model that describes information processing with the one-way quantum computer. To stress the importance of the cluster state for the scheme, we will use the abbreviation $QC_{\mathcal{C}}$ for "one-way quantum computer". The computational model underlying the $QC_{\mathcal{C}}$ has been described in a technical report in Ref. [15]. The purpose of the present paper is to give a summary of this model, concentrating on the concepts that we have introduced to describe computation with the $QC_{\mathcal{C}}$. We describe the objects that comprise the information processed with the $QC_{\mathcal{C}}$ and the temporal structure of this processing. The reader who is interested in the details of the derivations is referred to [15].

3.2 The $QC_{\mathcal{C}}$ as a Universal Simulator of Quantum Logic Networks

In this section, we give an outline of the universality proof [11] for the $QC_{\mathcal{C}}$. To demonstrate universality we show that the $QC_{\mathcal{C}}$ can simulate any quantum logic network efficiently. It shall be pointed out from the beginning that the network model does not provide the most suitable description for the $QC_{\mathcal{C}}$. Nevertheless, the network model is the most widely used form of describing a quantum computer and therefore the relation between the network model and the $QC_{\mathcal{C}}$ must be clarified.

For the one-way quantum computer, the entire resource for the quantum computation is provided initially in the form of a specific entangled state – the cluster state [14] – of a large number of qubits. Information is then written onto the cluster, processed, and read out from the cluster by one-particle measurements only. The entangled state of the cluster thereby serves as a universal "substrate" for any quantum computation. Cluster states can be created efficiently in any system with a quantum Ising-type interaction (at very low temperatures) between two-state particles in a lattice configuration. More specifically, to create a cluster state $|\phi\rangle_{\mathcal{C}}$, the qubits on a cluster $\mathcal{C}$ are at first all prepared individually in a state $|+\rangle = 1/\sqrt{2}(|0\rangle + |1\rangle)$ and then brought into a cluster state by switching on the Ising-type interaction H_{int} for an appropriately chosen finite time span T. The time evolution operator generated by this Hamiltonian which takes the initial product state to the cluster state is denoted by S.

The quantum state $|\phi\rangle_{\mathcal{C}}$, the cluster state of a cluster $\mathcal{C}$ of neighboring qubits, provides in advance all entanglement that is involved in the subsequent quantum computation. It has been shown [14] that the cluster state $|\phi\rangle_{\mathcal{C}}$ is characterized by a set of eigenvalue equations

$$\sigma_x^{(a)} \bigotimes_{a' \in ngbh(a)} \sigma_z^{(a')} |\phi\rangle_{\mathcal{C}} = (-1)^{\kappa_i} |\phi\rangle_{\mathcal{C}}, \tag{3.1}$$

where $ngbh(a)$ specifies the sites of all qubits that interact with the qubit at site $a \in \mathcal{C}$ and $\kappa_i \in \{0, 1\}$ for all $i \in \mathcal{C}$. The equations (3.1) are central for the proposed computation scheme. Cluster states specified by different sets $\{\kappa_i, i \in \mathcal{C}\}$ are local unitary equivalent, i.e. can be transformed into each other by local unitary rotations of single qubits, and are thus equally good for computation. In the following we will therefore confine ourselves to the case of

$$\kappa_i = 0 \;\; \forall i \in \mathcal{C}. \tag{3.2}$$

It is important to realize here that information processing is possible even though the result of every measurement in any direction of the Bloch sphere is completely random. The reason for the randomness of the measurement results is that the reduced density operator for each qubit in the cluster state is $\frac{1}{2}\mathbf{1}$. While the individual measurement results are irrelevant for the computation, the strict correlations between measurement results inferred from (3.1) are what makes the processing of quantum information on the $QC_{\mathcal{C}}$ possible.

For clarity, let us emphasize that in the scheme of the $QC_{\mathcal{C}}$ we distinguish between cluster qubits on $\mathcal{C}$ which are measured in the process of computation, and the logical qubits. The logical qubits constitute the quantum information being processed while the cluster qubits in the initial cluster state form an entanglement resource. Measurements of their individual one-qubit state drive the computation.

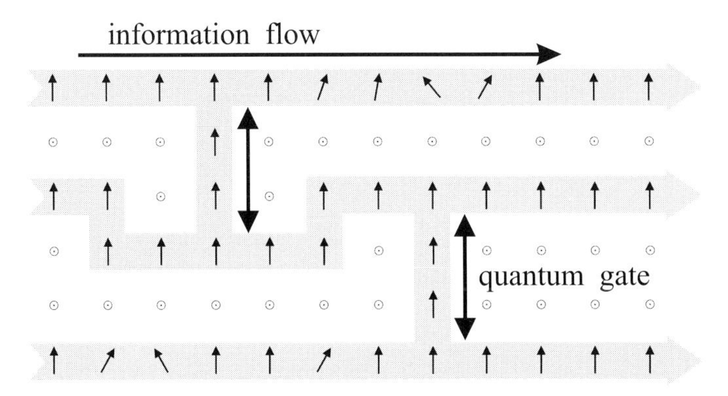

Figure 3.1: Simulation of a quantum logic network by measuring two-state particles on a lattice. Before the measurements the qubits are in the cluster state $|\phi\rangle_{\mathcal{C}}$ of (3.1). Circles $\odot$ symbolize measurements of σ_z, vertical arrows are measurements of σ_x, while tilted arrows refer to measurements in the x-y-plane.

To process quantum information with this cluster, it suffices to measure its particles in a certain order and in a certain basis, as depicted in Fig. 3.1. Quantum information is thereby propagated through the cluster and processed. Measurements of σ_z-observables effectively remove the respective lattice qubit from the cluster. Measurements in the σ_x- (and σ_y-) eigenbasis are used for "wires", i.e. to propagate logical quantum bits through the cluster, and for the CNOT-gate between two logical qubits. Observables of the form $\cos(\varphi)\,\sigma_x \pm \sin(\varphi)\,\sigma_y$ are measured to realize arbitrary rotations of logical qubits. For these cluster qubits, the basis in which each of them is measured depends on the results of preceding measurements. This introduces a temporal order in which the measurements have to be performed. The processing is finished once all qubits except a last one on each wire have been measured. The remaining unmeasured qubits form the quantum register which is now ready to be read out. At this point, the results of previous measurements determine in which basis these "output" qubits need to be measured for the final readout, or if the readout measurements are in the σ_x-, σ_y- or σ_z-eigenbasis, how the readout measurements have to be interpreted. Without loss of generality, we assume in this paper that the readout measurements are performed in the σ_z-eigenbasis.

To understand the $QC_{\mathcal{C}}$ in network model terms, in the same way as we decompose networks into gates, we would like to decompose a $QC_{\mathcal{C}}$-circuit as a simulator of a quantum logic network into simulations of quantum gates. This requires some adaption. First of all, we need to identify a quantum input and -output. To do so, we first modify the $QC_{\mathcal{C}}$-computation slightly and later remove this modification again. The modification is this: instead of creating a universal cluster state and subsequently measuring it we now allow for read-in of an arbitrary quantum input. Then, the modified procedure consists of the following steps. 1) Prepare a state $|\psi_{\text{in}}\rangle_I \otimes (\bigotimes_{j\in\mathcal{C}\setminus I} |+\rangle_j)$ where $|\psi_{\text{in}}\rangle$ is the input state prepared on a subset of the cluster qubits $I \subset \mathcal{C}$. 2) Entangle the state via the unitary evolution S generated by the Ising interaction. 3a) Measure all cluster qubits except for those of the output register $O \subset \mathcal{C}$. In

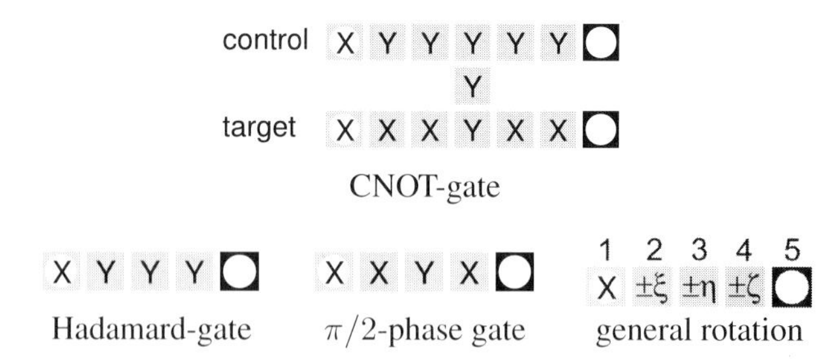

Figure 3.2: Realization of the required gates on the $QC_\mathcal{C}$. CNOT-gate between neighboring qubits, the Hadamard gate, the $\pi/2$ phase gate and the general rotation specified by its Euler angles ξ, η, ζ.

this way, the state $|\psi_{\text{in}}\rangle$ is teleported from I to O and at the same time processed. 3b) Measure the qubits in the output register O (readout).

Please note that for the "default input" $|\psi_{\text{in}}\rangle_I = \bigotimes_{i\in I} |+\rangle_i$ this modified procedure is equivalent to the original one. Then, steps 1 and 2 create a cluster state (which could as well be created by any other method) and steps 3a) and 3b) form the sequence of measurements. As will be discussed in Section 3.3, as long as the quantum input is known it is sufficient to consider $\bigotimes_{i\in I} |+\rangle_i$ and thus one-qubit measurements on cluster states.

Steps 1) - 3a) form the procedure to simulate some unitary network applied to the quantum register. It is decomposed into similar sub-procedures for gate simulation: loading the input, entangling operation, measurement of all but the output qubits. Provided the measurement bases have been chosen appropriately, a procedure of this type teleports a general input state from one part of the cluster to another and thereby also processes it.

We explain the $QC_\mathcal{C}$ as a succession of gate simulations, i.e. as repeated steps of entangling operations and measurements on sub-clusters. In reality, however, a different scheme is realized, namely first all the cluster qubits are entangled and second they are measured. These two ways to proceed are mathematically equivalent as has been demonstrated in [11]. The basic reason for this equivalence is that, in the sequential picture, later entangling operations commute with earlier measurements because they act on different particles. Therefore, the order of operations can be interchanged such that first all entangling operations and after that all measurements are performed.

In the following we review two points of the universality proof for the $QC_\mathcal{C}$: the realization of the arbitrary one-qubit rotation as a member of the universal set of gates, and the effect of the randomness of the individual measurement results and how to account for them. For the realization of a CNOT-gate see Fig. 3.2 and [11].

An arbitrary rotation $U_R \in SU(2)$ can be achieved in a chain of 5 qubits. Consider a rotation in its Euler representation

$$U_R(\xi, \eta, \zeta) = U_x(\zeta) U_z(\eta) U_x(\xi), \tag{3.3}$$

where the rotations about the x- and z-axis are $U_x(\alpha) = \exp\left(-i\alpha\frac{\sigma_x}{2}\right)$ and $U_z(\alpha) = \exp\left(-i\alpha\frac{\sigma_z}{2}\right)$. Initially, the first qubit is in some state $|\psi_{\text{in}}\rangle$, which is to be rotated, and the other qubits are in $|+\rangle$. After the 5 qubits are entangled by the time evolution operator S generated by the Ising-type Hamiltonian, the state $|\psi_{\text{in}}\rangle$ can be rotated by measuring qubits 1 to 4. At the same time, the state is also transfered to site 5. The qubits $1\dots4$ are measured in appropriately chosen bases, *viz.*

$$\mathcal{B}_j(\varphi_j) = \left\{ \frac{|0\rangle_j + e^{i\varphi_j}|1\rangle_j}{\sqrt{2}}, \frac{|0\rangle_j - e^{i\varphi_j}|1\rangle_j}{\sqrt{2}} \right\} \tag{3.4}$$

whereby the measurement outcomes $s_j \in \{0,1\}$ for $j = 1\dots4$ are obtained. Here, $s_j = 0$ means that qubit j is projected into the first state of $\mathcal{B}_j(\varphi_j)$. In (3.4) the basis states of all possible measurement bases lie on the equator of the Bloch sphere, i.e. on the intersection of the Bloch sphere with the x-y-plane. Therefore, the measurement basis for qubit j can be specified by a single parameter, the measurement angle φ_j. The measurement direction of qubit j is the vector on the Bloch sphere which corresponds to the first state in the measurement basis $\mathcal{B}_j(\varphi_j)$. Thus, the measurement angle φ_j is equal to the angle between the measurement direction at qubit j and the positive x-axis. For all of the gates constructed so far, the cluster qubits are either – if they are not required for the realization of the circuit – measured in σ_z, or – if they are required – measured in some measurement direction in the x-y-plane. In summary, the procedure to implement an arbitrary rotation $U_R(\xi,\eta,\zeta)$, specified by its Euler angles ξ,η,ζ, is this: 1. measure qubit 1 in $\mathcal{B}_1(0)$; 2. measure qubit 2 in $\mathcal{B}_2\left(-(-1)^{s_1}\xi\right)$; 3. measure qubit 3 in $\mathcal{B}_3\left(-(-1)^{s_2}\eta\right)$; 4. measure qubit 4 in $\mathcal{B}_4\left(-(-1)^{s_1+s_3}\zeta\right)$. In this way the rotation U'_R is realized:

$$U'_R(\xi,\eta,\zeta) = U_\Sigma U_R(\xi,\eta,\zeta). \tag{3.5}$$

The random byproduct operator

$$U_\Sigma = \sigma_x^{s_2+s_4}\sigma_z^{s_1+s_3} \tag{3.6}$$

can be corrected for at the end of the computation, as explained next.

The randomness of the measurement results does not jeopardize the function of the circuit. Depending on the measurement results, extra rotations σ_x and σ_z act on the output qubits of every implemented gate, as in (3.5), for example. By use of the propagation relations

$$U_R(\xi,\eta,\zeta)\,\sigma_z^s\sigma_x^{s'} = \sigma_z^s\sigma_x^{s'}\,U_R((-1)^s\xi,(-1)^{s'}\eta,(-1)^s\zeta), \tag{3.7}$$

$$\text{CNOT}\,{\sigma_z^{(t)}}^{s_t}{\sigma_z^{(c)}}^{s_c}{\sigma_x^{(t)}}^{s'_t}{\sigma_x^{(c)}}^{s'_c} = {\sigma_z^{(t)}}^{s_t}{\sigma_z^{(c)}}^{s_c+s_t}{\sigma_x^{(t)}}^{s'_c+s'_t}{\sigma_x^{(c)}}^{s'_c}\,\text{CNOT}, \tag{3.8}$$

these extra rotations can be pulled through the network to act upon the output state. There they can be accounted for by properly interpreting the σ_z-readout measurement results.

The propagation relations for the general rotations and for the CNOT gate, respectively, are different in the following respect. In the propagation relation for the CNOT gate the gate remains unchanged and the byproduct operator is modified. For rotations this is in general not possible if one demands that the byproduct operator must remain in the Pauli group, which is essential. Therefore, in the propagation relations for rotations the gate changes and the byproduct operator remains unmodified. This is the origin of adaptive measurement bases and thus the temporal structure of QC$_\mathcal{C}$-algorithms.

The reason why the propagation relation for the CNOT gate takes the form (3.8) is that it is in the Clifford group, the normalizer of the Pauli group, which means that a Pauli operator is mapped onto a Pauli operator under conjugation with any Clifford group element. There are also local rotations in the Clifford group, among them the Hadamard transformation and the $\pi/2$-phase gate. These rotations can be simulated more efficiently than by the procedure for general rotations described above. To see this, note that the measurement bases $\mathcal{B}(\varphi)$ and $\mathcal{B}(-\varphi)$ in (3.4) coincide for angles $\varphi = 0$ and for $\varphi = \pm\pi/2$. For $\varphi = 0$ the measurement basis $\mathcal{B}(\varphi)$ is the eigenbasis of σ_x, and for $\varphi = \pm\pi/2$ the measurement basis $\mathcal{B}(\varphi)$ is the eigenbasis of σ_y. In these cases, the choice of the measurement basis is not influenced by the results of measurements at other qubits. The Hadamard gate and the $\pi/2$ phase shift are such rotations. As displayed in Fig. 3.2, they are both realized by performing a pattern of σ_x- and σ_y-measurements on the cluster $\mathcal{C}$. The byproduct operators which are thereby created are

$$\begin{aligned} U_{\Sigma,H} &= \sigma_x^{s_1+s_3+s_4}\,\sigma_z^{s_2+s_3} \\ U_{\Sigma,U_z(\pi/2)} &= \sigma_x^{s_2+s_4}\,\sigma_z^{s_1+s_2+s_3+1}. \end{aligned} \tag{3.9}$$

Owing to the fact that the Hadamard- and the $\pi/2$-phase gate are in the Clifford group, the propagation relations for these rotations can also be written in a form resembling the propagation relation (3.8) for the CNOT-gate

$$\begin{aligned} H\,{\sigma_x}^{s_x}{\sigma_z}^{s_z} &= {\sigma_x}^{s_z}{\sigma_z}^{s_x}\,H, \\ U_z(\pi/2)\,{\sigma_x}^{s_x}{\sigma_z}^{s_z} &= {\sigma_x}^{s_x}{\sigma_z}^{s_x+s_z}\,U_z(\pi/2). \end{aligned} \tag{3.10}$$

As stated above, the measurement bases to implement the Hadamard- and the $\pi/2$-phase gate require no adjustment since only operators σ_x and σ_y are measured. The same holds for the implementation of the CNOT gate, see Fig. 3.2. Thus, all the Hadamard-, $\pi/2$-phase- and CNOT-gates of a quantum circuit can be implemented simultaneously in the first measurement round with no regard to their location in the network. In particular, quantum circuits which consist only of such gates, i.e. circuits in the Clifford group, can be realized in a single time step. As an example, many circuits for coding and decoding are in the Clifford group.

The fact that quantum circuits in the Clifford group can be realized in a single time step has previously not been known for networks. The best upper bound on the logical depth known so far scales logarithmically with the number of logical qubits [16]. One might therefore wonder whether the $\mathrm{QC}_\mathcal{C}$ is more efficient than a quantum computer realized as a quantum logic network. This is not the case in so far as both the quantum logic network computer and the $\mathrm{QC}_\mathcal{C}$ can simulate each other efficiently. The fact that each quantum logic network can be simulated on the $\mathrm{QC}_\mathcal{C}$ has been shown in [11]. The converse is also true because a resource cluster state of arbitrary size can be created by a quantum logic network of constant logical depth. Furthermore, the subsequent one-qubit measurements are within the set of standard tools employed in the network scheme of computation. In this sense, the operation of the $\mathrm{QC}_\mathcal{C}$ can be cast entirely in network language.

However, while the network model comprises the means that are used in a computation on a $\mathrm{QC}_\mathcal{C}$, it cannot describe *how* they have to be used. In particular, in the above construction – where a $\mathrm{QC}_\mathcal{C}$ simulating a quantum logic network is itself simulated by a more complicated network – the temporal order of the measurements and the rules to adapt the measurement bases are not provided with the network description. But without this additional information the network to simulate the $QC_\mathcal{C}$ is incomplete.

It should be noted that a link between the degree of parallelization of unitary operations and the logical depth of a quantum algorithm does not exist a priori. It is established only if quantum computation is identified with unitary evolution. The network model allows statements about how much one can parallelize networks composed of unitary gates. As an example, two unitary gates U_1, U_2 cannot be performed in parallel if they do not commute. For simulations of such gates with the $QC_\mathcal{C}$, however, this general restriction does not apply: The simulations of two non-commuting gates can still be parallelized if the gates are in the Clifford group.

With this observation we complete the survey of the universality proof [11] for the $QC_\mathcal{C}$. To summarize, for simulation of a quantum logic network on a one-way quantum computer, a set of universal gates can be realized by one-qubit measurements and the gates can be combined to circuits. Due to the randomness of the results of the individual measurements, extra byproduct operators occur. These byproduct operators specify how the readout of the simulated quantum register has to be interpreted. Also, they influence the bases of the one-qubit measurements.

In this section we have described the $QC_\mathcal{C}$ as a simulator of quantum logic networks. We adopted all the network notions such as the "quantum register" and "quantum gates". We have found an additional structure, the byproduct operator, which keeps track of the randomness introduced by the measurements. In a network-like description of the $QC_\mathcal{C}$, the byproduct operator appears as some unwanted extra complication that has to be and fortunately can be handled. In the next section we will point out in which respect the description of the $QC_\mathcal{C}$ as a network simulator is not adequate and in Section 3.4 we will present a different computational model for the $QC_\mathcal{C}$. For this model it will turn out that the byproduct operators form, in fact, the central quantities of information processing with the $QC_\mathcal{C}$, and that the "quantum register" and the "quantum gates" disappear.

3.3 Non-Network Character of the $QC_\mathcal{C}$

In the network model of quantum computation one usually regards the state of a quantum register as the carrier of information. The quantum register is prepared in some input state and processed to some output state by applying a suitable unitary transformation composed of quantum gates. Finally, the output state of the quantum register is measured by which the classical readout is obtained.

In this section we explain why the notions of "quantum input" and "quantum output" have no genuine meaning for the $QC_\mathcal{C}$ if we restrict ourselves to the situation where the quantum input state is *known*. Shor's factoring algorithm [5] and Grover's data base search algorithm [17] are both examples of such a situation. In these algorithms one always starts with the input state $\bigotimes_{i=1}^{n} 1/\sqrt{2}(|0\rangle_i + |1\rangle_i) = |++\ldots+\rangle$. Other scenarios are conceivable, e.g. where an unknown quantum input is processed and the classical result of the computation is retransmitted to the sender of the input state; or the unmeasured network output register state is retransmitted. These scenarios would lead only to minor modifications in the computational model. How to process an unknown quantum state has been briefly discussed in Section 3.2

but is not in the focus of this paper. So, let us assume that the quantum input is known. There it is sufficient to discuss the situation where an input state $|++..+\rangle_I$ is read in on some subset $I \subset \mathcal{C}$ of the cluster $\mathcal{C}$. Any other known input state can be created on the cluster from the standard quantum state $|++..+\rangle$, by some circuit preceding the main one.

Reading in an input state $\bigotimes_{i \in I} |+\rangle_i$ means to prepare the state $S[\bigotimes_{i \in I} |+\rangle_i \otimes (\bigotimes_{j \in \mathcal{C} \setminus I} |+\rangle_j)] = |\phi\rangle_{\mathcal{C}}$, i.e. to prepare nothing but a cluster state. The cluster state $|\phi\rangle_{\mathcal{C}}$ is a universal resource, no input dependent information specifies it. In this sense, the $\text{QC}_{\mathcal{C}}$ has no quantum input.

Similarly, the $\text{QC}_{\mathcal{C}}$ has no quantum output. Of course, the final result of any computation – including quantum computations – is a classical number, but for the quantum logic network the state of the output register before the readout measurements plays a distinguished role. For the $\text{QC}_{\mathcal{C}}$ this is not the case, there are just cluster qubits measured in a certain order and basis. The measurement outcomes contribute all to the result of the computation.

We have identified subsets I, O on the cluster $\mathcal{C} - I$ for the subset of the cluster which simulates the quantum register in its input state and O to simulate the quantum register in its output state – only to make the $\text{QC}_{\mathcal{C}}$ suitable for a description in terms of the network model. Such a terminology is not required for the $\text{QC}_{\mathcal{C}}$ a priori. It is not even appropriate: if, to perform a particular algorithm on the $\text{QC}_{\mathcal{C}}$, a quantum logic network is implemented on a cluster state there is a subset of cluster qubits which play the role of the output register. These qubits are not the final ones to be measured, but among the first (!).

As we have seen, the measurement outcomes from all the cluster qubits contribute to the result of the computation. The qubits from $O \subset \mathcal{C}$ simulate the output state of the quantum register and thus contribute directly. The cluster qubits in the set $I \subset \mathcal{C}$ simulate the input state of the quantum register and the outcomes obtained in their measurement contribute via the accumulated byproduct operator that is required to interpret the readout measurements on O. Finally, the qubits in the section $M \subset \mathcal{C}$ of the cluster by whose measurements the quantum gates are simulated also contribute via the byproduct operator.

Naturally there arises the question whether there is any difference in the way how measurements of cluster qubits in I, O or M contribute to the final result of the computation. As shown in [15], it turns out that there is none. This is why we abandon the notions of quantum input and quantum output altogether from the description of the $\text{QC}_{\mathcal{C}}$.

3.4 Computational Model

Quantum gates are not constitutive elements of the $\text{QC}_{\mathcal{C}}$; these are instead one-qubit measurements performed in a certain temporal order and in a spatial pattern of adaptive measurement bases. The most efficient temporal order of the measurements does not follow from the temporal order of the simulated gates in the network model. Therefore, a set of rules is required by which the optimal order of measurements can be inferred. Generally, for circuits which involve a vast number of measurements and subsequent conditional processing, it becomes essential to have an additional structure to process the classical information gained by the measurements. The $\text{QC}_{\mathcal{C}}$ provides such a structure - the information flow vector $\mathbf{I}(t)$ (see [15] and below). In fact, for the computational model of the $\text{QC}_{\mathcal{C}}$ this classical binary-valued quantity will turn out to be the central object for information processing.

So let us take a step back and look what the $QC_\mathcal{C}$ is. On the quantum level, the $QC_\mathcal{C}$ works by measuring quantum correlations of the initial universal cluster state. With the creation of the universal cluster state these quantum correlations are provided before the computation starts. In contrast to the network model, they are not created in a procedure specific to the computational problem. Therefore, for the $QC_\mathcal{C}$ there is no processing of information on the quantum level. In this sense, besides no quantum input and no quantum output, the $QC_\mathcal{C}$ has no quantum register either.

Central from the conceptual point of view but also vital for the practical realizability of the scheme is that the quantum correlations of the cluster state can be measured qubit-wise. This requires a temporal ordering of the measurements and adaptive measurement bases in accordance with previously obtained measurement results. Thus there is processing at the classical level.

The general view of a $QC_\mathcal{C}$-computation is as follows. The cluster $\mathcal{C}$ is divided into disjoint subsets $Q_t \subset \mathcal{C}$ with $0 \leq t \leq t_{\text{max}}$, i.e. $\bigcup_{t=0}^{t_{\text{max}}} Q_t = \mathcal{C}$ and $Q_s \cap Q_t = \emptyset$ for all $s \neq t$. The cluster qubits within each set Q_t can be measured simultaneously and the sets are measured one after another. The set Q_0 consists of all those qubits of which no measurement bases have to be adjusted, i.e. those of which the operator σ_x, σ_y or σ_z is measured. This comprises all the redundant qubits, the qubits to implement the Clifford part of the circuit and the qubits which simulate the network quantum output. In the subsequent measurement rounds only operators of the form $\cos\varphi\,\sigma_x \pm \sin\varphi\,\sigma_y$ are measured where $|\varphi| < \pi/2$, $\varphi \neq 0$. The measurement bases are adaptive in these rounds. The measurement outcomes from the qubits in Q_0 specify the measurement bases for the qubits in Q_1 which are measured in the second round, those from Q_0 and Q_1 together specify the bases for the measurements of the qubits in Q_2 which are measured in the third round, and so on. Finally, the result of the computation is calculated from the measurement outcomes in all the measurement rounds.

Now there arise two questions. First, "Given a quantum algorithm, how can one find the measurement pattern and in particular the temporal order in which the measurements are performed?". As for the measurement pattern, apart from a few exceptions such as the quantum Fourier transformation or the quantum adding circuit [18] presently we know no better than to straightforwardly simulate the network. Even then the optimal temporal order of the measurements is, as stated before, different from what one expects from the order of gates in the quantum logic network. The discussion of temporal complexity within the $QC_\mathcal{C}$ will lead us to objects such as the *forward cones*, the *byproduct images* and the *information flow vector* [15] which will be briefly introduced below. The second question is: "How complicated is the required classical processing?". In principle it could be that all the obtained measurement results had to be stored separately and the functions to compute the measurement bases were so complicated that one would gain no advantage over the classical algorithm for the considered problem. This is not at all the case. If the network algorithm runs on n qubits then the classical data that the $QC_\mathcal{C}$ has to keep track of is all contained in a $2n$-component binary valued vector, the information flow vector $\mathbf{I}(t)$. The update of $\mathbf{I}(t)$, the calculation to adapt the measurement bases of cluster qubits according to previous measurement outcomes and the final identification of the computational result are all elementary.

Let us first discuss the temporal ordering of the measurements. To understand how the sets Q_t of simultaneously measurable qubits are constructed we introduce the notion of forward cones. The forward cone $\text{fc}(k)$ of a cluster qubit $k \in \mathcal{C}$ is the set of all those cluster qubits

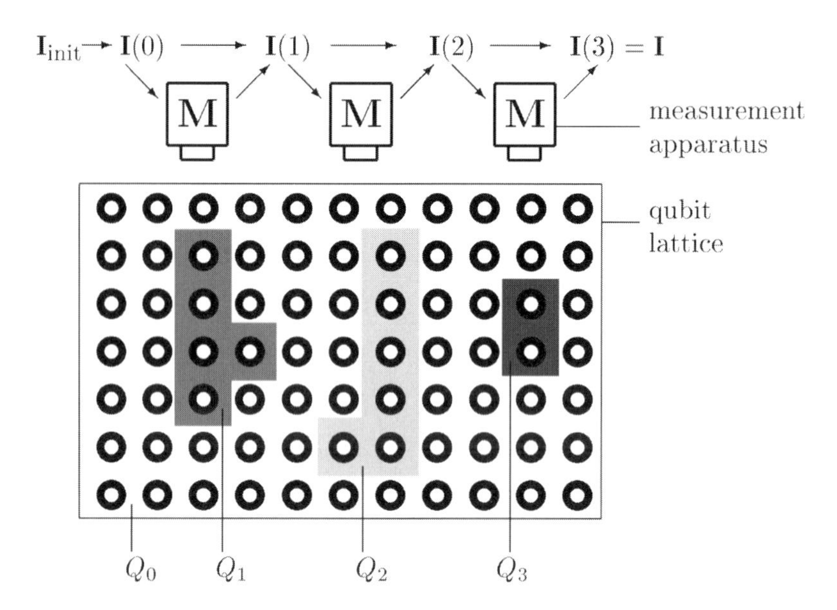

Figure 3.3: General scheme of the quantum computer via one-qubit measurements. The sets Q_t of lattice qubits are measured one after the other. The results of earlier measurements determine the measurement bases of later ones. All classical information from the measurement results needed to steer the $\text{QC}_\mathcal{C}$ is contained in the information flow vector $\mathbf{I}(t)$. After the last measurement round $t_{\rm max}$, $\mathbf{I}(t_{\rm max})$ contains the result of the computation.

$j \in \mathcal{C}$ whose measurement basis $\mathcal{B}(\varphi_{j,\rm meas})$ depends on the result s_k of the measurement of qubit k after the byproduct operator $(U_k)^{s_k}$ is propagated forward from the output side of the gate for whose implementation the cluster qubit was measured to the output side of the network. See Fig. 3.4. Similarly, the backward cone $\text{bc}(k)$ of a cluster qubit $k \in \mathcal{C}$ whose measurement bases depend upon the measurement result at qubit k when the byproduct operator is propagated backward to the input side of the network. The method to calculate the forward and backward cones follows immediately from their definitions. Quite surprisingly, it will turn out that only backward cones will appear in the computational model that finally emerges. Nevertheless, the forward cones are used to identify the sets of simultaneously measurable qubits as is explained below.

What does it mean that a cluster qubit j is in the forward cone of another cluster qubit k, $j \in \text{fc}(k)$? According to the definition, a byproduct operator created via the measurement at cluster qubit k influences the measurement angle $\varphi_{j,\rm meas}$ at cluster qubit j. To determine the measurement angle at j one must thus wait for the measurement result at k. Therefore, the forward cones generate a temporal ordering among the measurements. If $j \in \text{fc}(k)$, the measurement at qubit j is performed later than that at qubit k. This we denote by $k \prec j$

$$j \in \text{fc}(k) \Rightarrow k \prec j. \tag{3.11}$$

The relation "$\prec$" is a strict partial ordering, i.e. it is transitive and anti-reflexive. Anti-reflexivity is required for the scheme to be deterministic. Transitivity we use to generate "$\prec$" from the forward cones. This partial ordering can now be used to construct the sets $Q_t \subset \mathcal{C}$ of cluster qubits measured in measurement round t. Be $Q^{(t)} \subset \mathcal{C}$ the set of qubits which are to be measured in the measurement round t and all subsequent rounds. Then, Q_0 is the set of qubits which are measured in the first round. These are the qubits of which the observables σ_x, σ_y or σ_z are measured, so that the measurement bases are not influenced by other measurement results. Further, $Q^{(0)} = \mathcal{C}$. Now, the sequence of sets Q_t can be constructed using the following recursion relation

$$\begin{aligned} Q_t &= \left\{ q \in Q^{(t)} | \neg \exists p \in Q^{(t)} : p \prec q \right\} \\ Q^{(t+1)} &= Q^{(t)} \backslash Q_t . \end{aligned} \tag{3.12}$$

All those qubits which have no precursors in some remaining set $Q^{(t)}$ and thus do not have to wait for results of measurements of qubits in $Q^{(t)}$ are taken out of this set to form Q_t. The recursion proceeds until $Q^{(t_{\text{max}}+1)} = \emptyset$ for some maximal value t_{max} of t.

Let us now discuss the classical processing. The scheme that emerges is the following: The classical information gained by the measurements is processed within a flow scheme. The flow quantity is a classical $2n$-component binary vector $\mathbf{I}(t)$, where n is the number of logical qubits of a corresponding quantum logic network and t the number of the measurement round. This vector $\mathbf{I}(t)$, the information flow vector, is updated after every measurement round. That is, after the one-qubit measurements of all qubits of a set Q_t have been performed simultaneously, $\mathbf{I}(t-1)$ is updated to $\mathbf{I}(t)$ through the results of these measurements. In turn, $\mathbf{I}(t)$ determines which one-qubit observables are to be measured of the qubits of the set Q_{t+1}. The result of the computation is given by the information flow vector $\mathbf{I}(t_{\text{max}})$ after the last measurement round. From this quantity the result of the readout measurement on the quantum register in the corresponding quantum logic network can be read off directly without further processing.

In the following we briefly explain how this model arises. We already mentioned the accumulated byproduct operator $U_\Sigma|_\Omega$ which is the product of all the $(U_{\Sigma,k})^{s_k}$, the forward propagated byproduct operators randomly created by the measurement of qubits k with outcome s_k. $U_\Sigma|_\Omega$ determines how the readout has to be interpreted. Now note that the readout measurement results can themselves be expressed in terms of a byproduct operator. Let the quantum register be in the state $|\psi_{\text{out}}\rangle = \bigotimes_{i \in O} |s_i\rangle_{i,z}$ after readout. Then $|\psi_{\text{out}}\rangle$ can be written as $|\psi_{\text{out}}\rangle = U_R \bigotimes_{i \in O} |0\rangle_i$ with $U_R = \bigotimes_{i \in O} (\sigma_x^{(i)})^{s_i}$, i.e. as a byproduct operator U_R acting on some standard state $|0\rangle_O$. This standard state contains no information and can henceforth be discarded. The result of the computation is contained in the byproduct operators. It can be directly read off from the x-part of the operator $U_{\Sigma,R}|_\Omega = U_\Sigma|_\Omega U_R$.

If one discards the sign of these Pauli operators – an unphysical global phase – they can be mapped onto elements of a $2n$-dimensional discrete vector space $\mathcal{V}$. For this we use the isomorphism

$$\mathcal{I}:\ I \in \mathcal{V} \longrightarrow \mathcal{P}/\{\pm 1\} \ni U = \prod_{i=1}^{n} \left(\sigma_x^{(i)}\right)^{[I_x]_i} \left(\sigma_z^{(i)}\right)^{[I_z]_i}, \tag{3.13}$$

where $[I_x]_i, [I_z]_i \in \{0,1\}$ are the respective components of I_x, I_z and $\mathcal{P}$ denotes the Pauli group. In particular, the x-part of $\mathbf{I} = \mathcal{I}^{-1}(U_{\Sigma,R}|_\Omega)$ represents the result of the quantum computation, corresponding to the readout in the network model.

To establish the terminology in which the classical information processing with the $\mathrm{QC}_\mathcal{C}$ is described, we introduce the *byproduct images* and the *symplectic scalar product*. The byproduct image $\mathbf{F}_k$ of a cluster qubit $k \in \mathcal{C}$ is defined as $\mathbf{F}_k = \mathcal{I}^{-1}(U_{\Sigma,k}|_\Omega)$. Note that in the byproduct operators for the implementation of the CNOT- and the $\pi/2$-phase gate there are additional contributions which do not depend upon measurement results, i.e. the byproduct operators are not the identity for all measurement results being zero. These additional byproduct operators have their byproduct images as well. Since they cannot be related to a particular cluster qubit we attribute them to the gate g by whose implementation they are introduced and denote them by $\mathbf{F}_g$.

Byproduct images are easier to manipulate than the forward propagated byproduct operators to which they correspond via $\mathcal{I}$. There hold the relations $\mathcal{I}(\mathbf{F}_k + \mathbf{F}_l) = \pm\mathcal{I}(\mathbf{F}_k)\mathcal{I}(\mathbf{F}_l)$, $\mathcal{I}(s_k\mathbf{F}_k) = \pm(\mathcal{I}(\mathbf{F}_k))^{s_k}$. The symplectic scalar product of two byproduct images $\mathbf{F}_k$, $\mathbf{F}_l$ is defined as

$$(\mathbf{F}_k, \mathbf{F}_l)_S = \mathbf{F}_{k,x}^T\mathbf{F}_{l,z} + \mathbf{F}_{k,z}^T\mathbf{F}_{l,x} \bmod 2. \tag{3.14}$$

It is invariant under the Clifford group. A first application of the objects introduced above is the cone test [15],

$$\forall\, k \in \mathcal{C}, j \in Q^{(1)}: \;\; j \in \mathrm{fc}(k) \vee j \in \mathrm{bc}(k) \Longleftrightarrow (\mathbf{F}_j, \mathbf{F}_k)_S = 1. \tag{3.15}$$

Whether a qubit lies in some other qubits backward or forward cone can be read off from the respective byproduct images.

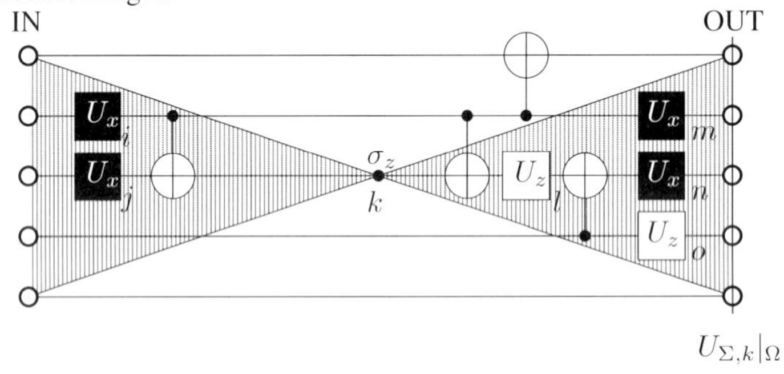

Figure 3.4: Forward and backward cones. The measurement of a cluster qubit k for the implementation of the shown quantum logic network may, depending on the measurement outcome, result in a byproduct operator σ_z (the underlying cluster and measurement pattern is not shown). This byproduct operator is propagated forward to act upon the output register as $U_{\Sigma,k}|_\Omega$. In forward propagation, it flips the measurement angles of the cluster qubits m, n by whose measurement one-qubit rotations are implemented. The cluster qubits m and n are thus in the forward cone of k, $m, n \in \mathrm{fc}(k)$, while l and o are not. Similarly, $i, j \in \mathrm{bc}(k)$.

We are now ready to discuss the process of computation. The goal is to collect all the byproduct images weighted with the measurement results, i.e. to finally obtain $\mathbf{I} =$

$\mathcal{I}^{-1}(U_{\Sigma,R}|_\Omega) = \sum_g \mathbf{F}_g + \sum_{k\in C} s_k \mathbf{F}_k$ with all the cluster qubits measured in the correct basis. Before the computation starts we propagate forward all the byproduct operators attributed to the gates since they do not depend on any measurement results. This reverses a number of angles specifying one-qubit rotations in the network to simulate. In this way, we obtain the algorithm angles $\{\varphi_{j,\text{algo}}^{\text{init}}\}$ from the network Euler angles. Further we collect the byproduct images of the gates, $\mathbf{I}_{\text{init}} = \sum_g \mathbf{F}_g$.

In the first measurement round we measure all the cluster qubits $k \in Q_0$ thereby removing the redundant qubits, implementing the Clifford gates of the circuit and measuring the "output register". This leaves us with byproduct operators scattered all over the place. These byproduct operators we propagate forward and include their byproduct images into the information flow vector at time $t = 0$, $\mathbf{I}(0) = \mathbf{I}_{\text{init}} + \sum_{k\in Q_0} s_k \mathbf{F}_k$. In forward propagation, the byproduct operators reverse some of the algorithm angles. In this way, we update the algorithm angles $\{\varphi_{j,\text{algo}}^{\text{init}}\}$ to the modified algorithm angles $\{\varphi_{j,\text{algo}}^{0}\}$, $\varphi_{j,\text{algo}}^{0} = (-1)^{\eta_j} \varphi_{j,\text{algo}}^{\text{init}}$ with $\eta_j = \sum_{k\in Q_0\,|\,j\in \text{fc}(k)} s_k$.

In subsequent measurement rounds we measure cluster qubits in adapted bases. This also produces byproduct operators in the middle of the network to simulate and we propagate them forward as before. The update of $\mathbf{I}(t)$ is just the same as in the first round. We find

$$\mathbf{I}(t) = \mathbf{I}(t-1) + \sum_{k\in Q_t} s_k \mathbf{F}_k = \mathbf{I}_{\text{init}} + \sum_{k\in \bigcup_{r=0}^{t} Q_r} s_k \mathbf{F}_k . \tag{3.16}$$

After the final update, $\mathbf{I}(t_{\text{max}}) = \mathbf{I}$ contains the result of the computation in its x-part.

As in the first round, the propagation of byproduct operators affects the angles that specify the measurement bases. The update of these angles in all the subsequent measurement rounds could be performed in the same way as in the first round providing one with a complete history $\{\varphi_{j,\text{algo}}^{t}\}$ of adapted angles. It is not necessary to generate and store this bulk of information. We only need to know the adapted angles $\varphi_{j,\text{meas}}$ at the time when the respective qubits are measured. These measurement angles $\varphi_{j,\text{meas}}$ can be obtained in a more compact procedure.

This leads to the question which measurement outcomes affect the choice of the measurement basis at qubit $j \in Q_t$, $t > 0$. All the measurement outcomes obtained at qubits $k \in \mathcal{C}$ with $j \in \text{fc}(k)$ contribute, i.e.

$$\begin{aligned} \varphi_{j,\text{meas}} &= \varphi_{j,\text{algo}}^{\text{init}} \, (-1)^{\sum_{k\in \mathcal{C}\,|\,j\in\text{fc}(k)} s_k} \\ &= \varphi_{j,\text{algo}}^{0} \, (-1)^{\sum_{k\in \mathcal{C}\backslash Q_0\,|\,j\in\text{fc}(k)} s_k} . \end{aligned} \tag{3.17}$$

In the second line of (3.17) we can now simplify the sign factor by use of the cone test (3.15). First note that the sum over $k \in \mathcal{C}\backslash Q_0$ reduces to a sum over $k \in \bigcup_{r=1}^{t-1} Q_r$ because otherwise $j \notin \text{fc}(k)$. Then, for $k \in Q_r, j \in Q_t$, $1 < r < t$ qubit j may only be in the forward cone of qubit k, but never in the backward cone $\text{bc}(k)$. Hence, the cone test simplifies to $j \in \text{fc}(k) \iff (\mathbf{F}_j, \mathbf{F}_k)_S = 1$ for such qubits, and we obtain $\sum_{k\in \mathcal{C}\backslash Q_0\,|\,j\in\text{fc}(k)} s_k = \sum_{k\in \bigcup_{r=1}^{t-1} Q_r} s_k (\mathbf{F}_k, \mathbf{F}_j)_S = (\mathbf{I}(t-1) - \mathbf{I}(0), \mathbf{F}_j)_S$, and thus

$$\varphi_{j,\text{meas}} = \varphi_{j,\text{algo}}' \, (-1)^{(\mathbf{I}(t-1),\mathbf{F}_j)_S} , \tag{3.18}$$

with $\varphi_{j,\text{algo}}' = \varphi_{j,\text{algo}}^{0} \, (-1)^{(\mathbf{I}(0),\mathbf{F}_j)_S}$. If one works this out one finds that the angles $\varphi_{j,\text{algo}}'$ are obtained from the corresponding Euler angles of the network by propagating the byproduct operators of the gates and of the qubits measured in the first round *backwards* to the input side of the network.

To sum up, the $2n$-component binary valued information flow vector represents the information that is processed with the $\mathrm{QC}_{\mathcal{C}}$. Although random in its numerical value after all measurement rounds but the final one, it has a meaning in every step of the computation. The rule for the adaption of measurement bases (3.18) invokes the random measurement results of qubits $k \in \mathcal{C} \backslash Q_0$ *only* via the information flow vector. The measurement results on qubits $k \in Q_0$ are absorbed into the angles $\varphi'_{j,\mathrm{algo}}$ and can be erased after these angles have been set. The angles $\varphi'_{j,\mathrm{algo}}$ remain unchanged in the further course of computation. After the final update at $t = t_{\mathrm{max}}$, when there are no measurement bases left to adjust, the information flow vector $\mathbf{I}(t_{\mathrm{max}})$ displays the result of the computation. The update of $\mathbf{I}(t)$ (3.16) and the rule to adjust the measurement angles (3.18) are very simple algebraic operations.

3.5 Conclusion

We have reviewed the computational model underlying the one-way quantum computer [15], which is very different from the quantum logic network model. The logical depth of certain algorithms is on the $\mathrm{QC}_{\mathcal{C}}$ lower than has so far been known for networks. As an example, on the $\mathrm{QC}_{\mathcal{C}}$ circuits composed of CNOT-, Hadamard- and $\pi/2$-phase gates have unit logical depth, independent of the number of gates or logical qubits. The best bound for networks known previously scales logarithmically. It therefore seems that the question of temporal complexity must be revisited.

The formal description of the $\mathrm{QC}_{\mathcal{C}}$ is based on primitive quantities of which the most important are the sets $Q_t \subset \mathcal{C}$ of cluster qubits defining the temporal ordering of measurements on the cluster state, and the binary valued information flow vector $\mathbf{I}(t)$ which is the carrier of the algorithmic information. Much of the terminology that one is familiar with from the network model has been abandoned since in case of the $\mathrm{QC}_{\mathcal{C}}$ no proper meaning can be assigned to these objects. In fact, the $\mathrm{QC}_{\mathcal{C}}$ has no quantum input, no quantum output, no quantum register and it does not consist of quantum gates.

The $\mathrm{QC}_{\mathcal{C}}$ is nevertheless quantum mechanical as it uses a highly entangled cluster state as the central physical resource. It works by measuring quantum correlations of the universal cluster state.

Acknowledgements

We would like to thank D. E. Browne and H. Wagner for helpful discussions.

References

[1] D. Deutsch, *Proc. R. Soc. London Ser. A* **1985**, 400, 97.

[2] E. Bernstein and U. Vazirani, *Proc. of the 25th Annual ACM Symposium on Theory of Computing* **1993**, 11.

[3] D. Deutsch, *Proc. R. Soc. London Ser. A* **1989**, 425, 73.

[4] A. Yao, *Proc. of the 34th Annual IEEE Symposium of Foundations of Computer Science* **1993**, 352.

[5] P. W. Shor, *SIAM J. Sci. Statist. Comput.* **1997**, 26, 1484.

[6] A. Barenco et al., *Phys. Rev. A* **1995**, 52, 3457.

[7] D. P. DiVincenzo, *Fortschritte der Physik* **2000**, 48, 771.

[8] R. B. Griffiths and C.-S. Niu, *Phys. Rev. Lett.* **1996**, 76, 3228.

[9] M. A. Nielsen and I. L. Chuang, *Phys. Rev. Lett.* **1997**, 79, 321.

[10] D. Gottesman and I. L. Chuang, *Nature (London)* **1999**, 402, 390.

[11] R. Raussendorf and H.-J. Briegel, *Phys. Rev. Lett.* **2001**, 86, 5188.

[12] E. Knill, R. Laflamme and G. J. Milburn, *Nature (London)* **2001**, 409, 46.

[13] M. A. Nielsen, *quant-ph/0108020* **2001**; A. Fenner and Y. Zhang, *quant-ph/0111077* **2001**; D. W. Leung, *quant-ph/0111122* **2001**.

[14] H.-J. Briegel and R. Raussendorf, *Phys. Rev. Lett.* **2001**, 86, 910.

[15] R. Raussendorf and H.-J. Briegel, Quant. Inf. Comp. **2002**, 6, 443.

[16] C. Moore and M. Nilsson, *quant-ph/9808027* **1998**.

[17] L. K. Grover, *Phys. Rev. Lett.* **1997**, 79, 325.

[18] R. Raussendorf, D. E. Browne and H.-J. Briegel: "Quantum computation via one-qubit measurements on cluster states", *In preparation.*

4 Quantum Correlations as Basic Resource for Quantum Key Distribution

M. Curty[1], *O. Gühne*[2,3], *M. Lewenstein*[3], *and N. Lütkenhaus*[1]

[1] Institut für Theoretische Physik
Universität Erlangen-Nürnberg
Germany

[2] Institut für Quantenoptik und Quanteninformation
Österreichische Akademie der Wissenschaften
Innsbruck, Austria

[3] Institut für Theoretische Physik
Universität Hannover
Germany

4.1 Introduction

Quantum key distribution (QKD) is a fascinating subject and is a prime example of the interdisciplinary character of the field of quantum information. Over recent years, it has seemed to be quite stagnant to people outside of the field. Recently, however, new developments have again put it into the focus of researchers with interest in practical applications. Moreover, also researchers interested in the fundamental limits and opportunities that quantum mechanics offers have turned their attention to this field.

The field of classical communication theory with its rich background in information theoretic security brings in new approaches and views that allow improvement of the performance of practical quantum key distribution devices. This improvement does not come from a tuning of the physical setup; instead, a mere change of classical communication protocols achieves this. These protocols, e.g. error correction and privacy amplification, always accompany quantum key distribution. An important tool in this context is two-way communication, which allows, e.g., the post-selection of data, as is used in its simplest form already in the so-called *sifting* in the Bennett–Brassard protocol. It is of fundamental and practical interest to explore the limits of these processes. While doing so, one can discover further principles on which quantum key distribution can be based and allow, e.g., the creation of a secret key from bound entangled states [21].

In Sec. 4.2 we will review the results of classical information science dealing with information theoretic security. In Sec. 4.3, we will link the ideas of security proofs in the quantum protocols to the classical results and give the general picture of achievable secure rates and upper bounds. Especially, we will show that a necessary precondition for successful QKD is that the data obtained experimentally allow us to verify some *effective* entanglement between sending and receiving parties. In Sec. 4.4 we will show how one can systematically search for this verifying proof of effective entanglement. In Sec. 4.5 we will illustrate this point for the 2-, 4-, and 6-state protocols. Moreover, in Sec. 4.6 we show how to apply the developed

Quantum Information Processing, 2nd Edition. Edited by Thomas Beth and Gerd Leuchs

ISBN: 3-527-40541-0

methods for these qubit-based protocols. In Sec. 4.7 we will briefly point out how to extend these results to a realistic implementation.

4.2 Background of Classical Information Theoretic Security

A secret key is a basic ingredient to many applications in cryptography. The most prominent example is the implementation of information theoretic secure communications via the one-time pad (Vernam cipher) [36]. The requirement on the secret key K, formed of n bits, for this use is that it is equiprobably distributed from the point of view of a potential eavesdropper. This implies that we have $H(K) = n$ where $H(K)$ is the Shannon entropy of the distribution of the key. Of course, in creating a secret key we will always have some imperfection. However, the key creation protocol should allow us to make the difference between the actual distribution and the ideal one arbitrarily small as some security parameter s of the key generation protocol grows. The measure for this difference has to be chosen such that the requirement of composability [1] holds and we can guarantee that the key generated in this manner can be used effectively as input in any protocol requiring an ideal secret key without compromising its security.

The problem of creating a secret key has been studied in depth in classical information theory. The usual starting point is to consider random variables A, B, and E that have a joint probability distribution $P(A, B, E)$ from which identical and independent draws are being made. The random variables are in the hands of three parties, commonly called Alice, Bob and Eve. Such joint distributions can come, e.g., from the Wyner wire tap model [39] where B is obtained from A via a symmetric binary channel with some error rate, and E is obtained from B via a second, independent symmetric binary channel with some other error rate. Wyner showed that in this scenario it is possible for Alice and Bob to communicate in secrecy, e.g. by creating a secret key between them and then using the one-time pad. We consider the more general scenario studied by Csiszár and Körner [11]. From the joint probability distribution $P(A, B, E)$ we can deduce the values of the pairwise mutual information, $I(A; B)$, $I(B; E)$ and $I(A; E)$. This is done by calculating the Shannon entropy of joint and marginal probability distributions, e.g. as

$$H(A, B) = \sum_{\binom{a \in A}{b \in B}} -P(a, b) \log_2 P(a, b).$$

Then the mutual Shannon information can be expressed, e.g., as $I(A; B) = H(A) + H(B) - H(A, B)$. Csiszár and Körner showed that it is possible for Alice and Bob to communicate in secrecy at a rate $S(A; B||E)$ of at least

$$S(A; B||E) \geq \max \{I(A; B) - I(A; E),\ I(A; B) - I(B; E)\}. \tag{4.1}$$

This rate can be achieved with one-way communication from Alice to Bob (Bob to Alice) if the maximum is obtained via $I(A; B) - I(A; E)$ ($I(A; B) - I(B; E)$). For this we perform one-way error correction[1] at the Shannon limit followed by generalized privacy amplification [4]. Even if the correlations do not allow one to generate a key according to this method, Maurer [28] showed that there are situations where nevertheless a secret key can be established by preprocessing the data in a so-called *advantage distillation step*. Maurer and Wolf [27] also provided an upper bound for the distillation rate of a secure key from correlated data via authenticated public communication, which is given by the *intrinsic information* $I(A;B\downarrow E)$. These authors proved that the rate of secret bits $S(A;B||E)$ that Alice and Bob can get by communicating with each other through a public authenticated channel satisfies

$$S(A;B||E) \leq I(A;B\downarrow E) = \min_{E\to\bar{E}} I(A;B|\bar{E}), \tag{4.2}$$

where the minimization runs over all possible channels $E \to \bar{E}$ characterized by the conditional probability $P(\bar{E}|E)$, and $I(A;B|\bar{E})$ is the mutual information between Alice and Bob given the public announcement of Eve's data based on the probabilities $P(A,B,\bar{E})$. This quantity is defined in terms of the Shannon entropy of conditional probability distributions, e.g.

$$H(X|\bar{E}) = \sum_{\binom{\bar{e}\in\bar{E}}{x\in X}} -p(x|\bar{e})\log_2 p(x|\bar{e})$$

as

$$I(A;B|\bar{E}) = \left[H(A|\bar{E}) + H(B|\bar{E}) - H(A,B|\bar{E})\right]. \tag{4.3}$$

Note that these results, e.g. the lower bound by Csiszár and Körner and the upper bound by Maurer and Wolf, were derived in the scenario where the three parties share *classical* correlations.

4.3 Link Between Classical and Quantum

Let us now consider generic quantum protocols of the following form: In a first phase, Alice and Bob exchange quantum signals and measure them. After that, they share classical random variables, while Eve, who interferes with the quantum signals, holds a quantum system that summarizes and contains all information about the signals that are available to her. She can act collectively by making joint measurements on quantum systems which have individually interacted with the quantum signals exchanged between Alice and Bob, or she can act coherently, i.e., she can prepare initially one quantum system of suitable dimension which interacts with all signals before being measured. The latter method allows a larger class of correlations between Alice and Bob and represents the most general strategy Eve can adopt.

Two types of schemes are used to create the correlated data in the first phase of QKD. In *entanglement-based* (EB) schemes an, in general, untrusted third party distributes a bipartite state to Alice and Bob. This party may even be Eve, who is in possession of a third subsystem

[1] Note that it is a non-trivial task to perform efficient one-way error correction. Current QKD implementations typically perform the efficiently implementable two-way error correction of Brassard and Salvail [7].

possibly entangled with those given to Alice and Bob. While the subsystems measured by Alice and Bob result in correlations described by $P(A, B)$, Eve can use her subsystem to obtain information about the data of the legitimate users.

In *prepare & measure* (P&M) schemes Alice prepares a random sequence of predefined non-orthogonal states $|\varphi_i\rangle$ that are sent to Bob through an untrusted quantum channel (possibly controlled by Eve). On the receiving side, Bob performs a positive operator value measure (POVM) on every signal he receives. Generalizing the ideas introduced by Bennett *et al.* [5], the signal preparation process in P&M schemes can be thought of as follows: Alice prepares an entangled bipartite state of the form $|\Psi\rangle_{AB} = \sum_i \sqrt{p_i}|\alpha_i\rangle|\varphi_i\rangle$, where the states $|\alpha_i\rangle$ form an orthonormal basis and $\{p_i\}_i$ represents the *a priori* probability distribution of the signal states $|\varphi_i\rangle$. If now Alice measures the first system in the basis $|\alpha_i\rangle$, she effectively prepares the (non-orthogonal) signal states $|\varphi_i\rangle$ with probabilities p_i. The action of the quantum channel on the state $|\Psi\rangle_{AB}$ leads to an effective bipartite quantum state shared by Alice and Bob. One important difference between P&M schemes with effective entanglement and EB schemes with real entanglement is that in the first case the reduced density matrix of Alice, $\rho_A = \mathrm{Tr}_B(|\Psi\rangle\langle\Psi|_{AB})$, is fixed and known and cannot be modified by Eve.

Once the quantum states have been exchanged and measured, the second phase of key generation protocols starts; Alice and Bob now use an authenticated public channel[2] to perform auxiliary protocols, such as error correction, privacy amplification and advantage distillation, to obtain a secret final key.

The advantage of QKD with respect to the classical scenario is that quantum mechanics allows us to say something about the quantum correlations Eve shares with Alice and Bob only by looking at the correlations shared by Alice and Bob. There are several situations where we can be sure that Eve is completely decoupled from Alice's and Bob's data. The first situation is where Alice and Bob share a maximally entangled state, which can be proven by local measurements and public discussion alone. A second situation is that where Alice sends a random sequence of non-orthogonal signal states to Bob who can verify that those signals arrive without change. These principles are exploited in security proofs.

As stressed before, unlike in the classical case, now Eve does not hold classical random variables, and therefore it is not sufficient to bound the quantities $I(A; B) - I(A; E)$ and $I(A; B) - I(B; E)$ in an attempt to utilize the Csiszár–Körner lower bound on the secret key rate. Recent results for collective [15] and coherent attacks [10] give rise to quantum versions of the Csiszár–Körner bounds. Another method to obtain provable secret key rates makes use of quantum error correction codes. Shor and Preskill [34] showed that the use of CSS quantum error correction codes is equivalent to a standard BB84 protocol [3] using one-way error correction and privacy amplification, so that this method is a quantum information theoretical analog to the Csiszár–Körner method. Also earlier proofs by Mayers [29, 30] contain these elements.

There are also methods that correspond to the advantage distillation method of Maurer. Consider that Alice and Bob share partially entangled qubits of such a form that an immediate measurement would not allow us to create a secret key, e.g., following Shor and Preskill, e.g.

[2]Note that the authentication can be obtained using the Wegman–Carter authentication method [38] that is unconditionally secure. It requires, however, an initial short secret key, thereby turning the key distribution protocol into an unconditional secure key growing protocol. This task cannot be achieved by purely classical means, so that the resulting protocol still demonstrates the power of quantum information theory.

the error rate is above 11 %. In this situation we know that we can still perform a quantum protocol that distills maximally entangled states [6, 14]. These can then be measured, and the correlations can be used in a Shor–Preskill type of fashion. However, this approach would require the use of coherent quantum manipulations by Alice and Bob. It turns out that some features of this approach can be transferred to the situation where Alice and Bob are allowed to perform only classical postprocessing of their data, which now explicitly uses two-way communication. Gottesman and Lo [18] presented a postselection procedure that can create a secret key even when the correlations between Alice and Bob are too noisy to obtain a secret key according to Shor and Preskill.

We see that the quantum domain has similar instruments as the classical case, and we would like to investigate what kind of correlations created in the first phase can lead to a secret key at all. Subsequently it is of interest to find bounds on key rates.

As a starting point for this consideration, note that the result of Maurer and Wolf can as well be adapted to the case where Alice, Bob and Eve start sharing a tripartite quantum state instead of a joint probability distribution. For this purpose, one can consider all possible tripartite states that Eve can establish using her eavesdropping method, and all possible measurements she could perform on her subsystem. This gives rise to a set of possible extensions $\mathcal{P}$ of the observable probability distribution $P(A, B)$ to $P(A, B, E)$. Now one can define the intrinsic information as

$$I(A; B \downarrow E) = \inf_{\mathcal{P}} I(A; B|E). \tag{4.4}$$

The main consequence of this fact is that, whenever the observable data $P(A, B)$ can be explained as coming from a tripartite state with a separable reduced density matrix for Alice and Bob, the intrinsic information vanishes and therefore no secret key can be established.

Observation [13] Assume that the observable joint probability distribution $P(A, B)$ together with the knowledge of the corresponding measurements performed by Alice and Bob can be interpreted as coming from a separable state σ_{AB}. Then the intrinsic information vanishes and no secret key can be distilled via public communication from the correlated data.

This observation has an important implication on the performance of quantum key distribution devices based on such a two-phase procedure. The generic picture is shown in Fig. 4.1. In such a picture we fix the signal states and the measurement devices of Alice and Bob. The observed correlations $P(A, B)$ are then simulated via a typical channel model. If we fix additionally the classical communication protocols, then we obtain security proofs that show that certain key rates can be obtained as a function of the transmission distance with some known upper bound on the distance. However, this upper bound is, typically, not of fundamental nature. For example, some proof techniques might be improved, or better classical protocols might be chosen to exploit the correlations $P(A, B)$. However, there are fundamental bounds that are based only on the correlations, not the protocols. At some distance, typically, we cannot exclude separable states as the origin of the observed correlations, and therefore no secret key can be extracted beyond that point.

Below that critical point it is still possible to put a bound on the achievable secret key rate. In [21] it has been shown that the secret key rate is bounded by the regularized relative

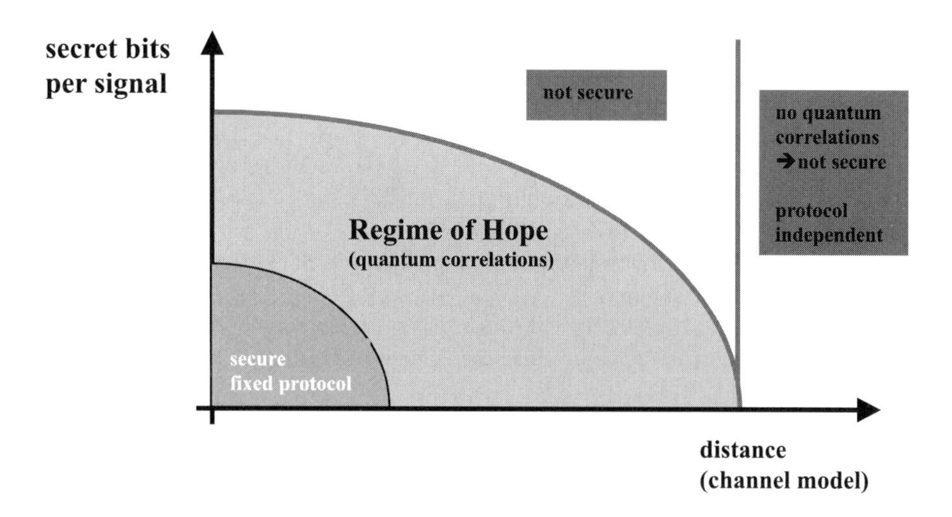

Figure 4.1: The generic performance diagram for practical quantum key distribution devices. The secret key rate per signal is plotted as a function of the distance in some channel model for the quantum signals. We see the domain of provable secret key rates for specified classical protocols, the upper bound on distance is determined by vanishing proven quantum correlations, and an upper bound on the secret key rate. The upper bounds are given only in terms of the observed correlations of the apparatus, and they are independent of the classical protocols chosen to extract the secret key from the correlated data.

entropy of entanglement. Here a minimization with respect to all density matrices compatible with the observable data can be performed. Another bound comes directly from the definition of intrinsic information and is given by the Shannon mutual information between Alice and Bob, $I(A;B)$.

One can therefore draw a diagram that shows, on the one hand, how much potential there might be in the correlations generated by a setup. The shown limitations cannot be shifted by more ingenious public discussion protocols. On the other hand it shows which rates can already be obtained by specified protocols. It is the goal of current research to divide up the "Regime of Hope" which is situated between the provably achievable secret key rates and the provable not achievable secret key rates to either of the two regions. Considering this type of diagram allows us to search for experimental set-ups that have at least the potential to meet given expectations on rate and distance.

4.4 Searching for Effective Entanglement

Given that quantum correlations are necessary for distilling a secure secret key, the question now is how to detect these quantum correlations in a given QKD scheme. More precisely, we have to answer the question whether the joint probability distribution $P(A,B)$, coming from the measurements performed by Alice and Bob during the protocol, allows them to conclude whether the effectively distributed state was entangled or not. In principle, any separability criteria might be employed to deliver this entanglement proof. The important question here is whether the chosen criterion can be used to provide a necessary and sufficient condition to detect entanglement when the knowledge about the state is not tomographically complete. As we will see below, it is a property of entanglement witnesses (EWs) that they allow one to obtain a necessary and sufficient criterion for separability even when the state cannot be completely reconstructed [13].

Let us first consider EB schemes. In these schemes, Alice and Bob perform some measurements on a bipartite quantum state distributed by an, in general, untrusted third party and retrieve the probability distribution $P(A, B)$ of the outcomes. Before showing that in this scenario EWs are especially appropriate to detect entanglement, let us recall some facts about witnesses [23, 35].

A witness is a Hermitian observable W with a positive expectation value on all separable states. So if a state ρ obeys $\mathrm{Tr}(\rho W) < 0$, the state ρ must be entangled. We say then that the state ρ is detected by W. In general, for every entangled state there exists a witness detecting it; however, this witness is in most cases very difficult to construct. Witnesses can be *optimized* in the following sense: A witness W_1 is called *finer* than another witness W_2 if W_1 detects all the states that are detected by W_2 and some states in addition. Finally, a witness W is called *optimal*, when there is no other witness that is finer than W [25]. Now we can state the following:

Theorem 1. [12, 13] *Assume that Alice and Bob can perform some local measurements with POVM elements $A_i \otimes B_i$, $i = 1, \ldots, n$, to obtain the probability distribution of the outcomes $P(A, B)$ on the distributed state ρ. Then the correlations $P(A, B)$ cannot originate from a separable state if and only if there is an EW of the form $W = \sum_i c_i\, A_i \otimes B_i$ which detects the effectively distributed state, i.e.,* $\mathrm{Tr}(W\rho) = \sum_i c_i\, P(A_i, B_i) < 0$.

We refer to witnesses that can be evaluated with the given POVM elements and the corresponding correlations $P(A, B)$ as *accessible*. According to Theorem 1, the set of all accessible witness operators gives rise to a necessary and sufficient condition for verifiable entanglement contained in the correlations $P(A, B)$: The joint probability distribution $P(A, B)$ can come exclusively from an entangled state if and only if at least one accessible witness in the set gives rise to a negative expectation value when it is evaluated with $P(A, B)$. Of course, in this set there is some redundancy. Typically, it contains witnesses that are finer than others, and therefore one can construct smaller sets of witnesses that are accessible and still have the property of being necessary and sufficient for verifying entanglement. Whenever this property holds, we refer to a set of witnesses $\mathcal{W}$ as being a *verification set*. The ultimate goal will be to obtain a *minimal verification set* in a compact description that contains no further redundancies to allow an efficient systematic search for verifiable entanglement by evaluating the members of this set. The rest of this paper is mainly concerned with the search for these minimal verification sets.

Before starting our quest for minimal verification sets, let us consider the case of P&M schemes, since in this section we have considered, so far, only EB schemes. As we mentioned previously, in this kind of scheme the reduced density matrix of Alice is fixed since Eve has no access to the state of Alice to try to modify it. However, this situation can also be incorporated in Theorem 1: We can add to the observables $A_i \otimes B_i$ other observables $C_i \otimes 1$ such that the observables C_i form a tomographic complete set of Alice's Hilbert space. Those witnesses that can be evaluated with this combined set of measurements can clearly be evaluated with the measurements $A_i \otimes B_i$ and the knowledge of the reduced density matrix of Alice.

Finally, let us emphasize again that there are many other separability criteria besides EWs that might be used for the detection of entanglement in quantum cryptographic schemes. For instance, the security of the first EB scheme proposed by Ekert in 1991 [16], the 4-state EB scheme, was based on the detection of quantum correlations by looking at possible violations of Bell inequalities. This criterion, or e.g. those based on uncertainty relations [19, 20], is directly linked to experimental data, which makes the implementation simple. Another interesting criterion that seems to be particularly suited for the case of P&M schemes, where the reduced density matrix of Alice is fixed and known, is, e.g., the reduction criterion [9, 22]. However, it is not clear whether these criteria guarantee to detect *all* entangled states that can be detected with the given set of measurements. In fact, in the case of the 4-state protocol, the knowledge of the performed measurements together with $P(A, B)$ allows the detection of entangled states beyond those that violate Bell-like inequalities.

4.5 Verification Sets

Now we illustrate the consequences of this point of view for some well-known qubit-based QKD protocols: the 6-state [8], the 4-state [3], and the 2-state [2] QKD schemes.

4.5.1 6-state Protocol

In the 6-state EB protocol, Alice and Bob perform projection measurements onto the eigenvectors of the three Pauli operators σ_x, σ_y, and σ_z on the bipartite qubit states distributed by Eve. In the corresponding P&M scheme, Alice prepares the eigenvectors of those operators by performing the same measurements on a maximally entangled two-qubit state. In both cases Alice and Bob have complete tomographic knowledge of the distributed state and therefore they can evaluate all EWs. In particular, the class of optimal witnesses (OEW) for two-qubit states is given by: $W = |\phi_e\rangle\langle\phi_e|^{T_P}$, where $|\phi_e\rangle$ denotes any entangled state of two-qubit systems [25]. Here T_P is the partial transposition, i.e., the transposition with respect to one of the subsystems. This means that in the 6-state protocol, for both EB and P&M schemes, all entangled states can be detected and the optimal witnesses form the minimal verification set. Note that, alternatively to the witness approach, in this case Alice and Bob can also employ the Peres–Horodecki criterion [23, 32].

4.5.2 4-state Protocol

In the 4-state EB scheme, Alice and Bob perform projection measurements onto two mutually unbiased bases, say the ones given by the eigenvectors of the two Pauli operators σ_x and σ_z. In the corresponding P&M scheme, Alice can use as well the same set of measurements but now on a maximally entangled state. Let us begin our analysis for the EB scheme [13]. In this case the accessible EWs, which we shall denote as W_4^{EB}, are of the form W_4^{EB} =

$\sum_{i,j=\{0,x,z\}} c_{ij}\, \sigma_i \otimes \sigma_j$, where $\sigma_0 = 1$ and c_{ij} are real numbers. This class of EWs can be characterized with the following observation:

Lemma. [13] *Given an entanglement witness W, we find $W \in W_4^{EB}$ iff $W = W^T = W^{T_P}$.*

It is straightforward to see that the OEW for two-qubit states introduced above do not fulfill this condition. This means that, in contrast to the case of the 6-state protocol, now there can be entangled states that give rise to correlations $P(A,B)$ that are not sufficient to prove the presence of entanglement. Next we present a set of optimal witnesses in the class W_4^{EB} which forms a minimal verification set of the 4-state EB protocol.

Theorem 2. [13] *Consider the family of operators $W = \frac{1}{2}(Q + Q^{T_P})$, where $Q = |\phi_e\rangle\langle\phi_e|$ and $|\phi_e\rangle$ denotes a real entangled state. The elements of this family are witness operators that are optimal in W_4^{EB} (OEW$_4^{EB}$) and detect all the entangled states that can be detected within W_4^{EB}.*

As we showed previously, for the case of a P&M scheme one can add to the set of observables measured in the protocol other observables such that they form a tomographically complete set of Alice's Hilbert space. So we have to add the operator $\sigma_y \otimes \sigma_0$ to the observables $\sigma_i \otimes \sigma_j$ with $i,j = \{0,x,z\}$. We obtain that the witnesses of the 4-state P&M protocol, denoted as $W_4^{P\&M}$, are of the form $W_4^{P\&M} = \sum_{i,j=\{0,x,z\}} c_{ij}\, \sigma_i \otimes \sigma_j + c_{y0}\, \sigma_y \otimes \sigma_0$. The analysis now is a bit more involved, but still one can prove that the set of OEW$_4^{EB}$ obtained in Theorem 2 is also sufficient to detect all entangled states that can be detected in the P&M scheme and still forms a verification set. We find then a *reduced verification set*, which still may contain some redundancies.

Theorem 3. [12] *The family of OEW$_4^{EB}$, $W = \frac{1}{2}(Q + Q^{T_P})$ with $Q = |\phi_e\rangle\langle\phi_e|$ and $|\phi_e\rangle$ a real entangled state, is sufficient to detect all entangled states that can be detected in the 4-state P&M scheme.*

As a special application of these results, as we will show below, we obtain that, for some asymmetric error patterns, the 6-state and the 4-state QKD protocols can detect entanglement even for error rates above $33\,\%$ and $25\,\%$, respectively [13].

4.5.3 2-state Protocol

The 2-state protocol [2] is based on the random transmission of only two non-orthogonal states, $|\varphi_0\rangle$ and $|\varphi_1\rangle$. On the receiving side, Bob measures each qubit he receives in a basis that is chosen, independently and at random, within the set $\{\{|\varphi_0\rangle, |\varphi_0^\perp\rangle\}, \{|\varphi_1\rangle, |\varphi_1^\perp\rangle\}\}$, with $|\langle\varphi_i|\varphi_i^\perp\rangle| = 0$. Equivalently, Alice can prepare an entangled state $|\Psi\rangle_{AB} = 1/\sqrt{2}\,(|0\rangle_A|\varphi_0\rangle_B + |1\rangle_A|\varphi_1\rangle_B)$, and then she measures her subsystem in the basis $\{|0\rangle, |1\rangle\}$. It can be shown that in this scheme the fact that the reduced density matrix of Alice is fixed is vital to detect quantum correlations in $P(A,B)$ [12].

We obtain that the EWs of the 2-state protocol, denoted as W_2, are of the form $W_2 = \sum_{i=\{0,z\},j=\{x,z\}} c_{ij}\, \sigma_i \otimes \sigma_j + \sum_{k=\{0,x,z,y\}} c_k\, \sigma_k \otimes \sigma_0$, where the second term includes a set of observables such that Alice can completely reconstruct the state of her subsystem. This set of witness operators can equivalently be rewritten as $W_2 = |0\rangle\langle 0| \otimes A + |1\rangle\langle 1| \otimes B + xC(\theta)$, where A and B represent two real symmetric operators, $A = A^T$ and $B = B^T$, and $xC(\theta)$ comprises the terms $c_x\, \sigma_x \otimes 1 + c_y\, \sigma_y \otimes 1$, with $x = |c_x + ic_y| \geq 0$ and $\theta = \tan^{-1}(c_y/c_x)$.

The first requirement that an operator W_2 must satisfy in order to qualify as an EW is that $\text{Tr}(W_2\sigma) \geq 0$ holds for all separable states σ. From this we obtain the requirements that $A \geq 0$, $B \geq 0$, and $x \leq \min_{|\psi\rangle} \sqrt{\langle\psi|A|\psi\rangle\langle\psi|B|\psi\rangle}$.

The second condition in order to guarantee that W_2 can detect entanglement is $W_2 \ngeq 0$. This last requisite implies that the operators A and B must be of full rank, and $x^2 > x^2_{\min} = \alpha/2 - \sqrt{\alpha^2/4 - \det(A)\det(B)}$, where $\alpha = \text{Tr}(AB)$. We then obtain a reduced set of EWs (Theorem 4) as verification set:

Theorem 4. [12] *The family of witness operators* $W_2 = |0\rangle\langle 0| \otimes A + |1\rangle\langle 1| \otimes B + xC(\theta)$, *with A and B being two real symmetric positive operators,* $A > 0$ *and* $B > 0$, *the operator* $C(\theta)$ *given above with* $\theta \in [0, 2\pi)$, *and such that* $x = \min_{|\psi\rangle} \sqrt{\langle\psi|A|\psi\rangle\langle\psi|B|\psi\rangle} > x_{\min}$ *is sufficient to detect all entangled states that can be detected in the 2-state protocol.*

4.6 Examples for Evaluation

Let us briefly analyze the implications of our results in the relationship between the bit error rate e in the 4-state and in the 6-state protocols and the presence of correlations of quantum mechanical nature [13]. Here the error rate e quantifies the rate of events where Alice and Bob obtain different results. It refers to the sifted key, i.e., considering only those cases where the signal preparation and detection methods employ the same polarization basis. In an intercept–resend attack Eve measures every signal emitted by Alice and prepares a new one, depending on the result obtained, that is given to Bob. This action corresponds to an entanglement-breaking channel, i.e., it is a channel Φ such that $I \otimes \Phi(\rho)$ is separable for any density matrix ρ on a tensor product space. Such a channel gives rise to $e \geq 25\,\%$ (4-state protocol) and $e \geq 33\,\%$ (6-state protocol), respectively [8, 17], which might seem to indicate that these values represent an upper bound for the tolerable error rate in the protocols (see also [31]). However, it turns out that, for some asymmetric error patterns, it is possible to detect the presence of quantum correlations even for error rates above $25\,\%$ ($33\,\%$) [13]. Let us illustrate this fact with an example that is motivated by the propagation of the polarization state of a single photon in an optical fiber.

We focus on the 4-state EB protocol only and we consider the particular joint probability distribution $P(A, B)$ given by Table 4.1, where the states $|\pm\rangle$ are defined as $|\pm\rangle = 1/\sqrt{2}\,(|0\rangle \pm |1\rangle)$. In principle, it is not straightforward to decide whether these correlations can be explained as coming exclusively from an entangled state or not. This is specially so since in this case the resulting bit error rate is given by $e \approx 35.4\,\%$.

To decide that question systematically, we can use the verification set defined by OEW_4^{EB}: $W = \frac{1}{2}(|\phi_e\rangle\langle\phi_e| + |\phi_e\rangle\langle\phi_e|^{T_P})$. The real states $|\phi_e\rangle$ can be parametrized as $|\phi_e\rangle = \cos\phi\,|00\rangle + \sin\phi\,[\cos\psi\,|01\rangle + \sin\psi\,(\cos\theta\,|10\rangle + \sin\theta\,|11\rangle)]$, with only three real parameters $\phi, \psi, \theta \in [0, 2\pi)$. From a practical point of view it is easier just to consider all angles ϕ, ψ, θ and allow the evaluation of some positive operators corresponding to states $|\phi_e\rangle$ that are not entangled. After expressing the witness operators as a *pseudo-mixture* [33], the condition $\text{Tr}(W\rho) < 0$ can be rewritten as $\sum_i f_i(\phi, \psi, \theta)\, P(A_i, B_i) < 0$, with $c_i = f_i(\phi, \psi, \theta)$

Table 4.1: Example of a $P(A, B)$ for the 4-state EB protocol, where $|\pm\rangle = 1/\sqrt{2}\,(|0\rangle \pm |1\rangle)$. The table is normalized such that $\sum_i P(A_i, B_i) = 1$.

$A \setminus B$	$\|0\rangle$	$\|1\rangle$	$\|+\rangle$	$\|-\rangle$
$\|0\rangle$	0.08058	0.04757	0.02106	0.10709
$\|1\rangle$	0.04623	0.07560	0.11349	0.00834
$\|+\rangle$	0.11808	0.01690	0.09319	0.04179
$\|-\rangle$	0.00873	0.10627	0.04136	0.07364

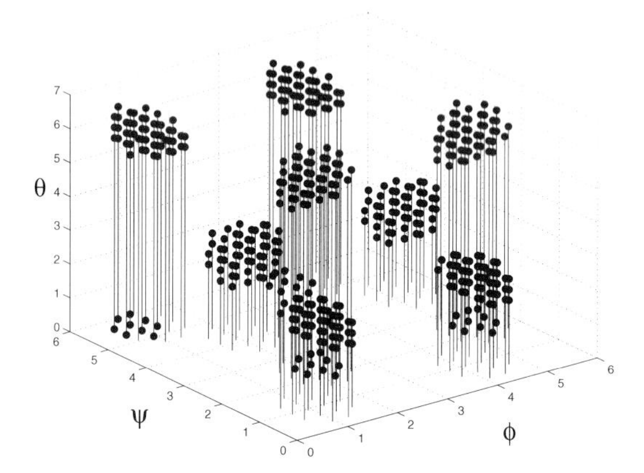

Figure 4.2: Illustration of several regimes of the parameters ϕ, ψ, θ leading to negative expectation values of the operators $\text{OEW}_4^{EB} = \frac{1}{2}(|\phi_e\rangle\langle\phi_e| + |\phi_e\rangle\langle\phi_e|^{T_P})$, with $|\phi_e\rangle = \cos\phi\,|00\rangle + \sin\phi\,[\cos\psi\,|01\rangle + \sin\psi\,(\cos\theta\,|10\rangle + \sin\theta\,|11\rangle)]$, when they are evaluated on the joint probability distribution $P(A, B)$ given in Table 4.1. The angles θ, ψ, ϕ are measured in radian.

for some functions f_i. Now it is easy to search numerically through the space of parameters ϕ, ψ, θ for quantum correlations in $P(A, B)$. This fact is illustrated in Fig. 4.2, where some combinations of these parameters detecting entanglement when they are evaluated on the $P(A, B)$ given in Table 4.1 are marked.

4.7 Realistic Experiments

The idea to check for quantum correlations in the observed data with the help of a verification set of witnesses applies also, in principle, to real implementations of QKD set-ups. One can incorporate any imperfection of the sources and the detection devices into the corresponding investigation within the framework of trusted devices. In that framework one characterizes detection devices by the use of an appropriate POVM description, e.g. on the infinite-dimensional Hilbert space of optical modes. For P&M schemes, one has to characterize additionally the given source via the reduced density matrix of the virtual internal preparation state, as described before. This idea then needs to be generalized to signal states that are described by mixed quantum states. For this purpose, one still uses a pure state as internal preparation, but Alice's signal preparation now no longer corresponds to a projection onto an orthogonal set of pure states, but to projections onto orthogonal subspaces, thereby effectively preparing mixed states.

In those general scenarios, it will be difficult to provide the minimal verification set of witnesses. Instead, one can fall back on the approach to search for just one accessible witness via numerical methods such as presented in the previous section. In this way, one can search through restricted classes of accessible witnesses at the price that the result of this search will not be conclusive, i.e. this search yields only a sufficient condition.

4.8 Conclusions

In this contribution we reported on current active research in QKD. The field is driven by the yearning for better understanding of the interaction between generation of quantum correlations and manipulations by local classical computation and two-way communication. From this we expect to find more efficient ways to turn quantum correlations into a secret key. There are limitations to the improvements that can be achieved, and these limitations help us to search through various implementations for quantum key distribution in order to evaluate their potential.

In this contribution we focused on the question of how to verify that the correlations we establish in a QKD setup can at all lead to secret keys when some channel noise and loss is added. This is an important first check in a situation where many different protocols are proposed while the corresponding security proofs are much harder to obtain.

Knowing the limitations for a given setup allows us to identify the bottleneck for performance. Then only a minor modification might be necessary to overcome that bottleneck, as has been demonstrated recently in the case of QKD with weak coherent pulses. By variation of the signal intensity [24,26,37], one can put tighter constraints on an eavesdropper and achieve significantly improved distances (more than 100 km) for unconditional secure QKD.

References

[1] M. Ben-Or, M. Horodecki, D. W. Leung, D. Mayers, and J. Oppenheim. The universal composable security of quantum key distribution. quant-ph/0409078, 2004.

[2] C. H. Bennett. Quantum cryptography using any two nonorthogonal states. *Phys. Rev. Lett.*, 68(21):3121–3124, May 1992.

[3] C. H. Bennett and G. Brassard. Quantum cryptography: Public key distribution and coin tossing. In *Proceedings of IEEE International Conference on Computers, Systems, and Signal Processing, Bangalore, India*, pages 175–179, New York, 1984. IEEE.

[4] C. H. Bennett, G. Brassard, C. Crépeau, and U. M. Maurer. Generalized privacy amplification. *IEEE Trans. Inf. Theory*, 41:1915–1923, 1995.

[5] C. H. Bennett, G. Brassard, and N. D. Mermin. Quantum cryptography without Bell's theorem. *Phys. Rev. Lett.*, 68(5):557–559, 1992.

[6] C. H. Bennett, G. Brassard, S. Popescu, B. Schumacher, J. A. Smolin, and W. K. Wootters. Purification of noisy entaglement and faithful teleportation via noisy channels. *Phys. Rev. Lett.*, 76(5):722–725, 1996.

[7] G. Brassard and L. Salvail. Secret-key reconciliation by public discussion. In Tor Helleseth, editor, *Advances in Cryptology - EUROCRYPT '93*, volume 765 of *Lecture Notes in Computer Science*, pages 410–423, Berlin, 1994. Springer.

[8] D. Bruß. Optimal eavesdropping in quantum cryptography with six states. *Phys. Rev. Lett.*, 81:3018–3021, 1998.

[9] N. Cerf, C. Adami, and R. M. Gingrich. Reduction criterion for separability. *Phys. Rev. A*, 60:898, 1999.

[10] M. Christandl, R. Renner, and A. Ekert. A generic security proof for quantum key generation. quant-ph/0402131, 2004.

[11] I. Csiszár and J. Körner. Broadcast channels with confidential messages. *IEEE Trans. Inf. Theory*, IT-24(3):339–348, May 1978.

[12] M. Curty, O. Gühne, M. Lewenstein, and N. Lütkenhaus. Detecting two-party quantum correlations in quantum key distribution protocols. quant-ph/0409047, 2004.

[13] M. Curty, M. Lewenstein, and N. Lütkenhaus. Entanglement as precondition for secure quantum key distribution. *Phys. Rev. Lett.*, 92:217903, 2004.

[14] D. Deutsch, A. Ekert, R. Josza, C. Macchiavello, S. Popescu, and A. Sanpera. Quantum privacy amplification and the security of quantum cryptography over noisy channels. *Phys. Rev. Lett.*, 77:2818–2821, 1996.

[15] I. Devetak and A. Winter. Relating quantum privacy and quantum coherence: an operational approach. *Phys. Rev. Lett.*, 93:080501, 2004.

[16] A. Ekert. Quantum cryptography based on Bell's theorem. *Phys. Rev. Lett.*, 67(6):661–663, 1991.

[17] A. Ekert and B. Huttner. Information gain in quantum eavesdropping. *J. Mod. Opt.*, 41:2455–2466, 1994.

[18] D. Gottesman and H. K. Lo. Proof of security of quantum key distribution with two-way classical communications. *IEEE Trans. Inf. Theory*, 49:457, 2003.

[19] O. Gühne. Characterizing entanglement via uncertainty relations. *Phys. Rev. Lett.*, 92:117903, 2004.

[20] H. F. Hofmann and S. Takeuchi. Violation of local uncertainty relations as a signature of entanglement. *Phys. Rev. A*, 68:032103, 2003.

[21] K. Horodecki, M. Horodecki, P. Horodecki, and J. Oppenheim. Secure key from bound entanglement. quant-ph/0309110, 2003.

[22] M. Horodecki and P. Horodecki. Reduction criterion of separability and limits for a class of distillation protocols. *Phys. Rev. A*, 59:4206, 1999.

[23] M. Horodecki, P. Horodecki, and R. Horodecki. Separability of mixed states: necessary and sufficient conditions. *Phys. Lett. A*, 223:1, 1996.

[24] W.-Y. Hwang. Quantum key distribution with high loss: Toward global secure communication. *Phys. Rev. Lett.*, 91:057901, 2003.

[25] M. Lewenstein, B. Kraus, J. I. Cirac, and P. Horodecki. Optimization of entanglement witnesses. *Phys. Rev. A*, 62:052310, 2000.

[26] H. K. Lo, X. Ma, and K. Chen. Decoy state quantum key distribution. quant-ph/0411004, 2004.

[27] U. Maurer and S. Wolf. Unconditionally secure key agreement and the intrinsic conditional information. *IEEE Trans. Inf. Theory*, 45:499–514, 1999.

[28] U. M. Maurer. Secret key agreement by public discussion from common information. *IEEE Trans. Inf. Theory*, 39(3):733–742, May 1993.

[29] D. Mayers. Quantum key distribution and string oblivious transfer in noisy channels. In *Advances in Cryptology—Proceedings of CRYPTO '96*, pages 343–357, Berlin, 1996. Springer. quant-ph/9606003.

[30] D. Mayers. Unconditional security in quantum cryptography. *JACM*, 43(3):351–406, May 2001.

[31] G. M. Nikolopoulos and G. Alber. Robustness of the BB84 quantum key distribution protocol against general coherent attacks. quant-ph/0403148, 2004.

[32] A. Peres. Separability criterion for density matrices. *Phys. Rev. Lett.*, 77:1413, 1996.

[33] A. Sanpera, R. Tarrach, and G. Vidal. Local description of quantum inseparability. *Phys. Rev. A*, 58:826–830, 1997.

[34] P. W. Shor and J. Preskill. Simple proof of security of the BB84 quantum key distribution protocol. *Phys. Rev. Lett.*, 85:441–444, 2000.

[35] B. Terhal. Bell inequalities and the separability criterion. *Phys. Lett. A*, 271:319, 2000.

[36] G. S. Vernam. Cipher printing telegraph systems. *J. AIEE*, 45:295, 1926.

[37] X. B. Wang. Beating the PNS attack in practical quantum cryptography. quant-ph/04110075, 2004.

[38] M. N. Wegman and J. L. Carter. New hash functions and their use in authentication and set equality. *J. Comp. Syst. Sci.*, 22:265–279, 1981.

[39] A. D. Wyner. The wire-tap channel. *Bell. Syst. Tech. J.*, 54(8):1355–1387, October 1975.

5 Increasing the Size of NMR Quantum Computers

S. J. Glaser[1], *R. Marx*[1], *T. Reiss*[1], *T. Schulte-Herbrüggen*[1], *N. Khaneja*[2], *J. M. Myers*[2], *and A. F. Fahmy*[3]

[1] Institut für Organische Chemie und Biochemie II
Technische Universität München
Garching, Germany

[2] Division of Engineering and Applied Sciences
Harvard University
Cambridge, USA

[3] Biological Chemistry and Molecular Pharmacology
Harvard Medical School
Boston, USA

5.1 Introduction

In this section, we discuss theoretical and experimental aspects of nuclear magnetic resonance (NMR) techniques for the implementation of quantum computing algorithms. Although we will focus on liquid state NMR techniques, many of the new theoretical approaches that are beeing developed in this field are expected to find applications not only in other fields of magnetic resonance such as solid state NMR or electron spin resonance (ESR), but also in different experimental approaches to quantum computing.

NMR is one of the most important methods for the identification of molecules and for the determination of their structure and dynamics, and it has found numerous applications in physics, chemistry, structural biology and medicine [1]. Sophisticated theoretical and experimental tools are available to manipulate the state of coupled nuclear spin systems, which gave NMR a head start in the experimental realization of fundamental concepts of quantum computing [2–4]. At present, NMR is most advanced in the practical implementation of quantum algorithms.

Nuclear spin systems are particularly well suited to act as qubits due to their isolation from the environment. Because tumbling of the molecules in a liquid sample decouples each molecule from all the others, the sample can be described by a density matrix for the n nuclear spins of the atoms of a single molecule [1] with only the spin-degrees of freedom corresponding to the desired Hilbert space of dimension 2^n.

A liquid NMR sample contains an ensemble of many identical spin systems, and it is not possible to manipulate or to detect individual spin systems. Hence, at the beginning of a computation, liquid-state NMR quantum computers are commonly prepared in a so-called pseudo-pure state [3] rather than in a pure state. Furthermore, ensemble-averaged expectation values are detected rather than observables of individual spin systems. Hence, NMR quantum

Quantum Information Processing, 2nd Edition. Edited by Thomas Beth and Gerd Leuchs

ISBN: 3-527-40541-0

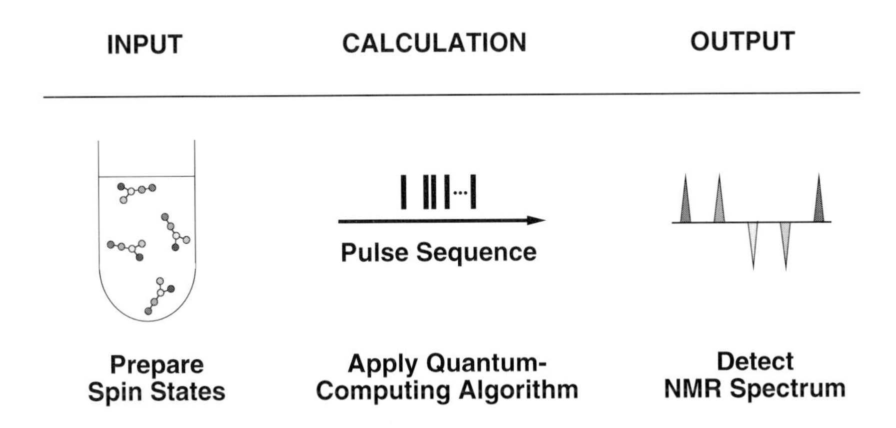

Figure 5.1: Schematic representation of a liquid state NMR experiment. The sample consists of a test tube with an ensemble of (about 10^{18}) molecules in an external magnetic field of about 10 Tesla. After preparing the density operator representing the ensemble of nuclear spin systems of the molecules in an initial state, the unitary transformations of a quantum computing algorithm are applied in the form of a sequence of radio-frequency pulses and delays. Finally, the outcome of the experiment is deduced from an acquired NMR spectrum.

computers are examples of expectation-value quantum computers (EVQC), which in place of an outcome of an occurrence of a measurement yields the expectation value. The expectation values representing the outcome of a given quantum algorithm can be extracted from the resulting NMR spectra. The basic computational steps of quantum algorithms can be realized with the help of radio-frequency pulses (see Fig. 5.1) and can always be broken down to spin-selective pulses and CNOT (Controlled NOT) gates between pairs of spins. Several reviews about many theoretical and practical aspects of NMR quantum computing can be found in the literature [5–13].

So far, many algorithms were implemented, such as the Deutsch-Jozsa algorithm for two [14, 15], three [16, 17] and five [18] qubits, variations of Grover's algorithm for two [19, 20] and three qubits [21], the period-finding algorithm for five qubits [22] and an implementation of Shor's algorithm for 7 qubits [23]. While it is unlikely that liquid state NMR quantum computing will ever be competitive with classical computers, it should be possible to extend the current technology well beyond ten qubits. Applications with up to twenty qubits may be feasible in the not too distant future. In the following, we will discuss theoretical and practical problems that need to be solved in order to approach this goal.

5.2 Suitable Molecules

From a chemical perspective, the number of qubits is mainly restricted by the availability of compounds with suitable spin systems. For the efficient implementation of an NMR quantum computer with n qubits, a molecule with n coupled spins 1/2 is required. In order to be able to perform a large number of basic computational steps (logic quantum gates) in a given algorithm, the time required for each quantum gate must be considerably smaller than the relaxation time of the nuclear spins.

Currently used sample preparations for liquid state NMR quantum computers result in nuclear spin relaxation times of up to several seconds. Characteristic spin–spin coupling constants are on the order of 10 to 10^2 Hz, resulting in a typical duration of two-qubit quantum gates between directly coupled spins of 10^{-2} seconds. Hence, sequences of up to 10^2 to 10^3 two-qubit quantum gates are feasible based on current liquid state NMR technology, and even more quantum gates may be possible by increasing the spin–spin coupling constants, e.g. by using dipolar couplings in liquid crystalline media [24] and by further increasing the relaxation times. Compared to two-qubit operations, single-spin quantum gates such as NOT or Hadamard gates are very short. For example, in heteronuclear spin systems, typical single-spin gate durations are on the order of 10^{-5} seconds. The minimum time required for a given single-spin quantum gate depends not only on the maximum amplitude of radio-frequency pulses but also on the smallest frequency difference of the nuclear spins in a given molecule [18] .

As the basic computational steps rely on the selective manipulation of individual spins, the frequency differences of nuclear spins should be as large as possible. In this respect, heteronuclear spin systems and homonuclear spins with a large range of chemical shifts are of advantage. In addition, large couplings should exist between the nuclear spins because the time required for the implementation of a direct logical combination of two qubits is inversely proportional to the size of the coupling between the corresponding nuclear spins. However, it is not required that all spins are mutually coupled because it is sufficient if the spins form a contiguous coupling network [25]. If the spins can be controlled individually, arbitrary coupling terms can be used, such as combinations of isotropic and dipolar couplings [26].

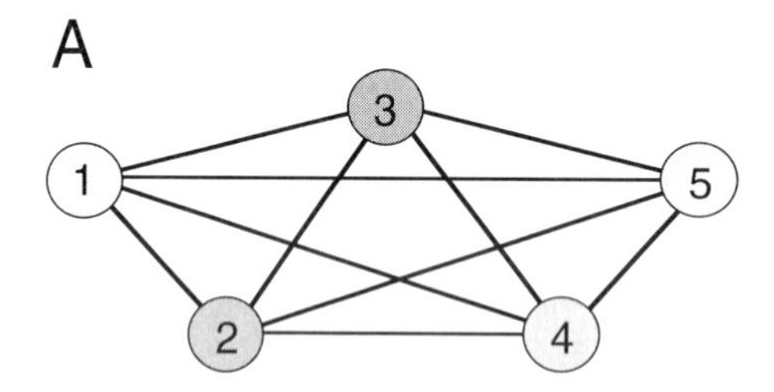

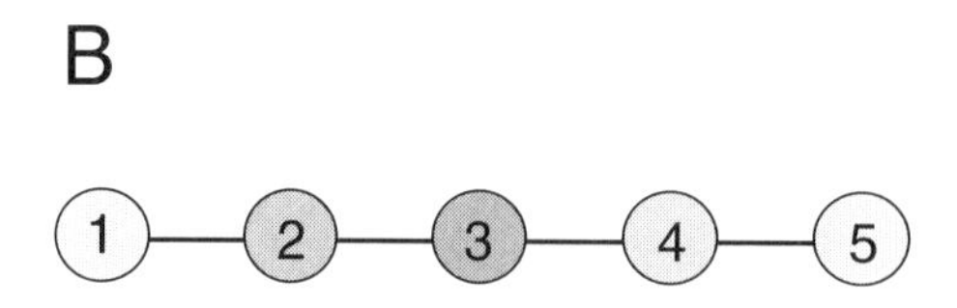

Figure 5.2: A: Schematic representation of a system consisting of $n = 5$ pair-wise coupled spins 1/2 (qubits). B: A spin chain is sufficient for an n-qubit NMR quantum computer.

For the first NMR quantum computers with up to three qubits, readily available compounds were used, such as 2,3-dibromothiophene [2,27], ^{13}C-chloroform [15], 2,3-dibromopropanoic acid [15,25], and $^{13}C_3$-alanine [28]. For the realization of the first five-qubit NMR quantum computer the compound BOC-($^{13}C_2$-^{15}N-$^{2}D_2^{\alpha}$-Gly)-F was synthesized [18,29] (see Fig. 5.3). If the deuterium spins are decoupled, the nuclear spins of $^1H^N$, ^{15}N, $^{13}C^{\alpha}$, $^{13}C'$ (i.e. C^O)

Figure 5.3: Schematic representation of BOC-$^{13}C_2$-^{15}N-$^{2}D_2^{\alpha}$-glycine-fluoride with the coupled five-spin system (represented schematically by white arrows) that forms the molecular basis of a five-qubit NMR quantum computer [18]. The atoms that form the spin system of interest are shown as balls. The rest of the molecule is shown in a stick representation.

and ^{19}F form a coupled spin system consisting of five spins 1/2. The ^1J coupling constants range between 13.5 Hz ($^1J_{N,C^\alpha}$) and 366 Hz ($^1J_{C',F}$), and for a magnetic field of 9.4 T, the smallest frequency differences is 12 kHz ($\nu_{C'} - \nu_{C^\alpha}$). Another synthetic five qubit system is a perfluorobutadienyl iron complex [22]. Here, the spin system is entirely homonuclear and consists of five coupled ^{19}F spins. A carbon-labelled analog of this compound has been used as a seven-qubit molecule for the implementation of Shor's algorithm [23]. Another seven-qubit molecule that has been suggested for NMR quantum computing applications is $^{13}C_4$-crotonic acid [30]. The design and synthesis of molecules with suitable spin systems for 10-20 qubits is not a trivial chemical challenge. In addition to the synthesis of molecules for a given number of qubits, an alternative approach for the realization of a molecular architecture with more than 10 coupled spins is the synthesis of polymers with a repetitive unit consisting of three or more spins [31]. This approach has the advantage that only a small number of resonances need to be selectively addressed. However, in such an architecture the implementation of quantum algorithms will require an additional overhead, which poses new challenges for the efficient implementation of quantum gates (*vide infra*).

5.3 Scaling Problem for Experiments Based on Pseudo-pure States

In addition to the practical problem to design and to synthesize suitable molecules, there are also a number of theoretical problems that have to be solved in order to increase the size of NMR quantum computers. As indicated in the introduction, NMR implementations of quantum algorithms are usually based on the creation of a so-called pseudo-pure state which results in a severe scaling problem:

So far, quantum algorithms have been developed for pure states and cannot be directly applied to mixed states. However, mixed states are typical of liquid state NMR experiments at about room temperature. The density matrix representing the mixed state of a spin system at thermal equilibrium is proportional to $\exp(-\mathcal{H}/kT)$, where $\mathcal{H}$ is the spin Hamiltonian for the n-spin molecule (in the liquid sample) used as a quantum register, k is Boltzmann's constant, and T is the temperature. In the high-temperature approximation the thermal density matrix is given adequately well by the first two terms in the Taylor expansion:

$$\rho_{eq} \approx 2^{-n}(\mathbf{1} - \mathcal{H}/kT) \approx N^{-1}\mathbf{1} - \frac{\hbar}{NkT}\sum_j \omega_j I_z^j \tag{5.1}$$

where $N = 2^n$, ω_j is the angular frequency of the j-th nucleus, and I_z^j is defined by a tensor product over all n spins in which all the factors are unit operators except for $\frac{1}{2}\text{Diag}(1, -1)$ as the j-th factor of the tensor product.

This mixed state can be transformed into a so-called pseudo-pure state [2, 3], resulting in a starting density matrix of the form

$$\rho = (1 - \epsilon)\mathbf{1}/N + \epsilon|\psi\rangle\langle\psi|, \tag{5.2}$$

for some (usually small) coeffient ϵ. As the identity operator $\mathbf{1}$ is invariant under (unitary) transformations and all observable operators of NMR are traceless, pseudo-pure states form convenient starting points for NMR implementations of quantum algorithms. However, this convenience comes at a high cost, because with an increasing number of qubits n, the coefficient ϵ decreases exponentially in n [32]. Hence, the available spectrometer signal decreases as well and severe signal-to-noise problems are expected for experiments with more than about 10 qubits.

This exponential signal loss is often thought to impose a fundamental limit on the scalability of liquid state NMR. Fortunately, this problem can be circumvented by several approaches that make it possible to avoid pseudo-pure states.

5.4 Approaching Pure States

The first approach is to prepare the sample not in a mixed state corresponding to the thermal equilibrium density operator at high temperature, but in a pure state corresponding to a low temperature close to 0 Kelvin. If the actual sample temperature were cooled down close to absolute zero, any sample would eventually freeze and no liquid-state NMR experiments would

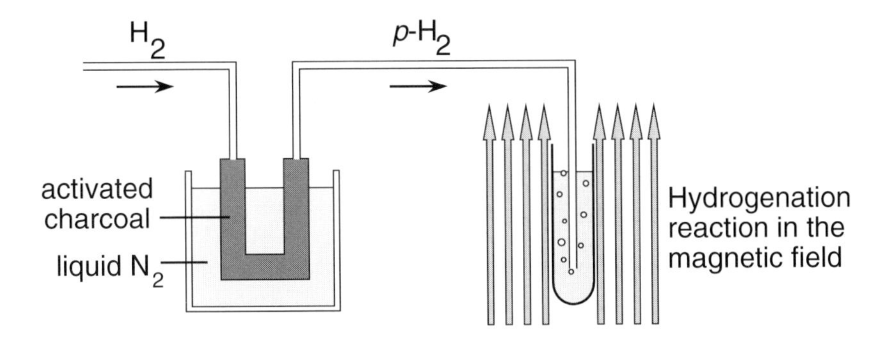

Figure 5.4: Hydrogen gas (H_2) is converted to para hydrogen gas ($p\text{-}H_2$), before it reacts with the host molecules in the magnetic field.

be possible. Fortunately it is possible to cool down only the spin temperature by increasing the nuclear spin polarization well beyond the thermal polarization, while leaving the sample at room temperature. Several spin polarization techniques are available to decrease the spin temperature. So far, the largest polarizations have been achieved experimentally using the spin order of parahydrogen (see Fig. 5.4) and applications to a simple quantum computing algorithm have been demonstrated [33]. Other techniques for polarization enhancement may be based on the use of laser-polarized Xenon [34]. Although promising, these techniques also pose a number of technical problems that currently limit their practical use in scaling up the number of qubits.

5.5 Scalable NMR Quantum Computing Based on the Thermal Density Operator

An attractive alternative approach to avoid the creation of pure or pseudo-pure states could be the design of NMR quantum computing algorithms that are based on the thermal density operator instead of a pure state. A premier example of this approach is a new scalable version of the Deutsch-Jozsa algorithm [35]. At the expense of an extra qubit and a more complex Oracle, balanced functions can be distinguished from constant functions using the starting state obtained merely by a hard 90° y-pulse applied to the thermal state, requiring neither the pseudo-pure state of Eq. (5.2) nor temporal averaging. One requires an Oracle for a function $f : \mathbf{Z}_{N/2} \rightarrow \mathbf{Z}_2$ that implements $U_{f'}$ not for f but for $f' : \mathbf{Z}_N \rightarrow \mathbf{Z}_2$, related to f by

$$f'(j) = \begin{cases} f(j) & \text{if } 0 \leq j \leq N/2 - 1 \\ 0 & \text{if } N/2 \leq j \leq N - 1 \end{cases} \tag{5.3}$$

It has been shown that given this more complex Oracle, an NMR quantum computer decides between balanced functions and constant functions. Furthermore, no exponentially growing demands are put on the resolution of the resulting output spectra [35].

For example, for the constant function $f_0(x_1, x_2, x_3) = 0$ and the balanced function $f_b(x_1, x_2, x_3) = x_1 \oplus x_2 x_3$, the scalable version of the Deutsch-Jozsa algorithm requires an

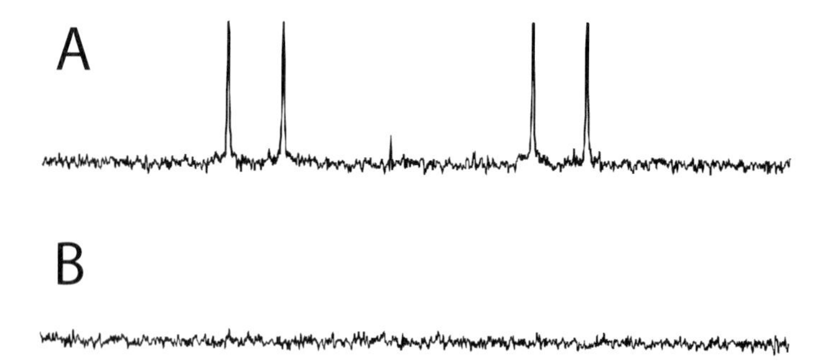

Figure 5.5: Experimental spectra [36] representing the result of the new version of the Deutsch-Jozsa algorithm [35] based on the thermal density operator for a constant (A) and a balanced (B) test function (see text).

additional qubit (x_0) and the implementation of $U_{f_0'}$ for $f_0'(x_0, x_1, x_2, x_3) = \overline{x}_0\, f_0(x_1, x_2, x_3) = 0$ and of $U_{f_b'}$ $f_b'(x_0, x_1, x_2, x_3) = \overline{x}_0\, f_b(x_1, x_2, x_3) = \overline{x}_0 x_1 \oplus \overline{x}_0 x_2 x_3$. For the 5-qubit system BOC-($^{13}C_2$-^{15}N-$^{2}D_2^{\alpha}$-Gly)-F [18, 29], the resulting spectra [36] of x_0 are shown in Fig. 5.5A and 5.5B for $U_{f_0'}$ and $U_{f_b'}$, respectively. Constant and balanced functions can be distinguished by the presence or absence of the signals .

With an increasing number of qubits, not only the synthetic requirements grow, but also the demands with respect to NMR instruments and pulse-sequence design, in order to keep the effects of experimental imperfections (such as the inhomogeneity of the radio-frequency fields) and spin relaxation under control.

5.6 Time-optimal Implementation of Quantum Gates

Although it is possible to decompose any quantum computing algorithm in a series of single-spin operations and two-spin gates between directly coupled spins, a fundamental question of both theoretical and practical interest is the minimum time required to realize a given unitary transformation in a given spin system and furthermore, what is the pulse sequence that achieves this minimum time. Geometric optimal control theory [37–40] provides powerful tools to address these issues as a question of time-optimal control of the unitary operator. An analytical characterization of such time-optimal pulse sequences can be given, and it has been shown that this problem can be reduced to finding shortest paths between certain cosets [38].

This approach is based on the fact that in many NMR applications there are two markedly different time scales that are characteristic for the evolution under radio frequency pulses (local single spin operations) and under couplings (two qubit gates). For example in heteronuclear spin systems, single spin operations are two orders of magnitude faster than evolution under typical spin–spin couplings. By making use of the Cartan decomposition of real semisimple Lie algebras $\mathfrak{g} = \mathfrak{k} \oplus \mathfrak{p}$ (where $[\mathfrak{k}, \mathfrak{k}] \subseteq \mathfrak{k}$; $[\mathfrak{k}, \mathfrak{p}] = \mathfrak{p}$; $[\mathfrak{p}, \mathfrak{p}] \subseteq \mathfrak{k}$), the goal of time-optimal realisations reduces to finding constrained shortest paths in the subgroup G/K, where in $n = 2$ spins-1/2, $G = SU(2^n)$ and $K = \big(SU(2)\big)^{\otimes n}$. Yet, also for $n > 2$, K corresponds to local actions on the individual spins, which proceed to good approximation

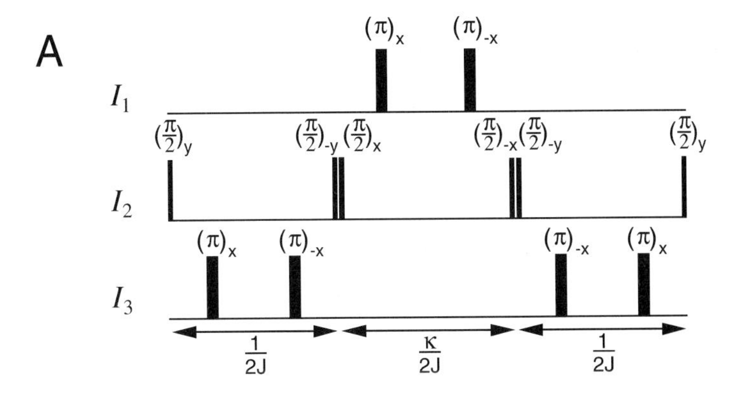

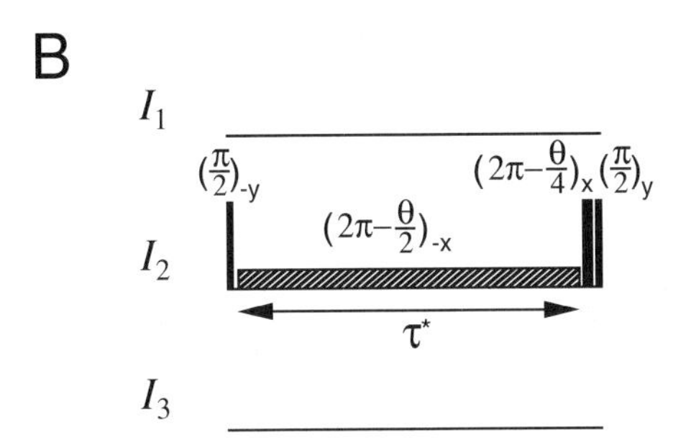

Figure 5.6: Conventional (A) and time-optimal (B) pulse sequences for synthesizing the propagator $U_{zzz}(\kappa) = \exp\{-i\theta I_{1z}I_{2z}I_{3z}\}$ with $\theta = 2\pi\kappa$. In the time-optimal (geodesic) pulse sequence, the radio-frequency amplitude ν_{rf} of the hatched pulse is $(2-\kappa)J/\sqrt{\kappa(4-\kappa)}$ [41].

arbitrarily fast (rf-pulses) as compared to evolution of couplings in G/K. Therefore, the time-optimal trajectories between points in G correspond to sub-Riemannian geodesics between cosets in G/K. Though using NMR as a paradigm, the formalism provides a more general approach to searching for time-optimal quantum computing gates.

A first example in a three-spin system is the time-optimal simulation of a three-spin-interaction Hamiltonian of the form

$$\mathcal{H}_{\alpha\beta\gamma} = 2\pi J_{\text{eff}}\, I_{1\alpha}I_{2\beta}I_{3\gamma} \tag{5.4}$$

where α, β, γ can be x, y or z and J_{eff} is an effective trilinear coupling constant. We considered a linear coupling topology consisting of a chain of three heteronuclear spins 1/2 with coupling constants $J_{12} = J_{23} = J$, $J_{13} = 0$ and the coupling term

$$\mathcal{H}_{\text{coup}} = 2\pi J I_{1z}I_{2z} + 2\pi J I_{2z}I_{3z}. \tag{5.5}$$

Based on ideas of geometric optimal control theory, the time-optimal realization of the trilinear coupling term $\mathcal{H}_{\alpha\beta\gamma}$ was derived [41], see Fig. 5.6.

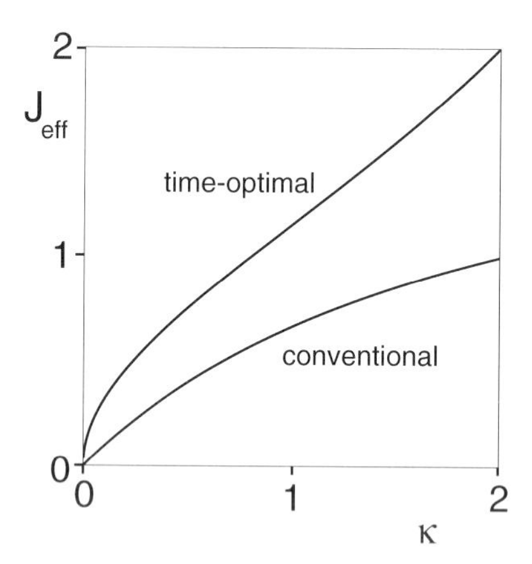

Figure 5.7: Scaling factor J_{eff}/J of the conventional and time optimal [41] simulation of the trilinear coupling Hamiltonian $\mathcal{H}_{zzz} = 2\pi J_{eff}\ I_{1z}I_{2z}I_{3z}$.

Compared to conventional approaches [42], the time-optimal synthesis of effective three-spin propagators of the form

$$\mathcal{U}_{\alpha\beta\gamma}(\kappa) = \exp\{-\mathrm{i}\kappa\ 2\pi\ I_{1\alpha}I_{2\beta}I_{3\gamma}\} \tag{5.6}$$

has a duration of only $\sqrt{\kappa(4-\kappa)}/2J$ [41] compared to the duration $(2+\kappa)/2J$ of conventional implementations [42]. For example, for $\kappa = 1, 0.5$ and 0.1, the time-optimal experiment requires only $1/\sqrt{3} = 57.7\%$, 52.9% and 29.7% of conventional implementations and hence provides significant time savings. The scaling factors J_{eff}/J are shown in Fig. 5.7.

Based on these trilinear effective coupling terms, it was possible to realize time-optimal implementations of indirect SWAP operation SWAP(1,3) between spins I_1 and I_3, which are not directly coupled by the following sequence of operations [41]:

$$\mathcal{U}_{\text{SWAP}(1,3)} = \mathcal{U}_{xzx}(1)\,\mathcal{U}_{yzy}(1)\,\mathcal{U}_{zzz}(1)\ \exp\left(\mathrm{i}\frac{\pi}{2}I_{2z}\right). \tag{5.7}$$

The time-optimal operation requires only 57.7 $\%$ of the duration that is required if the operation is decomposed in a sequence of three direct SWAP gates as SWAP(1,3) = SWAP(1,2) SWAP(2,3) SWAP(1,2).

Efficient transfer of arbitrary spin states along spin chains

Spin chains form potential architectures for quantum computing applications both in liquid state NMR and in solid state applications [43]. Here, a question of both theoretical and practical interest is how to efficiently transfer an unknown spin state of a given qubit efficiently along the chain to a given target qubit.

Based on a novel encoding scheme, a new approach has been developed to control such a transfer in spin chains of arbitrary length. The approach relies on the creation of a three

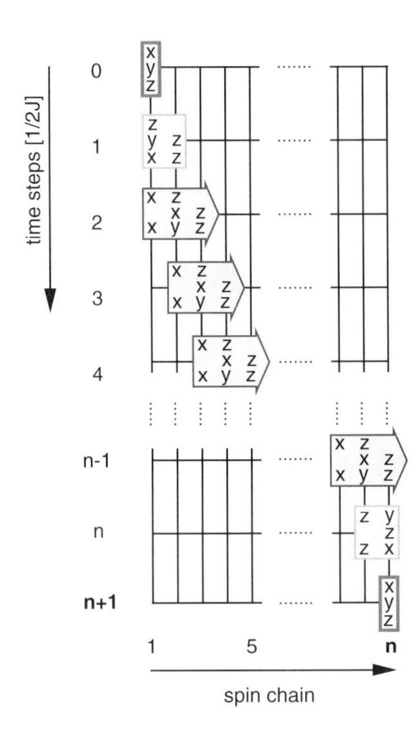

Figure 5.8: Schematic representation of the efficient transfer of an unknown spin state along the chain to a given target qubit. For clarity, spin operators such as I_{kx} are indicated by the letter x at position k. Similarly, bilinear or trilinear product operators are indicated only by the labels x, y, or z at the corresponding spin positions, omitting numerical prefactors and possible algebraic signs [44].

spin encoded state and efficient propagation of this encoded state through the spin state [44]. The transfer mechanism is schematically shown in Fig. 5.8. Compared to the conventional approach based on a sequence of direct SWAP gates, this novel transfer scheme requires only 50% for three-spin chains and converges to 33% as the chain length increases.

5.7 Conclusion

We have discussed several aspects that need to be considered for the realization of the next generation of NMR quantum computers with more than 10 qubits. We have not touched the question of entanglement, which has even led to claims that the NMR approach is no proper quantum computing at all [45, 46]. However, these discussions do not apply to NMR experiments in which pure states are created (see section 4). Even for NMR quantum computing experiments based on mixed or pseudo-pure states, the situation is far from clear, and it is argued that these experiments should in fact be regarded as quantum information processing [47]. No matter what the outcome of the discussion is, NMR certainly provides an ideal test bed to realize and test theoretical quantum algorithms for small spin systems. There are many practical problems for the scaling of NMR quantum computers to more than 10 qubits. However, none of these problems is fundamental in the sense that it cannot be circumvented in principle. The investigation of the limits of this approach is an exciting subject of current research which has already led to the development of new theoretical tools that may also prove useful for other experimental approaches to quantum computing.

References

[1] R. R. Ernst, G. Bodenhausen, and A. Wokaun, *Principles of Nuclear Magnetic Resonance in One and Two Dimensions* (Oxford University Press, Oxford, 1987).

[2] D. G. Cory, A. F. Fahmy, and T. F. Havel, *Proc. of the 4th Workshop on Physics and Computation* (New England Complex Systems Institute, Boston, MA, 1996), pp. 87–91.

[3] D. G. Cory, A. F. Fahmy, and T. F. Havel, Proc. Natl. Acad. Sci. USA **94**, 1634 (1997).

[4] N. A. Gershenfeld, and I. L. Chuang, Science **275**, 350 (1997).

[5] M. Mehring, Appl. Magn. Reson. **17**, 141 (1999).

[6] J. A. Jones, Fort. der Physik **48**, 909(2000).

[7] D. G. Cory, R. Laflamme, E. Knill, L. Viola, T. F. Havel, N. Boulant, G. Boutis, E. Fortunato, S. Lloyd, R. Martinez, C. Negrevergne, M. Pravia, Y. Sharf, G. Teklemariam, Y. S. Weinstein, W. H. Zurek, Fort. der Physik **48**, 875(2000).

[8] T. F. Havel, S. S. Somaroo, C. H. Tseng, D. G. Cory, Appl. Algebra Eng. Comm. **10**, 339(2000).

[9] J. A. Jones, Prog. NMR Spectrosc. **38**, 325(2001).

[10] J. A. Jones, Phys. Chem. Comm. **11**, 11 (2001).

[11] S. Glaser, Angew. Chem. 113, 151(2001); Angew. Chem. Int. Ed. **40**, 147(2001).

[12] L. M. K. Vandersypen, C. S. Yannoni, I. L. Chuang, quant-ph/0012108 (2001).

[13] T. F. Havel, D. G. Cory, S. Lloyd, N. Boulant, E. M. Fortunato, M. A. Pravia, G. Teklemariam, Y. S. Weinstein, A. Bhattacharyya, J. Hou, Am. J. Phys. **70**, 345(2002).

[14] J. A. Jones, and M. Mosca, J. Chem. Phys. **109**, 1648 (1998).

[15] I. L. Chuang, L. M. K. Vandersypen, X. L. Zhou, D. W. Leung, and S. Lloyd, Nature **393**, 143 (1998).

[16] N. Linden, H. Barjat, and R. Freeman, Chem. Phys. Lett. **296**, 61 (1998).

[17] K. Dorai, Arvind, A. Kumar, Chem. Phys. Rev. A, **61**, 042306 (2000).

[18] R. Marx, A. F. Fahmy, J. M. Myers, W. Bermel and S. J. Glaser, *Phys. Rev. A* **62**, 012310-1-8 (2000).

[19] I. L. Chuang, N. Gershenfeld, M. Kubinec, Phys. Rev. Lett. **80**, 3408 (1998).

[20] J. A. Jones, M. Mosca, R. H. Hansen, Nature **393**, 344 (1998)

[21] L. M. K. Vandersypen, M. Steffen, M. H. Sherwood, C. S. Yannoni, G. Breyta, I. L. Chuang, Appl. Phys. Lett. **76**, 468(2000).

[22] L. M. K. Vandersypen, M. Steffen, G. Breyta, C. Yannoni, R. Cleve, I. L. Chuang, Phys. Rev. Lett. **85**, 5452 (2000).

[23] L. M. K. Vandersypen, M. Steffen, G. Breyta, C. S. Yannoni, M. H. Sherwood, I. L. Chuang, Nature **414**, 883(2001).

[24] C. S. Yannoni, M. H. Sherwood, D. C. Miller, I. L. Chuang, L. M. K. Vandersypen, M. G. Kubinec, Appl. Phys. Lett. **75**, 3563 (1999).

[25] Z. L. Mádi, R. Brüschweiler, R. R. Ernst, J. Chem. Phys. **109**, 10603(1998).

[26] T. Schulte-Herbrüggen, K. Hüper, U. Helmke, S. J. Glaser, "Geometry of Quantum Computing by Hamiltonian Dynamics of Spin Ensembles", in: Applications of Geometric Algebra in Computer Science and Engineering, L. Dorst, C. Doran and J. Lasenby, Eds., Birkhäuser, Boston, 271 (2002).

[27] D. G. Cory, M. D. Price, T. F. Havel, Physica D **120**, 82(1998).

[28] D. G. Cory, W. Mass, M. Price, E. Knill, R. Laflamme, W. H. Zurek, T. F. Havel, S. S. Somaroo, Phys. Rev. Lett. **81**, 2152(1998).

[29] R. Marx, A. F. Fahmy, J. M. Myers, W. Bermel, S. J. Glaser, in *Quantum Computing*, Eds.: E. Donkor and A. R. Pirich, Proceedings of SPIE Vol. 4047, pp. 131-138 (2000).

[30] E. Knill, R. Laflamme, R. Martinez, C.-H. Tseng, Nature **404**, 368 (2000).

[31] S. Lloyd, Science **261**, 1569 (1993).

[32] W. S. Warren, Science **277**, 1688 (1997).

[33] P. Hübler, J. Bargon and S. J. Glaser, J. Chem. Phys. **113**, 2056-2059 (2000).

[34] A. S. Verhulst, O. Liivak, M. Sherwood, H.-M. Vieth, I. L. Chuang, Appl. Phys. Lett. **79**, 2480(2001).

[35] J. M. Myers, A. F. Fahmy, S. J. Glaser and R. Marx, Phys. Rev. A **63**, 032302/1(2001).

[36] R. Marx, Dissertation, Universität Frankfurt (2001).

[37] L. S. Pontryagin, V. Boltyanskii, R. Gamkrelidze, E. Meshecenko, The Mathematical Theory of Optimal Processes, Interscience (1962).

[38] N. Khaneja, R. Brockett, S. J. Glaser, Phys. Rev. A **63**, 032308/1(2001).

[39] N. Khaneja, S. J. Glaser, Chem. Phys. **267**, 11(2001).

[40] T. O. Reiss, N. Khaneja, S. J. Glaser, J. Magn. Reson. **154**, 192(2002).

[41] N. Khaneja, S. J. Glaser, R. Brockett, Phys. Rev. A **65**, 032301 (2002).

[42] C. H. Tseng, S. Saomaroo, Y. Sharf, E. Knill, R. Laflamme, T. F. Havel, D. G. Cory, Phas. Rev. A **61**, 012302 (2000).

[43] T. D. Ladd, R. J. Goldman, F. Yamaguchi, Y. Yamamoto, E. Abe, K. M. Itoh, Phys. Rev. Lett. **89**, 017901 (2002).

[44] N. Khaneja, S. J. Glaser, Phys. Rev. A **66**, 060301(R) (2002).

[45] K. Zyczkowski, P. Horodecki, A. Sanpera, M. Levenstein, Phys. Rev. A **58**, 883(1998).

[46] S. L. Braunstein, C. M. Caves, R. Jozsa, N. Linden, S. Popescu, R. Schack, Phys, Rev. Lett. **83**, 1054(1999).

[47] R. Laflamme, D. G. Cory, C. Negrevergne, L. Viola, Quantum Information & Computing **2**, 166 (2002).

6 On Lossless Quantum Data Compression and Quantum Variable-length Codes

Rudolf Ahlswede and Ning Cai

Universität Bielefeld
Bielefeld
Germany

6.1 Introduction

We survey results in the recently (in the late 90's) emerged area described in the title. The focus is on the compression rates, i.e. the average length of codewords.

In Shannon's Foundation of Information Theory ([27]) perhaps the most basic contributions are Source Coding Theorems (lossy and lossless) and the Channel Coding Theorem. In the most natural and simple source model DMS the source outputs a sequence $X_1, X_2, \ldots$ of independent, identically distributed random variables taking finitely many values. The Lossy Source Coding Theorem says that this sequence can be compresed by block coding with arbitrarily small error probability at a rate $H(P)$, the entropy of the common distribution P of the X_i's. (Later Shannon gave an extension replacing the probability of error criterion by general fidelity criteria and an ingenious formula for the rate-distortion function replacing $H(P)$.)

The Lossless Source Coding Theorem states that for variable-length codes the optimal data compression rate for an arbitrary source with distribution P is between $H(P)$ and $H(P) + 1$.

Whereas the beginning of Quantum Information Theory can be traced back for instance to Holevo's paper [15] from 1973 it started flourishing only in the midnineties. With Schumacher's Quantum Lossy Source Coding Theorem [24] one of Shannon's basic results could be carried over to the quantum world: a memoryless quantum source generating a sequence $|x_1, x_2, \ldots, x_n\rangle$ of pure states with probability $P^n(x^n)$, where $x^n = (x_1, \ldots, x_n)$ is a sequence of indices of the states from a finite index set $\mathcal{X}$, can be compressed at rate $S(P)$, the von Neumann entropy of the state $\sum_{x \in \mathcal{X}} P(x)|x><x|$ with arbitrary high fidelity.

Subsequent work on quantum lossy data compression can be found in [4, 5, 12–14, 18, 19, 22], and [28]. ([25] gives significant progess on channel coding; "single-letterisations" of the capacity formula are still not known.)

However, the extension of the Lossless Source Coding Theorem meets an obvious barrier: a measurement of the length of codewords will disturb the message. Thus quantum data cannot be compressed losslessly if only the quantum resource is available. This was pointed out by many authors, e.g. in [8, 9, 26].

Nevertheless some applications of quantum variable-length codes have been found. We report on these as follows:

In Section 6.2 we present basic concepts and results on quantum variable-length codes, mainly from [8] and [26].

In Section 6.3 it is explained how in [9] and [26] long quantum codes are build from quantum variable-length codes.

In Section 6.4 the model of Boström/Felbinger [8] (based on Bolström's [6,7]) for a quantum source with a classical *helper*, a classical channel informing the decoder about the lengths of codewords, is described.

In Sections 6.5 and 6.6 from [3] our recent results for this model and also our more general model are given.

In Section 6.7 for the first time in this survey we discuss a model and a result for *mixed* states, which are due to Koashi/Imoto [21].

In Section 6.8, finally we include a recent model for *lossy* quantum data compression, because it is related to the helper aspects.

6.2 Codes, Lengths, Kraft Inequality and von Neumann Entropy Bound

It is obvious that there is no way to compress losslessly classical nor quantum data by using block codes. So we always mean that a variable-length or in other words an indeterminate-length code (refered to in [26]) is employed, when we speak of lossless data compression.

6.2.1 The Codes

Quantum variable-length codes are defined by different authors, e.g. [8, 9] and [26]. The following is the definition by Boström and Felbinger in [8].

Let $\mathcal{H}$ be a Hilbert space of finite dimension d with an orthonormal basis

$$\mathcal{B}(\mathcal{H}) = \{|i>: i = 0, 1, 2, \ldots, d-1\}. \tag{6.1}$$

Denote by $\mathcal{H}^{\otimes n}$ the nth tensor power of the Hilbert space $\mathcal{H}$. For $\ell = 1, 2, \ldots, \ell_{\max}$ let $\mathcal{H}^{\otimes \ell}$ be a set of pairwise orthogonal (sub)spaces (in a sufficiently large Hilbert space). Then we can define the direct sum

$$\mathcal{H}^{\oplus \ell_{\max}} = \mathcal{H} \oplus \mathcal{H}^{\otimes 2} \oplus \cdots \oplus \mathcal{H}^{\otimes \ell_{\max}}, \tag{6.2}$$

a Hilbert space of dimension $\sum_{\ell=1}^{\ell_{\max}} d^\ell$. Now suppose we are given an information source space $\mathcal{S}$, a Hilbert space of finite dimension d', say. Then it was defined in [8] that a variable-length encoder $\mathcal{E}$ of maximal length $\ell_{\max}$ is a linear isometric operator $\mathcal{E}$ from $\mathcal{S}$ to a subspace $\mathcal{C} \subset \mathcal{H}^{\oplus \ell_{\max}}$ of dimension d' i.e., for all $|s\rangle, |s'\rangle \in \mathcal{S}$ $\langle \mathcal{E}(s)|\mathcal{E}(s')\rangle = \langle s|s'\rangle$, where $|\mathcal{E}(s'')\rangle = \mathcal{E}(|s''\rangle)$. $\mathcal{C}$ is called codeword space and (normalized) vectors (i.e., states) in it are called codewords.

To realize coding schemes, Schumacher and Westmoreland introduce zero-extended forms (zef) in [26]. For a codeword $\gamma^\ell \in \mathcal{H}^{\otimes \ell}$, its zef $|\gamma^\ell 0^{\ell_{\max}-\ell}\rangle$ is obtained by appending $(\ell_{\max} -$

ℓ)'s $|0\rangle$ to it. zef of a superposition of codewords $|\gamma^{\ell_i} >\in \mathcal{H}^{\otimes \ell_i}$ $i = 1, 2, \ldots, k$ (which in itself is a codeword) is the superposition of zefs of $|\gamma^{\ell_i}\rangle$'s. Similarly, to realize a coding scheme, $|0\rangle$'s are padded at the end in [9] and in front of codewords in [8].

6.2.2 Length Observable and Average Length of Codewords

In classical information theory the lengths of codewords in a variable-length code are determinate e.g., in the code $\{0, 10, 11\}$ the codewords $0, 10, 11$ have length $1, 2, 2$ respectively whereas the length of codewords in a quantum variable-length code are indeterminate because of superposition. Namely, for a vector $(a_1, a_2, \ldots, a_{\ell_{\max}}) \in \mathbb{C}^{\ell_{\max}}$, with $\sum_{\ell=1}^{\ell_{\max}} a_\ell^2 = 1$ and $|\gamma^\ell\rangle \in \mathcal{C} \cap \mathcal{H}^{\otimes \ell}$, $\sum_{\ell=1}^{\ell_{\max}} a_\ell |\gamma^\ell\rangle$ is a codeword (cf. (6.2)) because the encoder mapping is linear. So Schumacher and Westmoreland prefer to refer to the codes as "indeterminate codes" in [26]. One way to measure the lengths of codewords in this case is as follows ([26] and [8]). Let $\mathcal{H}^{\oplus \ell_{\max}}$ be the Hilbert space in (6.2) and let $\mathcal{P}_\ell$ be the projection of $\mathcal{H}^{\oplus \ell_{\max}}$ onto $\mathcal{H}^{\otimes \ell}$ for $\ell = 1, 2, \ldots, \ell_{\max}$. Then the observable $\mathcal{L} = \{\mathcal{P}_\ell\}$, where $\mathcal{P}_\ell$ corresponds to the outcome ℓ, is called length observable. Thus with the probability $\mathrm{tr}(|w\rangle\langle w|\mathcal{P}_\ell) = \langle w|\mathcal{P}_\ell|w\rangle = a_\ell^2$ the outcoming length of a codeword $|w\rangle = \sum_{\ell=1}^{\ell_{\max}} a_\ell |\gamma^\ell\rangle$, $|\gamma^\ell\rangle \in \mathcal{H}^{\otimes \ell}$, is ℓ when one measures the codeword by $\mathcal{L}$. Let

$$\wedge = \sum_{\ell=1}^{\ell_{\max}} \ell \mathcal{P}_\ell. \tag{6.3}$$

The expected outcoming length, also called the average length, of a codeword $|w\rangle$ is

$$\overline{L}(|w\rangle) = \mathrm{tr}(|w\rangle\langle w|\wedge) = \langle w| \wedge |w\rangle. \tag{6.4}$$

6.2.3 Kraft Inequality and von Neumann Entropy Bound

The quantum prefix codes i.e., a codes such that no codeword is a prefix of other codewords, are well studied by Schumacher and Westmoreland in [26]. In particular in similar ways as in Classical Information Theory they proved:

Quantum Kraft Inequality: *For all quantum prefix codes $\mathcal{C}$ and $D = \dim(\mathcal{H})$*

$$\sum_{\ell=1}^{\ell_{\max}} \dim(\mathcal{C} \cap \mathcal{H}^{\otimes \ell}) D^{-\ell} \leq 1. \tag{6.5}$$

Like in the classical case, it was shown in [26] that the quantum Kraft inequality holds for all uniquely decodable codes (cf. Section 6.3). For non-uniquely decodable codes, Boström and Felbinger in [8] extend them to prefix codes and then obtain a Kraft-type inequality with an additional term, which depends on the extension of the codes and therefore on the structure of the codes.

von Neumann Entropy Bound [26]: *Consider a quantum source which outputs a state* $|s\rangle$ *with probability* $P(s)$, *and* $\sigma = \sum_s P(s)|s\rangle\langle s|$. *Then for all uniquely decodable quantum codes, in particular for all quantum prefix codes in (6.4) the expected average length of codewords with respect to the probability* P *is lowerbounded by the von Neumann entropy* $S(\sigma)$ *of the state* σ.

6.2.4 Base Length

An important parameter, the base length $L(|w\rangle)$ of a codeword $|w\rangle$ in a quantum variable-length code was introduced in [8]:

$$L(|w\rangle) = \max\{\ell : \langle w|\mathcal{P}_\ell|w\rangle > 0\}. \tag{6.6}$$

That is, $L(|w\rangle)$ is the largest ℓ such that $a_\ell \neq 0$ if $|w\rangle$ is a superposition $|w\rangle = \sum_\ell a_\ell|\gamma^\ell\rangle$, $|\gamma^\ell\rangle \in \mathcal{C} \cap \mathcal{H}^{\otimes\ell}$. It is clear that for all codewords $|w\rangle$

$$\overline{L}(|w\rangle) \leq L(|w\rangle). \tag{6.7}$$

To store a codeword of base length ℓ, one needs a quantum register of length at least ℓ. So it is necessary for the decoder to know the base length of codewords.

6.3 Construct Long Codes from Variable-length Codes

When they observed that it is impossible to losslessly compress quantum data if only quantum resources are available (cf. next section), Braustein and Fuchs in [9] suggested to apply quantum variable-length codes to construct a long block code in the following way. First connect N codewords of a quantum variable-length code and then truncate the obtained codeword and keep the first $N(\tilde{L} + \delta)$ components, where $\tilde{L}$ is the expectation of the average lengths with respect to the source distribution. Then a block code of length $N(\tilde{L} + \delta)$ with high fidelity is obtained.

Algorithms for the purposes of storage and communication are also presented in [9]. It is shown that, in both cases, the computational complexity using quantum variable-length codes to construct long block codes is remarkably lower than the best known algorithms.

Constructing block codes from quantum variable-length codes is systematically analysed in [26]. A transformation, called condensation, is introduced. A code is said to be *condensable* if for all N there exists a unitary operator U (depending on N) such that for all $\gamma^{\ell_i} \in \mathcal{H}^{\otimes\ell_i}$, $i = 1, 2, \ldots, N$

$$U|\gamma^{\ell_1}0^{\ell_{\max}-\ell_1}\rangle|\gamma^{\ell_2}0^{\ell_{\max}-\ell_2}\rangle\ldots|\gamma^{\ell_N}0^{\ell_{\max}-\ell_N}\rangle = |\Psi^{\sum_{i=1}^N \ell_i}0^{N\ell_{\max}-\sum_{i=1}^N \ell_i}\rangle \tag{6.8}$$

for a $|\Psi^{\sum_{i=1}^N \ell_i}\rangle \in \mathcal{H}^{\otimes\sum_{i=1}^N \ell_i}$, and the process is called *condensation*. The code is called *simply* condensable and the condensation is said to be simple if for all γ^{ℓ_i} $i = 1, 2, \ldots, N$, $|\Psi^{\sum_{i=1}^N \ell_i}\rangle$ in (6.8) is $|\gamma^{\ell_1}\gamma^{\ell_2}\ldots\gamma^{\ell_N}\rangle$. Then a prefix code is simply condensable. Obviously simply condensable codes are analogous to uniquely decodable codes in Classical Information

Theory. So we also address simply condensable codes as uniquely decodable codes. Since unitary transformations are isoperimetric, a condensable code essentially is treated as a uniquely decodable code. In [26] quantum Kraft inequality and von Neumann entropy bound (see Section 6.2.3) for condensable codes are established. Based on them it is shown in [26] that the rate of asymptotically optimal codes of high fidelity constructed by condensation equals von Neumann entropy for pure state sources generating $|s\rangle$ with probability $P(s)$. More efficient ways to use quantum variable-length codes to build long block codes are also presented in [26].

6.4 Lossless Quantum Data Compression, if the Decoder is Informed about the Base Lengths

In this and the next two sections we consider lossless quantum data compression for pure state quantum sources. We emphasize again

Observation I: A length measurement performed at a codeword of a quantum variable-length code will destroy the codewords.

That means, there must be a way to inform the decoder about the base lengths of codewords in a procedure of lossless quantum data compression since to decode correctly the decoder must know the base length of the decoded codeword. Moreover it is noticed in [3]

Observation II: In general, there is no way to measure the base length of unknown codewords *without error*.

So to inform the decoder about the lengths of codewords the encoder should know the output of the quantum source i.e., the output should be visible (by the encoder).

Boström and Felbinger in [8] suggest to code the message in the following way.

1) Quantum Source Output Visible by Encoder: Suppose the encoder needs to encode the output states from the source space $\mathcal{S}$ of dimension d. He does this by a linear isometric operator from $\mathcal{S}$ to a subspace $\mathcal{C}$ of $\mathcal{H}^{\oplus \ell_{\max}}$ (c.f. Section 6.2.1). The output is visible, that is the encoder knows the output state of the quantum source and therefore the base length of the codeword to which the output state is encoded, say $\ell_{\mathcal{B}}$.

2) Classical Channel: Now the encoder knows $\ell_{\mathcal{B}}$ and has to inform the decoder about it. This is done via the classical channel. Thus the decoder may store and decode the codewords correctly.

We point out here that the classical channel in their model only is used to inform about the lengths of codewords. Under this assumption the authors of [8] proposed the following coding scheme for the discrete quantum source $\{(P(x), |x\rangle) : x \in \mathcal{X}\}$, which outputs the state $|x\rangle$, $x \in \mathcal{X}$ with probability $P(x)$, where $\mathcal{X}$ is a finite set,

(a) Choose a basis $\{|x_1\rangle, |x_2\rangle, \dots, |x_{d'}\rangle\}$ recursively as follows

(a$_1$) Choose an $|x_1\rangle$ such that $P(x_1) = \max_{x \in \mathcal{X}_1} P(x)$ for $\mathcal{X}_1 = \mathcal{X}$.

(a_i) Having chosen $|x_1\rangle, \dots, |x_{i-1}\rangle$, one first deletes all $|x'\rangle$ in the subspace spanned by $\{|x_1\rangle, \dots, |x_{i-1}\rangle\}$ from $\{|x\rangle : x \in \mathcal{X}\}$ and obtains a subset $\{|x''\rangle : x'' \in \mathcal{X}_i\}$, $\mathcal{X}_i \subset \mathcal{X}$. Then one chooses an $|x_i\rangle$ in $\mathcal{X}_i$ such that $P(x_i) = \max_{x'' \in \mathcal{X}_i} P(x'')$.

($a_{d'}$) The procedure is stopped at a vector $|x_{d'}\rangle$ such that $\mathcal{X}_{d'+1} = \varnothing$.

(b) Gram–Schmidt Orthonormalization: Obtain an orthonormal basis $\{|\beta_i\rangle : i = 1, 2, \dots, d'\}$ from $\{|x_i\rangle : i = 1, 2, \dots, d'\}$ by Gram–Schmidt orthonormalization.

(c) Encoding: Suppose $\dim(\mathcal{H}) = d$, and let $z_d(i)$ be the d-ary representation of number i and $w_d^{d'}(i)$ be the d-ary sequence of length d obtained by padding $d - \lceil \log_d i \rceil$'s 0 in front of $z_d(i)$ for $i = 1, 2, \dots, d'$. Then encode $|\beta_i\rangle$ to $|w_d^{d'}(i)\rangle$.

(d) Remove the redundancy and inform about the base length: Now assume a state $|x\rangle = \sum_{i=1}^{j} c_i|\beta_i\rangle$ for $c_j \neq 0$ as output. Then by the previous step and the linearity of the encoder, we know $|s\rangle$ is encoded to a codeword $\sum_{i=1}^{j} c_i|w_d^{d'}(i)\rangle$, a codeword starting with r zeros for $r = d - \lceil \log_d j \rceil$, say. Then the encoder, who knows $|s\rangle$ and consequently j, removes the r zeros to obtain a codeword of base length $\ell = \lceil \log_d j \rceil$, say, and inform the decoder about ℓ via a classical channel. Notice that the resulting codeword after removing the redudancy can be stored in a d-ary quantum register of length ℓ.

(e) Decoding: The decoder pads $d - \ell$ zeros in front of the received (quantum) codeword and recovers the state $|s\rangle$ by the inverse of the (isometric) encoder in Step (c).

6.5 Code Analysis Based on the Base Length

In our recent work [3] we systematically analyse the code of Boström and Felbinger [8], which is defined by the coding scheme 1), 2) in the previous section. To realize the coding scheme we expect that a codeword of base length ℓ can be stored in a quantum register of length ℓ. So we constrain the code $\mathcal{C}$ such that $\mathcal{C} \cap \mathcal{H}^{\otimes \ell}$ can be embedded in $\mathcal{H}^{\otimes \ell}$. Under this constraint we obtain a sufficient and necessary condition for the existence of codes of Section 6.4. For such a code $\mathcal{C}$ we denote by $\mathcal{C}_\ell$ the set of codewords of base lengths at most ℓ and by N_ℓ the number of codewords of base length ℓ. Then $\mathcal{C}_\ell$ is a linear subspace and the code exists iff for $\ell = 1, 2, \dots, \ell_{\max}$

$$\mathcal{C}_1 \subset \mathcal{C}_2 \subset \cdots \subset \mathcal{C}_{\ell_{\max}}, \tag{6.9}$$

$$\dim \mathcal{C}_\ell \leq d^\ell \tag{6.10}$$

or equivalently

$$\sum_{i=1}^{\ell} N_\ell \leq d^\ell. \tag{6.11}$$

To realize the coding scheme we may obtain its ref by appending $|0\rangle$'s to its codewords. Then we obtain the canonical codes introduced in [3].

Moreover we determine the optimal compression rate for an arbitrary quantum pure state source.

Let $\mathcal{S}$ be a Hilbert space of dimension d', which will serve as a source space. $\mathcal{F}$ is a σ-field on $\mathcal{S}$ and P is a probability distribution over $\mathcal{F}$, which is not necessary discrete. Suppose a quantum source outputs pure states in $F \in \mathcal{F}$ with probability $P(F)$.

We call a sequence of subspaces $L = \{L_\ell : \ell = 1, 2, \ldots, \ell_{\max} - 1\}$ for an $\ell_{\max}$ such that $d^{\ell_{\max}-1} < d' \leq d^{\ell_{\max}}$, where $d' = \dim \mathcal{S}$, d-nested if for all ℓ

$$\dim L_\ell = d^\ell, \tag{6.12}$$

$$L_1 \subset L_2 \subset \cdots \subset L_{\ell_{\max}-1}. \tag{6.13}$$

Denote by $\mathcal{L}_d(\mathcal{S})$ the set of d-nested sequences of subspaces of $\mathcal{S}$. Then we have

Theorem 6.1. (Ahlswede, Cai [3]) *The minimum achievable lossless compression rate of a quantum source, specified by a probability space $(S, \mathcal{F}, P)$, via a quantum variable-length code with a classical helping channel informing about base lengths i.e., the codes in Section 6.4, is*

$$R_0 \triangleq \ell_{\max} - \sup_{L \in \mathcal{L}_d(\mathcal{S})} \sum_{\ell=1}^{\ell_{\max}-1} P(L_\ell). \tag{6.14}$$

6.6 Lossless Quantum Data Compression with a Classical Helper

We have seen in Section 6.4 that in the codes introduced by Boström and Felbinger, the classical channel only transmits the base lengths of codewords. As the lengths of codewords actually carry information we naturally ask ourselves "Why do'nt we use the classical channel to send other information?" The following example in [3] shows that we can do better.

Example: Let $\dim \mathcal{H} = 2$, $\dim \mathcal{S} = 4$ and $\mathcal{S}_0$ and $\mathcal{S}_1$ be two orthogonal subspaces of $\mathcal{S}$ of dimension 2. P is a probability distribution over $\mathcal{S}$ such that $P(\mathcal{S}_1) = P(\mathcal{S}_2) = \frac{1}{2}$. Suppose the source outputs a state in $\mathcal{A} \subset \mathcal{S}$ with the probability $P(\mathcal{A})$. For a "continuous" source one may assume P is uniformly distributed on $\mathcal{S}_0 \cup \mathcal{S}_1$ and for the discrete quantum source one may assume P is uniformly distributed on a set of states

$$\{|u_i\rangle : i = 0, 1, 2, \ldots, m-1\} \cup \{|v_j\rangle : j = 0, 1, 2, \ldots, m-1\},$$

where $|u_i\rangle \in \mathcal{S}_0$, $|v_j\rangle \in \mathcal{S}_1$, and $m \geq 3$. But we shall see that the assumption for assigning the probabilities to the particular states makes no difference. Now $\ell_{\max} = 2$ and it is easy to see that the maximum probability of 2-dimensional subspaces of $\mathcal{S}$ is $\frac{1}{2}$. So by Theorem 6.1 the best quantum compression rate with classical helping channel informing the base length is $\frac{3}{2}$. Additionally the encoder has to send one bit to the decoder to inform him about the base length.

As in the current source the probability is concentrated on $\mathcal{S}_0 \cup \mathcal{S}_1$ the encoder has a more clever way to compress the quantum source. He can just simply choose arbitrary two unitrary operators U_0 and U_1, one mapping from $\mathcal{S}_0$ to $\mathcal{H}$ and the other from $\mathcal{S}_1$ to $\mathcal{H}$. In the case that a state $|s\rangle \in \mathcal{S}_i$ for $i = 0$ or 1, is output from the source, the encoder encodes it to $U_i|s\rangle$ by using operator U_i and sends i to the decoder via the classical channel. Then the decoder who knows i now decodes the quantum codeword by using U_i^{-1} and obtains $U_i^{-1}U_i|s\rangle = |s\rangle$. For this code the quantum compression rate is 1 and the encoder sends one bit via the classical channel. It is a better code.

This simple example motivated us to look for a more efficient way to use the classical helper. By Observations I and II in Section 6.4 the following assumptions are necessary.

(1) Visible encoding: The encoder knows the output state of the quantum source.

(2) The classical helper: There is a classical channel connecting the encoder and the decoder such that the encoder can send classical information to the decoder.

Under these assumptions we have the following coding scheme.

We let $\mathcal{H}$ and $\mathcal{S}$ be complex Hilbert spaces of dimensions d and d' respectively and P be a probability distribution with support set $\mathcal{U} \subset \mathcal{S}$.

Suppose a quantum source outputs a state $|u\rangle \in \mathcal{S}$ with probability $P(u)$. Without loss of generality we assume that $\mathcal{S} = \text{span}\{|u\rangle : u \in \mathcal{U}\}$, because otherwise we may replace $\mathcal{S}$ by $\text{span}\{|u\rangle : u \in \mathcal{U}\}$.

Coding Scheme:

(I) Partition $\mathcal{U}$ properly into $\{\mathcal{U}_j : j = 0, 1, \ldots, J-1\}$ for an integers J. For each j find the minimum ℓ_j such that there is an $d\ell_j$-dimensional subspace $\mathcal{S}_j$ of $\mathcal{S}$, containing $\text{span}\{|u\rangle : u \in \mathcal{U}_j\}$. We write $L_q(\mathcal{U}_j) = \ell_j$.

(II) For all $j \in \{0, 1, \ldots, J-1\}$, arbitrarily choose a unitrary operator U_j from $\mathcal{S}_j$ to $\mathcal{H}^{\otimes \ell_j}$.

(III) Suppose a $|u\rangle \in \mathcal{S}$ is output by the quantum source and assume that $|u\rangle \in \mathcal{S}_j$. Then the encoder encodes $|u\rangle$ to a codeword $|w(u)\rangle \triangleq \mathcal{U}_j|u\rangle \in \mathcal{H}^{\otimes L_q(\mathcal{U}_j)}$ by using the operator U_j. We say $|u\rangle$ is encoded to a quantum codeword $|w(u)\rangle$ of length $L_q(|w(u)\rangle) = L_q(\mathcal{U}_j)$. Then the encoder sends j by classical variable-length code e.g., Huffman code, for a classical source outputing $j \in \{0, 1, 2, \ldots, J-1\}$ with probability $Q(j) = P(\mathcal{U}_j)$, to the decoder via the classical channel.

(IV) Finally the decoder who has the quantum codeword $|w(u)\rangle = U_j|u\rangle$ and knows j from the classical channel, reconstructs the output state $|u\rangle$ by applying the operator U_j^{-1} to $|w(u)\rangle$.

It is not hard to see that this coding scheme is most general under the two assumptions, there is no better code than the best codes constructed by this coding scheme.

The key step is how to choose the partition in (I) and it is actually the most difficult part in the coding scheme.

We call a code constructed by coding scheme a quantum–classical variable-length code, or shortly a $q-c$ variable-length code and its two components, quantum and classical components respectively and speak of lossless quantum data compression with classical helper.

We denote by $L_c(\mathcal{U}_j)$ the length of the codeword to which the classical message is encoded by the classical variable-length code in step (III) of the coding scheme when $|u\rangle \in \mathcal{U}_j$. Then the classical and quantum components of the compression rate are

$$R_c = \sum_{j=0}^{J-1} P(\mathcal{U}_j) L_c(\mathcal{U}_j),$$

$$R_q = \sum_{j=0}^{J-1} P(\mathcal{U}_j) L_q(\mathcal{U}_j)$$

respectively. By Shannon's Lossless Source Coding Theorem ([27]; also in [10, 11]), with the notation $Q \triangleq \{Q(j) = P(\mathcal{U}_j) : j = 0, 1, \ldots, J-1\}$, R_c is bounded by

$$(\log a)^{-1} H(Q) \leq R_c < (\log a)^{-1} H(Q) + 1, \tag{6.15}$$

where a is the size of the alphabet of the classical code and H is Shannon's entropy.

To simplify notation, in the sequel we assume the size of the classical alphabet to be $a = d = \dim \mathcal{H}$. Then we have

Theorem 6.2. (Ahlswede, Cai [3]) *For any $q - c$ variable-length code,*

$$R_q + R_c \geq (\log d)^{-1} S(\rho) \tag{6.16}$$

where $S(\rho)$ is the von Neumann entropy of the state,

$$\rho \triangleq \sum_{u \in \mathcal{U}} P(u) |u><u|, \tag{6.17}$$

and equality holds iff the following conditions hold simultaneously.

(i) *For the probability Q in (6.15), i.e. $Q(j) = P(\mathcal{U}_j)$,*

$$R_c = (\log d)^{-1} H(Q). \tag{6.18}$$

(ii) *For all $j \neq j'$*

$$\mathcal{S}_j \perp \mathcal{S}_{j'}, \tag{6.19}$$

and

(iii) *for all $j \in \{0, 1, \ldots, J-1\}$*

$$P(\mathcal{U}_j)^{-1} \sum_{u \in \mathcal{U}_j} P(u) |u\rangle\langle u| = d_j'^{-1} \mathcal{P}_j, \tag{6.20}$$

where $d'_j = \dim \mathcal{S}_j$ and $\mathcal{P}_j$ is the projector onto subspace $\mathcal{S}_j$.

When the support set of the source distribution is a set of independent pure states, we have a sharper bound.

Proposition 6.1 (Ahlswede, Cai [3]). *Let $|u\rangle$, $u \in \mathcal{U}$, be a set of independent pure states, let P be a probability distribution on $\mathcal{U}$ and let a quantum source output $|u\rangle$ with probability $P(u)$, where $\mathcal{U}$ is a finite index set. Then for all $q - c$ variable-length codes for the source*

$$R_q + R_c \geq (\log d)^{-1} H(P)$$

with equality iff for all j $L_c(\mathcal{U}_j) = -(\log d)^{-1} \log P(\mathcal{U}_j), L_q(\mathcal{U}_j) = -(\log d)^{-1} \log |\mathcal{U}_j|$, and for all $u \in \mathcal{U}_j$ $P(u) = \frac{P(\mathcal{U}_j)}{|\mathcal{U}_j|}$.

We conclude this section with a few *Problems*, which we pose on lossless quantum data compression for pure state sources:

1. In [3] we showed that the gap between von Neumann entropy and the optimal compression rates in Theorem 6.2 may be arbitrary large. On the other hand by our knowledge successfully used quantum information measures are all in terms of von Neumann entropy. So we ask "Is there a quantity better to fit lossless quantum data compression than von Neumann entropy?"

2. For an arbitrary discrete memoryless quantum pure state source determine the optimal compression rate. For this problem we know that von Neumann entropy and Shannon entropy are lower and upper bounds and in general neither bound is tight.

3. Study other models of quantum data compression e.g., the quantum version of the identification problem treated in [2], which was introduced in the context of [1].

6.7 Lossless Quantum Data Compression for Mixed State Sources

We report here the work of Koashi and Imoto [21]. This is their model:

- A quantum source outputs for $i = 1, \ldots, I$ mixed states ρ_i in a Hilbert space $\mathcal{H}'_A$ with probability p_i.

- The measurement of lengths of codewords is performed in an auxiliary quantum system $\mathcal{H}_E$ so that it will (by assumption) not disturb the message.

More precisely the encoding–decoding operator is specified by a unitary operator U acting on $\mathcal{H}'_A \otimes \mathcal{H}_E$ such that for $i = 1, 2, \ldots, I$

$$\mathrm{tr}_E\big[U(\rho_i \otimes \Sigma_E)U'\big] = \rho_i, \tag{6.21}$$

where $\mathcal{H}_E$ is an auxiliary system initially prepared in a pure state Σ_E. They assume that there is an observable $\mathcal{L}$ acting on $\mathcal{H}_E$, which corresponds to the lengths of codewords such that the expected length of codewords for $\rho = \sum_i P_i \rho_i$,

$$\overline{\overline{L}} = \mathrm{tr}_E\big\{\mathcal{L}\ell_{\gamma_A}\big[U(\rho \otimes \Sigma_E)U'\big]\big\}. \tag{6.22}$$

The coding theorem is based on their previous work [20], where it was shown that for a set $\{\rho_i\}_i$ if mixed states a probability distribution $\{P_i\}_i$ and $\rho = \sum_i \rho_i P_i$, there is a unique decomposition of the support set $\mathcal{H}_A$ of ρ such that

$$\mathcal{H}_A = \oplus_\ell \mathcal{H}_J^{(\ell)} \otimes \mathcal{H}_K^{(\ell)} \tag{6.23}$$

and for all i

$$\rho_i = \oplus_\ell q^{(i,\ell)} \rho_J^{(i,\ell)} \otimes \rho_K^{(\ell)}, \tag{6.24}$$

where $\rho_J^{(i,\ell)}$ and $\rho_K^{(\ell)}$ are normalized density operators acting on $\mathcal{H}_J^{(\ell)}$ and $\mathcal{H}_K^{(\ell)}$, respectively, $q^{(i,\ell)}$ is the probability for the states to be in the subspace $\mathcal{H}_J^{(\ell)} \otimes \mathcal{H}_K^{(\ell)}$, $\rho_K^{(\ell)}$ is independent of i and $\{\rho_J^{(i,\ell)}\}_i$ cannot be expressed in a simultaneously block-diagonalized form. For a quantum source outputing mixed state ρ_i with probability P_i, we denote $P(\ell) = \sum_i P_i q^{(i,\ell)}$, $I_c = -\sum_\ell P(\ell) \log P(\ell) = H(P)$ and $P_{NC} = \sum_\ell P(\ell) \log \dim \mathcal{H}_J^{(\ell)}$.

Let $\overline{R}$ be the optimal compression rate for this model. Then the coding theorem in [21] says

$$I_c + D_{NC} \leq \overline{R} \leq I_c + D_{NC} + 2. \tag{6.25}$$

6.8 A Result on Tradeoff between Quantum and Classical Resources in Lossy Quantum Data Compression

In this last section we report a result on lossy quantum data compression due to Hayden, Jozsa, and Winter in [14], because it relates to the helper model, best briefly, because it concerns the lossy case whereas this survey primarily adresses the lossless case.

This is the model:

- A quantum DMS outputs a pure state $|u\rangle$, $u \in \mathcal{U}$ with probability $Q(u)$ and consequently outputs a sequence $|u^n\rangle = |u_1 u_2 \dots u_n\rangle$, $u^n \in \mathcal{U}^n$ with probability $Q^n(u^n) = \prod_{i=1}^n Q(u_i)$. In other words an ensembles $\mathcal{E} = \{|u\rangle, P(u)\}$ u is given.
- Assume that the encoder can send messages to the decoder via a classical channel at rate R bits per signa,.
- Then the trade-off function $Q^*(R)$ is defined as the asymptotically optimal compression rate (qbits per signal) with an arbitrarily high fidelity under the above assumptions.

To compute $Q^*(R)$ the authors Hayden, Jozsa, and Winter decompose the ensemble $\mathcal{E} = \{|u\rangle, P(u)\}$ into at most $|\mathcal{U}| + 1$ ensembles $\mathcal{E}_j = \{|u\rangle, W(u|j)\}$ with weight $P(j)$ and their union $\cup_j P(j)\mathcal{E}_j$ reporduces $\mathcal{E}$. This is equivalent to decompositing the probability distribution Q by an input distribution P over $\{0, 1, \dots, |\mathcal{U}|\}$ and using a classical channel $W : \{0, 1, \dots, |\mathcal{U}|\} \to \mathcal{U}$ such that for all $u \in \mathcal{U}$ $Q(u) = \sum_j P(j) W(u|j)$. Let $\mathcal{D}(R)$ be the set of decompositions with $I(P;W) = R$ and $\overline{S}(P) = \sum_j P(j)S(\mathcal{E}_j)$, where I is Shannon's mutual information and S is von Neumann entropy. Then

Theorem 6.3. (Hayden, Jozsa, Winter [13])

$$Q^*(R) = \min_{(P,W)\in\mathcal{D}(R)} \overline{S}(R).$$

Finally we remark that in addition to the difference between the models in Sections 6.4–6.6 and this section with respect to the property lossless versus property lossy, another difference is that we deal with general quantum sources in Sections 6.4–6.6 and with quantum DMS in this section.

Moreover, in Sections 6.4 and 6.5 the decoder allows only to send lengths of codewords via the classical channel and there is not such a restriction in Sections 6.6 and 6.7.

References

[1] R. Ahlswede, General theory of information transfer, Preprint 97–118, SFB 343 "Diskrete Strukturen in der Mathematik", Universität Bielefeld.

[2] R. Ahlswede, B. Balkenhol and C. Kleinewächter, Identification for sources, Preprint 00–120, SFB 343 "Diskrete Strukturen in der Mathematik", Universität Bielefeld, 2000.

[3] R. Ahlswede and N. Cai, On Lossless quantum data compression with a classical helper, submitted to IEEE Trans. Inf. Theory.

[4] H. Bornum, C. M. Caves, C. A. Fuchs, R. Jozsa, and B. Schumacher, On quantum coding for ensembles of mixed states, http://xxx.lanl.gov/abs/quant-ph/0008024v1,2000.

[5] H. Barnum, C. A. Fuchs, R. Jozsa, and B. Schumacher, General fidelity limit for quantum channels, Phys. Rev. A 54, 4707, 1996.

[6] K. Boström, Concept of a quantum information theory of many letters, http://xxx.lanl.gov/abs/quant-ph/0009052, 2000.

[7] K. Boström, Lossless quantum coding in many-letter space, http://xxx.lanl.gov/abs/quant-ph/0009073, 2000.

[8] K. Boström and T. Felbinger, Lossless quantum data compression and variable-length coding, preprint, 2002.

[9] S. L. Braunstein and C. A. Fuchs, A quantum analog of Huffman coding. http://xxx.lanl.gov/abs/quant-ph/9805080,1998.

[10] T. M. Cover and J. A. Thomas, Elements of Information Theory, Wiley and Sons, New York, 1991.

[11] I. Csiszár and J. Körner, Information Theory: Coding Theorems for Discrete Memoryless Systems, Academic Press, New York-San Francisco-London, 1981.

[12] I. Devetak and T. Berger, Quantum rate-distortion theory for memoryless sources, IEEE Trans. Inform. Theory, Vol. 48, 1580–1589, 2000.

[13] M. Hayashi, Exponents of quantum fixed-length pure state source coding, http://xxx.lanl.giv/abs/quant-ph/0202002, 2002.

[14] P. Hayden, R. Josza and A. Winter, Trading quantum for classical resourses in quantum data compression, http://xxx,Canl.gov.Ph/0204038, 2002.

[15] A. S. Holevo, Statistical problems in quantum physics, In Gisiro Maruyama and Jurii V. Prokhorov ed. Proceeding of 2nd Japan-USSR Sym. 104–119, Springer-Verlag Berlin, 1973.

[16] A. S. Holevo, The capacity of the quantum channel, with general signal states, IEEE Trans. Inf. Theory, 44 (1), 269–273, 1998.

[17] D. A. Huffman, A method for the construction of minimum redundancy codes, Proc. IRE 40, 1098–1101, 1952.

[18] R. Jozsa and B. Schumacher, A new proof of the quantum noiseless theorem, Mod. Opt. 14, 2343, 1994.

[19] M. Koashi and N. Imoto, Compressibility of mixed-state signals, http://xxx.lanl.giv/abs/quant-ph/0103128v1, 2001.

[20] M. Koashi and N. Imoto, What is possible without disturbing partially quantum states, http://xxx.lanl.gov/abs/quant-ph/01011444v2, 2001.

[21] M. Koashi and N. Imoto, Quantum information is incompressible without errors, http://xxx.lanl.gov/abs/quant-ph/0203045v1, 2002.

[22] M. A. Nielsen, Quantum Information Theory, PHD thesis, Univ. New Mexico, 1998.

[23] M. A. Nielsen and I. L. Chuang, Quantum Computation and Quantum Information, Cambridge, 2000.

[24] B. Schumacher, Quantum coding, Phys. Rev. A. 51, 2738–2747, 1995.

[25] B. Schumacher and M. P. Westmoreland, Sending classical information via noisy quantum channels, Phys. Rev. A, 56 (1), 131–138, 1997.

[26] B. Schumacher, M. P. Westmoreland, Indeterminate-length quantum coding, http://xxx.lanl.gov./abs/quant-ph/0011011, 2000.

[27] C. E. Shannon, A mathematical theory of communication, Bell. Syst. Tech. J. 27, 379–423, 1948.

[28] A. Winter, Coding theorems of quantum information theory, PhD. Thesis, Univ. Bielefeld, 1999.

7 Entanglement Properties of Composite Quantum Systems

K. Eckert, O. Gühne, F. Hulpke, P. Hyllus, J. Korbicz, J. Mompart, D. Bruß, M. Lewenstein, and A. Sanpera

Institut für Theoretische Physik
Universität Hannover
Hannover, Germany

We present here an overview of our work concerning entanglement properties of composite quantum systems. The characterization of entanglement, i.e. the possibility to assert if a given quantum state is entangled with others and how much entangled it is, remains one of the most fundamental open questions in quantum information theory. We discuss our recent results related to the problem of separability and distillability for distinguishable particles, employing the tool of witness operators. Finally, we also state our results concerning quantum correlations for indistinguishable particles.

7.1 Introduction

The processing of quantum information differs in a fundamental way from the processing of classical information: rather than allowing only boolean values "0" and "1" for a bit, a quantum bit or qubit is implemented by the quantum state of a two-level system, which can be in any superposition of "$|\,0\rangle$" and "$|\,1\rangle$", namely $|\,\psi\rangle = \alpha|\,0\rangle + \beta|\,1\rangle$. If several quantum states are involved, rather than dealing only with one string of bit values, e.g., "001011", the state of the composite system of qubits can be in a superposition of such strings, e.g., $|\,\Psi\rangle = a|\,001011\rangle + b|\,110100\rangle + c|\,010010\rangle + \ldots$. In general such a state cannot be written as a tensor product of states of its subsystems and, therefore, it is called *entangled*.

Entanglement is a key feature for most of the protocols used in quantum information such as, e.g., quantum teleportation, quantum cryptography, superdense coding, quantum algorithms and quantum error correction. Indeed the resources needed to implement a particular protocol of quantum information are closely linked to the entanglement properties of the states used in the protocol. Therefore, it is highly desirable to characterize the entanglement properties of quantum systems bearing in mind that this is a fundamental open problem of quantum theory but also that it is essential for the implementation of any possible task that relies on quantum bits.

In this manuscript we summarize our efforts in striving at some understanding of the properties of entanglement for composite quantum systems of distinguishable and indistinguishable particles. In the former case, one assumes that the involved quantum systems can be addressed separately. This happens either because the subsystems are located at different places so that their wavefunctions do not spatially overlap, or because they differ in some de-

Quantum Information Processing, 2nd Edition. Edited by Thomas Beth and Gerd Leuchs

ISBN: 3-527-40541-0

grees of freedom permitting thus to distinguish them. Until quite recently, this has been the most common approach considered in the framework of quantum information. However, the experimental progress in achieving quantum bits and quantum gates by means of solid state physics (quantum dots) and optical microtraps with neutral atoms, demands a new formalism which includes the statistical nature of the particles involved. In the last part of this manuscript we address this question.

In the frame of the DFG-Schwerpunkt on "Quanteninformationsverarbeitung" (quantum information processing) we have addressed these subjects in two different projects. The manuscript aims to give a comprehensive summary of our results for a reader familiar with the subject of separability and distillability. This summary, however, is by no means exhaustive. Important contributions to the above projects concerning entanglement measures, catalysis of entanglement and quantum game theory are presented elsewhere [1].

The manuscript is organized as follows: Sections 7.2–7.4 deal with the characterization of entanglement of composite quantum systems of distinguishable particles. In Section 7.2 we first briefly state the problem of separability, i.e., the question when the state of a composite quantum system does not contain any quantum correlations or entanglement, and then report our results concerning this question. In Section 7.3 we address the problem of distillability, i.e., the question when the state of a mixed composite quantum system can be transformed to a maximally entangled pure state by using local operations and classical communication. We report thereafter our progress concerning this subject. Section 7.4 deals with witness operators. We state our achievements in constructing, optimizing and implementing witness operators to detect entanglement. Also, a connection between a witness detecting a given state and the distillability and activability properties of the state is presented there. Finally, in Section 7.5 we address the study of quantum correlations in composite systems of identical particles and apply some of the formalisms previously developed for distinguishable particles to indistinguishable ones.

7.2 Separability of Composite Quantum Systems

In this section, before presenting our results, we define the problem of separability versus entanglement for a given quantum system. The reader interested in a tutorial description of the subject is addressed to references [2] and [3].

For simplicity, let us restrict ourselves here to the simplest case of composite systems: bipartite systems (traditionally denoted as Alice and Bob) of finite, but otherwise arbitrary dimensions. Physical states of such systems are, in general, mixed and are described by density matrices, i.e. hermitian, positive semi-definite linear operators of trace one (i.e. $\rho = \rho^\dagger$, $\rho \geq 0$, $\mathrm{Tr}\rho = 1$), acting in the Hilbert space of the composite system $\mathcal{H} = \mathcal{H}_A \otimes \mathcal{H}_B$. Without loosing generality we will assume that dim $\mathcal{H}_A = M \geq 2$ and dim $\mathcal{H}_B = N \geq M$.

Before proceeding further we introduce here some definitions that we will use throughout the paper. Given a density matrix ρ, we denote its kernel by $K(\rho) = \{|\phi\rangle : \rho|\phi\rangle = 0\}$, its range by $R(\rho) = \{|\phi\rangle : \exists|\psi\rangle \text{ s.t. } |\phi\rangle = \rho|\psi\rangle\}$, and its rank by $r(\rho) = \dim\ R(\rho) = NM - \dim K(\rho)$. Also the notion of *partial transposition* will be used throughout. The operation of partial transposition of a density matrix ρ means the transposition with respect to only one of the subsystems. If we express ρ in Alice's and Bob's orthonormal product basis,

$$\rho = \sum_{i,j=1}^{M} \sum_{k,l=1}^{N} \langle i,k\,|\rho|\,j,l\rangle |\,i,k\rangle\langle j,l\,| = \sum_{i,j=1}^{M} \sum_{k,l=1}^{N} \langle i,k\,|\rho|\,j,l\rangle |\,i\rangle_A\langle j\,| \otimes |\,k\rangle_B\langle l\,|, \tag{7.1}$$

then, the partial transposition with respect to Alice's system is given by:

$$\rho^{T_A} = \sum_{i,j=1}^{M} \sum_{k,l=1}^{N} \langle i,k\,|\rho|\,j,l\rangle |\,j\rangle_A\langle i\,| \otimes |\,k\rangle_B\langle l\,|. \tag{7.2}$$

Note that ρ^{T_A} is basis-dependent, but its spectrum is not. For the partial transpose ρ^{T_A} it might hold that $\rho^{T_A} \geq 0$, but this does not have to be true! As $(\rho^{T_A})^{T_B} = \rho^T$, and as $\rho^T \geq 0$ always holds, positivity of ρ^{T_A} implies positivity of ρ^{T_B} and vice versa. A density matrix ρ that fulfills $\rho^{T_A} \geq 0$ is termed *PPT state* for positive partial transpose, otherwise it is called *NPPT state* for non-positive partial transpose.

7.2.1 The Separability Problem

An essential step towards the understanding of entanglement is to first identify separable states, i.e, states that contain classical correlations only or no correlations at all. The mathematical definition of such states (separable states) in terms on convex combinations of product states was given by Werner in [4].

Definition 1 A given state ρ is *separable* iff

$$\rho = \sum_{i=1}^{k} p_i \rho_i^A \otimes \rho_i^B\,, \tag{7.3}$$

where $\sum_i p_i = 1$, and $p_i \geq 0$.

Notice that the above definition states that a separable state can be prepared by Alice and Bob by means of local operations (unitary operations, measurements, etc.) and classical communication (LOCC). However, the question whether a given state can be decomposed as a convex sum of product states like in Eq. (7.3) is by no means trivial – in fact there are no algorithms to check if such a decomposition for a given state ρ exist.

An entangled state is defined via the negation of the above definition. A given state ρ is *entangled* iff it cannot be decomposed as in Equation (7.3). Thus, the separability versus entanglement problem can be formulated as: *Given a composite quantum state described by ρ, can it be decomposed as a convex combination of product states or not?*

A major step in the answer of this problem and in the characterization of separability was done by Peres [5] and the Horodecki family [6] by providing a necessary condition for separability: the positivity of the partial transposition. Their results can be summarized in the following theorem:

Theorem 1 *If a density matrix ρ is separable then $\rho^{T_A} \geq 0$. If $\rho^{T_A} \geq 0$ in Hilbert spaces of dimensions 2×2 or 2×3 then ρ is separable.*

Notice that being PPT does not imply separability, except for low dimensional Hilbert spaces! Let us mention here, that also in [6], the problem of separability was rigorously reformulated in terms of the theory of positive maps. We will discuss about positive and completely positive maps in the forthcoming sections.

7.2.2 Results on The Separability Problem

An important tool for studying the properties of states with respect to their separability is the method of subtracting projectors onto product vectors from the given state. This method was developed in [7] and [8]: if there exists a product vector $|e, f\rangle \in R(\rho)$, the projector onto this vector (multiplied by some coefficient $\lambda > 0$) can be subtracted from ρ, such that the remainder is positive definite. A similar technique can be used for PPT states ρ: if there exists a product vector $|e, f\rangle \in R(\rho)$, such that $|e^*, f\rangle \in R(\rho^{T_A})$, the projector onto this vector (again multiplied by some $\lambda > 0$) can be subtracted from ρ, such that the remainder is positive definite and PPT. This observation allows to construct decompositions of a given ρ of the form

$$\rho = \lambda\sigma + (1 - \lambda)\delta, \tag{7.4}$$

where σ is separable, while δ is a so-called *edge state*, i.e. a state from which "nothing else" can be subtracted. In the case of decompositions for general entangled states, δ has no product vectors in the range. In the case of decompositions for PPT states δ can be taken as PPT edge state, i.e. a state that does not contain any product vector $|e, f\rangle \in R(\rho)$, such that $|e^*, f\rangle \in R(\rho^{T_A})$. The decompositions (7.4) can be optimized, by demanding λ to be maximal [7]. Such an optimal decomposition is illustrated in Figure 7.1. For separable states an important question (related to the optimization of the detection of entangled states [9]) concerns minimal decompositions, i.e. those containing minimal number of projectors on product vectors. In particular in Ref. [8] it has been shown that in 2×2 systems the minimal decomposition of separable states contains a number of projectors which is equal to the rank of the state. Minimal decompositions can also be considered in the form of pseudo-mixtures, where not all coefficients multiplying the projectors entering the decomposition are positive, see Section 7.4.2.

In the following we list the major results obtained by applying the decompositions (7.4) and constructing edge states in various systems, which has become a basic tool of the so-called Innsbruck-Hannover programme [10].

General properties of optimal separable approximations (decompositions) have been studied in Ref. [11] for the states ρ of bipartite quantum systems of arbitrary dimensions $M \times N$. For two qubit systems (M=N=2) the best separable approximation has a form of a mixture of a separable state and a projector onto a pure entangled state. We have formulated the necessary condition that the pure state in the best separable approximation is not maximally entangled. This result allowed Wellens and Kuś [12] to obtain an analytic form of the optimal decomposition in the 2×2 case and to relate the value of λ to the Wootters' concurrence [13]. We have demonstrated that the weight of the entangled state in the best separable approximation in arbitrary dimensions provides a good entanglement measure. We have proven that in general, for arbitrary M and N, the best separable approximation corresponds to a mixture of a

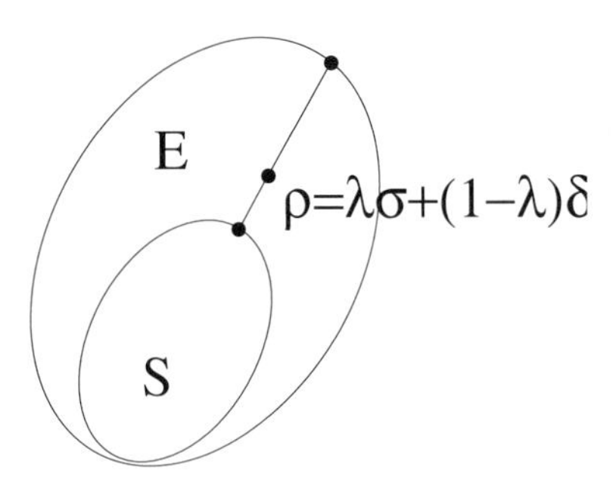

Figure 7.1: Illustration of the decomposition of ρ into a separable state σ and an edge state δ: $\rho = \lambda\sigma + (1-\lambda)\delta$

separable and an entangled state which are both unique. We have developed also a theory of optimal separable approximations for states with positive partial transpose, and discussed procedures of constructing such decompositions.

The decomposition techniques and investigations of edge states have then be applied to $2 \times N$ systems in [14] and [15]. We have analyzed the separability properties of PPT density operators supported on $\mathcal{C}^2 \otimes \mathcal{C}^N$. We have shown that if $r(\rho) = N$, then it is separable, and that bound entangled states have rank larger than N. We have also solved the separability problem for low rank states: we have given a separability criterion for a generic density operator such that the sum of its rank and the one of its partial transpose does not exceed $3N$. If it exceeds this number we show that one can subtract projectors onto product vectors until decreasing it to $3N$, while keeping the positivity of ρ and its partial transpose. This automatically gives us a sufficient criterion for separability for general density operators. We also prove that all density operators that remain invariant (or, more generally close to being invariant) after partial transposition with respect to the first system are separable. Finally, in Ref. [14] we have also presented a simple elementary proof of the Peres-Horodecki separability criterion in 2×2 – dimensional systems.

The results for $2 \times N$ systems were then generalized to $M \times N$ systems [16], where we have also been able to solve the separability problem for low rank states. We have considered low rank density operators ϱ supported on a $M \times N$ Hilbert space for arbitrary $M \leq N$ and with a positive partial transpose $\varrho^{T_A} \geq 0$. For rank $r(\varrho) \leq N$ we have proven that having a PPT is necessary and sufficient for ϱ to be separable; in this case we have also provided its minimal decomposition in terms of pure product states. It follows from this result that there are no bound entangled states of rank 3 having a PPT. We have also presented a necessary and sufficient condition for the separability of generic density matrices for which the sum of the ranks of ϱ and ϱ^{T_A} satisfies $r(\varrho) + r(\varrho^{T_A}) \leq 2MN - M - N + 2$. This separability condition has the form of a constructive check, providing thus also a pure product state decomposition for separable states, and it works in those cases where a system of coupled polynomial equations has a finite number of solutions, as expected in the generic case.

The same research programme can also be applied to $2 \times 2 \times N$ systems [17]. We have investigated separability and entanglement of mixed states in $\mathcal{C}^2 \otimes \mathcal{C}^2 \otimes \mathcal{C}^N$ three-party quantum systems. We have shown that all states ρ with positive partial transposes that have rank

$r(\rho) \leq N$ are separable. For the three-qubit case ($N = 2$) we have proven that all PPT states ρ with rank $r(\rho) = 3$ are separable. We provided also constructive separability checks for the states ρ that have the sum of the rank of ρ and the ranks of partial transposes with respect to all subsystems smaller than $15N - 1$.

We have studied also the problem of separability and entanglement properties of completely positive maps acting on operators acting in the composite Hilbert space of Alice and Bob [18, 19]. We have studied when a physical operation can produce entanglement between two systems that are initially disentangled. The formalism that we have developed allows to show that one can perform certain non-local operations with unit probability by performing local measurements on states that are very weakly entangled. This formalism is a generalization of the Jamiołkowski isomorphism, that connect maps with operators, to the case of maps acting on tensor product spaces. We have associated with every completely positive map (CPM) acting on states of Alice and Bob, an operator acting on two copies of Alice's and Bob's space. The isomorphism connects separable CPMs with separable states, PPT CPMs with PPT entangled states, and so one. It provides a powerful tool to classify CP maps, using results known for states.

Last, but not least, we have applied our methods and techniques to study states in infinite-dimensional Hilbert spaces, i.e. continuous variable states. A particularly important class of such states, that is very frequently used in experiments with photons, is formed by the so-called $Gaussian$ states. Gaussian states can be defined by the requirement that their associated Wigner function has a Gaussian form. We have been able to solve the separability problem of Gaussian states for two parties each having an arbitrary number of photon (harmonic oscillator) modes ([20], for a review see [21]), and for three parties each having one harmonic mode [22]. For bipartite systems of arbitrarily many modes the necessary and sufficient condition consists in an iterative transformation of the correlation matrix of a given state and provides an operational criterion, since it can be checked by a simple computation with arbitrary accuracy. Moreover, it allows us to find a pure product-state decomposition of any given separable Gaussian state. Our criterion is independent of the one based on partial transposition, and obviously, since it detects all entangled states, it is strictly stronger than the PPT criterion. We have also derived a necessary and sufficient condition for the separability of tripartite three mode Gaussian states, that is easy to check for any such state. We have given a classification of the separability properties of those systems and have shown how to determine for any state to which class it belongs. We have also shown that there exist genuinely tripartite bound entangled states (see 7.3) and have pointed out how to construct and prepare such states.

7.3 The Distillability Problem

For many applications in quantum information processing one needs a maximally entangled state of two parties, i.e. a state in $M \times N$ dimensions of the form

$$| \Psi_{\max} \rangle = \frac{1}{\sqrt{M}} \sum_{i=1}^{M} | i, i \rangle \ . \tag{7.5}$$

However, even if an experimental source that creates such a state is available, during storage or transmission along a noisy channel the state will interact with the environment and evolve into a mixed state, thus loosing the property of being maximally entangled.

The idea of distillation and purification, i.e. enhancement of the entanglement of a given mixed state by local operations and classical communication (LOCC) was proposed by Bennett *et al.* [23], Deutsch *et al.* [24] and Gisin [25]. Again, for Hilbert spaces of composite systems with dimension lower or equal to 6, any mixed entangled state can always be distilled to a pure maximally entangled state. Since for such systems entanglement is equivalent to non-positivity of the partial transpose, we conclude that for systems in 2×2 and 2×3 dimensions all NPPT states are distillable [26]. It was shown by the Horodecki family [27] that the PPT property implies undistillability. Somehow surprisingly, in higher dimensions there exist states that are entangled but cannot be distilled. These states, namely PPT entangled states, are called bound entangled states, contrary to free entangled states which can be distilled. In general, the distillability problem can be formulated as: *Given a composite quantum state described by* ρ*, is it distillable or undistillable?*

The problem of distillability can be rigorously formulated [27] so that it reduces to the following theorem:

Theorem 2 *ρ is distillable iff there exists a state $|\psi\rangle$ from a 2×2-dimensional subspace, $|\psi\rangle = a|e_1\rangle|f_1\rangle + b|e_2\rangle|f_2\rangle$, such that*

$$\langle\psi|(\rho^{T_A})^{\otimes K}|\psi\rangle < 0 \tag{7.6}$$

for some K.

7.3.1 Results on the Distillability Problem

As mentioned above, in finite dimensions there exist so-called *bound entangled* states, which are entangled, but their entanglement cannot be distilled. One possibility of the construction of such states was given in [28]. There we have presented a family of bound entangled states in 3×3 dimensions. Their density matrix depends on 7 independent parameters and has 4 different non-vanishing eigenvalues. This construction can e.g. be useful when testing whether some new entanglement criterion detects bound entanglement.

Apart from several examples for bound entangled states with positive partial transpose we have some evidence that also bound entanglement, i.e., entanglement that cannot be distilled, of states with non-positive partial transpose exists: in [29] we study the distillability of a certain class of bipartite density operators which can be obtained via depolarization starting from an arbitrary one. This class is a one-parameter family of states that consist of a weighted sum of projectors onto the symmetric and the antisymmetric subspace. Our results suggest that non-positivity of the partial transpose of a density operator is not a sufficient condition for distillability, when the dimension of both subsystems is higher than two. This conjecture has been found independently in [30], and is still an open problem. The present understanding of the decomposition of mixed states into separable, undistillable entangled and distillable entangled states is shown in Figure 7.2.

We have also addressed the distillability and bound entanglement question in the contexts of infinite dimensional Hilbert spaces [31]. We have introduced and analyzed the definition

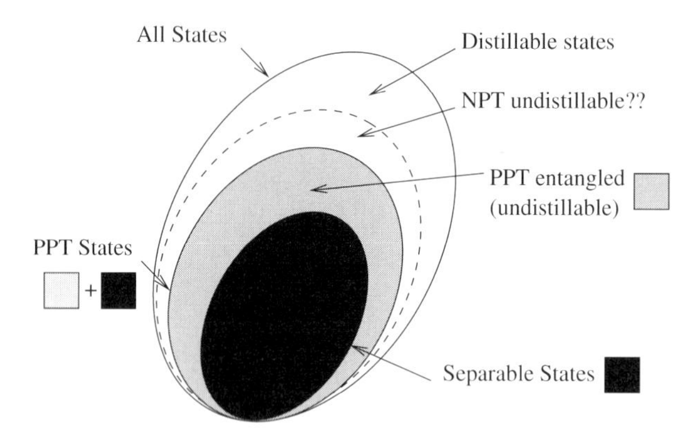

Figure 7.2: Schematic representation of the set of all states, decomposed into the various subsets explained in the text

of generic bound entanglement for the case of continuous variables. We have provided some examples of bound entangled states for that case, and discussed their physical sense in the context of quantum optics. We have raised the question of whether the entanglement of these states is generic. As a byproduct, we have obtained a new many-parameter family of bound entangled states with positive partial transpose in Hilbert spaces of arbitrary finite dimension. We have also pointed out that the "entanglement witnesses" (see Section 7.4) and positive maps revealing the corresponding bound entanglement can be easily constructed.

Furthermore, we have studied how rare separable and non-distillable states of continuous variables [18] are. In finite dimensional Hilbert spaces, we have earlier demonstrated that the volumes of the set of separable states and PPT entangled states are both non-zero, and that there exists a vicinity of the identity operator that contains separable states only [32]. This turned out not to be the case for continuous variable systems. Also we have proven that the set of non–distillable continuous variable states is nowhere dense in the set of all states, i.e., the states of infinite–dimensional bipartite systems are generically distillable. This automatically implies that the sets of separable states, entangled states with positive partial transpose, and bound entangled states are also nowhere dense in the set of all states. All these properties significantly distinguish quantum continuous variable systems from the spin like ones. The aspects of the definition of bound entanglement for continuous variables has also been analyzed in the context of the theory of Schmidt numbers. In particular, the main result was generalized to the set of states of arbitrary Schmidt number and to the single copy regime.

7.4 Witness Operators for the Detection of Entanglement

7.4.1 Definition and Geometrical Interpretation of Witness Operators

A very useful tool to detect entanglement is the so-called entanglement witness. An entanglement witness is an observable ($\mathcal{W}$) which reveals the entanglement (if any) of a given state ρ. This concept, which was introduced and studied in [6, 33], reformulates the problem of separability in terms of witness operators:

Theorem 3 *A density matrix ϱ is entangled iff there exists a Hermitian operator $\mathcal{W}$ with $Tr(\mathcal{W}\varrho) < 0$ and $Tr(\mathcal{W}\sigma) \geq 0$ for any separable state σ.*

We say that the witness $\mathcal{W}$ "detects" the entanglement of ϱ. The existence of entanglement witnesses is just a consequence of the Hahn-Banach theorem, that states: *Let S be a convex, compact set, and let $\varrho \notin S$. Then there exists a hyper-plane that separates ϱ from S.*

Figure 7.3 illustrates the concept of an entanglement witness $\mathcal{W}$, represented by a hyper–plane (dashed line) that separates the state ρ from the convex compact set S. We have also depicted in the figure, an optimal entanglement witness $\mathcal{W}_{opt}$ (represented by straight line) together with other optimal witnesses. Optimal witnesses are tangent to the set of separable states (The concept of optimization will be explained in the next subsection). One can immediately grasp from the figure, that, in order to completely characterize the set of separable states S one should find all the witnesses tangent to S. Unfortunately, infinitely many witnesses are needed for such a task!

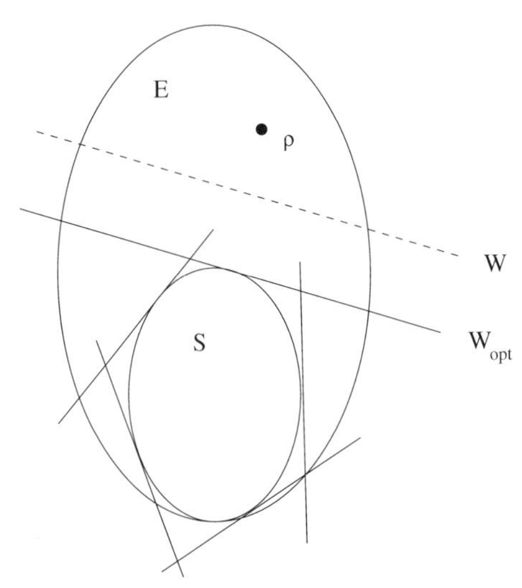

Figure 7.3: Geometrical picture of entanglement witnesses and their optimization.

Witness operators are also related to maps. Indeed, there is an isomorphism that connects maps with operators known as Jamiołkowski isomorphism: each entanglement witness $\mathcal{W}$ on an $M \times N$ space defines a positive map $\mathcal{E}$ that transforms positive operators on an M or N–dimensional Hilbert space into positive operators on an M or N–dimensional space [34]. The maps corresponding to entanglement witnesses are positive, but not completely positive (i.e. there is an extension $\mathbb{1} \otimes \mathcal{E}$ which is not positive), and thus allow to "detect" the entanglement of ρ.

Entanglement witnesses for PPT states and the corresponding maps have the property of being *non-decomposable*. A witness is called decomposable iff it can be written in the form

$\mathcal{W} = P + Q^{T_A}$ with both P and Q positive. Otherwise it is non-decomposable. Correspondingly a map is decomposable iff it can be represented as a combination of positive maps and partial transposition and it is non-decomposable otherwise.

7.4.2 Results on Witness Operators

How does one construct an entanglement witness? In [35] we provide a canonical form of mixed states in bipartite quantum systems in terms of a convex combination of a separable state and an edge state, as defined in Section 7.2. We construct entanglement witnesses for all edge states, and present a canonical form of *non-decomposable* entanglement witnesses and the corresponding positive maps. We present a characterization of separable states using a special class of entanglement witnesses.

An entanglement witness $\mathcal{W}$ is called *optimal*, if there exists no entanglement witness that detects states in addition to the ones detected by $\mathcal{W}$. Geometrically, this corresponds to the hyperplane defined by the witness being tangent to the set of separable states, see Figure 7.3. In [36] we give necessary and sufficient conditions for entanglement witnesses to be optimal. We show how to optimize a general witness, and then we particularize our results to witnesses that can detect PPT entangled states, i.e. non-decomposable witnesses. This method also permits the systematic construction of non-decomposable positive maps.

The tool of witness operators can be applied to give a finer classification of entangled states by detecting their so-called *Schmidt number*. The Schmidt number of a mixed state was introduced in [37] as a generalization of the Schmidt rank for pure states: it characterizes the maximal Schmidt rank of the pure states in the "most simple" decomposition of ρ, i.e. the one that needs the lowest maximal Schmidt rank. The definition of the Schmidt number k is given by

$$\varrho = \sum_i p_i | \Psi_i^{r_i} \rangle \langle \Psi_i^{r_i} | , \quad k = \min_{\{dec\}} (r_{\max}) , \tag{7.7}$$

where r_i denotes the Schmidt rank of the state $| \Psi_i \rangle$, the minimization is done over all possible decompositions of ρ, and $r_{\max} = \max_i(r_i)$ is the maximal Schmidt rank of a given decomposition. In [38] we investigate the Schmidt number of an arbitrary mixed state by constructing a Schmidt number witness that detects it. We present a canonical form of such witnesses and provide constructive methods for their optimization. In this context we also find strong evidence that all bound entangled states with positive partial transpose in two qutrit systems have Schmidt number two.

In the articles summarized above, we were considering bipartite systems only. Can one use the method of entanglement witnesses for systems of more than two particles? This question was addressed in [39], where we introduce a classification of mixed three-qubit states. The case of pure three-qubit states was studied in [40]: here the authors show that there exist two inequivalent classes of states with genuine tripartite entanglement, the so-called GHZ- and W-states. In [39] we define the classes of mixed separable, biseparable, W- and GHZ-states, which are successively embedded into each other. We show that contrary to pure W-type states, the mixed W-class is not of measure zero. We construct witness operators that detect the class of a mixed state, and discuss the conjecture that all PPT entangled states belong to the

W-class. The classification of three-qubit states into the sets mentioned above is schematically shown in Figure 7.4.

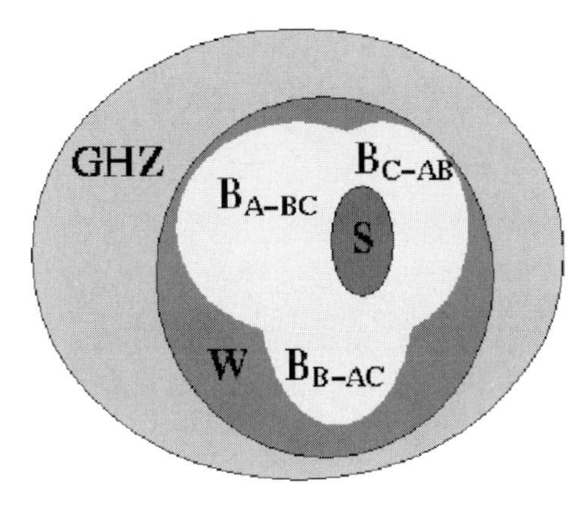

Figure 7.4: Schematic view of the set of three-qubit states.

Although the concept of entanglement witnesses is born from a mathematical background, it is by no means purely academic. We have been studying the possibility of implementing a witness with only few *local* projection measurements in [9]. If some prior knowledge of the density matrix is given, which is usually the case in a realistic experiment, one can construct a suitable entanglement witness and find its minimal decomposition into a *pseudo-mixture* (i.e. a mixture that contains at least one negative coefficient) of local projectors, i.e.

$$W = \sum_i c_i | a_i \rangle \langle a_i | \otimes | b_i \rangle \langle b_i | , \tag{7.8}$$

where the coefficients c_i are real and fulfill $\sum_i c_i = 1$. As local projection measurements can be performed with present day technology, some simple measurements then tell the experimentalist whether his given state is indeed entangled. The general solution to the optimization problem of finding the *minimal* number of measurements is yet unknown. We discuss a realistic example for two qubits, and suggest the first method for the detection of bound entanglement with local measurements.

The tool of witness operators is not only useful for addressing the separability problem, but also for studying the distillability problem: In [41] we introduce a formalism that connects entanglement witnesses and the distillation and activation properties of a state. We apply this formalism to two cases: First, we rederive the results presented in [42], namely that one copy of any bipartite state with non–positive partial transpose is either distillable, or activable. Second, we show that there exist three–partite NPPT states, with the property that two copies can neither be distilled, nor activated.

Finally, an overview of our programme that investigates quantum correlations and entanglement in terms of convex sets is given in [10]. There we present a unified description of optimal decompositions of quantum states and the optimization of witness operators that detect whether a given state belongs to a given convex set. We illustrate this abstract formulation with several examples, and discuss relations between optimal entanglement witnesses and n-copy non-distillable states with non-positive partial transpose.

7.5 Quantum Correlations in Systems of Fermionic and Bosonic States

The notion of entanglement discussed in the previous sections applies to situations where the parties are separated by macroscopic distances. Various mechanisms to create entanglement or to perform quantum gate operations, e.g. in the context of quantum dots [43] or neutral atoms in optical microtraps [44], however require a direct interaction at short distances between indistinguishable particles. We have developed a framework to study quantum correlations in such situations where the bosonic or fermionic character of indistinguishable particles become important. We have furthermore described a possible implementation of a quantum logic gate for neutral atoms in optical microtraps and studied bosonic correlations in this case.

7.5.1 What is Different with Indistinguishable Particles?

To illustrate the consequences of indistinguishability consider two fermions located in a double well potential as a schematic model of electrons in quantum dots and assume the qubit to be implemented in the spin degree of freedom. Let the initial situation be such that each well contains one electron. Even if they are prepared completely independently, their pure quantum state has to be written in terms of Slater determinants in order to respect the indistinguishability. Operator matrix elements between such Slater determinants contain terms due to the antisymmetrization, but if the spatial wavefunctions of electrons located in different wells have only vanishingly small overlap, then the matrix elements will tend to zero for any physically meaningful operator. This situation is generically realized if the supports of the single-particle wavefunctions are essentially centered around locations being sufficiently apart from each other, or the particles are separated by a sufficiently large energy barrier. In this case the antisymmetrization has no physical effect and for all practical purposes it can be neglected.

If the two wells are moved closer together, or the energy barrier is lowered, such that the electrons are no longer completely localized in one well, then the fermionic statistics is clearly essential and the two-electron wave-function has to be antisymmetrized. Note that in this situation the space of states written in terms of single-particle states no longer has a tensor product structure because the actual state space is just a subspace of the complete tensor product. As a consequence of this fact any antisymmetrized state formally resembles an entangled state although these correlations are not useful as individual particles cannot be accessed. To emphasize this fundamental difference between distinguishable and indistinguishable particles, we will use the term *quantum correlations* to characterize *useful* correlations in systems of indistinguishable particles as opposed to correlations arising purely from their statistics.

We remark that there are different possible ways to quantify quantum correlations. An approach which can be seen as complementary to the one which we will describe here was discussed by Zanardi [45] who ignored the original tensor product structure through partitioning of the physical space into subsystems and introducing a tensor product structure in terms of modes. The entangled entities then are no longer particles but modes.

7.5.2 Results on Quantum Correlations for Indistinguishable Particles

Let us consider the case of two identical fermions sharing an N-dimensional single-particle space $\mathcal{H}_N$. The total Hilbert space is $\mathcal{A}(\mathcal{H}_N \otimes \mathcal{H}_N)$ where $\mathcal{A}$ denotes the antisymmetrization operator. A general state vector can be written as

$$|w\rangle = \sum_{i,j=1}^{N} w_{ij} f_i^\dagger f_j^\dagger |\Omega\rangle \tag{7.9}$$

with fermionic creation operators $f_i^\dagger$ acting on the vacuum $|\Omega\rangle$. The antisymmetric coefficient matrix w_{ij} fulfills the normalization condition $\mathrm{tr}\,(w^* w) = -1/2$. Under a unitary transformation of the single-particle space, $f_i^\dagger \mapsto \sum_j U_{ji} f_j^\dagger$, w transforms as $w \mapsto U w U^T$.

Theorem 4 For every pure two-fermion state $|\, w\rangle$ there exists a unitary transformation of the single particle space such that in the new basis of creation operators $f_i^\dagger$ the state is of the form

$$|\, w\rangle = 2 \sum_{k=1}^{m} z_k f_{2k}^\dagger f_{2k-1}^\dagger |\, \Omega\rangle \tag{7.10}$$

with $2 \cdot m \leq N$ and z_k real and positive.

Each term in this decomposition corresponds to an elementary Slater determinant which is an analogue of a product state in systems consisting of distinguishable parties. Thus, when expressed in such a basis, $|w\rangle$ is a sum of elementary Slater determinants where each single-particle basis state enters at most one term. In this basis the number m of Slater determinants is furthermore minimal and these Slater determinants are thus the analogues of the products states occurring in the Schmidt decomposition of a bi-partite state of distinguishable particles. Therefore we call m the *fermionic Slater rank* of $|w\rangle$ [46], and an expansion of the form (7.10) a *Slater decomposition* of $|w\rangle$. For bosons there exists a similar expansion in terms of elementary two-boson Slater permanents representing doubly occupied states [47].

For two fermions the smallest single-particle space allowing for non-trivial correlations is four-dimensional. In this case a quantity analogous to the concurrence introduced by Wootters as an entanglement measure for two distinguishable qubits [13] can be constructed in the following way:

Theorem 5 Let $|\, w\rangle$ be a two-fermion state in a four-dimensional single-particle space. Then the concurrence $\mathcal{C}(|\, w\rangle)$, defined as

$$\mathcal{C}(|\, w\rangle) = \left| \frac{1}{2} \sum_{i,j,k,l=1}^{4} \epsilon^{ijkl} w_{ij} w_{kl} \right|, \tag{7.11}$$

(ϵ is the fully antisymmetric unit tensor) has the following properties: (i) $\mathcal{C}(|\, w\rangle)$ is invariant under unitary transformations of the single-particle space, (ii) $0 \leq \mathcal{C}(|\, w\rangle) \leq 1$ and (iii)

$\mathcal{C}(|w\rangle) = 0$ iff $|w\rangle$ has Slater rank one and $\mathcal{C}(|w\rangle) = 1$ iff $|w\rangle$ has maximal Slater rank, i.e. Slater rank two.

The concurrence $\mathcal{C}(|w\rangle)$ thus fully characterizes quantum correlations in the case of pure states and for $N = 4$. For mixed two-fermion states in a four-dimensional single-particle space characterized by a density matrix ρ a *Slater number* can be defined similar to the Schmidt number for mixed states of two qubits as the maximal Slater rank of a decomposition of ρ into pure states minimized over all decompositions. Also we can define the mixed state concurrence as

$$\mathcal{C}(\rho) = \inf_{\{p_i, |w_i\rangle\}} \left\{ \sum_i p_i \mathcal{C}(|w_i\rangle) \right\} \tag{7.12}$$

where the infimum is taken over all decompositions of ρ. With this definition we find:

Theorem 6 Let $\rho = \sum_i p_i |w_i\rangle\langle w_i|$ be a mixed two-fermion state (for N=4). Define $|\widetilde{w}_i\rangle = \sum_{i,j,k,l=1}^{4} \epsilon^{ijkl} w_{kl} f_i^\dagger f_{jl}^\dagger |\Omega\rangle$ and $\tilde{\rho} = \sum_i p_i |\widetilde{w}_i\rangle\langle \widetilde{w}_i|$ and let λ_i be the real and non-negative eigenvalues of $\rho\tilde{\rho}$ in descending order of magnitude. Then

$$\mathcal{C}(\rho) = \max(0, \lambda_1 - \sum_{i=2}^{6} \lambda_i) \tag{7.13}$$

and ρ has Slater number one iff $\mathcal{C}(\rho) = 0$, i.e iff $\lambda_1 \leq \sum_{i=2}^{6} \lambda_i$.

Notice that in a similar way the concurrence can be defined and calculated for pure and mixed states of two bosons in a two-dimensional single-particle space.

For higher-dimensional single-particle spaces there exist necessary and sufficient criteria to determine the Slater rank of pure fermionic and bosonic states by contracting their coefficient matrix w with the ϵ-tensor [47]. These become only necessary criteria when applied to mixed states and apparently a full and explicit characterization of higher-dimensional two-boson and two-fermion mixed states is not possible. Furthermore for the case of more than two particles a straight-forward generalization of the Slater decomposition cannot be given. This is again similar to the case of more than two qubits where a Schmidt decomposition of a general pure state does not exist [48]. Consider for example states of three fermions in a six-dimensional Hilbert space. It is in general not possible to find a unitary transformation of the single-particle space that brings a given state to a form $|w\rangle \propto z_1 f_1^\dagger f_2^\dagger f_3^\dagger |\Omega\rangle + z_2 f_4^\dagger f_5^\dagger f_6^\dagger |\Omega\rangle$, which would be the analogue of the two-fermion Slater decomposition. There however exist criteria to identify pure uncorrelated states, i.e. states that can be written as a single Slater determinant [47].

Finally we notice that for the case of two fermions or bosons in higher-dimensional single particle spaces ($N > 4$ for fermions, $N > 2$ for bosons) the concepts of witnesses can be applied [46, 47]. As explained in Section 7.4 for the case of distinguishable particles, k-edge states can be introduced as states that become non-positive when $\epsilon |w^{<k}\rangle\langle w^{<k}|$ is subtracted for some state $|w^{<k}\rangle$ of Slater rank $< k$. Then fermionic and bosonic k-Slater witnesses can be defined that detect states of Slater number k. These witnesses can furthermore be optimized as demonstrated in Section 7.4.

7.5.3 Implementation of an Entangling Gate with Bosons

In [44] we investigate quantum computation with bosonic neutral atoms in optical microtraps [49]. In contrast to other methods with the qubit being implemented in an internal degree of freedom, we study the case where the qubit is implemented in the motional state of the atoms, i.e., in the two lowest vibrational states of each trap. The quantum gate operation is performed by adiabatically approaching two traps each containing one particle such that tunneling and cold collisions occur and thus the bosonic character of the atoms is important. We especially address the implementation of a $\sqrt{\mathrm{SWAP}}$-gate, i.e., a two-qubit gate that transforms states $|0\rangle_A|1\rangle_B$ and $|1\rangle_A|0\rangle_B$ to maximally entangled states while leaving $|0\rangle_A|0\rangle_B$ and $|1\rangle_A|1\rangle_B$ unchanged. The fidelity of the gate operation is evaluated as a function of the degree of adiabaticity in moving the traps and for rubidium atoms in state-of-the-art optical microtraps we obtain gate durations in the range of a few tens of milliseconds. Taking into account error mechanisms like spontaneous scattering of photons we calculate error rates of the gate operation and show that proof-of-principle experiments should be possible.

7.6 Summary

The characterization and classification of entangled states is a very challenging open problem of modern quantum theory. We have presented some approaches and partial solutions to this problem. The methods we used are the optimal decomposition of a given state into a separable and an entangled state, and the tool of witness operators. We have summarized various advances in the separability and distillability problem, and addressed the question of experimental implementation of witness operators. However, many open questions still remain to be solved.

Acknowledgments

We wish to thank A. Acín, G. Birkl, I. Cirac, W. Dür, A. Ekert, G. Giedke, P. Horodecki, S. Karnas, B. Kraus, M. Kuś, D. Loss, C. Macchiavello, A. Peres, J. Schliemann, G. Vidal and X. Yi for collaboration and discussions.

References

[1] J. Eisert and M. Wilkens, *Mod. Opt.* 2000,*47*, 2543; J. Eisert and M. Wilkens, *Phys. Rev. Lett.* 2000, *85*, 437; J. Eisert, T. Felbinger, P. Papadopoulos, M. B. Plenio, and M. Wilkens *Phys. Rev. Lett.* 2000, *84*, 1611; K. Bostroem and T. Felbinger *Phys. Rev. A.* 2002 *65*, 032313; K. Audenaert, J. Eisert, E. Jane, M.B. Plenio, S. Virmani and B. De Moor *Phys. Rev. Lett.* 2001, *87*, 217902.

[2] M. Lewenstein, D. Bruß, J. I. Cirac, B. Kraus, M. Kus, J. Samsonowicz, A. Sanpera and R. Tarrach, *Proceedings of the Conference "Quantum Optics Kuhtai 2000", J. Mod. Opt.* 2000, *47*, 2481.

[3] D. Bruß, *Proceedings of the ICQI Rochester conference* 2001, *J. Math. Phys.* 2002, *43*, 4237, lanl e-print quant-ph/0110078.

[4] R. F. Werner, *Phys. Rev. A* 1989, *40*, 4277.

[5] A. Peres, *Phys. Rev. Lett.* 1996, *77*, 1413.

[6] M. Horodecki, P. Horodecki and R. Horodecki, *Phys. Lett. A* 1996, *223*, 8.

[7] M. Lewenstein and A. Sanpera, *Phys. Rev. Lett.* 1998, *80*, 2261, lanl e-print quant-ph/9707043.

[8] A. Sanpera, R. Tarrach und G. Vidal, *Phys. Rev. A* 1998, *58*, 826, lanl e-print quant-ph/9801024.

[9] O. Gühne, P. Hyllus, D. Bruß, A. Ekert, M. Lewenstein, C. Macchiavello, and A. Sanpera, *Phys. Rev. A* 2002, *66*, 062305, lanl e-print quant-ph/0205089.

[10] D. Bruß, J. I. Cirac, P. Horodecki, F. Hulpke, B. Kraus, M. Lewenstein, and A. Sanpera, *Proceedings of the ESF QIT conference Gdansk, July 2001, J. Mod. Opt.* 2002, *49*, 1399, lanl e-print quant-ph/0110081.

[11] S. Karnas and M. Lewenstein, *J. Phys. A* 2001, *34*, 6919, lanl e-print quant-ph/0011066.

[12] T. Wellens and M. Kuś, *Phys. Rev. A* 2001, *64*, 052302, lanl e-print quant-ph/0104098.

[13] W. K. Wootters, *Phys. Rev. Lett.* 1998, *80*, 2245.

[14] M. Lewenstein, J. I. Cirac und S. Karnas, lanl e-print quant-ph/9903012.

[15] B. Kraus, J. I. Cirac, S. Karnas und M. Lewenstein, *Phys. Rev. A* 2000, *61*, 062302, lanl e-print quant-ph/9912010.

[16] P. Horodecki, M. Lewenstein, G. Vidal und J. I. Cirac, *Phys. Rev. A* 2000, *62*, 032310, lanl e-print quant-ph/0002089.

[17] S. Karnas and M. Lewenstein, *Phys. Rev. A* 2001, *64*, 042313, lanl e-print quant-ph/0102115.

[18] P. Horodecki, J. I. Cirac, and M. Lewenstein, *to appear in "Multiparticle Entanglement", Eds. S. L. Braunstein and A. K. Pati (Springer, Heidelberg, 2002)*, lanl e-print quant-ph/0103076.

[19] B. Kraus, W. Dür, G. Vidal, J. I. Cirac, M. Lewenstein, N. Linden, and S. Popescu, *Z. Naturforsch.* 2001, *56*, 91.

[20] G. Giedke, B. Kraus, M. Lewenstein, and J. I. Cirac, *Phys. Rev. Lett.* 2001, *87*, 167904, lanl e-print quant-ph/0104050.

[21] G. Giedke, B. Kraus, L. M. Duan, P. Zoller, J. I. Cirac, and M. Lewenstein, *Fortsch. Phys.* 2001, *49 (10-11)*, 973.

[22] G. Giedke, B. Kraus, M. Lewenstein, and J. I. Cirac, *Phys. Rev. A* 2001, *64*, 052303, lanl e-print quant-ph/0103137.

[23] C. H. Bennett, G. Brassard, S. Popescu, B. Schumacher, J. A. Smolin and W. K. Wootters, *Phys. Rev. Lett.* 1996, *76*, 722; C. H. Bennett, H. J. Bernstein, S. Popescu and B. Schumacher, *Phys. Rev. A* 1996, *53*, 2046.

[24] D. Deutsch, A. Ekert, R. Jozsa, Ch. Macchiavello, S. Popescu and A. Sanpera, *Phys. Rev. Lett.* 1996, *77*, 2818.

[25] N. Gisin, *Phys. Lett. A* 1996, *210*, 151.

[26] M. Horodecki, P. Horodecki and R. Horodecki, *Phys. Rev. Lett.* 1997, *78*, 574.

[27] M. Horodecki, P. Horodecki and R. Horodecki, *Phys. Rev. Lett.* 1998, *80*, 5239.

[28] D. Bruß und A. Peres, *Phys. Rev. A* 2000, *61*, 30301, lanl e-print quant-ph/9911056.

[29] W. Dür, J. I. Cirac, M. Lewenstein und D. Bruß, *Phys. Rev. A* 2000, *61*, 062313, lanl e-print quant-ph/9910022.

[30] D. DiVincenzo, P. Shor, J. Smolin, B. Terhal, and Ashish V. Thapliyal, *Phys. Rev. A* 2000, *61*, 062312.

[31] P. Horodecki und M. Lewenstein, *Phys. Rev. Lett.* 2000, *85*, 2657, lanl e-print quant-ph/0001035.

[32] K. Życzkowski, P. Horodecki, A. Sanpera und M. Lewenstein, *Phys. Rev. A* 1998, *58*, 883, lanl e-print quant-ph/9804024.

[33] B. Terhal, *Phys. Lett. A* 2000, *271*, 319.

[34] A. Jamiołkowski, *Rep. Math. Phys.* 1972, *3*, 275.

[35] M. Lewenstein, B. Kraus, P. Horodecki und J. I. Cirac, *Phys. Rev. A* 2001, *63*, 044304, lanl e-print quant-ph/0005112.

[36] M. Lewenstein, B. Kraus, J. I. Cirac und P. Horodecki, *Phys. Rev. A* 2000, *62*, 052310, lanl e-print quant-ph/ 0005014.

[37] B. Terhal and P. Horodecki, *Phys. Rev. A* 2000, *61*, 040301(R).

[38] A. Sanpera, D. Bruß und M. Lewenstein, *Phys. Rev. A* 2001, *63*, 050301, lanl e-print quant-ph/0009109.

[39] A. Acín, D. Bruß, M. Lewenstein, and A. Sanpera, *Phys. Rev. Lett.* 2001, *87*, 040401, lanl e-print quant-ph/0103025.

[40] W. Dür, G. Vidal, and J. I. Cirac, *Phys. Rev. A* 2000, *62*, 062314.

[41] B. Kraus, M. Lewenstein, and J. I. Cirac, *Phys. Rev. A* 2002, *65*, 042327, lanl e-print quant-ph/0110174.

[42] T. Eggeling, K.-G. Vollbrecht, R. Werner, and M. Wolf, *Phys. Rev. Lett.* 2001, *87*, 257902, lanl e-print quant-ph/0104095.

[43] J. Schliemann, D. Loss, and A. H. MacDonald, *Phys. Rev. B* 2001, *63*, 085311.

[44] K. Eckert, J. Mompart, X. X. Yi, J. Schliemann, D. Bruß, G. Birkl, and M. Lewenstein, *Phys. Rev. A* 2002, *66*, 042317, lanl e-print quant-ph/0206096.

[45] P. Zanardi, *Phys. Rev. A* 2002, *65*, 042101, lanl e-print quant-ph/0104114.

[46] J. Schliemann, J. I. Cirac, M. Kus, M. Lewenstein, and D. Loss, *Phys. Rev. A* 2001, *64*, 022303, lanl e-print quant-ph/0012094.

[47] K. Eckert, J. Schliemann, D. Bruß, and M. Lewenstein, *Ann. of Phys.* 2002, *299*, 88, lanl e-print quant-ph/0203060.

[48] A. Acin, A. Andrianov, L. Costa, E. Jane, J. I. Latorre, and R. Tarrach, *Phys. Rev. Lett.* 2000, *85*, 1560.

[49] G. Birkl, F. Buchkremer, R. Dumke, M. Volk, W. Ertmer, *Optics Comm.* 2001, *191*, 67, lanl e-print quant-ph/0012030.

8 Non-Classical Gaussian States in Noisy Environments

Stefan Scheel[1] *and Dirk–Gunnar Welsch*[2]

[1] Quantum Optics and Laser Science
Blackett Laboratory
Imperial College
London, UK

[2] Theoretisch-Physikalisches Institut
Friedrich-Schiller-Universität Jena
Jena, Germany

8.1 Introduction

Research in the field of quantum information theory nowadays focusses more and more on the possibilities of practical implementation of quantum information processes. One of the most promising attempts has seemed to be the usage of entangled Gaussian states since they are rather easy to produce with optical means. In this article we will review properties of Gaussian states and describe operations on them. The interaction of the electromagnetic field with an absorbing dielectric as a special type of environmental interaction will serve as the basis for the understanding of decoherence and entanglement degradation of Gaussian states of light propagating through fibers.

In Section 8.2 we shortly review the definition of Gaussian states and operations that can be performed on them. These include symplectic (unitary) transformations, completely positive (CP) trace-preserving maps, and projective measurements. The important definitions of classicality, separability, and entanglement are defined and appropriate measures are given for them. In Section 8.3 we review some results on entanglement degradation in dielectric environments. We discuss the decoherence of Gaussian quantum states of light transmitted through absorbing dielectric objects such as fibers. Section 8.4 is devoted to the study of quantum teleportation in noisy environments. Special emphasis is put onto the question of choosing the correct displacement on the receiver's side.

8.2 Gaussian States and Gaussian Operations

To begin with, we define Gaussian states and Gaussian operations and discuss some important properties such as non-classicality, separability, and entanglement. A quantum state is called Gaussian if its characteristic function, and hence an appropriately chosen phase-space function, is an exponential form that is at most quadratic in its canonical variables. Most importantly, an N-mode Gaussian state is fully characterized by the first and second moments of the $2N$ canonical variables $\hat{\boldsymbol{\xi}}^{\mathrm{T}} = (\hat{x}_1, \hat{p}_1, \ldots, \hat{x}_N, \hat{p}_N)$ obeying the canonical commutation

Quantum Information Processing, 2nd Edition. Edited by Thomas Beth and Gerd Leuchs

ISBN: 3-527-40541-0

relations ($\hbar=1$)

$$\left[\hat{\xi}_i, \hat{\xi}_j\right] = i\Sigma_{ij}, \tag{8.1}$$

where $\mathbf{\Sigma}$ is the $(2N)$-dimensional symplectic matrix

$$\mathbf{\Sigma} = \begin{pmatrix} 0 & 1 \\ -1 & 0 \end{pmatrix}^{\oplus N}. \tag{8.2}$$

This definition seems to suggest that higher moments do not play any rôle. However, this is not correct, since all higher moments can be derived from first and second moments. The first moments describe the position $\boldsymbol{\kappa}$ of the 'centre-of-mass' of the Gaussian function in phase-space, whereas the second moments describe the fluctuations of the canonical variables or equivalently, the width of the Gaussian in phase-space. They can be collected in the positive symmetric $(2N \times 2N)$-dimensional covariance matrix $\mathbf{\Gamma}$, viz.

$$\Gamma_{ij} = 2\mathrm{Tr}\left[\left(\hat{\xi}_i - \langle\hat{\xi}_i\rangle\right)\left(\hat{\xi}_j - \langle\hat{\xi}_j\rangle\right)\hat{\varrho}\right] - i\Sigma_{ij}. \tag{8.3}$$

The associated charateristic function is the Fourier transform of the Wigner function. Note that without the term $-i\Sigma_{ij}$ one would obtain the covariance matrix corresponding to the density operator in normal order. The fluctuations, however, are not independent from each other. This means that not every positive $(2N \times 2N)$-dimensional matrix is an admissible covariance matrix, i.e. describes a physical state. The reason for this behaviour is the simple fact that the fluctuations have to obey the Heisenberg uncertainty relations which in matrix form read as

$$\mathbf{\Gamma} + i\mathbf{\Sigma} \geq 0. \tag{8.4}$$

With the above definitions we can write the characteristic function of a general Gaussian state as

$$\chi(\boldsymbol{\lambda}) = \exp\left[-\tfrac{1}{4}\boldsymbol{\lambda}^{\mathrm{T}}\mathbf{\Gamma}\boldsymbol{\lambda} + i\boldsymbol{\lambda}^{\mathrm{T}}\mathbf{\Sigma}\boldsymbol{\kappa}\right]. \tag{8.5}$$

For the following discussion we will focus on the second moments only. We can transform Gaussian states into each other by transforming the covariance matrix as $\mathbf{\Gamma}\rightarrow\mathbf{\Gamma}'=\mathbf{S}\mathbf{\Gamma}\mathbf{S}^{\mathrm{T}}$, where the matrix $\mathbf{S}$ has to be chosen such that the canonical commutation relations (8.1) are fulfilled. This requirement leads to the constraint $\mathbf{S}\mathbf{\Sigma}\mathbf{S}^{\mathrm{T}}=\mathbf{\Sigma}$. On the level of states, these matrices generate unitary transformations $\hat{\varrho}\rightarrow\hat{\varrho}'=\hat{U}(\mathbf{S})\hat{\varrho}\hat{U}^{\dagger}(\mathbf{S})$. In turn, any unitary transformation which is bilinear in the canonical variables can be described by a symplectic matrix. All symplectic matrices $\mathbf{S}$ form the (non-compact) symplectic group Sp($2N$,$\mathbb{R}$). As for every group, it has certain normal forms associated with it. One of the most important ones is the generalized Euler decomposition [1]: Every symplectic matrix $\mathbf{S}$ can be decomposed into three factors, $\mathbf{S} = \mathbf{O}_1\mathbf{D}\mathbf{O}_2$, where the $\mathbf{O}_i$ belong to the N-dimensional (compact) orthogonal group and $\mathbf{D}$ is a diagonal matrix with entries $(k_1, 1/k_1, \ldots, k_N, 1/k_N)$. This decomposition has an immediate physical interpretation: The $\mathbf{O}_i$ generate N-mode rotations (generated by beam splitter networks and phase shifters) and the matrix $\mathbf{D}$ describes single-mode squeezings.

8.2.1 Classicality

A very special feature of a quantum state is its possible non-classical character, which means that its inherent statistics cannot be modelled by a classical probability distribution. A widely used definition has been given by Titulaer and Glauber [2] saying that a state is classical (or 'predetermined' in their words) if and only if its P function is positive semi-definite and sufficiently smooth, hence $P \in L_2(\mathbb{R}^+)$. This definition has recently been translated into a measurable criterion which is based on quadrature distributions [3].

Let us discuss the implications for a general Gaussian N-mode state. The main result is that we regard a Gaussian state as non-classical whenever one or more eigenvalues of its covariance matrix drop below 1. The reasoning behind the argument is that in such a case the characteristic function does not possess a Fourier transform since it increases exponentially at infinity. To answer the question of whether a Gaussian state is classical or not, we recall that any Gaussian state (with zero mean) can be generated by acting on a thermal state (being classical) with a symplectic matrix $\mathbf{S}$. That is, we can start with a covariance matrix of the form $\mathbf{\Gamma}_{\mathrm{th}}=2\mathrm{diag}(n_1,n_1,\ldots,n_N,n_N)+\mathbf{1}$, with n_i being the mean thermal excitation number in the i-th mode. Next we act on $\mathbf{\Gamma}_{\mathrm{th}}$ with a symplectic matrix $\mathbf{S}$ and compute the eigenvalues of the transformed covariance matrix $\mathbf{S}\mathbf{\Gamma}_{\mathrm{th}}\mathbf{S}^{\mathrm{T}}$ with $\mathbf{S}=\mathbf{O}_1\mathbf{D}\mathbf{O}_2$.

To give a simple example, let us consider a single-mode Gaussian state. The sought eigenvalues are simply the eigenvalues of the covariance matrix ($k=e^{\zeta}$)

$$\mathbf{D}\mathbf{\Gamma}_{\mathrm{th}}\mathbf{D} = \mathrm{diag}\left[e^{2\zeta}(2n+1), e^{-2\zeta}(2n+1)\right] \tag{8.6}$$

(the orthogonal matrices $\mathbf{O}_i$ do not play a rôle here), from which we immediately find the well-known relation $|\zeta_{\max}|=\frac{1}{2}\ln(2n+1)$ for the maximal squeezing value such that the state stays classical [4]. We see that a state can be classical, though it is squeezed. Needless to say that multimode Gaussian states can be treated in the same way. Clearly, the number of parameters one can choose increases rapidly.

Furthermore, a non-classicality measure $C(\hat{\varrho})$ based on the relative entropy can be defined as

$$C(\hat{\varrho}) = \min_{\hat{\sigma}_{\mathrm{class.}}} \mathrm{Tr}\left[\hat{\varrho}\left(\ln\hat{\varrho} - \ln\hat{\sigma}\right)\right]. \tag{8.7}$$

The advantage of using this entropic measure for Gaussian states is that it can be easily calculated [5]. Note that this measure can be equally well used as the measure based on Bures' metric defined in [6].

8.2.2 CP Maps and Partial Measurements

Until now we have described unitary operations on the level of states. Decoherence processes, however, correspond to operations in a larger class, the completely positive maps. Physically, they are unitary operations in a larger Hilbert space (Naimark theorem [7]). A typical example is the interaction with a dissipative environment which is traced out afterwards. On the level of covariance matrices, CP maps act as

$$\mathbf{\Gamma} \to \mathbf{\Gamma}' = \boldsymbol{A}\mathbf{\Gamma}\boldsymbol{A}^{\mathrm{T}} + \boldsymbol{G}, \tag{8.8}$$

where G is a positive symmetric matrix and A an arbitrary matrix, with the only constraint being that the resulting matrix $\mathbf{\Gamma}'$ should be a valid covariance matrix, i.e. fulfils Eq. (8.4). The non-orthogonality of the matrix A is a direct consequence of dissipation, and G is the additional noise as required by the dissipation-fluctuation theorem. As a simple example, G can represent pure thermal noise. In this case, one can produce a thermal state with covariance matrix $\mathbf{\Gamma}_{\text{th}}=(2n+1)\mathbf{1}$ from the vacuum with $\mathbf{\Gamma}_{\text{vac}}=\mathbf{1}$ with the help of the matrices $A=\mathbf{1}$ and $G=2n\mathbf{1}$.

Another important class of operations is generated by (homodyne) measurements on subsystems. For that, let us write the covariance matrix of a bipartite system in block form $\mathbf{\Gamma}=\begin{pmatrix} \mathbf{C}_1 & \mathbf{C}_3 \\ \mathbf{C}_3^{\mathrm{T}} & \mathbf{C}_2 \end{pmatrix}$ where $\mathbf{C}_1$ and $\mathbf{C}_2$ are the principal submatrices with respect to the subsystems 1 and 2, respectively, and $\mathbf{C}_3$ is the correlation matrix between the subsystems. In general, the dimension of all these matrices can be different, e.g., $\mathbf{C}_1$ is $2N \times 2N$, $\mathbf{C}_2$ is $2M \times 2M$, and $\mathbf{C}_3$ is $2N \times 2M$. Projection measurements on subsystem 2 onto a Gaussian state with covariance matrix $\mathbf{D}$ leads to the Schur complement [8] with respect to the leading principal submatrix $\mathbf{C}_1$,

$$\mathbf{\Gamma} \to \mathbf{\Gamma}' = \mathbf{C}_1 - \mathbf{C}_3 \left(\mathbf{C}_2 + \mathbf{D}^2\right)^{-1} \mathbf{C}_3^{\mathrm{T}} . \tag{8.9}$$

The probability of measuring a certain outcome is $\left[\det \left(\mathbf{C}_2+\mathbf{D}^2\right)\right]^{-1/2}$. Note that the dimension of the matrix $\mathbf{\Gamma}'$ is now reduced. When $\mathbf{D}$ is the identity matrix, then the transformation (8.9) describes vacuum projection. For $\mathbf{D}=\lim_{d\to 0}(1/d, d, \ldots, 1/d, d)$ it describes homodyne detection, in which case the inverse has to be thought of as the Moore–Penrose (MP) inverse [9]. An analogous treatment has to be made for the corresponding determinant. Combining the transformations (8.8) and (8.9) gives the most general (not necessarily trace-preserving) Gaussian operation where the name means that they leave the Gaussian character of a state invariant.

8.2.3 Separability and Entanglement

So far, we have discussed general features of Gaussian states and Gaussian operations. In what follows, we will focus on bipartite Gaussian states and their entanglement properties. To be more precise, we restrict ourselves to the class of (1×1)-mode states, in which one mode of the electromagnetic field is entangled with another one. The covariance matrix $\mathbf{\Gamma}$ is thus a (4×4)-matrix. In order to check whether a given bipartite Gaussian state is entangled or not, a necessary and sufficient separability criterion has been developed [10, 11]. It is equivalent to the Peres–Horodecki criterion for entangled qubits [12]. The test consists of checking whether the *partial* transpose of the density operator is again a valid density operator. In this case, the state is separable and therefore not entangled. On the level of the covariance matrix, the partial transpose consists of pre- and post-multiplying $\mathbf{\Gamma}$ by the diagonal matrix $\mathbf{P}=\text{diag}(1,1,1,-1)$, viz. $\mathbf{\Gamma}^{\text{P.T.}}=\mathbf{P}\mathbf{\Gamma}\mathbf{P}^{\mathrm{T}}$. Hence, if $\mathbf{\Gamma}^{\text{P.T.}}+i\mathbf{\Sigma}\geq 0$ the state is separable. In the block notation used in Section 8.2.2, this criterion translates as [11]

$$\det \mathbf{C}_1\mathbf{C}_2 + \left(1 - |\det \mathbf{C}_3|\right)^2 - \mathrm{Tr}\left[\mathbf{C}_1\mathbf{\Sigma}\mathbf{C}_3\mathbf{\Sigma}\mathbf{C}_2\mathbf{\Sigma}\mathbf{C}_3^{\mathrm{T}}\mathbf{\Sigma}\right] \geq \det \mathbf{C}_1 + \det \mathbf{C}_2 . \tag{8.10}$$

Once one has checked for inseparability, one may ask how much entangled the state is. This answer is given by computing the logarithmic negativity [13], so far the only computable measure for states in infinite-dimensional Hilbert spaces. It is defined as the sum of the symplectic eigenvalues of $\mathbf{\Gamma}^{\mathrm{P.T.}}$, which translates as [8]

$$E_N(\hat{\varrho}) = \begin{cases} -(\log \circ f)(\mathbf{\Gamma}), & \text{if } f(\mathbf{\Gamma}) < 1, \\ 0 & \text{otherwise}, \end{cases} \tag{8.11}$$

where

$$f(\mathbf{\Gamma}) = \left(\tfrac{1}{2}\left(\det \mathbf{C}_1 + \det \mathbf{C}_2\right) - \det \mathbf{C}_3 - \sqrt{\left[\tfrac{1}{2}(\det \mathbf{C}_1 + \det \mathbf{C}_2) - \det \mathbf{C}_3\right]^2 - \det \mathbf{\Gamma}} \right)^{1/2} . \tag{8.12}$$

An immediate consequence is that it is impossible to distill Gaussian states with Gaussian operations [8, 14]. We are now in the position to look at practically important situations such as entanglement degradation and quantum teleportation in noisy environments.

8.3 Entanglement Degradation

Let us consider the influence of passive optical devices on entangled two-mode quantum states. We have seen in the previous section that Gaussian states are fully characterized by their (first and) second moments. In order to describe the influence of dielectric objects on the quantum state we therefore need to know only how these moments transform. The easiest way to see this is to look at the quantum-optical input-output relations [15, 16]. They relate the amplitude operators of the electromagnetic field impinging on the dielectric object to the amplitude operators of the field leaving the device. In the following we will assume that the dielectric material shows absorption as it is generically the case. This is due to the fact that the dielectric permittivity (as well as the magnetic susceptibility) has to fulfil the Kramers–Kronig relations, which connect the real part of the permittivity (responsible for dispersion) to the imaginary part (responsible for absorption). That said one recognizes that quantization of the electromagnetic field in absorbing matter has to be performed.

Here we will use the macroscopic description [15,17] that is most suitable for the situation we have in mind. The electromagnetic field is quantized by expanding it in terms of a complete set of bosonic vector fields $\hat{\mathbf{f}}(\mathbf{r},\omega)$ that describe collective excitations of the electromagnetic field, the matter polarization, and the reservoir responsible for absorption, viz.

$$\hat{\mathbf{E}}(\mathbf{r}) = i\sqrt{\frac{\hbar}{\pi\varepsilon_0}} \int d^3s \int\limits_0^\infty d\omega \, \frac{\omega^2}{c^2} \sqrt{\operatorname{Im}\varepsilon(\mathbf{s},\omega)}\, \boldsymbol{G}(\mathbf{r},\mathbf{s},\omega)\, \hat{\mathbf{f}}(\mathbf{s},\omega) + \text{H.c.}, \tag{8.13}$$

with $\boldsymbol{G}(\mathbf{r},\mathbf{s},\omega)$ being the classical Green tensor. Expanding the electromagnetic field in terms of amplitude operators outside the dielectric device where no matter is present, one can derive the input-output relation [16]

$$\hat{\mathbf{b}}(\omega) = \mathbf{T}(\omega)\hat{\mathbf{a}}(\omega) + \mathbf{A}(\omega)\hat{\mathbf{g}}(\omega) \tag{8.14}$$

in which the $\hat{a}_i(\omega)$ and $\hat{b}_i(\omega)$ $(i=1,2)$ denote the amplitude operators of the incoming and outgoing fields at frequency ω, and the $\hat{g}_i(\omega)$ denote the noise operators associated with absorption inside the device. The matrices $\mathbf{T}(\omega)$ and $\mathbf{A}(\omega)$ are the transformation and absorption matrices, respectively. They are determined by the Green tensor and satisfy the (energy-) conservation relation $\mathbf{T}(\omega)\mathbf{T}^{+}(\omega)+\mathbf{A}(\omega)\mathbf{A}^{+}(\omega)=\mathbf{1}$. With regard to Gaussian states, we need to know how second moments of the amplitude operators transform. Assuming that the incoming field modes and the device are not correlated, we derive, for example,

$$\begin{aligned}\langle\hat{b}_1^\dagger(\omega)\hat{b}_1(\omega)\rangle &= |T_{11}(\omega)|^2\langle\hat{a}_1^\dagger(\omega)\hat{a}_1(\omega)\rangle + |T_{12}(\omega)|^2\langle\hat{a}_2^\dagger(\omega)\hat{a}_2(\omega)\rangle \\ &\quad + |A_{11}(\omega)|^2\langle\hat{g}_1^\dagger(\omega)\hat{g}_1(\omega)\rangle + |A_{12}(\omega)|^2\langle\hat{g}_2^\dagger(\omega)\hat{g}_2(\omega)\rangle. \end{aligned} \tag{8.15}$$

Let us now look at the most generic two-mode entangled Gaussian state, the two-mode squeezed vacuum (TMSV). In the Fock basis it reads ($\zeta\in\mathbb{R}^+$)

$$|\text{TMSV}\rangle = \exp\zeta(\hat{a}_1\hat{a}_2-\hat{a}_1^\dagger\hat{a}_2^\dagger)|00\rangle = \sqrt{1-q^2}\sum_{n=0}^{\infty}q^n|nn\rangle\,, \quad q=\tanh\zeta\,. \tag{8.16}$$

It is separable only if $q=0$, otherwise it is entangled, where the entanglement content is $E=-\ln(1-q^2)-q^2/(1-q^2)\ln q^2 \lesssim 2\zeta = E_N$. In the language used in Section 8.2, the TMSV translates into a covariance matrix

$$\mathbf{\Gamma} = \begin{pmatrix} c & 0 & s & 0 \\ 0 & c & 0 & -s \\ s & 0 & c & 0 \\ 0 & -s & 0 & c \end{pmatrix}, \quad c=\cosh 2\zeta, \quad s=\sinh 2\zeta. \tag{8.17}$$

We are interested how the entanglement changes when the two modes are transmitted through *two* fibers with equal transmission coefficients $|T|$, reflection coefficients $|R|$, and mean thermal photon numbers n_{th}. Applying the input-output relations (8.14), the second moments transform as

$$c\to c|T|^2+|R|^2+(2n_{\text{th}}+1)(1-|T|^2-|R|^2)\,, \quad s\to s|T|^2, \tag{8.18}$$

which translates into a transformation of the covariance matrix as

$$\mathbf{\Gamma}\to|T|^2\mathbf{\Gamma}+\left[|R|^2+(2n_{\text{th}}+1)(1-|T|^2-|R|^2)\right]\mathbf{1}, \tag{8.19}$$

from which we can easily identify the matrices $\boldsymbol{A}$ and $\boldsymbol{G}$ in the transformation (8.8). Note that we have not taken care about local phases since they do not play any rôle in determining separability and entanglement. Applying the criterion (8.10) to the covariance matrix (8.19), we can easily rewrite the condition of separability as [18]

$$n_{\text{th}} \ge \frac{|T|^2(1-e^{-2\zeta})}{2(1-|R|^2-|T|^2)}\,. \tag{8.20}$$

Restricting ourselves to vanishing incoupling losses ($R=0$) and imposing the Lambert–Beer law of extinction $|T|=e^{-l/l_A}$, with l_A being the characteristic absorption length of the fibers, from Eq. (8.20) we derive for the fiber length after which the TMSV becomes separable

$l_S = \frac{1}{2} l_A \ln[1 + (1 - e^{-2\zeta})/2n_{\text{th}}]$. These results say that for fibers at zero temperature entanglement will always be present (although possibly arbitrarily small). For finite n_{th} the separability length l_S is a function of the squeezing parameter ζ of the input state.

Let us now look at the actual entanglement content in terms of the logarithmic negativity. Again, we only need to insert the transformed covariance matrix (8.19) into Eq. (8.11). Neglecting again the mean thermal photon numbers of the fibers ($n_{\text{th}} = 0$), we derive for the logarithmic negativity

$$E_N = -\ln\left[1 - |T|^2(1 - e^{-2\zeta})\right], \tag{8.21}$$

which is independent of the reflection coefficient R. For perfect transmission ($T=1$), we obtain the familiar result $E_N = 2\zeta$. Another immediate consequence of Eq. (8.21) is that the transmittable entanglement saturates. In the limit of infinite initial squeezing ($\zeta \to \infty$) the maximal entanglement that can pass through fibers of given length l is just $E_{N,\max} = -\ln\left[1 - |T|^2\right] = -\ln\left[1 - e^{-2l/l_A}\right]$. In other words, no matter how strong the initial squeezing was, the amount of transmittable entanglement is limited by the fiber properties (in this simple model the fiber length normalized to the characteristic absorption length). Moreover, the entanglement depends in a simple manner on the overall losses (accounting for reflection and absorption, since $1 - |T|^2 = |R|^2 + |A|^2$) during transmission. Figure 8.1 shows the dependence of the maximal entanglement on the fiber length. It says that in order to transmit the quantum information equivalent to $E_N = 1$, the transmission distance must not exceed the value $l = l_A/2 \ln 2 \approx 0.35\, l_A$. Equivalently, the transmission coefficient must not be smaller than $|T|^2 = 0.5$.

Let us apply this result to quantum dense coding considered, where a classical bit is encoded in a coherent shift of one mode of a TMSV. The result in [19] is that classical coding is superior to quantum dense coding whenever the transmission coefficient $|T|^2$ drops below the value of 0.75, which corresponds, according to Eq. (8.21), to a log-negativity of 2. The example clearly shows the limits of the performance of quantum information processes due to the unavoidably existing entanglement degradation.

8.4 Quantum Teleportation in Noisy Environments

A widely discussed process in quantum information theory has been quantum teleportation [21]. Its purpose is to transport a substantial amount of information about an unknown (single) quantum state spatially from one location to another without transmitting the state itself. The idea is the following. A signal mode that is prepared in the unknown quantum state is mixed with one mode of a (maximally) entangled two-mode state, and the resulting output quantum state is subsequently measured by homodyne detection. The (single-)measurement result is submitted classically to the receiver who, depending on the measurement result, performs a coherent displacement of the quantum state in which his mode (of the two-mode entangled system) has been prepared owing to the measurement.

In order to make contact with the theory developed in the previous sections, we restrict ourselves to the case in which the unknown state is a Gaussian (with zero mean). That is, the initial state in mode 0 can be described by a (2×2) covariance matrix $\boldsymbol{\Gamma}_{\text{in}}$. This class covers all squeezed and thermal states and combinations of them. Accordingly, the source of the entangled state in modes 1 and 2 is assumed to produce a TMSV with a covariance matrix $\mathbf{\Gamma}_{\text{TMSV}}$ as in Eq. (8.17). The covariance matrix of the total state at the beginning of the teleportation process is then the direct sum $\mathbf{\Gamma}_{\text{in}} \oplus \mathbf{\Gamma}_{\text{TMSV}}$.

As already mentioned, the transmission lines from the entanglement source to the sender and the receiver, which are realized, e.g., by fibers, are not perfect in practice. The question we would like to address is to what extent the teleportation protocol has to be modified when decoherence is present. Especially, we will try to answer the question where the source of the entangled state has to be located to obtain an optimal teleportation protocol.

8.4.1 Imperfect Teleportation

Let us slightly generalize the situation, compared to the previous section, in that we consider the possibility of having two different fibers connecting the entanglement source with the sender and the receiver. The matrix $\mathbf{\Gamma}_{\text{TMSV}}$ will thus change to [20]

$$\mathbf{\Gamma}_{\text{TMSV}} \to \mathbf{\Gamma}_{\text{dec}} = \begin{pmatrix} a & 0 & c_1 & c_2 \\ 0 & a & c_2 & -c_1 \\ c_1 & c_2 & b & 0 \\ c_2 & -c_1 & 0 & b \end{pmatrix} \equiv \begin{pmatrix} \mathbf{A} & \mathbf{C} \\ \mathbf{C}^{\text{T}} & \mathbf{B} \end{pmatrix} \tag{8.22}$$

with

$$a = c|T_1|^2 + |R_1|^2 + (2n_{\text{th},1}+1)(1-|T_1|^2-|R_1|^2), \tag{8.23}$$
$$b = c|T_2|^2 + |R_2|^2 + (2n_{\text{th},2}+1)(1-|T_2|^2-|R_2|^2), \tag{8.24}$$

$$c_1 = s\text{Re}(T_1T_2), \tag{8.25}$$
$$c_2 = s\text{Im}(T_1T_2). \tag{8.26}$$

Here, the T_i and R_i ($i=1,2$) are the transmission and reflection coefficients of the ith fiber.

The operation of mixing one mode of the transmitted TMSV (covariance matrix $\mathbf{\Gamma}_{\text{dec}}$) with the signal mode at a symmetric beam splitter is described by a symplectic matrix of the form

$$\mathbf{S}_{\text{BS}} = \frac{1}{\sqrt{2}} \begin{pmatrix} \mathbf{1} & \mathbf{1} \\ -\mathbf{1} & \mathbf{1} \end{pmatrix}. \tag{8.27}$$

Remembering that $\mathbf{S}_{\text{BS}}$ acts only on modes 0 and 1, the covariance matrix of the tripartite state reads as $\mathbf{\Gamma}_{012} = (\mathbf{S}_{\text{BS}} \oplus \mathbf{1}_2)\,(\mathbf{\Gamma}_{\text{in}} \oplus \mathbf{\Gamma}_{\text{dec}})(\mathbf{S}_{\text{BS}} \oplus \mathbf{1}_2)^{\text{T}}$, where $\mathbf{1}_2$ is the identity operation on mode 2, the receiver's side. In block matrix notation, $\mathbf{\Gamma}_{012}$ reads

$$\mathbf{\Gamma}_{012} = \begin{pmatrix} (\mathbf{\Gamma}_{\text{in}}+\mathbf{A})/2 & (\mathbf{A}-\mathbf{\Gamma}_{\text{in}})/2 & \mathbf{C}/\sqrt{2} \\ (\mathbf{A}-\mathbf{\Gamma}_{\text{in}})/2 & (\mathbf{\Gamma}_{\text{in}}+\mathbf{A})/2 & \mathbf{C}/\sqrt{2} \\ \mathbf{C}^{\text{T}}/\sqrt{2} & \mathbf{C}^{\text{T}}/\sqrt{2} & \mathbf{B} \end{pmatrix} := \begin{pmatrix} \mathbf{M} & \mathbf{N} \\ \mathbf{N}^{\text{T}} & \mathbf{B} \end{pmatrix}. \tag{8.28}$$

As shown in the Appendix, homodyne detection (with respect to the variables x_0 and p_1 after the beam splitter) is equivalent to a partial Fourier transform [Eqs. (A.1) and (A.4)]. Let us denote the measurement outcomes corresponding to the variables x_0 and p_1 by X_{x_0} and X_{p_1}, respectively. The characteristic function at the receiver's side will then be

$$\chi_{\mathrm{rec}}(\boldsymbol{\lambda}_2) = p(X_{x_0}, X_{p_1}) \exp\left[-\frac{1}{4}\boldsymbol{\lambda}_2^{\mathrm{T}}\boldsymbol{\Gamma}_{\mathrm{rec}}\boldsymbol{\lambda}_2 + i\,\boldsymbol{\lambda}_2^{\mathrm{T}}\mathbf{N}^{\mathrm{T}}(\boldsymbol{\pi}\mathbf{M}\boldsymbol{\pi})^{\mathrm{MP}}\begin{pmatrix} X_{x_0} \\ -X_{p_1}\end{pmatrix}\right]. \quad (8.29)$$

The probability of obtaining a certain pair of measurement outcomes is

$$p(X_{x_0}, X_{p_1}) = \frac{1}{\pi\sqrt{\det(\boldsymbol{\pi}\mathbf{M}\boldsymbol{\pi})}} \exp\left[-\begin{pmatrix} X_{x_0} & -X_{p_1}\end{pmatrix}(\boldsymbol{\pi}\mathbf{M}\boldsymbol{\pi})^{\mathrm{MP}}\begin{pmatrix} X_{x_0} \\ -X_{p_1}\end{pmatrix}\right], \quad (8.30)$$

and the covariance matrix of the receiver's quantum state obtains the form

$$\boldsymbol{\Gamma}_{\mathrm{rec}} = \mathbf{B} - \mathbf{N}^{\mathrm{T}}(\boldsymbol{\pi}\mathbf{M}\boldsymbol{\pi})^{\mathrm{MP}}\mathbf{N}, \quad (8.31)$$

where $\boldsymbol{\pi} = \mathrm{diag}(0, 1, 1, 0)$ and the superscript MP denotes the Moore–Penrose inverse. The matrix $\boldsymbol{\pi}$ is, in fact, a projector onto the (p_0, x_1)-plane. It removes all entries of $\mathbf{M}$ except the square block matrix specified by the 1-entries. The Moore–Penrose inverse of a matrix containing a square block on the diagonal and zeros otherwise is nothing but a matrix with the inverse of the square block at the same position (and zero everywhere else).

If we denote the elements of the covariance matrix of the unknown signal state by $\boldsymbol{\Gamma}_{\mathrm{in}} = \begin{pmatrix} x & z \\ z & y \end{pmatrix}$, the involved matrices read

$$\mathbf{N} = \begin{pmatrix} c_2 & -c_1 \\ c_1 & c_2 \end{pmatrix}, \quad (\boldsymbol{\pi}\mathbf{M}\boldsymbol{\pi})^{\mathrm{MP}} = \frac{1}{(a+x)(a+y) - z^2}\begin{pmatrix} a+x & z \\ z & a+y \end{pmatrix}, \quad (8.32)$$

and the covariance matrix (8.31) thus takes the form of

$$\boldsymbol{\Gamma}_{\mathrm{rec}} = \begin{pmatrix} b & 0 \\ 0 & b \end{pmatrix} - \frac{1}{(x+a)(y+a) - z^2} \times \begin{pmatrix} c_2^2(x+a) + c_1^2(y+a) + 2c_1c_2z & c_1c_2(y-x) - z(c_1^2 - c_2^2) \\ c_1c_2(y-x) - z(c_1^2 - c_2^2) & c_1^2(x+a) + c_2^2(y+a) - 2c_1c_2z \end{pmatrix}. \quad (8.33)$$

In the limit $\zeta \to \infty$, perfect transmission of the TMSV, and the phase adjusted such that $c_2 = 0$, the covariance matrix on the receiver's side becomes $\boldsymbol{\Gamma}_{\mathrm{rec}} = \boldsymbol{\Gamma}_{\mathrm{in}}$. In particular, we recover that for perfect teleportation an infinitely squeezed TMSV (with $\zeta \to \infty$) is needed. Note that $\boldsymbol{\Gamma}_{\mathrm{rec}}$ does not depend on the measurement result, even for non-perfect transmission. That means that the information about the quantum features of a Gaussian state are not transported via the classical channel. What we see, though, is that the covariance matrix is sensitive to the properties of the channel itself.

8.4.2 Teleportation Fidelity

Generically, the TMSV is neither infinitely squeezed nor are the transmission lines from the entanglement source to sender and receiver perfect. Hence, one needs a measure for the success rate of the teleportation protocol. For teleporting pure quantum states, one commonly

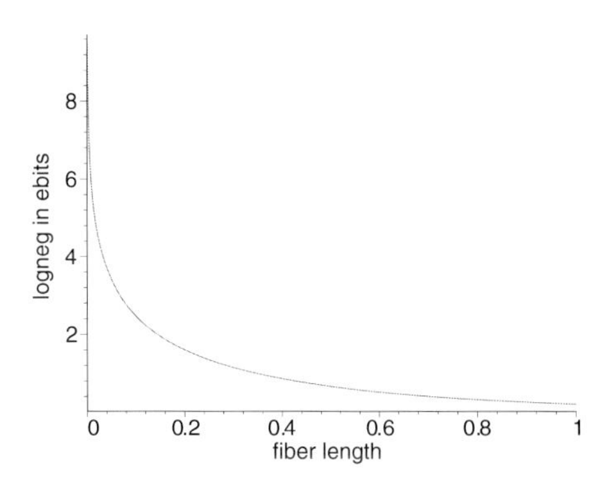

Figure 8.1: Maximally transmittable entanglement as a function of the fiber length (in units of the absorption length) for the case of a TMSV input.

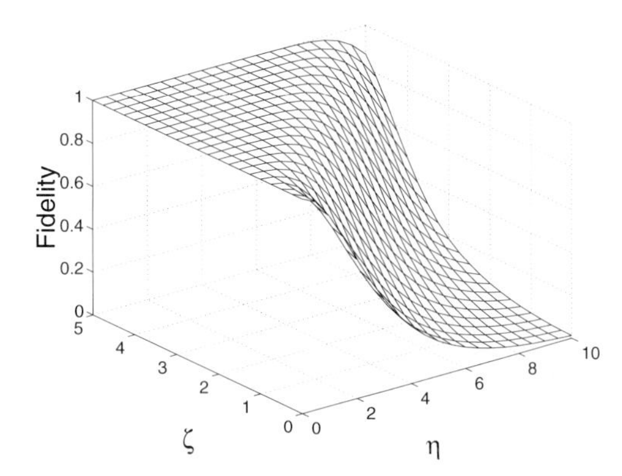

Figure 8.2: Teleportation fidelity F_{qu} of a (pure) squeezed state (squeezing parameter η) by means of an unperturbed TMSV (squeezing parameter ζ).

uses the fidelity defined as the overlap between the quantum state to be teleported and the quantum state at the receiver. In terms of their characteristic functions, the fidelity is given as a double integral over the (complex) parameter $\boldsymbol{\lambda}$ as

$$F = \frac{1}{\pi}\int d^2\boldsymbol{\lambda}\,\chi_{\mathrm{in}}(\boldsymbol{\lambda})\chi_{\mathrm{rec}}(-\boldsymbol{\lambda}). \tag{8.34}$$

For Gaussian states with zero mean the negative argument in Eq. (8.34) does not matter since the characteristic function is an exponential quadratic form of $\boldsymbol{\lambda}$. As we have seen, the homodyne detection introduces a coherent displacement, but does not influence the covariance matrix. With regard to the quantum properties, it is therefore sufficient to compute the overlap between characteristic functions with zero mean. In that way, from Eq. (8.34) we find that

$$F \to F_{\mathrm{qu}} = \frac{2}{\sqrt{\det(\boldsymbol{\Gamma}_{\mathrm{in}} + \boldsymbol{\Gamma}_{\mathrm{rec}})}}\,. \tag{8.35}$$

For perfect teleportation, the covariance matrix of the receiver's state is identical to the covariance matrix of the signal state to be teleported, and the fidelity becomes $F_{\mathrm{qu}} = (\det\boldsymbol{\Gamma}_{\mathrm{in}})^{-1/2} = 1$. Even if entanglement degradation can be disregarded, perfect teleportation cannot be realized, because of finite squeezing. Let us consider a (pure) squeezed signal state and disregard entanglement degradation, so that the entangled state is simply the unchanged (pure) TMSV state. The parameters to look at are $a = b = c = \cosh 2\zeta$, $c_1 = s = \sinh 2\zeta$, $c_2 = 0$, $x = y = \cosh\eta$, $z = \sinh\eta$. The teleportation fidelity according to Eq. (8.35) is then

$$F_{\mathrm{qu}} = \sqrt{1 - \frac{\sinh^2\eta}{(\cosh\eta + \cosh 2\zeta)^2}} \equiv \sqrt{1 - \frac{\sinh^2\eta}{(\cosh\eta + \cosh E_N)^2}} \tag{8.36}$$

[cf. Eq. (8.21)], whose behaviour is illustrated in Fig. 8.2. We see that for chosen squeezing parameter η of the signal state the fidelity is a monotonous function of the inserted entanglement E_N. In particular, the classical teleportation fidelity (obtained for $\zeta = 0$) is read off as $\sqrt{2/(1+\cosh\eta)}$.

8.4.3 Choice of the Coherent Displacement

There have been some discussions recently about what kind of coherent displacement should be applied [22, 23]. Let us look at Eq. (8.29). In the limit of an infinitely entangled TMSV the product $\mathbf{N}^{\mathrm{T}}(\boldsymbol{\pi}\mathbf{M}\boldsymbol{\pi})^{\mathrm{MP}}$ becomes the symplectic matrix $\boldsymbol{\Sigma}$ which is independent of the initial state. Thus, in the ideal case the coherent displacement is solely determined by the measurement results. This is the standard teleportation protocol, where the coherent displacement to be performed on the receivers's side is exactly $X_{x_0} - iX_{p_1}$.

Even if an infinitely entangled TMSV were available, entanglement degradation would prevent one from observing this simple result. On recalling Eq. (8.32) together with Eqs. (8.23) and (8.25), the relevant matrix product $\mathbf{N}^{\mathrm{T}}(\boldsymbol{\pi}\mathbf{M}\boldsymbol{\pi})^{\mathrm{MP}}$ now reads

$$\lim_{\zeta\to\infty} \mathbf{N}^{\mathrm{T}}(\boldsymbol{\pi}\mathbf{M}\boldsymbol{\pi})^{\mathrm{MP}} = \frac{1}{|T_1|^2} \begin{pmatrix} \mathrm{Im}\,(T_1T_2) & \mathrm{Re}\,(T_1T_2) \\ \mathrm{Re}\,(T_1T_2) & -\mathrm{Im}\,(T_1T_2) \end{pmatrix}. \tag{8.37}$$

If we write the transmission coefficients as $T_i = |T_i|e^{i\varphi_i}$, Eq. (8.37) becomes

$$\lim_{\zeta\to\infty} \mathbf{N}^{\mathrm{T}}(\boldsymbol{\pi}\mathbf{M}\boldsymbol{\pi})^{\mathrm{MP}} = \boldsymbol{\Sigma}\left|\frac{T_2}{T_1}\right| \begin{pmatrix} \cos(\varphi_1+\varphi_2) & -\sin(\varphi_1+\varphi_2) \\ \sin(\varphi_1+\varphi_2) & \cos(\varphi_1+\varphi_2) \end{pmatrix}. \tag{8.38}$$

We see that, apart from the (irrelevant) phase shift $-(\varphi_1+\varphi_2)$, the standard teleportation protocol must be modified in so far that the absolute value of the coherent displacement has to be scaled by $|T_2/T_1|$. This result is in agreement with [22]. Note that the displacement (in this limit) is still independent of the unknown state and no averaging has to be performed. We like to emphasize that this limit is actually the only situation in which the word 'teleportation' makes sense.

In practice, an infinitely entangled state is not available. Nevertheless, one might want to apply the displacement (8.38), which is optimal for infinite entanglement. This, however, introduces further loss in fidelity. The modification is an exponential function that depends on the displacement of the signal state and the state at the receiver's side after performing the displacement. These results show that understanding decoherence and its associated processes is important for quantum information processing in the presence of absorption.

Acknowledgements

We would like to thank K. Audenaert, A.V. Chizhov, J. Eisert, P.L. Knight, L. Knöll, M.B. Plenio, and W. Vogel for many fruitful discussions on the subject. This work has been funded in parts by a Feodor-Lynen fellowship of the Alexander von Humboldt foundation (S. Scheel), the QUEST programme of the European Union (HPRN-CT-2000-00121), and the Deutsche Forschungsgemeinschaft (DFG-Schwerpunkt Quanteninformationsverarbeitung).

Appendix: Homodyne detection

Perfect homodyne detection is a projective quadrature component measurement. Thus, we have

$${}_2\langle X,\varphi|\hat{\varrho}|X,\varphi\rangle_2 = \frac{1}{\pi^2}\int d^2\boldsymbol{\lambda}_1 d^2\boldsymbol{\lambda}_2\, \hat{D}^\dagger(\boldsymbol{\lambda}_1)\chi(\boldsymbol{\lambda}_1,\boldsymbol{\lambda}_2)\,{}_2\langle X,\varphi|\hat{D}^\dagger(\boldsymbol{\lambda}_2)|X,\varphi\rangle_2, \tag{A.1}$$

where the $|X, \varphi\rangle$ are the quadrature-component eigenstates with eigenvalues X for chosen phase parameter φ ($\varphi=0$ corresponds to an x measurement, $\varphi=\pi/2$ to a p measurement). It can be expanded as [24]

$$|X, \varphi\rangle = \pi^{-1/4} e^{-X^2/2} \exp\left[-\tfrac{1}{2}\left(\hat{a}^\dagger e^{i\varphi}\right)^2 + 2^{1/2} X \hat{a}^\dagger e^{i\varphi}\right] |0\rangle, \tag{A.2}$$

and hence the diagonal-matrix elements of the coherent displacement operator read

$$_2\langle X, \varphi|\hat{D}^\dagger(\boldsymbol{\lambda}_2)|X, \varphi\rangle_2 = \delta(x_2 \cos\varphi + p_2 \sin\varphi) \exp\left[iX\left(x_2 \sin\varphi - p_2 \cos\varphi\right)\right] \tag{A.3}$$

[$\boldsymbol{\lambda}_2^{\mathrm{T}} = (x_2, p_2)/\sqrt{2}$]. For $\varphi=0, \pi/2$, Eq. (A.3) reduces to

$$_2\langle X, \varphi|\hat{D}^\dagger(\boldsymbol{\lambda}_2)|X, \varphi\rangle_2 = \begin{cases} \delta(x_2) e^{-iXp_2} & (\varphi = 0), \\ \delta(p_2) e^{iXx_2} & (\varphi = \pi/2). \end{cases} \tag{A.4}$$

Equivalently, one can use the relation for the projector onto a quadrature-component eigenstate in the form [24]

$$|X, \varphi\rangle\langle X, \varphi| = \frac{1}{2\pi} \int dy\, e^{-iyX} \hat{D}\left(\frac{iye^{i\varphi}}{\sqrt{2}}\right) \tag{A.5}$$

and use the completeness of the quadrature-component states $|X, \varphi\rangle$ and the relation $\mathrm{Tr}[\hat{D}(\lambda)\hat{D}(\lambda')] = \pi\delta(\lambda + \lambda')$. On the level of the covariance matrices, the δ function leads to the projection matrix $\boldsymbol{\pi}$ in Eq. (8.31) since it deletes all matrix elements associated with the measured variables. Performing the remaining Gaussian integration, which is equivalent to a Fourier transform, introduces the Schur complement with respect to the leading submatrix corresponding to the subsystem 1.

References

[1] Arvind, B. Dutta, N. Mukunda, and R. Simon, *quant-ph/9509002*.

[2] U.M. Titulaer and R.J. Glauber, Phys. Rev. **140**, B676 (1965).

[3] W. Vogel, Phys. Rev. Lett. **84**, 1849 (2000).

[4] P. Marian and T.A. Marian, Phys. Rev. A **47**, 4474 (1993); 4487 (1993).

[5] S. Scheel and D.-G. Welsch, Phys. Rev. A **64**, 063811 (2001).

[6] P. Marian, T.A. Marian, and H. Scutaru, Phys. Rev. Lett. **88**, 153601 (2002).

[7] M.A. Naimark, C.R. Acad. Sci. USSR **41**, 359 (1943).

[8] J. Eisert, S. Scheel, and M.B. Plenio, Phys. Rev. Lett. **89**, 137903 (2002).

[9] R.A. Horn and C.R. Johnson, *Matrix Analysis* (Cambridge University Press, Cambridge, 1985).

[10] L.M. Duan, G. Giedke, J.I. Cirac, and P. Zoller, Phys. Rev. Lett. **84**, 2722 (2000).

[11] R. Simon, Phys. Rev. Lett. **84**, 2726 (2000).

[12] A. Peres, Phys. Rev. Lett. **77**, 1413 (1996); P. Horodecki, Phys. Lett. A **232**, 333 (1997).

[13] J. Eisert, PhD thesis (Potsdam, February 2001); G. Vidal and R.F. Werner, Phys. Rev. A **65**, 032314 (2002).

[14] J. Fiurášek, Phys. Rev. Lett. **89**, 137904 (2002); G. Giedke and J.I. Cirac, Phys. Rev. A **66**, 032316 (2002).
[15] L. Knöll, S. Scheel, and D.-G. Welsch, in *Coherence and Statistics of Photons and Atoms*, ed. J. Peřina (Wiley, New York, 2001).
[16] T. Gruner and D.-G. Welsch, Phys. Rev. A **54**, 1661 (1996).
[17] T. Gruner and D.-G. Welsch, Phys. Rev. A **53**, 1818 (1996); Ho Trung Dung, L. Knöll, and D.-G. Welsch, Phys. Rev. A **57**, 3931 (1998); S. Scheel, L. Knöll, and D.-G. Welsch, Phys. Rev. A **58**, 700 (1998).
[18] S. Scheel, PhD thesis (Jena, July 2001).
[19] M. Ban, J. Opt. B: Quantum Semiclass. Opt. **2**, 786 (2000).
[20] S. Scheel, L. Knöll, T. Opatrný, and D.-G. Welsch, Phys. Rev. A **62**, 043803 (2000).
[21] C.H. Bennett et al., Phys. Rev. Lett. **70**, 1895 (1993); S.L. Braunstein and H.J. Kimble, Phys. Rev. Lett. **80**, 869 (1998).
[22] D.-G. Welsch, S. Scheel, and A.V. Chizhov, Proceedings of the 7th International Conference on Squeezed States and Uncertainty Relations, Boston, 2001 (www.wam.umd.edu/∼ys/boston.html).
[23] M.S. Kim, Phys. Rev. A **64**, 012309 (2001); A.V. Chizhov, L. Knöll, and D.-G. Welsch, Phys. Rev. A **65**, 022310 (2002).
[24] W. Vogel, S. Wallentowitz, and D.-G. Welsch, *Quantum Optics, An Introduction* (Wiley-VCH, Berlin, 2001).

9 Quantum Estimation with Finite Resources

Thorsten C. Bschorr, Dietmar G. Fischer, Holger Mack, Wolfgang P. Schleich, and Matthias Freyberger

Abteilung für Quantenphysik
Universität Ulm
Ulm, Germany

Abstract

We discuss the quantum estimation of characteristic parameters that describe quantum devices. It turns out that for the estimation of a Pauli channel entangled qubits can serve as a nonclassical resource. Moreover, we analyze a generalized estimation scheme that involves the distribution of entangled particles via separate channels.

9.1 Introduction

Quantum mechanics has completely changed our understanding of nature. It is an operator–based theory that abandons all classical premises that seem to be intuitively clear. The fundamental change in the transition from classical physics to quantum physics can be seen in the respective concepts of the state of a physical object. A quantum state is fully described by a density operator which is neither an observable nor can it be assigned to a single system. It rather formulates the catalogue of expectations, that is of the statistics that can be observed when we make measurements on an ensemble of microscopic systems.

The most important consequence of this transition is that we cannot determine characteristic properties of a quantum object by adopting classical strategies [1, 2]. In particular, the full reconstruction of a quantum state [3] needs in general infinitely many measurements. Since each measurement changes the state, it furthermore requires an infinite ensemble of identically prepared quantum objects. Starting from finite resources we can never measure but only *estimate* the underlying state [1, 2, 4–6]. Thus quantum resources are valuable and we have to treat them carefully.

This fact has immediate consequences for quantum information processing. For a given task sophisticated schemes have to be invented in order to squeeze out maximum performance. Quantum algorithms [7] provide excellent examples for this sophistication since they perform even better than any classical approach for a given resource. In the present work we shall describe estimation methods for a quantum channel. In general such a channel can be considered as a black box quantum device that transforms an input state to an output state according to rules enforced by quantum mechanics. Our aim is to extract the characteristic parameters of the device by using only a finite number of measurements. In particular, we show that

Quantum Information Processing, 2nd Edition. Edited by Thomas Beth and Gerd Leuchs

ISBN: 3-527-40541-0

entanglement can strongly enhance this estimation process and therefore acts like a purely nonclassical resource.

The structure of the paper is as follows. We briefly review the essential concepts of quantum devices and quantum channels in Sec. 9.2. In Sec. 9.3 we introduce the reader to the basic idea of channel estimation. We describe the estimation scheme with single qubits in Sec. 9.4.1 and proceed with the entangled estimation scheme in Sec. 9.4.2. A comparison between the single qubit and the entangled version shows that the entangled one enhances the estimation quality. We generalize our estimation scheme in Sec. 9.5 by introducing a second channel which is used for entanglement distribution by Alice and Bob. Finally, we give a short outlook in Sec. 9.6.

9.2 Quantum Devices and Channels

A quantum device can be viewed as a black box, which transforms an input quantum state $\hat{\rho}$ into an output state $\hat{\rho}'$ according to the rules enforced by quantum mechanics. Hereby, the transformation can be as simple as a unitary operation, e. g., a free time evolution. However, also measurements—including the selection of subensembles—and wanted or unwanted couplings to other quantum systems or environments are possible. Examples of quantum devices are quantum computers, information transfer channels or any measurement apparatus.

Karl Kraus [8] has formulated the quantum-mechanical description and the corresponding restrictions of such general state changes[1]. He considers the most general possible transformation of a quantum system which is initially described by a density operator $\hat{\rho}$. This transformation $\mathcal{C}$ can be written in terms of Kraus operators $\hat{A}_i$ and yields the final density operator

$$\hat{\rho}' = \mathcal{C}\hat{\rho} = \sum_i \lambda_i \hat{A}_i \hat{\rho} \hat{A}_i^\dagger \,. \tag{9.1}$$

The parameters λ_i are the occurrence probabilities of the operators $\hat{A}_i$ and hence we need $\sum_i \lambda_i = 1$. If the equality $\sum_i \lambda_i \hat{A}_i^\dagger \hat{A}_i = \hat{1}$ holds, then the transformation is trace preserving. In this case Eq. (9.1) represents the most general form of a completely positive (CP) linear map [9]. On the other hand we might find $\mathrm{Tr}\hat{\rho}' < \mathrm{Tr}\hat{\rho} = 1$. This means that some of the quantum systems are lost inside the quantum device. This could happen due to absorption processes—e. g., a photon is absorbed in a crystal—or due to the selection of subensembles, which is also known as filtering. An simple example for such a filtering would be a Stern–Gerlach device that blocks one atomic beam and therefore selects a subensemble with a certain spin direction.

Any quantum device can be interpreted as a *quantum channel* in the sense of Eq. (9.1) through which the quantum states are sent and we will use this latter notion in the following. Furthermore, we will concentrate on channels for qubits as the fundamental carriers of quantum information.

[1] We are considering devices without memory. Their action on the nth quantum system does not depend on the sequence of previous actions on the quantum systems $1, \ldots, n-1$. In other words, the device has always to be prepared in the same initial state before the next quantum system is sent.

9.3 Estimating Quantum Channels

We now focus our attention on the problem of determining the properties of a quantum channel that connects two parties, say Alice and Bob. In particular, we consider the case when Alice and Bob use a finite amount N of quantum systems for that purpose. This is certainly the realistic case. Alice and Bob have no or only partial knowledge about the properties of the quantum channel. Within a specific class each quantum channel $\mathcal{C}$ can be parameterized by a vector $\vec{\lambda}$, cf. Eq. (9.1), with L independent components. We will therefore denote these channels by $\mathcal{C}_{\vec{\lambda}}$ throughout this chapter. The parameters λ_i are the quantities that Alice and Bob want to determine using only finite resources.

Alice and Bob can apply the following protocol in order to estimate the properties of a quantum channel. They agree on a finite set of N quantum states $\hat{\rho}_i$, which are prepared by Alice and then sent through the quantum channel $\mathcal{C}_{\vec{\lambda}}$. The channel disturbs the state and Bob receives according to Eq. (9.1) the N quantum states $\hat{\rho}'_i = \mathcal{C}_{\vec{\lambda}}\,\hat{\rho}_i$. He now performs suitable measurements on these N quantum systems. From the results he deduces an estimated parameter vector $\vec{\lambda}^{(est)}$. Of course, a good estimation scheme should result in an estimated parameter vector close to the actual parameter vector $\vec{\lambda}$ of the quantum channel.

To quantify this notion of "closeness" we have to introduce a quantitative measure for it. We emphasize that several choices are possible. In this paper we choose the *statistical cost function*

$$c(N,\vec{\lambda}) \equiv \sum_{\ell=1}^{L} \left(\lambda_\ell - \lambda_\ell^{(est)}(N)\right)^2 , \tag{9.2}$$

which is derived from the variance of actual and estimated parameters. Note that the elements of the estimated parameter vector $\vec{\lambda}^{(est)}$ strongly depend on the available resources, i. e. the number N of systems prepared by Alice. We also emphasize that c describes the error for one single run of an estimation protocol, that is, for one channel estimation based on a sequence $\mathcal{J}$ of N measurement results. However, we are not interested in the single run error, but in the average error of a given protocol. Therefore, we average over all possible sequences of measurement outcomes $\mathcal{J}$ to obtain the *mean statistical cost function*

$$\overline{c}(N,\vec{\lambda}) \equiv \langle c(N,\vec{\lambda})\rangle_{\mathcal{J}}, \tag{9.3}$$

while keeping the number N of resources fixed.

Note that the cost function $\overline{c}$ depends on the parameterization $\vec{\lambda}$ of the quantum channel. That is, the estimation errors change when we change the parameterization of the channel despite the fact that it physically remains the same channel. Nevertheless, we can use $\overline{c}$ to compare the performance of different estimation protocols.

9.4 Entanglement and Estimation

Up to now we have basically described the estimation principle based on sending single qubits through the channel. However, we are not restricted to such separable estimation schemes. Instead of sending single qubits we could use entangled qubit pairs, also called ebits [10]. In this section we will demonstrate the superiority of such nonclassical methods for the estimation of the so–called Pauli channel.

The Pauli channel is widely discussed in the literature, especially in the context of quantum error correction [11]. Its name originates from the error operators $\hat{A}_i$ of the channel. These error operators are the unitary Pauli operators $\hat{A}_1 = \hat{\sigma}_x \equiv |0\rangle\langle 1| + |1\rangle\langle 0|$, $\hat{A}_2 = \hat{\sigma}_y \equiv i(|1\rangle\langle 0| - |0\rangle\langle 1|)$ and $\hat{A}_3 = \sigma_z \equiv |0\rangle\langle 0| - |1\rangle\langle 1|$. The operators define the quantum mechanical analogue to bit errors in a classical communication channel since $\hat{\sigma}_x$ causes a bit flip, i. e. $|0\rangle$ transforms into $|1\rangle$ and vice versa. Analogously, $\hat{\sigma}_z$ leads to a phase flip, that is, $|0\rangle + |1\rangle \leftrightarrow |0\rangle - |1\rangle$. Moreover, $\hat{\sigma}_y$ results in a combined bit and phase flip.

Three probabilities $\vec{\lambda} \equiv (\lambda_1, \lambda_2, \lambda_3)^T$ for the occurrence of these errors describe the Pauli channel completely. If an initial quantum state $\hat{\rho}$ is sent through the channel, the Pauli channel transforms $\hat{\rho}$ into

$$\hat{\rho}' = \mathcal{C}_{\vec{\lambda}}\hat{\rho} = (1 - \lambda_1 - \lambda_2 - \lambda_3)\hat{\rho} + \lambda_1\hat{\sigma}_x\hat{\rho}\hat{\sigma}_x + \lambda_2\hat{\sigma}_y\hat{\rho}\hat{\sigma}_y + \lambda_3\hat{\sigma}_z\hat{\rho}\hat{\sigma}_z \tag{9.4}$$

in accordance with Eq. (9.1). Thus the density operator $\hat{\rho}$ remains unchanged with probability $1 - \lambda_1 - \lambda_2 - \lambda_3$, whereas with probabilities λ_1, λ_2 and λ_3 the corresponding Pauli operators rotate the original state $\hat{\rho}$. Moreover the transformation, Eq. (9.4), is trace preserving.

9.4.1 Estimation using Single Qubits

We will first describe the separable qubit estimation scheme for the Pauli channel, before we analyze the one based on entanglement. Following the general estimation scheme presented in Sec. 9.3, the protocol to estimate the parameters of the Pauli channel, Eq. (9.4), is depicted in Fig. 9.1 (a). It requires the preparation of three different quantum states pointing along three orthogonal Bloch-sphere directions. Alice sends (i) $N/3$ qubits in the state $|0\rangle$, (ii) $N/3$ qubits in the state $(|0\rangle + |1\rangle)/\sqrt{2}$, and (iii) $N/3$ qubits in the state $(|0\rangle + i|1\rangle)/\sqrt{2}$. Bob measures these states in their corresponding eigenbasis.

The reason why this choice of prepared input states is optimal is that they are eigenstates of the channel error operators σ_z, σ_x and σ_y. Thus the resulting output states only depend on two instead of three channel parameters λ_i. The measurements are then designed in such a way that the variation in measured probabilities P due to a change of channel parameters is maximal. The measurement probabilities for finding the result "1_j" in the z, x and y basis are given by $P[1_z] = \lambda_1 + \lambda_2$, $P[1_x] = \lambda_2 + \lambda_3$ and $P[1_y] = \lambda_1 + \lambda_3$. Note that the probabilities depend on more than one channel parameter.

By inverting this system of linear equations we can express the channel parameters λ_i by the probabilities $P[1_j]$. However, in an experiment with finite resources we have to replace the probabilities $P[1_j]$ by the corresponding relative frequencies $n_j/(N/3)$ for the measurement result "1_j". The estimated parameter values then read

$$\vec{\lambda}^{(est)} = \frac{1}{2N/3}\begin{pmatrix} -n_x + n_y + n_z \\ n_x - n_y + n_z \\ n_x + n_y - n_z \end{pmatrix} . \tag{9.5}$$

We note that, although the probabilities λ_i are positive or vanish, their estimated values $\lambda_i^{(est)}$ may be negative. Nonetheless, the average cost function $\bar{c}$ defined by Eq. (9.3) can always be

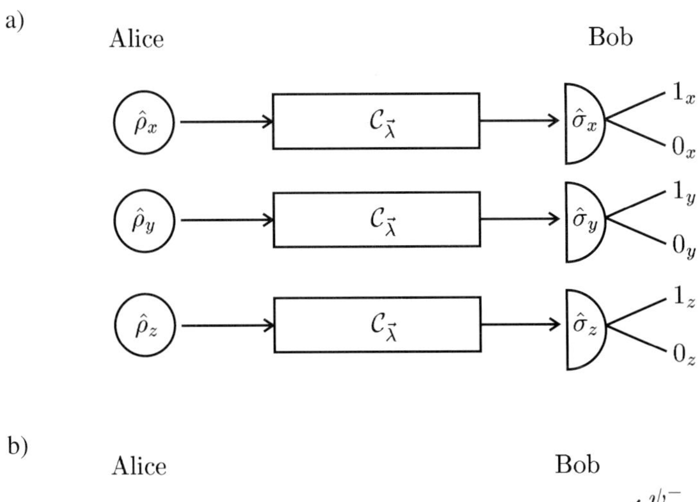

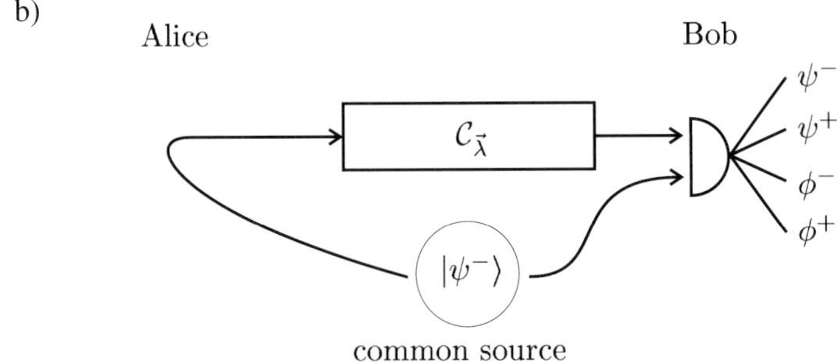

Figure 9.1: Two schemes for estimating the Pauli channel $C_{\vec{\lambda}}$ from an initial supply of N qubits. In scheme (a) Alice prepares single qubits in three different eigenstates $\hat{\rho}_i$ ($N/3$ qubits in each state) of the error operators $\hat{\sigma}_i$ and sends them through the quantum channel. After receiving the qubits, Bob measures the operator $\hat{\sigma}_i$ for each qubit. Thus he finally possesses N measurement results from which he estimates the parameters $\vec{\lambda}$. In scheme (b) Alice and Bob share $N/2$ entangled pairs of qubits prepared by a common source. Alice sends her qubit through the quantum channel to Bob who then performs a Bell measurement. In this scheme Bob records only $N/2$ measurement results.

evaluated. For separable qubits we find

$$\begin{aligned}
\bar{c}^{\,\mathrm{sep}}(N,\vec{\lambda}) &= \sum_{n_x=0}^{N/3}\sum_{n_y=0}^{N/3}\sum_{n_z=0}^{N/3}\binom{N/3}{n_x}\binom{N/3}{n_y}\binom{N/3}{n_z} \\
&\quad \times P[1_x]^{n_x}\left(1-P[1_x]\right)^{N/3-n_x} \times P[1_y]^{n_y}\left(1-P[1_y]\right)^{N/3-n_y} \\
&\quad \times P[1_z]^{n_z}\left(1-P[1_z]\right)^{N/3-n_z} \\
&\quad \times \left[\left(\lambda_1-\lambda_1^{(est)}\right)^2+\left(\lambda_2-\lambda_2^{(est)}\right)^2+\left(\lambda_3-\lambda_3^{(est)}\right)^2\right] \\
&= \frac{9}{2N}\left[\lambda_1(1-\lambda_1)+\lambda_2(1-\lambda_2)+\lambda_3(1-\lambda_3)\right. \\
&\qquad \left.-\lambda_1\lambda_2-\lambda_2\lambda_3-\lambda_3\lambda_1\right]\,, \qquad (9.6)
\end{aligned}$$

where we have averaged over all possible runs $\mathcal{J}$ weighted by the corresponding binomial distribution.

For fixed N the average error has a maximum at $\lambda_1 = \lambda_2 = \lambda_3 = 1/4$, where all acting error operators and the identity operation occur with the same probability. On the other hand, the average error vanishes when we have a faithful transmission or when one of the errors occurs with certainty.

9.4.2 Estimation using Entangled States

The protocol that we have considered so far was based on single qubits prepared in pure states. These qubits are then sent through the channel one after another. However, one can envisage estimation schemes with different features.

An interesting and powerful alternative scheme is based on the use of entangled states [10] as a nonclassical resource. The scheme is depicted in Fig. 9.1 (b). It requires Alice and Bob to share N pairwise entangled qubits prepared by a common source. We assume that each pair is described by the Bell state

$$|\psi^-\rangle \equiv \frac{1}{\sqrt{2}}\left(|0\rangle|1\rangle - |1\rangle|0\rangle\right) . \tag{9.7}$$

Alice sends her $N/2$ qubits of the entangled pairs (ebits) through the Pauli channel, which transforms the entangled state into the mixed state

$$\begin{aligned}\hat{\rho}' = \mathcal{C}_{\vec{\lambda}}(|\psi^-\rangle\langle\psi^-|) &= \lambda_1|\phi^-\rangle\langle\phi^-| + \lambda_2|\phi^+\rangle\langle\phi^+| + \lambda_3|\psi^+\rangle\langle\psi^+| \\ &\quad + (1-\lambda_1-\lambda_2-\lambda_3)\,|\psi^-\rangle\langle\psi^-|,\end{aligned} \tag{9.8}$$

where $|\phi^\pm\rangle \equiv \left(|0\rangle|0\rangle \pm |1\rangle|1\rangle\right)/\sqrt{2}$ and $|\psi^\pm\rangle \equiv \left(|0\rangle|1\rangle \pm |1\rangle|0\rangle\right)/\sqrt{2}$ form the complete Bell basis.

Bob performs $N/2$ Bell measurements, i. e., projects each of the $N/2$ ebits onto one of the four Bell states with probabilities

$$P[\phi^-] = \lambda_1, \quad P[\phi^+] = \lambda_2, \quad P[\psi^+] = \lambda_3, \quad P[\psi^-] = 1-\lambda_1-\lambda_2-\lambda_3. \tag{9.9}$$

Consequently, the estimated parameter values are directly given by $\lambda_k^{(est)} = n_k/(N/2)$ with the relative frequency $n_k/(N/2)$ with numbering $k = 1, 2, 3$ or 4 for the measurement result ϕ^-, ϕ^+, ψ^+ or ψ^-.

The average error

$$\begin{aligned}\bar{c}^{\,\mathrm{ent}}(N,\vec{\lambda}) &= \sum_{n_1+n_2+n_3+n_4=N/2} \frac{(N/2)!}{n_1!n_2!n_3!n_4!} \\ &\times \quad P[\phi^-]^{n_1} P[\phi^+]^{n_2} P[\psi^+]^{n_3} P[\psi^-]^{n_4} \sum_{k=1}^{3}\left(\lambda_k - \lambda_k^{(est)}\right)^2\end{aligned} \tag{9.10}$$

of the entangled estimation based on the statistical cost function, Eq. (9.2), now reads

$$\bar{c}^{\,\mathrm{ent}}(N,\vec{\lambda}) = \frac{1}{N/2}\left[\lambda_1(1-\lambda_1) + \lambda_2(1-\lambda_2) + \lambda_3(1-\lambda_3)\right]. \tag{9.11}$$

We can now compare the average errors of separable and entangled estimation schemes. As emphasized above, for a fair comparison we have to consider the same number N of available qubits for both schemes. We find that the difference

$$\begin{aligned}\Delta(N,\vec{\lambda}) &\equiv \bar{c}^{\,\text{sep}}(N,\vec{\lambda}) - \bar{c}^{\,\text{ent}}(N,\vec{\lambda}) \\ &= \frac{1}{2N}\left[5(1-\lambda_1-\lambda_2-\lambda_3)(\lambda_1+\lambda_2+\lambda_3)\right. \\ &\qquad \left. + \lambda_1\lambda_2 + \lambda_1\lambda_3 + \lambda_2\lambda_3\right] \geq 0 \end{aligned} \tag{9.12}$$

defined by Eqs. (9.6) and (9.10) is nonnegative for all possible parameter values $\vec{\lambda}$. This result clearly shows an enhancement of the estimation quality due to the nonclassical use of entangled qubit pairs.

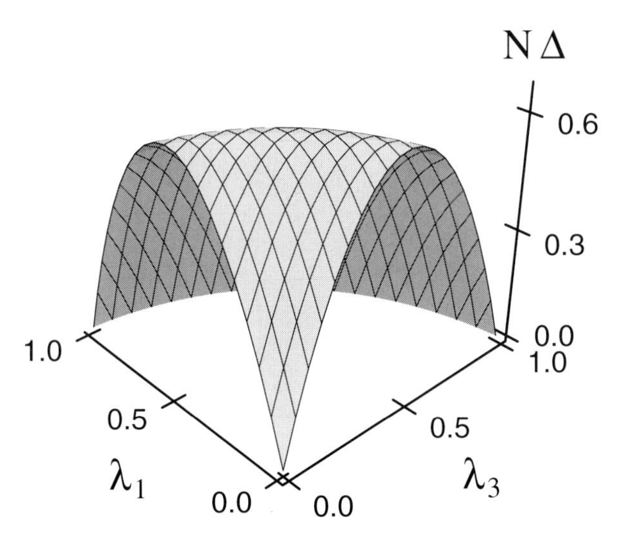

Figure 9.2: Gain due to the use of entangled qubit pairs as compared to separable ones. For the specific example of $\lambda_2 = 0$ we have plotted $N \cdot \Delta$, Eq. (9.12), versus the channel probabilities λ_1 and λ_3 in the allowed parameter range $0 \leq \lambda_1 + \lambda_3 \leq 1$. This quantity, which is independent of N, is always nonnegative. For $\lambda_2 = 0$ the maximum gain of $N \cdot \Delta^{max} = 25/38 \approx 0.66$ is reached for $\lambda_1 = \lambda_3 = 5/19$. In the general case ($\lambda_2 \geq 0$) the maximum value of $N\Delta^{max} = 75/112 \approx 0.67$ is found for $\lambda_1 = \lambda_2 = \lambda_3 = 5/28$.

The enhancement is illustrated in Fig. 9.2 for the case $\lambda_2 = 0$ which brings out clearly the essential features of $\Delta(N,\vec{\lambda})$, Eq. (9.12). It is always positive except for the extremal points $\vec{\lambda} = (0,0,0)^T$, $\vec{\lambda} = (1,0,0)^T$ and $\vec{\lambda} = (0,0,1)^T$, where Δ vanishes.

Instead of comparing the errors for the same number of available qubits one could also compare the estimation quality for the same number K of channel applications. For the entangled case we have $K = N/2$ and for the separable case we set $K = N$. Hence, the enhancement

$$\begin{aligned}\tilde{\Delta}(K,\vec{\lambda}) &\equiv \bar{c}^{\,\text{sep}}(N=K,\vec{\lambda}) - \bar{c}^{\,\text{ent}}(N/2=K,\vec{\lambda}) \\ &= \frac{1}{2K}\left[7(1-\lambda_1-\lambda_2-\lambda_3)(\lambda_1+\lambda_2+\lambda_3)\right. \\ &+ \left. 5\lambda_1\lambda_2+5\lambda_1\lambda_3+5\lambda_2\lambda_3\right] \geq 0 \end{aligned} \tag{9.13}$$

due to entanglement is even larger.

9.5 Generalized Estimation Schemes

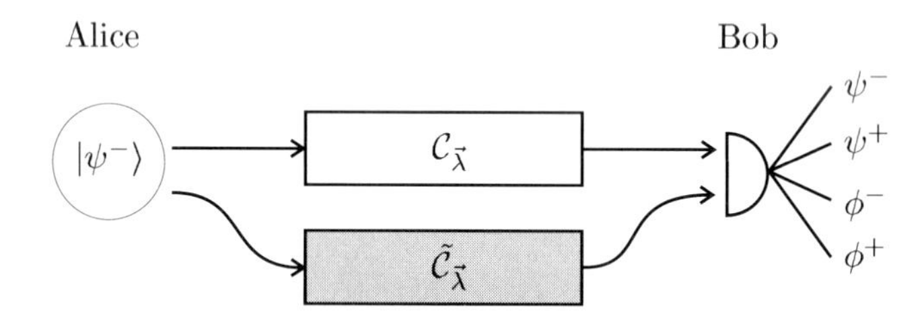

Figure 9.3: Generalized estimation scheme. We assume that the parameters of the channel $\tilde{\mathcal{C}}_{\vec{\lambda}}$ are known. In this approach $\tilde{\mathcal{C}}_{\vec{\lambda}}$ is used as a reference to estimate the unknown channel $\mathcal{C}_{\vec{\lambda}}$.

The nonclassical estimation scheme of the preceeding section did not take into account that in practice Alice and Bob might be located far from each other. Therefore, also the transfer of the second particle of the ebit must be accomplished with the help of a channel that introduces errors. This means that Alice and Bob are now connected to each other by two quantum channels as shown in Fig. 9.3. The quantum channel labeled $\mathcal{C}_{\vec{\lambda}}$ is the one whose parameter vector $\vec{\lambda}$ we want to estimate. The other quantum channel, a calibrated reference quantum channel $\tilde{\mathcal{C}}_{\vec{\lambda}} \equiv \mathcal{C}_{\tilde{\vec{\lambda}}}$, is used to distribute the second particle of the entangled state between Alice and Bob. We assume, that its parameter vector $\tilde{\vec{\lambda}}$ is known.

9.5.1 Estimation with Two Channels

Alice can prepare ebits in a perfect $|\psi^-\rangle$ state. After passing one particle through the reference channel $\tilde{\mathcal{C}}_{\vec{\lambda}}$, Alice and Bob share a Bell diagonal state, that is a mixed state which is represented by the density operator

$$\begin{aligned}\hat{\rho}' = \tilde{\mathcal{C}}_{\vec{\lambda}}(|\psi^-\rangle\langle\psi^-|) &= \tilde{\lambda}_1|\phi^-\rangle\langle\phi^-| + \tilde{\lambda}_2|\phi^+\rangle\langle\phi^+| + \tilde{\lambda}_3|\psi^+\rangle\langle\psi^+| \\ &+ \left(1-\tilde{\lambda}_1-\tilde{\lambda}_2-\tilde{\lambda}_3\right)|\psi^-\rangle\langle\psi^-|\,. \end{aligned} \tag{9.14}$$

This state $\hat{\rho}'$ has a fidelity $F \equiv \langle\psi^-|\hat{\rho}'|\psi^-\rangle = 1-\tilde{\lambda}_1-\tilde{\lambda}_2-\tilde{\lambda}_3 < 1$ and therefore is no longer maximally entangled. Hence, we now have to discuss the problem of channel estimation with a non–perfect Bell diagonal state.

In complete analogy to the method of the previous section, Bob performs a Bell measurement after Alice has also passed the second qubit through the unknown channel $C_{\vec{\lambda}}$. From the measured frequencies n_k after the measurement of $N/2$ qubits, one can then again calculate the estimated parameter vector $\vec{\lambda}^{(est)}$. We do not present the corresponding expression for $\vec{\lambda}^{(est)}$ here, because it is lengthy and does not provide much insight [12]. The corresponding average estimation error $\bar{c}^{\,\mathrm{ent}}(N, \vec{\lambda}, \vec{\tilde{\lambda}})$ is again defined according to Eq. (9.10) and can be calculated too. Instead of presenting these expressions explicitly we rather concentrate on a different question.

9.5.2 What is the Optimal Reference Channel?

Since we cannot avoid the influence of the reference channel $\tilde{C}_{\vec{\lambda}}$ in a realistic scheme, we have to ask, which one results in the smallest average estimation error? We expect, of course, $\tilde{C}_{\vec{\lambda}} = \hat{1}$ because we then have the previously discussed case where we can use maximally entangled states for our estimation. But, in general, Alice and Bob cannot establish an ideal quantum channel for entanglement distribution. Therefore we ask, which parameters $\vec{\tilde{\lambda}} \neq (0,0,0)^{\mathrm{T}}$ should the reference channel $\tilde{C}_{\vec{\lambda}}$ have in order to enable estimation with the smallest average estimation error?

To answer the question, we look at the average estimation error for an unknown quantum channel $C_{\vec{\lambda}}$, that is we consider the average estimation error

$$\langle\, \bar{c}^{\,\mathrm{ent}}(N, \vec{\lambda}, \vec{\tilde{\lambda}})\,\rangle_{\vec{\lambda}} = \iiint\limits_{0 \le \lambda_1+\lambda_2+\lambda_3 \le 1} \bar{c}^{\,\mathrm{ent}}(N, \vec{\lambda}, \vec{\tilde{\lambda}})\, \mathrm{d}\lambda_1\, \mathrm{d}\lambda_2\, \mathrm{d}\lambda_3 \tag{9.15}$$

averaged over all possible parameter vectors $\vec{\lambda}$. Reference [12] shows that this quantity has a global minimum for

$$\tilde{\lambda}_1 = \tilde{\lambda}_2 = \tilde{\lambda}_3 \equiv \tilde{\lambda}\,. \tag{9.16}$$

Inserting this set of parameters into Eq. (9.14) we get

$$\hat{\rho}' = \left(1 - 3\tilde{\lambda}\right) |\psi^-\rangle\langle\psi^-| + \tilde{\lambda}\left(|\phi^-\rangle\langle\phi^-| + |\phi^+\rangle\langle\phi^+| + |\psi^+\rangle\langle\psi^+|\right). \tag{9.17}$$

States of this form are called Werner states [13] with fidelity $F \equiv \langle\psi^-|\hat{\rho}'|\psi^-\rangle = 1 - 3\tilde{\lambda}$. We therefore obtain the result that the reference channel $\tilde{C}_{\vec{\lambda}}$ should produce a Werner state in order to get the best estimation results for an arbitrary Pauli channel $C_{\vec{\lambda}}$.

When we rewrite Eq. (9.17) in the form

$$\hat{\rho}' = \left(1 - 4\tilde{\lambda}\right) |\psi^-\rangle\langle\psi^-| + 4\tilde{\lambda} \cdot \frac{1}{4}\hat{1} \tag{9.18}$$

we see that it is in fact a depolarizing reference channel $\tilde{C}_{\vec{\lambda}}$ that leads—on average—to our desired estimation with a small average estimation error.

9.5.3 Estimation with Werner States

Let us briefly recall important properties of a Werner state, Eq. (9.17) with fidelity $F = 1-3\tilde{\lambda}$. In the case $F = \frac{1}{4}$, one obtains a totally mixed state $\hat{\rho}' = \frac{1}{4}\hat{1}$, which certainly does not yield any information about the channel $\mathcal{C}_{\vec{\lambda}}$. For $F > \frac{1}{2}$ the Werner state has a nonzero negativity of its partial transpose [14] and is therefore nonseparable that is, for $F > \frac{1}{2}$ there is a chance that a Werner state improves the parameter estimation compared to the separable case. For $F > \frac{1}{8}(3\sqrt{2}+2) \approx 0.78$ the Werner state violates the Bell–CHSH inequality [15, 16] and for $F = 1$ we obtain the maximally entangled Bell state $|\psi^-\rangle$, which we already used in Sec. 9.4.2. In the following we therefore restrict ourselves to the domain $\frac{1}{2} < F \leq 1$ where $\hat{\rho}'$ is nonseparable and may lead to a nonclassical estimation enhancement.

Alice and Bob share $N/2$ qubits in a Werner state, Eq. (9.17). The relation between our estimated channel parameter $\vec{\lambda}^{(est)}$ and the measured frequencies n_k of Bob's Bell measurement now reads

$$\lambda_k^{(est)} = \frac{3\frac{n_k}{N} + F - 1}{4F-1}\,, \tag{9.19}$$

where again n_1 denotes the number of ϕ^- results, n_2 the number of ϕ^+ results and n_3 the number of ψ^+ results.

The average estimation error, Eq. (9.10), can still be evaluated analytically and we arrive at

$$\begin{aligned}\bar{c}^{\,\mathrm{ent}}(N,\vec{\lambda},F) &= \frac{1}{N/2}\left(\frac{3}{4F-1}\right)^2 \\ &\times \sum_{i=1}^{3}\Big\{\lambda_i(1-\lambda_i) + \tfrac{4}{3}(1-F)(\lambda_i - \tfrac{1}{4})(2\lambda_i - 1) \\ &\qquad -\tfrac{16}{9}(1-F)^2(\lambda_i - \tfrac{1}{4})^2\Big\}\,. \end{aligned} \tag{9.20}$$

Instead of comparing the average error $\bar{c}^{\,\mathrm{sep}}$, Eq. (9.6), of the separable scheme to the error $\bar{c}^{\,\mathrm{ent}}$, Eq. (9.20), of the non–perfect entangled estimation scheme, as we did in Sec. 9.4.2, we now focus on the resources needed to obtain a desired estimation error. We require

$$\bar{c}^{\,\mathrm{sep}}(N^{\mathrm{sep}},\vec{\lambda}) \leq \gamma \tag{9.21}$$

and

$$\bar{c}^{\,\mathrm{ent}}(N^{\mathrm{ent}},\vec{\lambda},F) \leq \gamma \tag{9.22}$$

for a given maximum estimation error γ and ask for the minimum number $N^{\mathrm{sep}}(\vec{\lambda},\gamma)$ and $N^{\mathrm{ent}}(\vec{\lambda},\gamma,F)$ of resources that we need to fulfill these inequalities. If the difference $\Delta N(\vec{\lambda},\gamma,F) \equiv N^{\mathrm{sep}}(\vec{\lambda},\gamma) - N^{\mathrm{ent}}(\vec{\lambda},\gamma,F)$ is positive, the entangled estimation scheme with a given fidelity F requires less resources than the separable one. We therefore have the case of entanglement serving as a nonclassical resource.

For $F = 1$, that is the ideal case with an ideal reference channel $\tilde{\mathcal{C}}_{\vec{\lambda}}$, we always get $\Delta N(\vec{\lambda}, F = 1, \gamma) \geq 0$, which is in fact no surprise since according to Eq. (9.12) we have always $\Delta(N,\vec{\lambda}) \geq 0$.

In Fig. 9.4 we discuss the difference $\Delta N(\vec{\lambda}, F, \gamma = 0.01)$ in needed resources for two different channels. Whereas in (a) we show ΔN for a depolarizing channel $\mathcal{C}_{\vec{\lambda}}$ with $\vec{\lambda} = (\lambda, \lambda, \lambda)^{\mathrm{T}}$, (b) displays this difference for the same parameter for a Pauli channel $\mathcal{C}_{\vec{\lambda}}$ with $\vec{\lambda} = (\lambda, 0.1, 0.2)^{\mathrm{T}}$. The range in which entanglement serves as a nonclassical resource increases with F. Moreover, for $F = 1$ we always have a nonclassical enhancement.

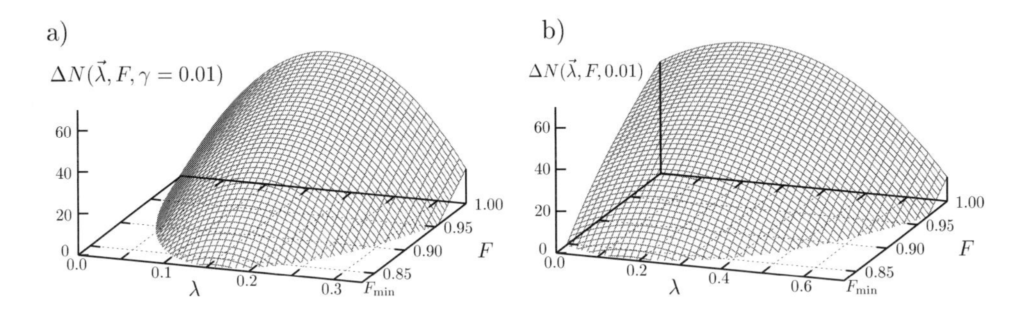

Figure 9.4: Difference $\Delta N(\vec{\lambda}, F, \gamma)$ in needed resources to obtain a given estimation accuracy $\gamma = 0.01$ as function of $\vec{\lambda}$ and F. In (a) we show the case of a depolarizing channel, that is $\vec{\lambda} = (\lambda, \lambda, \lambda)^{\mathrm{T}}$. In (b) we have plotted the difference in resources for a Pauli channel with $\vec{\lambda} = (\lambda, 0.1, 0.2)^{\mathrm{T}}$. There is a wide range where we obtain $\Delta N(\vec{\lambda}, F, \gamma = 0.01) > 0$. In this range we need less resources (qubits) if we estimate our channel with entangled states as compared to an estimation with separable states. However, we also find a range for which $\Delta N(\vec{\lambda}, F, \gamma = 0.01) < 0$ (points not plotted). In this case estimation with separable states is the better choice.

9.6 Outlook

Quantum estimation problems arise naturally when we encode and decode information using a finite resource of quantum objects. We emphasize that it is the structure of quantum mechanics, and in particular, the measurement process that leads inevitably to these questions. This situation is in contrast to classical estimation theory which is basically motivated by insufficient measurement accuracies. From this point of view it is most interesting to see that at the same time quantum theory itself provides the nonclassical means like entanglement to enhance the quality of an estimation procedure.

Although we have discussed estimation for the specific example of a Pauli channel, the principle can certainly be applied to characterize other quantum devices. Moreover, further estimation techniques, for example maximum likelihood schemes, have to be analyzed too. And finally all these studies will certainly lead to a deeper understanding of quantum mechanics when applied to finite ensembles.

References

[1] C.W. Helstrom, *Quantum Detection and Estimation Theory* (Academic Press, New York, 1976).

[2] A.S. Holevo, *Probabilistic and Statistical Aspects of Quantum Theory* (North–Holland, Amsterdam, 1982).

[3] See the reviews: D. Leibfried, T. Pfau, and D. Monroe, Phys. Today **51**, 22 (1998); M. Freyberger, P. Bardroff, C. Leichtle, G. Schrade, and W. Schleich, Phys. World **10** (11), 41 (1997).

[4] A. Peres, and W.K. Wootters, Phys. Rev. Lett. **66**, 1119 (1991); S. Massar, and S. Popescu, Phys. Rev. Lett. **74**, 1259 (1995); R. Derka, V. Bužek, and A.K. Ekert, Phys. Rev. Lett. **80**, 1571 (1998); J.I. Latorre, P. Pascual, and R. Tarrach, Phys. Rev. Lett. **81**, 1351 (1998); D.G. Fischer, S.H. Kienle, and M. Freyberger, Phys. Rev. A **61**, 032306 (2000); Z. Hradil, J. Summhammer, G. Badurek, and H. Rauch, Phys. Rev. A **62**, 014101 (2000); K. Banaszek, and I. Devetak, Phys. Rev. A **64**, 052307 (2001); E. Bagan, M. Baig, A. Brey, and R. Muñoz-Tapia, Phys. Rev. A **63**, 052309 (2001).

[5] K.R.W. Jones, Phys. Rev. A **50**, 3682 (1994).

[6] M. Paris and J. Řeháček (Eds.), *Quantum State Estimation*, Lecture Notes in Physics **649** (Springer, Berlin, 2004).

[7] D. Bouwmeester, A. Ekert, and A. Zeilinger (Eds.), *The Physics of Quantum Information: Quantum Cryptography, Quantum Teleportation, Quantum Information* (Springer, Berlin, 2000); M.A. Nielsen, and I.L. Chuang, *Quantum Computation and Quantum Information* (Cambridge University Press, Cambridge, 2000); G. Alber, T. Beth, M. Horodecki, P. Horodecki, R. Horodecki, M. Rötteler, H. Weinfurter, R. Werner, and A. Zeilinger, *Quantum Information: An Introduction to Basic Theoretical Concepts and Experiments*, Springer Tracts in Modern Physics **173** (Springer, Berlin, 2001).

[8] K. Kraus, Annals of Phys. **64**, 311 (1971); K. Kraus, *States, Effects and Operations*, Lecture Notes in Physics **190** (Springer, Berlin, 1983).

[9] A. Fujiwara, and P. Algoet, Phys. Rev. A **59**, 3290 (1999).

[10] D.G. Fischer, H. Mack, M.A. Cirone, and M. Freyberger, Phys. Rev. A **64**, 022309 (2001); T.C. Bschorr, D.G. Fischer, and M. Freyberger, Phys. Lett. A **292**, 15 (2001).

[11] P.W. Shor, Phys. Rev. A **52**, 2493 (1995); A.M. Steane, Phys. Rev. Lett. **77**, 793 (1996); R. Laflamme, C. Miquel, J.P. Paz, and W.H. Zurek, Phys. Rev. Lett. **77**, 198 (1996); A.R. Calderbank, and P.W. Shor, Phys. Rev. A **54**, 1098 (1996); A.M. Steane, Proc. Roy. Soc. London A **452**, 2551 (1996); D. Gottesman, Phys. Rev. A **54**, 1862 (1996); A.R. Calderbank, E.M. Rains, P.W. Shor, and N.J.A. Sloane, Phys. Rev. Lett. **78**, 405 (1997); E. Knill, and R. Laflamme, Phys. Rev. A **55**, 900 (1997); A. Steane, Rep. Prog. Phys. **61**, 117 (1998).

[12] T.C. Bschorr, *Reinigung von verschränkten Quantenbits mit endlichen Ressourcen und die Anwendung beim Schätzen von Quantenkanälen* (Diploma Thesis, University of Ulm, 2002).

[13] R.F. Werner, Phys. Rev. A **40**, 4277 (1989).

[14] G. Vidal, and R.F. Werner, Phys. Rev. A **65**, 032314 (2002).

[15] J.F. Clauser, M.A. Horne, A. Shimony, and R.A. Holt, Phys. Rev. Lett. **23**, 880 (1969).

[16] A. Peres, Phys. Rev. A **54**, 2685 (1996).

10 Size Scaling of Decoherence Rates

Christopher S. Maierle and Dieter Suter

Universität Dortmund
Fachbereich Physik – Experimentelle Physik III
Dortmund, Germany

10.1 Introduction

As quantum computers grow to include more and more qubits, it becomes important to understand how decoherence rates scale with this increase of system size. In general, it is believed that decoherence rates increase linearly with number of qubits [1,2]. Success probabilities of the form $e^{-\gamma t}$ where γ is a decoherence rate therefore scale exponentially with size. If, however, the interactions which give rise to decoherence display permutation symmetry so that groups of two or more qubits are coupled identically to the bath modes, it is then possible to construct decoherence free subspaces which are immune to the effects of decoherence [3–5]. Any realistic system will of course contain terms which break the permutation symmetry. Nevertheless, if the symmetry breaking terms are small, the formerly decoherence free subspaces will still be sub-decoherent in the sense that states in these subspaces will tend to decohere more slowly than states chosen at random from the full Hilbert space [2, 3].

In this work we consider how decoherence rates scale with system size when the qubits are acted upon by bath modes which are correlated, but not perfectly correlated. We will mainly consider relaxation in the Markovian regime where the resulting exponential relaxation allows for a simple and unambiguous characterization of relaxation rates. For the sake of simplicity, we will consider relaxation which can be modeled by fluctuating random fields. Previous work, however, has mainly focused on decoherence arising from qubits linearly coupled to a bosonic bath [1,2,4,5] and so by way of connection we show in the first section how a bosonic bath can function as a random field and then briefly review the special cases of independent and collective decoherence. We conclude our introductory material by arguing that one can use an average decoherence rate as a measure of decoherence.

The main results of this work follow from analyzing the size scaling of average decoherence rates for a number of encodings, all of which are decoherence free in the limit where each physical qubit sees the same fluctuating field. We show that imperfectly correlated fields give rise to the expected linear scaling of decoherence rates with system size but that the scaling rate varies depending on the encoding and the correlations. In particular, we show how the strength of the correlations determines which encodings give rise to the most favorable decoherence rate scaling.

Quantum Information Processing, 2nd Edition. Edited by Thomas Beth and Gerd Leuchs

ISBN: 3-527-40541-0

10.2 Decoherence Models

We will consider relaxation which can be understood as arising from a fluctuating random field. In this model, the Hamiltonian is,

$$H = \sum_i \frac{\omega_i}{2}\sigma_z^{(i)} + \sum_i \gamma_i \mathbf{B}^{(i)}(t) \cdot \sigma^{(i)} \tag{10.1}$$

where $\mathbf{B}^{(i)}(t)$ is the fluctuating field seen by the ith qubit and the σ are Pauli spin operators. In order that the above Hamiltonian give rise to relaxation we must consider an ensemble of quantum computers, each of which is subject to a different set of fields. We then define an average density matrix, $\rho_{\text{avg}} = \overline{U(t,0)\rho(0)U^\dagger(t,0)}$ where $U(t,0)$ is the time evolution operator generated by the Hamiltonian in Eq. (10.1) and the bar indicates an average over the ensemble of random fields. Physically, the ensemble averaging arrises either because we perform many repeated experiments and average the results, or because each measurement corresponds to an average over many individual quantum computers, as is the case in ensemble NMR quantum computing. Regardless of the underlying physical nature, there exists, in general, no closed form solution for ρ_{avg}. However, in the Markovian limit, where the fields fluctuate on a timescale fast compared to the dynamics which they generate, one can write the approximate equation of motion [6],

$$\dot{\rho}_{\text{avg}} = -i\left[\sum_i \frac{\omega_i}{2}\sigma_z^{(i)}, \rho_{\text{avg}}\right] - \Gamma\rho_{\text{avg}} \tag{10.2}$$

where the relaxation superoperator, Γ, is given by

$$\Gamma\rho_{\text{avg}} = \sum_{i,j}\sum_{q,q'} \frac{1}{2} J_{qq'}^{ij}(q\omega_i)\left[\sigma_{q'}^{(j)}, [\sigma_q^{(i)}, \rho_{\text{avg}}]\right] \tag{10.3}$$

and the spectral densities $J_{qq'}^{ij}(\omega)$ are

$$J_{qq'}^{ij}(\omega) = \gamma_i\gamma_j \int_{-\infty}^{\infty} \overline{B_q^{(i)}(t)B_{q'}^{(j)}(t+\tau)} e^{i\omega\tau} d\tau \tag{10.4}$$

where the indices i and j label qubits and q and q' take on the values 0 and ± 1 corresponding to σ_z and $\frac{1}{\sqrt{2}}\sigma_\pm$ respectively.

If the random fields do not fluctuate fast enough to induce transitions between the Zeeman levels of $\sum_i \frac{\omega_i}{2}\sigma_z^{(i)}$, then the dominant source of relaxation is the $[\sigma_z^{(i)}, [\sigma_z^{(j)}, \rho_{\text{avg}}]]$ term in Eq. (10.3). In this case, often known as pure dephasing, relaxation can be understood as arising from fluctuations in energy levels caused by the z component of the random field.

As mentioned in the introduction, previous investigations of decoherence in quantum computers have often utilized the spin boson model as a starting point, rather than the semi-classical fluctuating random field model outlined above. When using the spin boson model, one starts with the Hamiltonian, $H_{SB} = H_S + H_B + H_I$ where $H_S = \sum_i \frac{\omega_i}{2}\sigma_z^{(i)}$ is the

Hamiltonian for the qubits, $H_B = \sum_k \omega_k^B (b_k^\dagger b_k + \frac{1}{2})$ is the Hamiltonian for the bath modes, and the interaction between the bath and the qubits is given by,

$$H_I = \sum_k g_k \sigma_+ b_k + f_k \sigma_+ b_k^\dagger + h_k \sigma_z b_k + \text{h.c.} \tag{10.5}$$

One then calculates a reduced density matrix $\rho_{red} = \text{Tr}_B[\rho]$ by tracing over the bath degrees of freedom in the full density matrix. As with the semi-classical random field model, there exists, in general, no closed form solution for the reduced density matrix. One can, again, consider the Markovian limit where the bath fluctuates on a timescale fast compared to the dynamics that it generates. If one assumes, as is usually done, an initial state for the system and bath of the form $\rho_S(0) \otimes \rho_B(0)$ so that the system and bath are not initially entangled, then Eq. (10.2) also holds for the reduced density matrix, ρ_{red} when the classical correlation functions $\overline{B_q^{(i)}(t) B_{q'}^{(j)}(t+\tau)}$ are replaced by their quantum mechanical analogs $\text{Tr}_B[\hat{B}_q(t)\hat{B}_{q'}(t+\tau)\rho_B(0)]$ where the operators $\hat{B}_q(t)$ are constructed from the bath operators in Eq. (10.5) written in the interaction representation.

There are, of course, important differences between the fully quantum mechanical spin boson model and the semi-classical random field model. Unless ad-hoc modifications are made, semi-classical relaxation models generally lead to infinite temperature equilibrium states where all energy levels are equally populated whereas fully quantum mechanical models can yield the correct thermodynamic equilibrium states [6]. Additionally, fully quantum mechanical models account for entanglement between the system and bath and can therefore lead to decoherence even at zero temperature where the ensemble of bath states contains only one member [2]. Nevertheless, assuming that the system and bath are initially unentangled, the equations of motion for ρ_{avg} and ρ_{red} depend identically on correlation functions of the fields and field operators. We will not prove this statement here, but it holds generally regardless of whether the dynamics are Markovian or not.

10.3 Collective and Independent Decoherence

We now briefly review the special cases of independent and collective decoherence. In the Markovian limit, relaxation rates have a very simple dependence on the spectral densities $J_{qq'}^{ij}(\omega)$, which are Fourier transforms of correlation functions for the random fields seen by the ith and jth qubits. If there exists no correlation between the random fields seen by different qubits, then the spectral densities $J_{qq'}^{ij}(\omega)$ vanish unless $i = j$ and we can rewrite the relaxation superoperator Γ as a sum of superoperators each of which determines the relaxation of a single physical qubit. This case is known as independent decoherence because each qubit relaxes independently of the state of the others [1,2].

If, on the other hand, each qubit is affected identically by the fluctuating field, then the Hamiltonian in Eq. (10.1) can be rewritten in the form,

$$H = \sum_i \omega_i \sigma_z^{(i)} + \gamma B(t) \cdot \sigma^{\text{coll}} \tag{10.6}$$

where σ^{coll} is a collective operator with components $\sigma_q^{\text{coll}} = \sum_i \sigma_q^{(i)}$. In this case, known as collective decoherence [4,5], one can construct a decoherence free subspace out of the states which are annihilated by the collective operators σ_q^{coll}. These subspaces are decoherence free [3] because, within the subspace all matrix elements of the interaction term $\gamma B(t) \cdot \sigma^{\text{coll}}$ are zero so that the fluctuating field effectively does not exist so long as we stay within the subspace.

Since it is hard to imagine why a large number of qubits would all be subject to the exact same bath fluctuations, previous workers have proposed encoding logical qubits in small clusters of physical qubits [7, 8]. The idea is that it may be reasonable to expect a small group of qubits to see nearly the same random field fluctuations if the qubits are located physically close to one another. In other words, when it is not possible to write the Hamiltonian, even approximately, in terms of collective operators involving all the physical qubits, one hopes for a Hamiltonian which can be written in the form,

$$H = \sum_i \omega_i \sigma_z^{(i)} + \sum_c \gamma B^c(t) \cdot \sigma^c \tag{10.7}$$

where the σ^c are collective operators for small clusters of spins and have the components $\sigma_q^c = \sum_{i \in c} \sigma_q^{(i)}$ with the sum running over physical qubits in the cluster labeled by c. Again, the states which are annihilated by the cluster operators σ_q^c can be used to form a decoherence free subspace. Our primary goal is to consider decoherence which arrises inside of these subspaces due to the presence of imperfectly correlated fields. We will find it useful to have a simple measure of decoherence and so to this end, we argue in the next section that the average decoherence rate represents a good measure of 'overall' decoherence.

10.4 Average Decoherence Rate as a Measure of Decoherence

A system of n qubits inhabits a Liouville space of dimension 4^n and, rather than focusing on the decay of individual matrix elements, we would like to consider the overall loss of quantum information due to decoherence. In order to quantify this we can consider the overlap $F(t)$ between the actual state of the quantum computer at time t and the state that would result in the absence of decoherence. Taking the operations of the ideal quantum computer to be contained in the Liouville operator $\mathcal{L}(t)$ and the decoherence to be represented by the superoperator Γ, we write,

$$F(t) = \langle\langle \rho_0 | \mathcal{U}_{\mathcal{L}}^{\dagger} \mathcal{U}_{\mathcal{L}+\Gamma} | \rho_0 \rangle\rangle = \langle\langle \rho_0 | \mathcal{U}_{\Gamma(t)} | \rho_0 \rangle\rangle \tag{10.8}$$

where $\mathcal{U}_{\mathcal{L}}$, $\mathcal{U}_{\mathcal{L}+\Gamma}$, and $\mathcal{U}_{\Gamma(t)}$ are the Liouville space time evolution operators generated by $\mathcal{L}(t)$, $\mathcal{L}(t) + \Gamma$, and $\mathcal{U}_{\mathcal{L}} \Gamma \mathcal{U}_{\mathcal{L}}^{-1}$ respectively and ρ_0 is the initial state of the quantum computer. The double brackets in the above expression indicate a scalar product in Liouville space which corresponds to the quantity $\text{Tr}[\rho_0 U_{\Gamma(t)} \rho_0 U_{\Gamma(t)}^{\dagger}]$ in the usual Hilbert space notation [9]. From a physical point of view, one expects that as the quantum computer functions, it will transfer states through various Liouville space pathways. If this transfer is ergodic in the sense that each pathway tends to equally sample the various possible relaxation rates, then we expect that $F(t)$ will decay with a rate which is an average of the decoherence rates of the system.

We can present a somewhat firmer derivation of this concept by using the projection operator technique [10] to write the following equation of motion for $F(t)$:

$$\dot{F}(t) = -\langle\langle\rho_0|\Gamma(t)|\rho_0\rangle\rangle F(t) - \int_0^t K(t,\tau)F(\tau)d\tau. \tag{10.9}$$

The kernel $K(t,\tau)$ is given by

$$K(t,\tau) = \langle\langle\rho_0|\Gamma(t)\mathcal{U}_{(1-P)\mathcal{U}_{\mathcal{L}}\Gamma\mathcal{U}_{\mathcal{L}}^{-1}}(1-P)\Gamma(\tau)|\rho_0\rangle\rangle, \tag{10.10}$$

where P is the projection operator $|\rho_0\rangle\rangle\langle\langle\rho_0|$ and we have assumed the normalization condition $\langle\langle\rho_0|\rho_0\rangle\rangle = 1$. If we assume that the computer functions on a timescale τ_{QC} much faster than the timescale for decoherence ($\tau_{QC}|\Gamma| \ll 1$), as it must if the quantum computer is able to successfully calculate, then we expect $K(t,\tau)$ to go to zero for times $t-\tau \gg \tau_{QC}$ so that the integral $\int_0^t K(t,\tau)F(\tau)d\tau$ is of the order $|\Gamma(t)|^2\tau_{QC}F(t)$. The second term is therefore small compared to the first (which is of order $|\Gamma(t)|F(t)$) and so we drop it. Solving for $F(t)$ then yields

$$F(t) = e^{-\int_0^t\langle\langle\rho_0|\Gamma(\tau)|\rho_0\rangle\rangle d\tau}F(0). \tag{10.11}$$

The eigenvalues of Γ give the decoherence rates of the system and since $\Gamma(t) = \mathcal{U}_{\mathcal{L}}\Gamma\mathcal{U}_{\mathcal{L}}^{-1}$ is a unitary transformation of Γ, it must have the same eigenvalues. Defining $|\psi_n(t)\rangle\rangle$ as the normalized Liouville space eigenvectors of $\Gamma(t)$ with eigenvalues γ_n, we can write

$$\int_0^t \langle\langle\rho_0|\Gamma(\tau)|\rho_0\rangle\rangle d\tau = \int_0^t \sum_n |c_n(\tau)|^2\gamma_n d\tau, \tag{10.12}$$

where the $c_n(\tau)$ are given by $\langle\langle\rho_0|\psi_n(\tau)\rangle\rangle$. The $|c_n(\tau)|^2$ correspond to 'how much' of the quantum computer's state is in the nth Liouville space decay mode.

We now make two crucial assumptions: First, we assume that the quantum computer gates represented by $\mathcal{L}(t)$ do not take amplitude out of whatever subspace we are using and that our initial state $|\rho_0\rangle\rangle$ is located entirely within our subspace. This then implies the normalization condition $\sum_{n\in S}|c_n(\tau)|^2 = 1$ where the sum is over states $|\psi_n(\tau)\rangle\rangle$ which lie in our subspace. Secondly, we assume that the evolution of our quantum computer is sufficiently complex so that the time average $\frac{1}{t}\int_0^t|c_n(\tau)|^2d\tau$ can be replaced by $\frac{1}{N_S}$ where N_S is the dimension of our subspace. This then leads to $F(t) = e^{-\overline{\gamma}t}$ where $\overline{\gamma} = \frac{1}{N_S}\sum_{n\in S}\gamma_n$ is the average decoherence rate in our subspace. Physically, the replacement of the time average over $|c_n(\tau)|^2$ by $\frac{1}{N_S}$ means that the operations of the quantum computer transfer amplitude ergodically through Liouville space such that each of the decay modes of Γ are equally sampled.

The above argument has hopefully demonstrated the plausibility of using an average decoherence rate as a measure of overall decoherence time. In this work we will not consider the issue further and now turn to the calculation of average decoherence rates in the presence of imperfectly correlated fields.

10.5 Decoherence Rate Scaling due to Partially Correlated Fields

In what follows, it will be useful to make a number of simplifying assumptions about the spectral densities which appear in the relaxation superoperator Γ given in Eq.(10.3). First, we will assume that the qubits all have identical gyromagnetic ratios and that the root mean square field seen by each qubit is the same. This assumption is in no way essential and will merely serve to simplify our results by allowing us to express the spectral densities as $J^{ij}_{qq'}(\omega) = C_{ij}J_{qq'}(\omega)$ where the correlation coefficient C_{ij} is 1 for $i = j$. For $i \neq j$, C_{ij} is some number between 0 and 1 which represents the degree of correlation between the fields seen by the ith and jth qubits. Negative correlation coefficients corresponding to anti-correlated fields are of course also possible but we will not consider this case. Finally, we assume that there exists no correlation between the z component of the random field and the x and y components. With these assumptions we can now write the relaxation superoperator in the simplified form,

$$\begin{aligned}\Gamma\rho_{\text{avg}} &= \frac{1}{2}J_{zz}(0)\sum_{i,j}C_{ij}\left[\sigma_z^{(i)},[\sigma_z^{(j)},\rho_{\text{avg}}]\right] \\ &\quad +\frac{1}{2}J_{+-}(\omega)\sum_{i,j}C_{ij}\left(\left[\sigma_+^{(i)},[\sigma_-^{(j)},\rho_{\text{avg}}]\right]+\left[\sigma_-^{(i)},[\sigma_+^{(j)},\rho_{\text{avg}}]\right]\right) \qquad (10.13)\end{aligned}$$

where ω is the Larmor frequency of the qubits which are all equal now due to the assumption of identical gyromagnetic ratios.

In order to evaluate the average decoherence rate in a given subspace, we simply need to trace Γ over the states in the subspace. If our subspace has d dimensions in Hilbert space, then we can write the average decoherence rate $\overline{\gamma}$ as

$$\overline{\gamma} = \frac{1}{d^2}\sum_{n,m}\text{Tr}[|\psi_n\rangle\langle\psi_m|\Gamma|\psi_m\rangle\langle\psi_n|], \qquad (10.14)$$

where the $|\psi_n\rangle$ are any set of orthonormal basis states for our subspace. We will consider the average decoherence rate in subspaces where all matrix elements of the cluster operators given in Eq. (10.7) vanish. As discussed before, these subspaces are decoherence free in the limit where all qubits in each cluster see exactly the same field fluctuations.

We now derive two simple identities which will be useful in evaluating Eq. (10.14). First, we show that

$$\sum_n\langle\psi_n|\sigma_z^{(i)}\sigma_z^{(j)}|\psi_n\rangle = \frac{-d_c}{n_c-1}, \qquad (10.15)$$

where $i \neq j$ label two qubits in a cluster of n_c physical qubits and the sum runs over the d_c states $|\psi_n\rangle$ which are annihilated by the collective operator $\sigma_z^c = \sum_{i\in c}\sigma_z^{(i)}$ for the cluster. In this case, d_c is given by $\frac{n_c!}{(n_c/2)!^2}$ but for reasons of generality we will continue to write d_c. We begin by noting that σ_z^c commutes with all permutation operators which exchange two qubits in the cluster. Thus $\sum_n \langle\psi_n|\sigma_z^{(i)}\sigma_z^{(j)}|\psi_n\rangle$ is independent of the indices i and j for $i \neq j$ and we can write

$$\begin{aligned}\sum_n \langle\psi_n|\sigma_z^{(i)}\sigma_z^{(j)}|\psi_n\rangle &= \frac{1}{n_c(n_c-1)}\sum_{i\neq j}\sum_n \langle\psi_n|\sigma_z^{(i)}\sigma_z^{(j)}|\psi_n\rangle \\ &= \frac{1}{n_c(n_c-1)}\left(\sum_n \langle\psi_n|\sigma_z^c\sigma_z^c|\psi_n\rangle\right. \\ &\quad \left. -\sum_i\sum_n \langle\psi_n|(\sigma_z^{(i)})^2|\psi_n\rangle\right), \end{aligned} \tag{10.16}$$

from which Eq. (10.15) follows by first noting that $\sum_n \langle\psi_n|\sigma_z^c\sigma_z^c|\psi_n\rangle$ is equal to zero and $\sum_i\sum_n \langle\psi_n|(\sigma_z^{(i)})^2|\psi_n\rangle = n_c d_c$. We also have the identity

$$\sum_n \langle\psi_n|\sigma_z^{(i)}|\psi_n\rangle = 0 \tag{10.17}$$

which can be shown by an argument similar to that given above.

We now return to the evaluation of the average decoherence rate $\overline{\gamma}$ given in Eq. (10.14). For simplicity, we will first consider the case of pure dephasing for a cluster of n_c qubits where Γ is given simply by

$$\Gamma\rho_{\text{avg}} = \frac{1}{2}J_{zz}(0)\sum_{i,j}C_{ij}\left[\sigma_z^{(i)},[\sigma_z^{(j)},\rho_{\text{avg}}]\right]. \tag{10.18}$$

If the qubits all saw exactly the same field fluctuations then the subspace of states annihilated by the collective cluster operator $\sigma_z^c = \sum_{i\in c}\sigma_z^{(i)}$ would be decoherence free. We will refer to this subspace as the $m_z = 0$ subspace. Armed with the identities given in Eqs. (10.15) and (10.17) it is a simple matter to show that in the $m_z = 0$ subspace, Eq. (10.14) yields an average decoherence rate of

$$\overline{\gamma} = n_c J_{zz}(0)(1 - C_{\text{avg}}) \tag{10.19}$$

where C_{avg} is the average cross correlation coefficient given by

$$C_{\text{avg}} = \frac{1}{n_c(n_c-1)}\sum_{i,j}C_{ij}. \tag{10.20}$$

Since a Hilbert space of dimension d_c can be used to encode $n_{LQ} = \text{Log}_2 d_c$ logical qubits, we can write the average decoherence rate per logical qubit as

$$\overline{\gamma}/n_{LQ} = \frac{n_c}{\text{Log}_2 d_c}J_{zz}(0)(1 - C_{\text{avg}}). \tag{10.21}$$

A scalable quantum computer would then be constructed by concatenating many of these clusterized subspaces together to create basis states of the form $|\psi_n\rangle \otimes |\psi'_{n'}\rangle \otimes \cdots$ where the primes indicate that $|\psi_n\rangle$ and $|\psi'_{n'}\rangle$ are the basis states for separate clusters of qubits. Using

Eqs. (10.17) and (10.14) one can show that the average decoherence rate is unaffected by correlations in the fields seen by different clusters. Thus, Γ can be written as a sum of terms, each of which acts on a single cluster of qubits and the average decoherence rate therefore scales linearly with the number of clusters. Furthermore, the number of logical qubits which we can encode also scales linearly with the number of clusters and we therefore have that Eq. (10.21) is valid not only for a single cluster of physical qubits but also for an arbitrary number of such clusters used together to create a quantum computer. Simply put, we have shown that for a clusterized encoding where logical qubits are encoded in the states annihilated by the collective cluster operators $\sigma_z^c = \sum_{i \in c} \sigma_z^{(i)}$, decoherence rates scale linearly with the number of logical qubits with a proportionality constant given in Eq. (10.21).

Examining Eq. (10.21) which we now understand to represent decoherence rate scaling, we see that it contains the factor $\frac{n_c}{\mathrm{Log}_2 d_c}$ which is nothing more than one over the efficiency of our encoding. Now on the one hand, larger clusters lead to higher encoding efficiencies [5] and thereby to an *decrease* of the factor $\frac{n_c}{\mathrm{Log}_2 d_c}$ but in order to determine the effect of cluster size on decoherence rate scaling, we must also consider the factor $(1 - C_{\mathrm{avg}})$ which we generally expect to *increase* for larger clusters.

In order to further investigate the dependence of decoherence rate scaling on cluster geometry, it will be useful to specify a model for the C_{ij}. For the sake of simplicity, we consider only the two cases where the correlation coefficients have either an exponential or gaussian dependence on the physical separation between the ith and jth qubits so that we have $C_{ij} = e^{-\alpha r_{ij}}$ or $C_{ij} = e^{-\beta r_{ij}^2}$. In the limit where the intra-cluster correlation coefficients are all close to 1, we can then write the factor $(1 - C_{\mathrm{avg}})$ as $\alpha \bar{r}$ or $\beta \overline{r^2}$ where the bar indicates an average over qubit pairs. Combining this with Eq. (10.21) gives the decoherence scaling rate as $\frac{n_c}{\mathrm{Log}_2 d_c} J_{zz}(0) \alpha \bar{r}$ or $\frac{n_c}{\mathrm{Log}_2 d_c} J_{zz}(0) \beta \overline{r^2}$ for the case of exponential or gaussian spatial field correlations respectively. Written in this form, one can see clearly that the beneficial effect of increasing the encoding efficiency by using larger cluster sizes must be weighed against the accompanying increase in the average inter-qubit separation. Physically this simply reflects the fact that the random field fluctuations are more strongly correlated over shorter distances which is what led us to consider the clusterized encodings in the first place.

In Table (10.1) we evaluate the above expressions for decoherence rate scaling for a number of cluster geometries. The physical arrangement of the qubits in the clusters are chosen so that all geometries have the same nearest neighbor inter-qubit separation. For the $m_z = 0$ subspaces, which protect only against pure dephasing, the results are expressed in terms of the rate scaling that is obtained for the pairwise encoding, and we see that the tetrahedral and octahedral geometries that allow for a low average inter-qubit separation give rise to the best scaling. The 2×2 and 2×3 square encodings on the other hand do not utilize three dimensional space as efficiently as the tetrahedral and octahedral geometries do. For the case of gaussian spatially correlated fields which give a higher penalty for increased inter-qubit separation, these encodings are outperformed by the simple pairwise encoding.

It would be nice to make a general statement as to which cluster sizes or geometries give rise to the most favorable decoherence rate scaling. This is, however, complicated by the fact that the encoding efficiency as a function of cluster size varies depending on the type of encoding. We can, for example, consider subspaces spanned by the singlet states which are annihilated not only by the collective cluster operators σ_z^c, but also by the collective raising

Table 10.1: Comparison of decoherence rate scaling for different cluster geometries using the exponential and gaussian models for the spatial dependence of the field correlations. For the $m_z = 0$ subspaces, rates are listed in terms of the rate obtained for the pairwise encoding, for the singlet subspaces, rates are in terms of that obtained for the tetrahedral encoding.

Geometry	# Phys.	$m_z = 0$		Singlet	
	Qubits	Exponential	Gaussian	Exponential	Gaussian
Pair	2	1.00	1.00		
Tetrahedral	4	0.77	0.77	1.00	1.00
Square 2×2	4	0.88	1.03	1.14	1.33
Square 2×3	6	0.98	1.53	0.91	1.42
Octahedral	6	0.75	0.83	0.70	0.78
Cubic $2 \times 2 \times 2$	8	0.83	1.12	0.67	0.90
Cubic $2 \times 3 \times 3$	18	1.01	1.94	0.64	1.23
Cubic $4 \times 4 \times 4$	64	1.37	4.02	0.75	2.19

and lowering operators $\sigma_{\pm}^{c} = \sum_{i \in c} \sigma_{\pm}^{(i)}$. These singlet subspaces therefore protect not only against pure dephasing but also against amplitude damping induced by T_1 processes [4, 11]. By carrying out an analysis similar to that in Eqs. (10.15-10.20), one can show that the singlet subspaces give rise to average decoherence rates per logical qubit of

$$\overline{\gamma}/n_{LQ} = \frac{n_c}{\mathrm{Log}_2 d_c}(J_{zz}(0) + 2J_{+-}(\omega))(1 - C_{\mathrm{avg}}), \tag{10.22}$$

which is very similar to the rate scaling that we derived earlier except that the dimension of the singlet subspace for a cluster of n_c qubits is not equal to $\frac{n_c!}{(n_c/2)!^2}$ but is instead given by $n_c![(n_c/2)!(n_c/2+1)!]^{-1}$ [4]. In Table (10.1) we also compare the decoherence rate scaling of the singlet subspaces for a number of cluster sizes and geometries. Since a minimum of four physical qubits are needed to encode one logical qubit, we give the decoherence rate scaling factors for the singlet subspaces in terms of the rate obtained for a cluster of four physical qubits in a tetrahedral geometry.

Again examining the results in Table (10.1), we see that of the clusterizations considered, the cubic $2 \times 3 \times 3$ cluster with 18 physical qubits gives the best rate scaling when the random field fluctuations have an exponential spatial correlation whereas the octahedral cluster with 6 physical qubits wins out if the fields have a gaussian spatial correlation. These results reflect the fact that the efficiency of encoding using the singlet states of 18 physical qubits ($\frac{1}{18}\mathrm{Log}_2 \frac{18!}{9!10!} = .68$) represents a nearly three-fold increase over that obtained by using only four qubits ($\frac{1}{4}\mathrm{Log}_2 \frac{4!}{2!3!} = .25$). For the case of the $m_z = 0$ encodings, the increased efficiency obtained upon going from a cluster of four qubits to a cluster of 18 qubits is not nearly so great and the rate scaling is actually slightly worse than one obtains using pairs of physical qubits. Thus, we see that which cluster geometry gives the best decoherence rate scaling depends both upon the type of encoding used and also the form of the spatial field correlations. Note also that the results in Table (10.1) do not depend upon the actual amount of spatial field correlation but only upon the form of the spatial correlations as a function of distance. In other words, the optimal cluster size does not depend upon how small the parameter α is in the expression $C_{ij} = e^{\alpha r_{ij}}$ but only on the functional form of the C_{ij} as a function of inter-qubit separation.

10.6 Conclusion

We have analyzed the scaling of average decoherence rates with system size and have seen that in order to obtain an encoding which optimizes decoherence rate scaling, one must find a cluster size and geometry which balances encoding efficiency against the degree of correlation in the random field fluctuations seen by the physical qubits.

Note that we have made no attempt to define in detail how individual logical qubits should be encoded, but have only considered how decoherence rates scale for subspaces which could in principle be used to encode qubits. Except for the pairwise clustering, all of the encodings considered above would require at least some of the logical qubits to be defined in terms of states involving entanglements between physical qubits from separate clusters which of course further complicates the already difficult task of designing gates. Nevertheless, one can still obtain improved scaling rates even without using all of the sub-decoherent states of the individual clusters. For example, a $2 \times 2 \times 2$ cubic cluster of physical qubits yields 14 singlet states and even if we only use 8 of these to encode 3 qubits, the encoding still gives a scaling rate which is improved over that for a tetrahedral geometry with 4 physical qubits. Moreover, the logical qubits are now defined such that single qubit gates can be implemented by manipulating single clusters of physical qubits.

As an additional possibility, one could also consider how scaling rates are affected by incorporating the use of states from subspaces which are not decoherence free in the limit of perfect collective decoherence but which are still sub-decoherent in the presence of spatially correlated field fluctuations. Whatever the details of the encoding, however, we expect that the principle of optimizing decoherence rate scaling by balancing the encoding efficiency against the average spatial correlations of the bath will remain important.

References

[1] W. G. Unruh, Phys. Rev. A **51**, 992 (1995).
[2] G. M. Palma, K. A. Suominen, and A. K. Ekert, Proc. R. Soc. London A **452**, 567 (1996).
[3] D. A. Lidar, I. L. Chuang, and K. B. Whaley, Phys. Rev. Lett. **81**, 2594 (1998).
[4] P. Zanardi and M. Rasetti, Phys. Rev. Lett. **79**, 3306 (1997).
[5] L. Duan and G. Guo, Phys. Rev. A **57**, 737 (1998).
[6] A. Abragam, *Principles of Nuclear Magnetism* (Oxford Univ. Press, Oxford, 1961).
[7] L. Duan and G. Guo, Phys. Rev. Lett. **79**, 1953 (1997).
[8] P. Zanardi, Phys. Rev. A **57**, 3276 (1998).
[9] S. Mukamel, *Principles of Nonlinear Optical Spectroscopy* (Oxford Univ. Press, New York, 1995).
[10] R. Zwanzig, J. Chem. Phys. **33**, 1338 (1960).
[11] D. Bacon, K. R. Brown, and K. B. Whaley, Phys. Rev. Lett. **87**, 247902 (2001).

11 Reduced Collective Description of Spin-Ensembles

Mathias Michel, Harry Schmidt, Friedemann Tonner, and Günter Mahler

Institute of Theoretical Physics I
Stuttgart, Germany

We investigate the structure of Liouville-space with respect to Hamilton- and density-operators. Starting from available resources and limited information we study classes of unitary transformations and mixing strategies for quantum-networks. Scaling properties with respect to ergodicity, storage capacity and separability are obtained.

11.1 Introduction

Any machine – including a computer – is eventually defined by its repertoire of transformations acting on the space of accessible states. In the quantum domain these characteristics are conveniently described within Liouville-spaces. To be useful, i.e. to implement a specific function, the Liouville-space for the density-operator must severely be constrained, as we would otherwise have a hard time to block out all the undesired pathways (cf. "axles and connecting rods" within a mechanical machine). It goes without saying that those constraints have to be realized physically; typically, they result from symmetry and/or design-efforts (structure etc.) built into the Hamiltonian [1].

It is the purpose of this investigation to explore such pathways in the face of incomplete knowledge about some quantitative features of those transformations. (We consider such control limits as generic for large-scale quantum networks.) The resulting statistical approach allows to discuss not only specific realizations but also "typical" features. Optimally, one would expect to find various classes of systems with characteristic behavior; in the worst case few, if any, predictable properties will survive. Of main concern will be the scalability of such networks [2].

Our reasoning will be based on some sort of "quantum game": Just as the famous Jaynes' principle [3], which is able to give a logical foundation of thermodynamical equilibrium properties, such games imply a chain of logical inferences based on available resources, limited information, and the rules of quantum mechanics. They may be formulated as a way "to win bets" under given preconditions.

11.2 Operator Representations

Quantum-mechanical modeling has to refer to specific representations; only these allow to define concrete model- and state-parameters. Though their choice is, in principle, a matter of convenience, some representations are often better suited for a particular purpose than others.

Quantum Information Processing, 2nd Edition. Edited by Thomas Beth and Gerd Leuchs

ISBN: 3-527-40541-0

As a basis in Liouville-space for an n-level system, $\mathcal{L}$, we introduce the set of unitary basis operators [4, 5]

$$\widehat{U}_{a,b} := \sum_{p=0}^{n-1} \omega^{bp} |\underline{p+a}\rangle\langle p| \quad \text{with} \quad \omega = \mathrm{e}^{\frac{2\pi \mathrm{i}}{n}}, \tag{11.1}$$

where $\underline{x} = (x \mod n)$ and $a, b = 0, \ldots, n-1$. Note that the operator $\widehat{U}_{0,0}$ represents the identity operator. Based on the orthonormality relation $\mathrm{Tr}\{\widehat{U}_{a,b}\, \widehat{U}^{\dagger}_{a',b'}\} = n\, \delta_{aa'}\, \delta_{bb'}$ any operator $\widehat{A}$ within $\mathcal{L}$ can be expanded as

$$\widehat{A} = \frac{1}{n} \sum_{a=0}^{n-1} \sum_{b=0}^{n-1} u_{a,b}(\widehat{A})\, \widehat{U}_{a,b} \quad \text{with} \quad u_{a,b}(\widehat{A}) = \mathrm{Tr}\{\widehat{U}^{\dagger}_{a,b}\, \widehat{A}\}. \tag{11.2}$$

For $n = 2$ the unitary operators are directly related to the hermitian SU(2)-operators, the well-known Pauli-operators $\widehat{\sigma}_i$ ($i = 0, x, y, z$):

$$\widehat{U}_{0,0} = \widehat{\sigma}_0 = \widehat{1}, \quad \widehat{U}_{0,1} = \widehat{\sigma}_z, \quad \widehat{U}_{1,0} = \widehat{\sigma}_x, \quad \widehat{U}_{1,1} = \mathrm{i}\,\widehat{\sigma}_y. \tag{11.3}$$

Consider now a (N, n) quantum network, i.e. a set of N subsystems, n levels each. As a basis in the n^{2N}-dimensional Liouville-space $\mathcal{L}$ one can use products of unitary basis operators:

$$\hat{C}_{\{\boldsymbol{a},\boldsymbol{b}\}} = \bigotimes_{\mu=1}^{N} \widehat{U}_{a_\mu, b_\mu}, \tag{11.4}$$

$\boldsymbol{a}$ and $\boldsymbol{b}$ are index vectors of N components each. Because of the orthonormality relation $\mathrm{Tr}\{\hat{C}_{\{\boldsymbol{a},\boldsymbol{b}\}}\, \hat{C}^{\dagger}_{\{\boldsymbol{a}',\boldsymbol{b}'\}}\} = n^N\, \delta_{\boldsymbol{a}\boldsymbol{a}'}\, \delta_{\boldsymbol{b}\boldsymbol{b}'}$ these *product operators* form a complete orthogonal set for the expansion of any network operator:

$$\widehat{A} = \frac{1}{n^N} \sum_{\{\boldsymbol{a},\boldsymbol{b}\}} u_{\{\boldsymbol{a},\boldsymbol{b}\}}(\widehat{A})\, \hat{C}_{\{\boldsymbol{a},\boldsymbol{b}\}} \quad \text{with} \quad u_{\{\boldsymbol{a},\boldsymbol{b}\}}(\widehat{A}) = \mathrm{Tr}\{\hat{C}^{\dagger}_{\{\boldsymbol{a},\boldsymbol{b}\}}\, \widehat{A}\}. \tag{11.5}$$

For later reference we define the measure

$$\mathrm{Tr}\{\widehat{A}^2\} = \frac{1}{n^N} \sum_{\{\boldsymbol{a},\boldsymbol{b}\}} |u_{\{\boldsymbol{a},\boldsymbol{b}\}}(\widehat{A})|^2 \tag{11.6}$$

which is invariant under any unitary transformation.

To describe $(N, 2)$-networks it is more convenient to use Pauli-operators instead of the unitary operators $\widehat{U}_{a,b}$. We introduce the alternative index notation $\widehat{C}_{\alpha\beta\gamma,p}$ [5], where α, β and γ specify the "multiplicity" of $\widehat{\sigma}_x$, $\widehat{\sigma}_y$ and $\widehat{\sigma}_z$, respectively, index p denotes the permutation of these operators among the N nodes. The number of such permutations and hence the index range for $p \in [0, \Omega - 1]$ is

$$\Omega(N; \alpha, \beta, \gamma) = \binom{N}{\alpha}\binom{N-\alpha}{\beta}\binom{N-\alpha-\beta}{\gamma} = \frac{N!}{\alpha!\beta!\gamma!(N-\alpha-\beta-\gamma)!}. \quad (11.7)$$

$c = \alpha+\beta+\gamma \leq N$ is the number of subsystems the operator $\widehat{C}_{\alpha\beta\gamma,p}$ acts on (c-body-operator).

Finally, collective operators can be defined as linear combinations of the $\widehat{C}_{\alpha\beta\gamma,p}$, see [5],

$$\widehat{E}_{\alpha\beta\gamma,b} = \sum_{p=0}^{\Omega-1} \omega_\Omega^{pb}\, \widehat{C}_{\alpha\beta\gamma,p} \quad \text{with} \quad \omega_\Omega = \mathrm{e}^{2\pi\mathrm{i}/\Omega}, \quad b \in [0, \Omega - 1]. \quad (11.8)$$

They form an alternative complete orthogonal basis for the Liouville-space, with the relation $\mathrm{Tr}\{\widehat{E}_{\alpha\beta\gamma,b}\,\widehat{E}^\dagger_{\alpha'\beta'\gamma',b'}\} = \Omega 2^N\, \delta_{\alpha\alpha'}\, \delta_{\beta\beta'}\, \delta_{\gamma\gamma'}\, \delta_{bb'}$. The subclass $b = 0$ is permutation-symmetric, any permutation-symmetrical operator can be expanded in terms of $\widehat{E}_{\alpha\beta\gamma} \equiv \widehat{E}_{\alpha\beta\gamma,0}$. The space of these operators is denoted by $\mathcal{L}_{\text{sym}}$, its dimension

$$\dim(\mathcal{L}^{(N)}_{\text{sym}}) = \frac{(N+1)(N+2)(N+3)}{6} \quad (11.9)$$

grows only polynomially in N. The space of permutation-symmetrical operators with at most c-body interaction will be denoted by $\mathcal{L}^{(c)}_{\text{sym}}$, its dimension can be evaluated from Eq. (11.9) replacing N by c. In this symmetrical subspace the representation of some operator $\widehat{A}$ reads

$$\widehat{A} = \frac{1}{2^N} \sum_{\substack{\{\alpha\beta\gamma\}\\ \alpha+\beta+\gamma\leq c}} e_{\alpha\beta\gamma}(\widehat{A})\, \widehat{E}_{\alpha\beta\gamma} \quad \text{with} \quad e_{\alpha\beta\gamma}(\widehat{A}) = \frac{1}{\Omega}\, \mathrm{Tr}\{\widehat{A}\,\widehat{E}_{\alpha\beta\gamma}\} \quad (11.10)$$

$$\text{and} \quad \mathrm{Tr}\{\widehat{A}^2\} = \frac{\Omega}{2^N} \sum_{\substack{\{\alpha\beta\gamma\}\\ \alpha+\beta+\gamma\leq c}} |e_{\alpha\beta\gamma}(\widehat{A})|^2. \quad (11.11)$$

The local collective operators $\widehat{E}_{100}$, $\widehat{E}_{010}$ and $\widehat{E}_{001}$ coincide with the many-particle Pauli-operators $\widehat{S}_i = \sum_\mu \widehat{\sigma}_i(\mu)$, $i = x, y, z$. All operators $\widehat{E}_{\alpha\beta\gamma}$ can be expressed as linear combinations of products of these operators and the identity $\widehat{1}$, e.g.

$$\widehat{E}_{200} = \frac{1}{2}\widehat{S}_x^2 - \frac{N}{2}\widehat{1}. \quad (11.12)$$

This notation is especially convenient for operators with at most two-body terms, $c = 2$. Independent of N, these can be written in the "non-linear" form

$$\widehat{A} = a_0\widehat{1} + \boldsymbol{a} \cdot \widehat{\boldsymbol{S}} + \widehat{\boldsymbol{S}}^t \cdot \mathsf{A} \cdot \widehat{\boldsymbol{S}}, \quad (11.13)$$

with exactly ten independent parameters $\{a_0, a_j, A_{ij} = A_{ji};\ i,j = x,y,z\}$, corresponding to $\dim(\mathcal{L}^{(2)}_{\text{sym}})$. $\widehat{\boldsymbol{S}}^t$ denotes the transpose of the vector operator $\widehat{\boldsymbol{S}}$. Eq. (11.13) is a pertinent example of a reduced collective description. The various parameter sets, $u_{\{\boldsymbol{a},\boldsymbol{b}\}}$, $e_{\alpha\beta\gamma}$ etc. are connected via linear transformations.

11.3 Hamilton Models

In this section we restrict ourselves to spin-network-Hamiltonians ($n = 2$). We will take here the position that the respective model parameters, though constrained, e.g., by symmetry, will not be completely known. This implies a statistical approach for the unitary transformations generated by these Hamiltonians.

11.3.1 Symmetry-constrained Networks

Any Hamilton-model $\widehat{H} \in \mathcal{L}^{(2)}_{\text{sym}}$ for a permutation-symmetrical $(N,2)$-network with up to 2-body interactions can be written as (cf. Eq. (11.10))

$$\widehat{H} = \hbar\delta_0\widehat{E}_{000} + \hbar(\overbrace{\delta_1\widehat{E}_{100} + \delta_2\widehat{E}_{010} + \delta_3\widehat{E}_{001}}^{\text{local terms}} + \\ + \underbrace{\gamma_1\widehat{E}_{200} + \gamma_2\widehat{E}_{020} + \gamma_3\widehat{E}_{002} + \gamma_4\widehat{E}_{110} + \gamma_5\widehat{E}_{101} + \gamma_6\widehat{E}_{011}}_{\text{interaction terms}}). \tag{11.14}$$

The following two examples are based on the representation (11.13):

The many-body Jaynes–Cummings-Model with dipole-interaction is defined by

$$\widehat{H} = \tfrac{\hbar h_x}{2}\widehat{S}_x + \tfrac{\hbar h_z}{2}\widehat{S}_z + \tfrac{\hbar H_{zz}}{2}\widehat{S}_z^2 = \tfrac{\hbar h_x}{2}\widehat{E}_{100} + \tfrac{\hbar h_z}{2}\widehat{E}_{001} + \hbar H_{zz}\left(\widehat{E}_{002} + \tfrac{N}{2}\widehat{E}_{000}\right). \tag{11.15}$$

The most remarkable feature of this model is its ability to create maximally entangled N-body-states, so called Cat-states

$$|\text{Cat}_\varphi\rangle = \frac{1}{\sqrt{2}}\left(|00\dots0\rangle + \mathrm{e}^{\mathrm{i}\varphi}|11\dots1\rangle\right), \tag{11.16}$$

out of the product states $|00\dots0\rangle$ and $|11\dots1\rangle$, in which all particles are in the same local state $|0\rangle$ or $|1\rangle$, respectively. φ is an arbitrary phase factor. The time needed for the transition from $|00\dots0\rangle$ or $|11\dots1\rangle$ to $|\text{Cat}_\varphi\rangle$ depends on the number of particles N.

Another example is the Hamiltonian

$$\widehat{H} = \hbar H_{xx}\widehat{S}_x^2 = \hbar H_{xx}(2\widehat{E}_{200} + \widehat{E}_{000}). \tag{11.17}$$

It describes N two-level systems in $\widehat{\sigma}_x \otimes \widehat{\sigma}_x$-interaction without local terms. This system also allows to create the states $|\text{Cat}_\varphi\rangle$ for N even, but the time needed is independent of N.

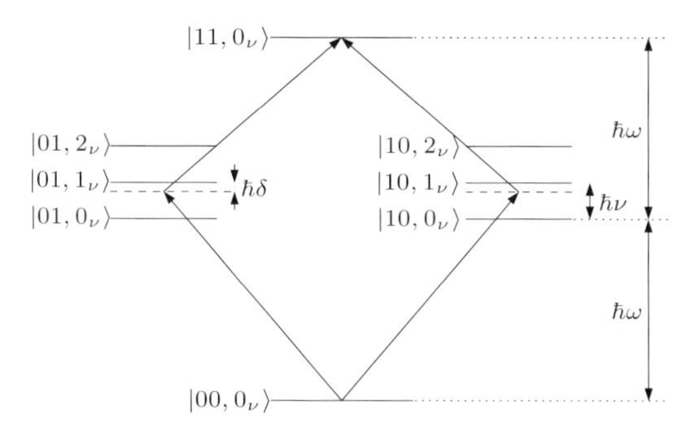

Figure 11.1: Scheme for creating the Hamiltonian (11.4) for two particles.

A permutation-symmetrical interaction as in (11.17) can be realized by two-level charged particles confined in a harmonic potential – e.g. ions in a trap [6,7]. For a rough understanding of how it works we consider only two spin-$1/2$ particles with internal energy splitting $\hbar\omega$, ν is the frequency of a particular collective oscillatory mode of the particles in the trap.

The particles are initially prepared in their internal spin-down state $|0\rangle$ and in the ground state of their collective oscillation $|0_\nu\rangle$: $|\psi(0)\rangle = |00, 0_\nu\rangle$. Now two fields with frequencies $\omega + \nu - \delta$ and $\omega - \nu + \delta$, respectively, drive the two-step transition from $|00, 0_\nu\rangle$ to $|11, 0_\nu\rangle$. For sufficiently large detuning δ, the intermediate states $|10, 1_\nu\rangle$ and $|01, 1_\nu\rangle$ are negligibly occupied, see Figure 11.1. The resulting effective Hamiltonian is

$$\hat{H} = \hbar\gamma(|11\rangle\langle 00| + |00\rangle\langle 11| + |10\rangle\langle 01| + |01\rangle\langle 10|) = \frac{\hbar\gamma}{2}\hat{S}_x^2 - \hbar\gamma\hat{1}, \tag{11.18}$$

which is just the Hamiltonian (11.17), except for a global energy shift.

This second system is but one example of a more general case: To create a totally permutation-symmetrical interaction between an arbitrary number of nodes of a quantum network one can use a so-called "bus-mode" mediating the interaction. In the case of the ion trap the collective motion acts as the bus-mode. Other systems based on this principle have been proposed, e.g. inductively coupled Josephson-qubits, see [8]. Generalization to multiple ensembles have also been considered [9].

11.3.2 Topology-constrained Networks

Design efforts for a (N, n)-network are typically aimed at a specific connectivity between subsystems. This connectivity can be specified by a $N \times N$ matrix, $T_{\mu\nu}$, where $T_{\mu\nu} = 1$, if the subsystems are connected, and $T_{\mu\nu} = 0$ otherwise. This topological structure can be formulated as a graph $G(V, E)$, vertex set $V = \{1, 2, \ldots, N\}$, edge set E containing tuples $\{\mu, \nu\}$ with $\mu, \nu \in V$ (cf. inset of Figure 11.4). The structure can then be taken as a basis for a matrix-model

$$\widehat{H} = \sum_{\mu,\nu} H_{\mu\nu}|\psi_\mu\rangle\langle\psi_\nu| \tag{11.19}$$

with $H_{\mu\nu} = h_{\mu\nu} T_{\mu\nu}$, $h_{\mu\nu}$ defines the quantitative features of the model. The operators $\widehat{P}_{\mu\nu} = |\psi_\mu\rangle\langle\psi_\nu|$ form another orthogonal basis in Liouville-space. Examples for such localized states $|\psi_\mu\rangle$ will be given in Section 11.4.2.

11.4 State Models

The representation (11.5) can also be applied to the density operator $\widehat{\rho}$. The expansion coefficients $u_{\{a,b\}}$ are entries of the so-called *coherence vector* (see [4]). The coherence vector specifies local as well as correlation properties of any (N, n)-network state. Here we focus on specific sections of the whole state space. Firstly we want to investigate pure states with special properties: the family of 1-particle excitations (see Section 11.4.2) and the family of total permutation-symmetrical states (see Section 11.4.1). Secondly we consider 1-parameter families of mixed states (see Section 11.4.3, 11.4.4).

11.4.1 Totally Permutation-symmetric Subspace

Again, we restrict ourselves to $(N, 2)$-networks. Under the action of permutation-symmetrical operators $\widehat{E}_{\alpha\beta\gamma}$ all subspaces of the Hilbert-space $\mathcal{H}$ of given permutation symmetry are invariant. This is in particular true for the totally symmetrical subspace $\mathcal{H}_{\text{sym}}$. As a basis in this subspace the states $|s\rangle$ can be used, where s is the number of spin-up and $(N - s)$ is the number of spin-down particles,

$$|s\rangle = \widehat{\mathcal{S}}_{\text{sym}} |\underbrace{0 \ldots 0}_{N-s} \underbrace{1 \ldots 1}_{s}\rangle \propto \widehat{E}_{s00} |00 \ldots 0\rangle. \tag{11.20}$$

$\widehat{\mathcal{S}}_{\text{sym}}$ denotes the symmetrisation-operator with proper normalization. There obviously exist $N + 1$ such states, the dimension of $\mathcal{H}_{\text{sym}}$ therefore is $N + 1$ and scales only linearly with N!

11.4.2 Collective 1-particle Excitations

For a $(N, 2)$-network there are N 1-particle excitations of the product-form

$$|\mu\rangle = |0, \ldots, 0, 1_\mu, 0, \ldots, 0\rangle \quad \mu = 1, \ldots, N. \tag{11.21}$$

These subsystem-selective states obviously form a subset out of the 2^N-dimensional state-space. A family of collective 1-particle excitations can thus be specified by

$$|y\rangle = \sum_\mu y_\mu |\mu\rangle \tag{11.22}$$

where y_μ is an effective wave function.

11.4.3 1-parameter Families of Non-pure States

Let us now consider a class of non-pure states of a (N, n) quantum network, which will be called "generalized Werner-states". These states are defined by the 1-parameter-family of density operators:

$$\widehat{\rho}(\epsilon, |\psi\rangle) = \frac{1}{n^N}(1-\epsilon)\widehat{1} + \epsilon\,|\psi\rangle\langle\psi| \tag{11.23}$$

with $|\psi\rangle$ denoting a given pure state and ϵ is a real mixture parameter with $0 \le \epsilon \le 1$. It is possible to expand this state in terms of product operators as shown in Eq. (11.5):

$$\widehat{\rho}(\epsilon, |\psi\rangle) = \frac{1}{n^N}\left(\widehat{1} + \sum_{\{\boldsymbol{a},\boldsymbol{b}\}\neq\{\boldsymbol{0},\boldsymbol{0}\}} u_{\{\boldsymbol{a},\boldsymbol{b}\}}(\widehat{\rho}(\epsilon, |\psi\rangle))\, \hat{C}_{\{\boldsymbol{a},\boldsymbol{b}\}}\right). \tag{11.24}$$

Obviously, $u_{\{\boldsymbol{a},\boldsymbol{b}\}}(\widehat{\rho}(\epsilon, |\psi\rangle)) = \epsilon u_{\{\boldsymbol{a},\boldsymbol{b}\}}(\widehat{\rho}(1, |\psi\rangle))$, i.e. all coherence vector components of this generalized Werner-state are scaled down by ϵ compared to the respective pure state value. Because of this fact one may say, that these mixed states are "pale" states.

A special case of the generalized Werner-states is obtained if the pure state $|\psi\rangle$ is taken as the special Cat-state $|\mathrm{Cat}_0\rangle$, with

$$|\mathrm{Cat}_0\rangle = \frac{1}{\sqrt{n}}\sum_{i=0}^{n-1}|i, \ldots, i\rangle, \tag{11.25}$$

the original Werner-state [10].

11.4.4 Families of Separable States: "Modules"

Another interesting state family for (N, n)-networks are special separable states, which we call *modules*. They are defined by their product operator expansion (cf. Eq. (11.5)):

$$\hat{\rho}_{\mathrm{mod}}(\{\boldsymbol{a}, \boldsymbol{b}\}, \phi, \Xi) = \frac{1}{n^N}\left(\widehat{1} + \Xi\left(\mathrm{e}^{\mathrm{i}\phi}\,\hat{C}_{\{\boldsymbol{a},\boldsymbol{b}\}} + \mathrm{e}^{-\mathrm{i}\phi}\,\hat{C}^{\dagger}_{\{\boldsymbol{a},\boldsymbol{b}\}}\right)\right) \tag{11.26}$$

with an arbitrary phase ϕ and a real positive constant Ξ. For positivity of these states to hold, we must require $\Xi \le \frac{1}{2}$. The coherence vector of each module contains the minimum number of non-zero entries: Two elements are non-zero, because of the hermiticity of the modules. Modules, in turn, can be generated by mixing pure product states, in this way we have shown that modules are separable states [11].

11.5 Ensembles

For N, $n \gg 1$, Hilbert-space becomes a large space indeed. If nothing was known beforehand, some general statements could be drawn from Hilbert-space statistics [12, 13]. If everything was known (initial state, $\widehat{H}$-model), the whole state trajectory would completely be determined. In between, limited information can be cast into appropriate ensembles.

Based on the preceding sections we will now study specific $\widehat{H}$-ensembles (Section 11.5.1, 11.5.2) and state ensembles (Section 11.5.3, 11.5.4).

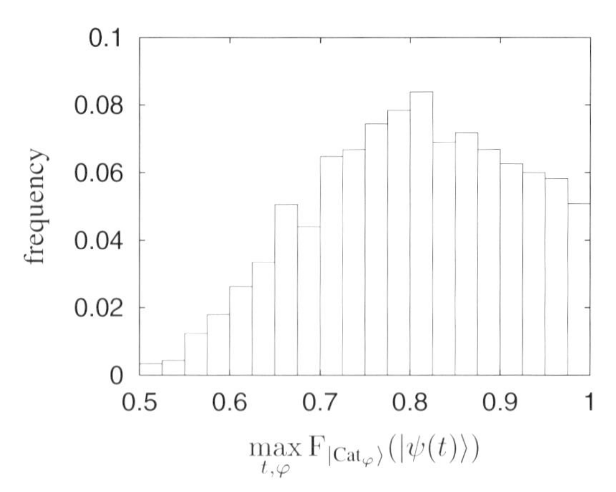

Figure 11.2: Fidelity (11.17) maximized with respect to t and φ for 6 qubits.

11.5.1 Trajectories and Ergodicity

We restrict ourselves to permutation-symmetrical Hamilton-models within $\mathcal{L}_{\mathrm{sym}}^{(2)}$ (see Section 11.3.1) acting on the totally permutation-symmetrical subspace of Hilbert-space (see Section 11.4.1). In this subspace, there exists no operator, apart from the identity $\widehat{1}$, that commutes with the whole set as given in (11.14). One may wonder, whether this fact may lead to an "ergodic behavior" or whether there are regions in this subspace that are disconnected.

Ergodicity will be probed for special transitions only, namely for those from the product state $|00\ldots0\rangle$ to the product state $|11\ldots1\rangle$ and to the maximally entangled states $|\mathrm{Cat}_\varphi\rangle$ according to (11.16). Since for $\delta_3 \neq 0$ energy conservation prohibits these transitions, we will only consider Hamiltonians with $\delta_3 = 0$, i.e. with vanishing Zeeman-Terms. The other parameters entering Eq. (11.14) are assumed to be unknown; they will be taken from identical Gaussian distributions

$$P(x) = \frac{1}{\sqrt{2\pi\sigma}} \mathrm{e}^{x^2/2\sigma^2} \tag{11.27}$$

with variance $\sigma^2 = 1$, and $x = \delta_1, \delta_2, \ldots, \gamma_6$. Their product $P(\delta_1)\cdot P(\delta_2)\cdots \equiv P(\delta_1, \delta_2, \ldots)$ is then $\propto \exp(-a\,\mathrm{Tr}\{\widehat{H}^2\})$ and as such invariant (cf. Eq. (11.11)). In this way we have built in the prior knowledge that our Hamiltonian has the form Eq. (11.14) though with respect to an unknown basis. Note that a different choice of σ only means a rescaling of all the Hamilton-parameters and thus a rescaling of the time in the unitary evolution generated by $\widehat{H}$.

Typically, these systems mediate the transition from $|00\ldots0\rangle$ to $|11\ldots1\rangle$, at least approximately. However, for the transition from $|00\ldots0\rangle$ to $|\mathrm{Cat}_\varphi\rangle$ this is not the case. There exist systems which never reach any of the states $|\mathrm{Cat}_\varphi\rangle$, although energy conservation would permit this transition for at least some φ.

To measure the "distance" between the state $|\psi(t)\rangle$ under the time evolution generated by one of the Hamiltonians (11.14) (with $\delta_3 = 0$) to the states $|\mathrm{Cat}_\varphi\rangle$, we use the fidelity

$$\mathrm{F}_{|\mathrm{Cat}_\varphi\rangle}(|\psi(t)\rangle) = |\langle \mathrm{Cat}_\varphi|\psi(t)\rangle|^2, \tag{11.28}$$

which is maximized over all t and $\varphi \in [0, 2\pi)$. So, no prediction whatsoever can be made about the time needed for this transition for those systems that actually make it. It can only be determined how close the trajectory passes by the family of states $|\mathrm{Cat}_\varphi\rangle$ *at any time*.

Figure 11.2 shows the relative frequency of different maximized fidelities for six qubits. The coefficients were randomly chosen as explained above with $\delta_3 = 0$. Obviously, for some systems the state never gets closer to $|\mathrm{Cat}_\varphi\rangle$ than already the initial state was:

$$\mathrm{F}_{|\mathrm{Cat}_\varphi\rangle}(|00\ldots0\rangle) = 1/2. \tag{11.29}$$

Only few realizations (11.14) allow for the transition from $|00\ldots0\rangle$ to one of the states $|\mathrm{Cat}_\varphi\rangle$ with good accuracy. A lot of ensemble members therefore are not ergodic. The results become even more dramatic, when N is increased. In this case, the fraction of Hamilton-models decreases which reach the maximum entangled state $|\mathrm{Cat}_\varphi\rangle$ starting from the product state $|00\ldots0\rangle$. Already for $N = 60$ qubits almost no model is left with a trajectory approaching $|\mathrm{Cat}_\varphi\rangle$. More than half of the class members even don't get closer to $|\mathrm{Cat}_\varphi\rangle$ than initially, see Figure 11.3.

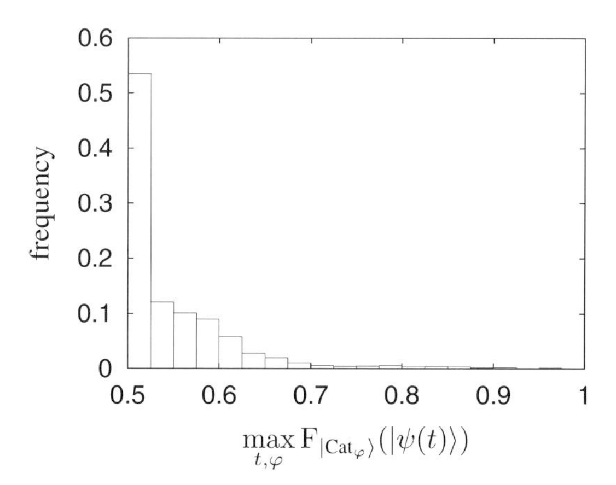

Figure 11.3: Relative frequency of different values of the fidelity (11.17) for 60 qubits.

It is clear that there exist special Hamilton-models of the present type mediating the transition exactly for any N, e.g. $\widehat{H} = \hbar H_{xx}\widehat{S}_x^2$ according to Eq. (11.17). But it can be suspected that these systems form only a subset of measure zero within $\mathcal{L}_{\mathrm{sym}}^{(2)}$. In fact, even the systems in their close vicinity show a behavior similar to Figure 11.3 as N is increased.

If confirmed in the $N \to \infty$ limit, the present $\widehat{H}$-ensemble could be regarded as (asymptotically) "non-ergodic" in symmetrical Hilbert-space. This would have a direct impact on the controlability of quantum systems in the limit of large numbers of subsystems. Since Hamiltonians are never implemented perfectly, there will be regions in Hilbert-space that tend to become inaccessible. In this sense, no permutation-symmetrical system designed to produce maximally entangled states would be scaleable.

11.5.2 Leakage and Storage Capacity

We consider the topology-constrained network, Eq. (11.19), in the N-dimensional space of 1-particle excitations $|\psi_\mu\rangle = |\mu\rangle$ (cf. Eq. (11.21)). This is our prior knowledge. The Hamilton-parameters, $h_{\mu\nu}$, are assumed to be unknown: Choosing for the diagonal elements, $x = h_{\mu\mu}$ a distribution with variance $\sigma^2 = 2$, the real and imaginary part of the off-diagonal elements $x = h_{\mu\nu}$ ($\mu < \nu$) are distributed with variance $\sigma^2 = 1$. The product of these functions is then, again, a function of the invariant $\mathrm{Tr}\{\widehat{H}^2\}$ only (cf. Section 11.3.2). This is reminiscent of the uniform distribution of hermitian matrices (Gaussian unitary ensemble, GUE) in the complete Hilbert-space (cf. [12, 14]).

Given a realization of $\hat{H}$ the time evolution is unitary:

$$\hat{U}(t) = \mathrm{e}^{-\mathrm{i}\hat{H}t/\hbar} = \sum_{j=1}^{N} e^{-\mathrm{i}k_j t/\hbar} |K_j\rangle\langle K_j| \tag{11.30}$$

where $|K_j\rangle$ and k_j are the eigenvectors and eigenvalues of $\hat{H}$, respectively.

In order to analyze the system behavior, we probe the system by exciting one of the base states $|\psi_\mu\rangle$ and measuring another base state $|\psi_\nu\rangle$. The pertinent conditional probability is given by the fidelity

$$F(\nu|\mu;t) = \left|\langle\psi_\nu|\hat{U}(t)|\psi_\mu\rangle\right|^2 = \left|\sum_{j=1}^{N} \mathrm{e}^{-\mathrm{i}k_j t/\hbar} \langle\psi_\nu|\, K_j\rangle \langle K_j|\, \psi_\mu\rangle\right|^2$$

We restrict ourselves to the time independent case and take the implicit mean over all times

$$F(\nu|\mu) = \sum_{j=1}^{N} |\langle\psi_\nu|\, K_j\rangle|^2 \, |\langle K_j|\, \psi_\mu\rangle|^2 \tag{11.31}$$

which holds in the case of non-degenerate eigenvalues.

For the discrete distance $d = d(G, \mu, \nu)$ between the base states μ and ν on the graph G (d =minimum number of steps, cf. Figure 11.4) we get

$$F^*(d|\mu) = \sum_{\nu;d=d(G,\mu,\nu)} F(\nu|\mu) \tag{11.32}$$

It gives us the conditional probability of finding the excitation at a distance d from where we know to have started and thus a measure of leakage. In case of all states μ being equivalent we simply write $F(d)$ instead of $F^*(d|\mu)$.

The usage of the system as a quantum register for storing base states would require the retrieval of information with a small error probability $\mathcal{E}$.

We assume that the bits are stored in states each centered at hyperspheres of radius r. With the number of states $n(d)$ at a distance d we are using $N_r = \sum_{d=0}^{r} n(d)$ states for each bit. Then we can define the storage capacity

$$C = 1/N_r \tag{11.33}$$

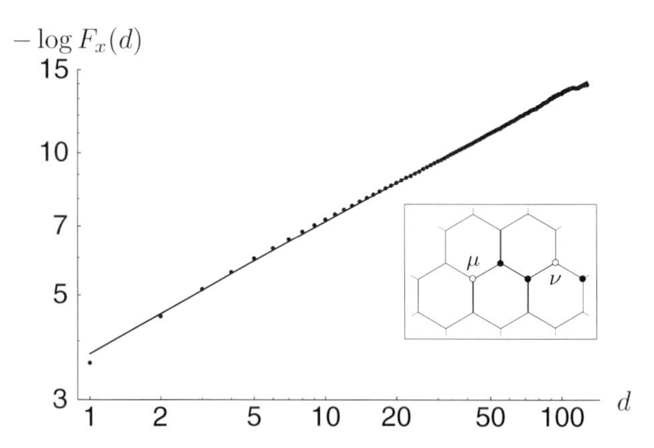

Figure 11.4: Probability $F_x(d)$ of measuring the system in a state at distance d from the initial state on the x-axis of a 256×128 honeycomb lattice with periodic boundary conditions, averaged over all initial states. The "stretched exponential" fit to the data points is depicted as a line. The inset shows how the distance d between μ and ν is defined (here $d = 3$).

($1/C$ might be interpreted as the "redundancy") and express the error probability as

$$\mathcal{E} = \sum_{d>r} F(d). \tag{11.34}$$

$\mathcal{E}$ gives the probability to find the given state outside its assigned sphere and thus outside the region, where the state can uniquely be identified.

In the following we limit our discussion to a 2D honeycomb lattice, which provides the lowest number of nearest neighbors of 3. We numerically estimate $F(d)$ for one realization on a honeycomb lattice with 256×128 vertices (largest distances $(d_x, d_y) = (128, 128)$) and periodic boundary conditions, calculating $F_x(d)$ along the x-direction and averaging over all initial states (Figure 11.4). The best fit to the data is a "stretched exponential" $(0 < \lambda < 1)$

$$F(d) \approx \alpha\, \mathrm{e}^{-\beta d^{\lambda}} \tag{11.35}$$

with fit parameters $\alpha = 1.590, \beta = 4.221$ and $\lambda = 0.2566$. The deviation between fit and data is characterized by the standard deviation $\sigma = 0.082$. Note that the data could also be approximated by a power law for small distances or a power law times an exponential (with small exponential prefactor) for the full range.

Assuming isotropy we expect $F(d) = 3dF_x(d) + F_x(0)$, and $\sum_d F(d)$ should be 1. Instead we find $\sum_d F(d) = 0.856$, which we interpret to result from residual anisotropy. After renormalizing $F(d)$ we can evaluate the resulting memory capacity in this system. The level of $\mathcal{E} = 10^{-2}$ is reached at a capacity $C = 6 \cdot 10^{-5}$, corresponding to a circle radius of $r = 104$ (see Figure 11.5). Extrapolating the fit to $F(d)$ to larger distances, we reach the error levels $\mathcal{E} = \{10^{-3}, 10^{-4}, 10^{-5}\}$ at $C = \{4.9 \cdot 10^{-6}, 1.4 \cdot 10^{-6}, 5.1 \cdot 10^{-7}\}$ and $r = \{369, 686, 1141\}$, respectively.

For our discussion of leakage, the behavior of $F(d)$ already for small d is relevant. The slow decay of $F(d)$ severely limits the storage of bits with an acceptable error probability.

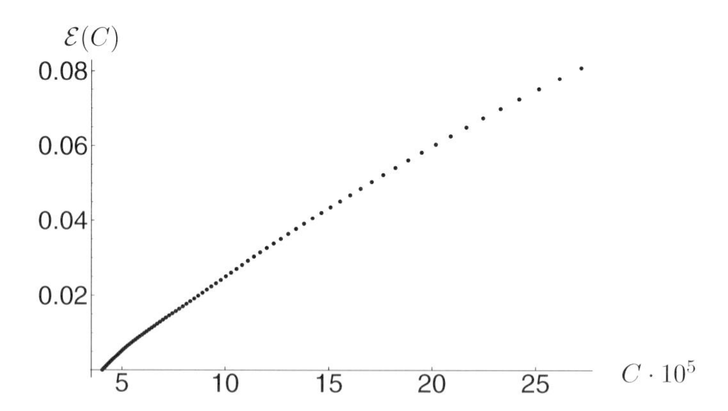

Figure 11.5: Error probability $\mathcal{E}(C)$ over memory capacity C extracted from the data $F_x(d)$ as of Figure 11.4. Because of the finite size of the system the error $\mathcal{E} = 0$ is reached.

Distinguishing the possible leakage cases by the fit parameter λ (cf. Eq. (11.35)) into exponential leakage ($\lambda = 1$, weak leakage), "stretched exponential" leakage ($0 < \lambda < 1$, strong leakage) and delocalization/full leakage ($\lambda = 0$), we are left with the case in between.

This type of study can be extended to other topological structures ("full coupling", one- and two-dimensional structures based on different lattices etc.). Our model could be interpreted as an Anderson model (for a review see [15]) but with diagonal and off-diagonal disorder and a fixed ratio between both disorder types. As is the case in the 2D Anderson model, we have found localization behavior for our 2D model. Note that we discuss leakage and localization not in the energy-controlled but in the local position-controlled picture: The initial states are prepared as $|\psi_\mu\rangle$ state with some energy uncertainty.

11.5.3 Mixing Strategies

We now consider families of stationary mixed states as our resources and ask ourselves, what other states can be reached by specific mixing processes. During such a mixing the coherence vector length (cf. Eq. (11.6))

$$L^2 = \sum_{\{\boldsymbol{a},\boldsymbol{b}\}\neq\{\boldsymbol{0},\boldsymbol{0}\}} |u_{\{\boldsymbol{a},\boldsymbol{b}\}}|^2 = n^N \operatorname{Tr}\{\hat{\rho}^2\} - 1 \tag{11.36}$$

will further decrease.

In general, it is very hard to decide whether an ad-hoc coherence vector (for a (N, n)-network) corresponds to an allowed state in Hilbert-space. To the best of our knowledge the geometrical shape generated by the set of all pure-state coherence vectors is unknown; it is anything but a simple hypersphere (the Bloch-sphere, applicable to the 2-dimensional Hilbert-space, is an exception).

However, restrictions on the vector components fade away as we decrease the coherence-vector length. With mixtures based on the separable modules of Section 11.4.4, it is possible to construct any coherence vector $u_{\{\boldsymbol{a},\boldsymbol{b}\}}$ of down-scaled length L. So, for N, n finite any direction of this vector is allowed, if only its length is taken to be small enough (though finite). The set of accessible states is given by

$$\hat{\rho}_{\text{mix}} = \frac{1}{N_{\text{mod}}} \sum_{\{\boldsymbol{a},\boldsymbol{b}\}\neq\{\boldsymbol{0},\boldsymbol{0}\}} K(\{\boldsymbol{a},\boldsymbol{b}\},\phi)\, \hat{\rho}_{\text{mod}}(\{\boldsymbol{a},\boldsymbol{b}\},\phi,\Xi). \tag{11.37}$$

with the mixture coefficients $K(\{\boldsymbol{a},\boldsymbol{b}\},\phi)$ constrained by

$$\sum_{\{\boldsymbol{a},\boldsymbol{b}\}\neq\{\boldsymbol{0},\boldsymbol{0}\}} K(\{\boldsymbol{a},\boldsymbol{b}\},\phi) = N_{\text{mod}} \tag{11.38}$$

By construction all these states are separable.

11.5.4 State Construction and Separability

To what extent can we reach a given state by means of this mixing strategy? A simple answer is obtained for the generalized Werner-states: We merely have to compare their product-operator expansion according to Eq. (11.24) with the above form (11.37). In this way we get an upper bound for ϵ: [11]

$$\epsilon \leq \epsilon_{\text{mod}} \equiv \frac{2\Xi}{N_{\text{mod}}} = \frac{1}{N_{\text{mod}}}. \tag{11.39}$$

Clearly, this mixing process is a sufficient criterion for separability. Every module is separable and also their mixture cannot lead to entanglement. Unfortunately, it is not possible to evaluate the distance to the separability boundary because of the lack of a general criterion for separability.

However, for the original Werner-state (see Section 11.4.3), it has been shown that the Peres-criterion is a necessary *and* sufficient criterion for separability [16]. A Werner-state is separable if and only if

$$\epsilon \leq \frac{1}{n^{N-1}+1} \equiv \epsilon_{\text{opt}}. \tag{11.40}$$

For this special case we have been able to find an optimized mixing process for any (N,n)-network with n prime [11]. This mixing algorithm leads to a Werner-state directly at the separability boundary, as given above. Modules are not optimal for this process, i.e. they produce a much lower bound $\epsilon_{\text{mod}} < \epsilon_{\text{opt}}$.

The "complexity" of the mixture (defined as the approximate number of different product states needed for the mixture) in both cases increases exponentially with the number of subsystems N.

11.6 Summary and Outlook

The main motivation of this investigation has been to contribute to the challenging scalability problem of quantum networks. For this purpose we have discussed ensembles of states and of

Hamilton-models for spin-networks (N subsystems). The reduced collective description has turned out to be very useful for the treatment of symmetry- or topology-constrained Hamiltonians, irrespective of their size. The ensemble concept has been introduced to formalize limited information and control (considered inevitable for realistic large networks). We have discussed state evolutions under the influence of unitary transformations and mixing strategies. We have been able to show

- that systems constrained by permutation symmetry tend to be non-ergodic in the large-N limit,
- that any down-scaled coherence vector can be generated in the separable domain,
- how any Werner-state can be mixed out of product states though with exponentially growing resources; (there are procedures even working at the separability border),
- how leakage within typical quantum networks severely limits the maximum number of storage bits (this number also depends on the underlying topology).

References

[1] Mahler, G., Grappling with qubits, Science **292**, 57 (2002).

[2] Fitzgerald, R., What really gives a quantum computer its power?, Phys. Today Jan. 2000, 20.

[3] Jaynes, E. T., Information theory and statistical mechanics, Phys. Rev. **108**, 171 (1957).

[4] Mahler, G., Weberruß, V. A., Quantum Networks: Dynamics of Open Nanostructures, SPRINGER, Berlin (1998).

[5] Otte, A., Mahler, G., Adapted-operator representations: Selective versus collective properties of quantum networks, Phys. Rev. A **62**, 012303 (2000).

[6] Mølmer, K., Sørensen, A., Multiparticle Entanglement of Hot Trapped Ions, Phys. Rev. Lett. **82**, 1835-1838 (1999).

[7] Sackett, C. A., Kielpinski, D.,King, B. E., Langer, C.,Mayer, V., Myatt, C. J., Rowe, M., Turchette, Q. A., Itano, W. M., Wineland, D. J., Monroe, C., Experimental entanglement of four particles, Nature **404**, 256-259 (2000).

[8] Makhlin, Y., Schön, G., Shnirman, A., Quantum-state engineering with Josephson-junction devices, Rev. Mod. Phys. **73**, 357-400 (2001).

[9] Lukin, M. D., Fleischhauer, M., Cole, R., Duan, L. M., Jaksch, D., Cirac, J. I., Zoller, P., Dipole Blockade and Quantum Information Processing in Mesoscopic Atomic Ensembles, Phys. Rev. Lett. **87**, 037901 (2001).

[10] Werner, R. F., Quantum states with Einstein-Podolsky-Rosen Correlations admitting a hidden variable model, Phys. Rev. A **40**, 4277 (1989).

[11] Michel, M., Mahler, G., Modular construction of special mixed quantum states, in preparation (2002).

[12] Otte, A., Separabilität in Quantennetzwerken, Universität Stuttgart, (PhD-Thesis, Stuttgart, 2001, unpublished).

[13] Mahler, G., Gemmer, J., Stollsteimer, M., Quantum Computer as a thermodynamical machine, Superlattices and Microstructures (invited), in press (2002).

[14] Haake, F., Quantum Signature of Chaos, SPRINGER, Berlin (1991).

[15] Lee, P. A., Ramakrishnan, T. V., Disordered electronic systems, Rev. Mod. Phys. **57**, No. 2, 287 (1985).

[16] Pittenger, A. O., Rubin, M. H., Separability and Fourier representations of density matrices, Phys. Rev. A **62**, 032313 (2000).

12 Quantum Information Processing with Defects

F. Jelezko and J. Wrachtrup

University of Stuttgart
3. Institute of Physics
Stuttgart, Germany

12.1 Introduction

Defects in wide-band-gap semiconductors and insulators may have a large impact on solid state quantum physics in general and quantum information processing and communication in particular. The reason for this is two-fold. First, defects in wide-band-gap semiconductors and dielectric host material are optically accessible, which makes them of use in quantum communication. At the same time such defects can be wired up, i.e. electrical readout of quantum states has been shown. This may enable easy scaling up and integration into existing electronic circuits. Second, defects can be considered as point-like structures with respect to delocalization of electron wave functions. As a consequence, transitions remain atomic-like and electron–phonon as well as spin–phonon coupling may be small. The largest dephasing times for spins and electronic transitions found in solids are a consequence of this. While this all sounds promising, nanostructuring of wide-band-gap semiconductors and dielectrics as well as nanodeposition of defects is still in its infancy. This, in part, is the reason for the limited number of applications in quantum information processing so far. This contribution will highlight recent results on a color center in diamond, the nitrogen-vacancy center. The defect is one of the best characterized so far. Its spin states are well suited for precise state manipulation, and nanodeposition of defects is an emerging technique.

12.2 Properties of Nitrogen-vacancy Centers in Diamond

Diamond is well known for its extreme mechanical, electrical, and optical properties, such as its hardness, wide band gap (5.5 eV), high carrier mobility, and large number of optically active defects. Interest in diamond has increased in the past decade as a consequence of the developments that have occurred in its synthesis through the chemical vapor deposition process. This technique permits the synthesis of diamond in a reproducible way, regarding its dopants and morphology. Nowadays, it is possible to generate optically active paramagnetic defects via ion implantation.

Although many such defect centers in diamond have been reported in the literature, most of them cannot be detected as single centers because they show a very weak oscillator strength and/or the presence of metastable states, which limits the magnitude of the fluorescence signal. The nitrogen-vacancy (NV) defect has up to now been the only paramagnetic system that

Quantum Information Processing, 2nd Edition. Edited by Thomas Beth and Gerd Leuchs

ISBN: 3-527-40541-0

can be detected optically as a single center. The NV defect is a naturally occurring defect in diamond with nitrogen impurities. This defect center is found particularly often in type IB synthetic diamond, in which it can be produced by irradiation and subsequent annealing at temperatures above 550 °C. Radiation damage creates vacancies in the diamond lattice. Annealing treatment leads to migration of vacancies toward nitrogen atoms, creating NV defects. The NV center can also be produced in type IIA diamonds by N^+ ion implantation. The center shows a linear Stark effect, which is due to the absence of inversion symmetry. Figure 12.1 shows the generally accepted model of the NV center. Based on neutron irradiation experiments, Mita and co-workers have reported that the NV center is negatively charged [1].

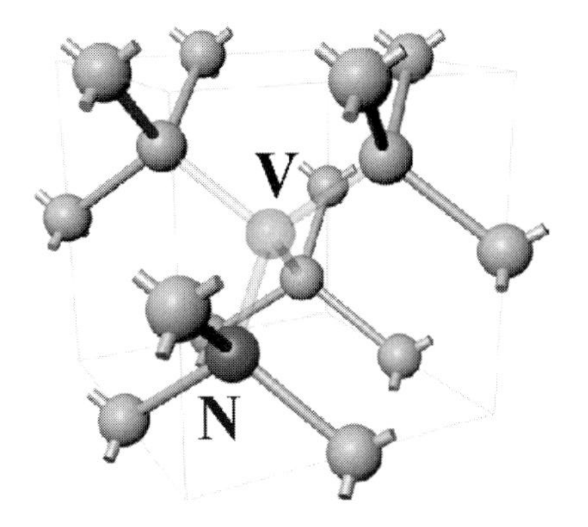

Figure 12.1: The structure of the nitrogen-vacancy defect in diamond.

The fluorescence spectrum of the center consists of a sharp zero-phonon line at 638 nm (1.945 eV) [2, 3]. The oscillator strength of the optical transition of the NV center is comparable with that of the GRI center (neutral vacancy in diamond). The fluorescence quantum yield is 0.99 [4]. The electron–phonon coupling is strong (Huang–Rhys factor $S = 3.65$). The fluorescence lifetime of the center is 11.6 ns and 13.3 ns at 77 K and 700 K, respectively [4,5].

The energy level scheme of the NV center is shown in Fig. 12.2. The ground state of the center is a spin triplet (T_0). At zero magnetic field, the ground triplet state of the center is split by the coupling of two unpaired electron spins in the diamond crystal field into three sublevels X, Y ($m_S = \pm 1$) and Z ($m_S = 0$), separated by 2.88 GHz [6, 7]. The spin–lattice relaxation time is 1.17 ms. Through low-temperature studies, the mechanism for ground-state spin–lattice relaxation has been determined to be a two-phonon process involving 63 meV phonons. The first parity-allowed transition occurs to a triplet excited state at 1.944 eV (T_1). The excited state has a more complicated fine structure, originating from spin–spin and spin–orbit interactions and from strain in the crystal. It consists of two groups, each with three sublevels separated by a few GHz within each group. The energy separation between the groups is about 40 wavenumbers [8]. The substates show a fast decay to lower-lying excited-state level [6, 9–11]. Only the lowest three spin sublevels of the excited triplet state are shown here. Satellite lines in persistent hole-burning spectra indicate a zero-field spin splitting in the excited triplet state of 0.6 GHz. The first metastable singlet excited state (S_1) is 200 cm^{-1} below the first excited singlet state [12].

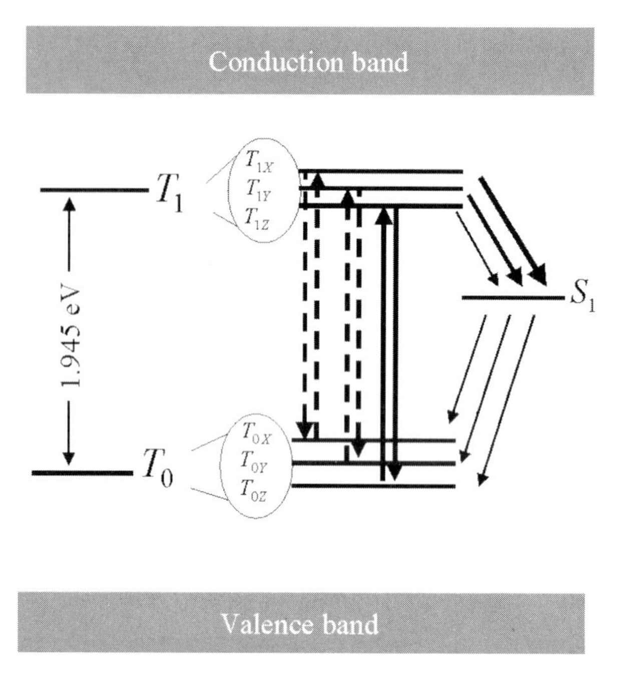

Figure 12.2: The energy level scheme of the NV defect center.

12.3 Readout of Spin State via Site-selective Excitation

The splitting between sublevels of two triplet states is larger than the homogeneous linewidth of the optical transition at $T = 2$ K. Hence, site-selective excitation can be used as a tool to determine the state of the defect center spin. When a narrow-band laser is tuned in resonance with the transition between a specific pair of spin sublevels T_{0i} and T_{1j}, the defect center can be excited to the T_1 state. Here we assume that the sublevel T_{0i} has nonzero initial population and that the transition $T_{0i} \rightarrow T_{1j}$ is allowed by selection rules. The probability of transition is proportional to $|\langle S_i | S_j \rangle|^2$, where S_i and S_j are spin wave functions of the T_{0i} and T_{1j} substates. In general, the principal spin axes X, Y, Z of the ground triplet state may be different from those of the excited triplet level due to different distributions of the electronic density of the two unpaired electrons. Hence, it can be expected that electronic transitions from every spin sublevel of the ground state to every spin sublevel of the excited state are possible. On the other hand, the symmetry of a defect often gives the invariant direction of the principal spin axes [13]. Therefore, taking into account the C_{3V} symmetry of the NV defect, the conservation of principal spin axes can be assumed. In this case, the only allowed electronic transitions are those between sublevels of the same spin components: T_{0i}, T_{1i}. Thus, there are three optically allowed transitions, which are shown in Fig. 12.2 by pairs (for excitation and emission) of vertical lines. Due to different zero-field splittings in the ground and excited states [8], those three optical resonances will appear at different spectral positions when the laser is scanned through the T_0–T_1 absorption band and three peaks can be expected in the fluorescence excitation spectra.

The fluorescence excitation spectrum of a single NV center is shown in Fig. 12.3. Surprisingly only a single excitation line appears in the excitation spectrum. The single center absorption line has a Lorentzian shape. The linewidth of around 280 MHz is larger than the limit imposed by radiative decay of the excited triplet state. The cause of line broadening can be attributed to a spectral diffusion processes. Since the spectral width of a single center is much narrower than the zero-field splitting in the ground state $\Delta E = 2.88$ GHz, the excitation line shown in Fig. 12.3 marks a *specific spin configuration of the defect.* The absence of satellites related to transitions between other sublevels has been explained by sublevel-specific intersystem crossing parameters [14]. When the excitation laser is in resonance with the T_0–T_1 transition, the fluorescence exhibits a telegraph behavior. This is shown in the inset of Fig. 12.3, where the fluorescence of a single NV defect is plotted as a function of time.

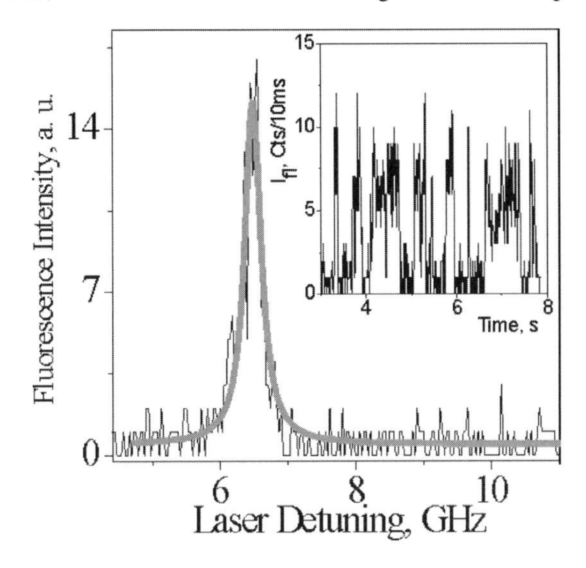

Figure 12.3: The fluorescence excitation spectrum of an NV defect in diamond at $T = 1.6$ K. The fitted curve is a Lorentzian function. The inset shows the time trace of the fluorescence signal when the excitation laser was tuned to resonance of the transition.

A detailed analysis of the excitation–emission pathway is necessary for an explanation of blinking behavior [14]. There are three different excitation pathways of a single defect center. First, as long as an optically excited defect center remains in the T_{1i}–T_{0i} channels, it emits fluorescence. The emission rate is rather high, typically on the order of tens of megahertz. Second, the defect center can also undergo intersystem crossing from T_{1i} to the singlet state S, which is characterized by the rate k_{iS} specific for this T_{1i} sublevel. As soon as an intersystem crossing (ISC) to S takes place, the single defect center transition is out of resonance with the driving laser field, and no fluorescence is expected to occur. Third, the defect center can return back to the ground triplet T_0 within the lifetime of the metastable singlet state, which typically is much longer than the radiative lifetime of the excited triplet state. Within this picture, there are three possibilities for this transition, differing by the final substate of the ground triplet state T_0. Only one of them, the ISC transition $S \rightarrow T_{0i}$, brings the molecule back into resonance with the excitation laser, thus resulting in the termination of the dark interval in the fluorescence and restoration of fluorescence emission. Note that the complete pathway $T_{1i} \rightarrow S \rightarrow T_{0i}$ conserves spin projection. The other ISC transitions, involving a change in spin projection, do not return the center to resonance with the laser because of the large zero-field splitting in the ground triplet state. To restore fluorescence in this case, a spin relaxation transition must take place. This thermally activated process is known to be slow (on the order of seconds) at liquid-helium temperatures. Another mechanism is the coupling of the triplet electronic spin of the studied single impurity in the T_0 state to electronic spins T_0 of other defects of the same type in a host lattice. These spin–spin interactions result in flip-flop processes between different centers (cross-relaxation), nearly equalizing the T_0 substate populations of the optically excited center. However, the latter mechanism is effective only at large concentrations of defect centers, while experiments with single centers are performed on samples with a low NV center concentration. If some of the pathways $T_{1i} \rightarrow S \rightarrow T_{0n}$ $(n \neq i)$ become more active, this will increase the durations of dark periods in the fluorescence.

The appearance of a single line in the excitation spectrum can be understood as follows. Let us consider in more detail the case when the laser is tuned in resonance with the fine-structure transition between the T_{0Z} and T_{1Z} substates, as shown in Fig. 12.2 with solid vertical lines. In this case the system can be described in terms of the general optical Bloch equations. The average fluorescence intensity emitted by a single molecule can be written as [14]:

$$I_{Fl} = \frac{AB}{3(A + B + k_S) + B\left(1 + \frac{k_S}{R}\frac{k_X + k_Y + R}{k_Z}\right)},$$

where A and B are the rates of spontaneous emission and absorption corresponding to transition $T_{1Z} \rightarrow T_{0Z}$; k_S is the singlet population rate corresponding to transition $T_{1Z} \rightarrow S$; k_X, k_Y, and k_Z are the singlet depopulation rates to the ground-state sublevels X, Y, and Z, respectively; and R is the spin–lattice relaxation rate. If the spin–lattice relaxation rate R is much lower than the rate of decay of the excited singlet state, the fluorescence signal under complete optical saturation can be written as $I_{Fl}^{\infty} = AR/k_S$. By substituting here the expected rates $R \sim 1\,\mathrm{s}^{-1}$ and $A \sim 10^8\,\mathrm{s}^{-1}$ and $k_S \sim A \sim 10^3\,\mathrm{s}^{-1}$, and taking into account a detection efficiency of the setup $\sim 1\,\%$, the fluorescence signal can be estimated to be on the order of a few thousands photocounts per second, which is in good agreement with experimental data. Note that faster shelving rates have recently been observed from correlation measurements on single NV centers [15–17]. These are attributed to k_S of the other transitions. If one of those other transitions is pumped, i.e. $T_{0X;Y} \rightarrow T_{1X;Y}$, then the intersystem crossing rate k_S is three orders of magnitude larger ($k_S \sim 10^6\,\mathrm{s}^{-1}$) and roughly 0.1–1 detectable photocounts per second is calculated. The intersystem crossing rate is thus essential for the possibility to detect the resonance lines of individual defects. Spectral hole-burning experiments show a $k_S = 12.4$ kHz, which can be attributed to shelving from Z sublevel [18]. The intersystem crossing process for the X and Y sublevels of the triplet excited state is probably faster. The conclusion is that those resonant lines corresponding to sublevels other than the $T_{1Z} \rightarrow T_{0Z}$ are not observable because of the low fluorescence intensity.

12.4 Magnetic Resonance on a Single Spin at Room Temperature

The manipulation of single spin states is of great importance for quantum computing applications. Single spin coherence in solids has been observed for organic systems [19–21]. However, coherent experiments on the NV center are of particular interest for two reasons. First, spin coherence time is not limited by the electronic lifetime of the excited triplet state. Second, the magnetic resonance signal can be detected at room temperature.

In general, optical excitation rates B from different spin substates T_{0ii} as well as the spontaneous emission rates A_T to these substates are different for different substates T_1. In particular, different spin sublevels of the excited electronic state are characterized by spin-selective rates for ISC transitions to and from the singlet state S_1 resulting in their different populations. A theoretical model of the photokinetics of the NV defect under room-temperature non-selective excitation [22] predicts optical alignment of the center in which most of the population is in the Z substate. The Z substate of the excited triplet state shows the lowest ISC rate. Hence, the resonant microwave decreases the population of this state, resulting in a decrease of the steady-state fluorescence signal (negative ODMR effect). Since no selective excitation is required, magnetic resonance experiments can be performed under ambient conditions.

The simplest coherence spin resonance experiment is the detection of transient nutations. In this experiment, a resonant microwave field is applied to the sample. This microwave field induces transitions between ground-state sublevels, resulting in a spin precession in a direction perpendicular to the direction of the applied field in the rotating frame. This is equivalent to a periodic change of populations of the spin sublevels.

Several experimental aspects related to measurements on a single spin must be pointed out. First, the measurement of a single electron spin is projective, i.e. readout always projects the spin state into one of the eigenstates. Therefore, coherent experiments require several measurements for detection of coherent oscillations. Second, it is necessary to initialize the state of the spin in the beginning of the experiment. This can be achieved by a strong non-selective optical excitation, which projects the NV center into the Z sublevel of the triplet ground state. Third, the projective nature of the spin detection leads to the fact that a continuous measurement of the spin state induces decoherence [23].

Figure 12.4 shows the transient nutation experiment on a single NV center. The system was polarized into the $m_S = 0$ sublevel by strong unselective optical illumination. After that, the laser was switched off and a microwave pulse of variable duration was applied. The readout of the spin state was achieved by monitoring the fluorescence intensity. The experimental data clearly demonstrate a periodic modulation of the fluorescence signal. This corresponds to the coherent oscillation of the electronic spin between the $m_S - 0$ and $m_S - 1$ sublevels. The fluorescence intensity starts at a high level, corresponding to population of the $m_S = 0$ sublevel. Upon increasing the pulse length, the fluorescence intensity decreases, reaching the case of population inversion at a pulse duration of 16 ns (π-pulse). Note that the coherent nutation experiment is equivalent to realization of a NOT gate.

The decay of coherent oscillations is determined by the electron spin dephasing time T_2, which typically ranges from 1.5 to 3 μs, depending on the defect center under study. Recently, ensemble experiments have shown that T_2 in this system can reach values larger than 60 µs, with increasing T_2 in samples with low nitrogen content [23]. The most important dephasing mechanism is spin flip-flop processes, either directly between electron spins of the NV center and residual nitrogen impurities in the diamond lattices (P-center, $S = 1/2$) or via hyperfine coupling to the nitrogen nuclear spin. In both cases, the dephasing rate is strongly distance-dependent ($1/r^3$). The differences among the T_2 values of different NV centers are possibly due to a change in the distance between the electron spin of the center and the nearest-neighbor nitrogen in the lattice.

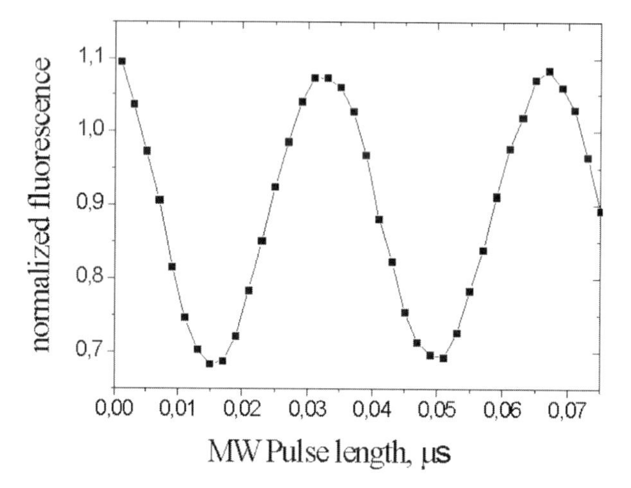

Figure 12.4: Transient nutation of the electron spin between ground-state sublevels of a single NV defect (ambient conditions). The spin states have been initialized by non-selective optical pulse. Subsequently, a resonant microwave pulse of variable duration was applied. The state was read out optically. 106 measurements have been accumulated to obtain a smooth curve.

12.5 Magnetic Resonance on a Single ^{13}C Nuclear Spin

The nuclear spin I of the most abundant carbon isotope ^{12}C is $I = 0$ and the ^{13}C isotope has a nuclear spin $I = 1/2$. The ^{13}C nuclear spins in the first coordination shell could be used as the second qubit. Approximately 30 % of the spin density (the probability of finding the electron at the place of the nuclei) of both unpaired electron spins is at the next-nearest-neighbor carbons [24]. According to the McConnell relation, the spin density is proportional to the hyperfine splitting. As a result of the exponential decrease of the spin density distribution, the hyperfine coupling of the more distant carbon to the electron spin is negligible.

If an electron spin is interacting with a paramagnetic nuclear spin, the spin Hamiltonian describing the coupled system is

$$\hat{H} = g_e\beta_e\hat{S}\hat{B} + \hat{S}\overleftrightarrow{D}\hat{S} + \hat{S}\overleftrightarrow{A}\hat{I} - g_n\beta_n\hat{I}\hat{B}.$$

Here $\overleftrightarrow{D}$ is the fine-structure tensor due to the interaction of the two uncoupled electron spins, and $\overleftrightarrow{A}$ is the hyperfine interaction tensor related to coupling between the electron and the nuclear spin. The hyperfine coupling of a ^{13}C ($I = 1/2$) nucleus in the first coordination shell around the defect center is known to be 130 MHz [25]. The natural abundance of ^{13}C in the samples used is 1.1 %. Hence, one out of 30 defect centers should have a ^{13}C in either of the positions 1 to 3. The CW ESR spectrum of single NV defect coupled to a single ^{13}C nucleus is shown in Fig. 12.5.

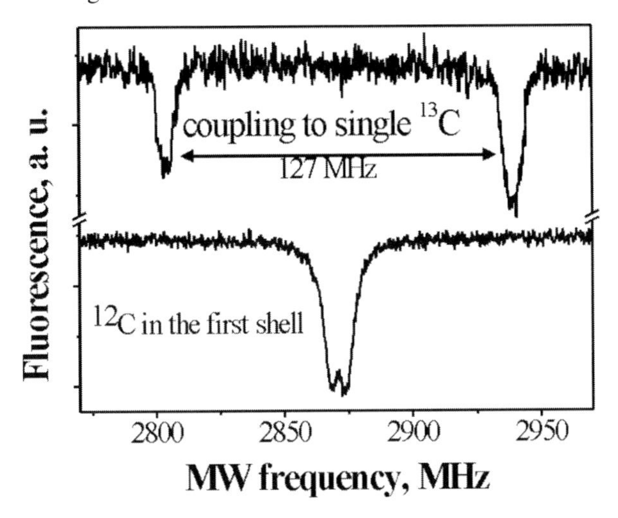

Figure 12.5: The CW ODMR spectrum of a single NV defect with single ^{13}C nucleus in the first shell. The lower spectrum represents the case in which all carbon of the first shell is the ^{12}C isotope.

A scheme describing the spin levels relevant in this situation is shown in Fig. 12.6A. Among the spin levels, four transitions are allowed in first order. A and B are electron spin resonance transitions ($\Delta m_S = \pm 1$, $\Delta m_I = 0$) and C and D are nuclear magnetic resonance transitions ($\Delta m_I = \pm 1$, $\Delta m_S = 0$). The splitting between states 1 and 2 is determined by the hyperfine coupling of the ^{13}C nucleus ($\approx$ 130 MHz), whereas the splitting between states 3 and 4 is given by the nuclear Zeeman interaction (2–10 MHz).

Since our readout scheme is only sensitive to the electron spin state, all changes in the nuclear spin states need to be detected via the electron spin during pulsed NMR experiments. This is equivalent to an optically detected electron nuclear magnetic double resonance (ENDOR) experiment [26]. Figure 12.6C shows Rabi nutations of the nuclear spin measured by this technique. In the experiment the electron spin is initialized first by a laser pulse (duration 3 μs). After initialization (see Fig. 12.6B), the system is found either in state 3 or 4. If the system is in state 4, a new initialization is started until state 3 is populated. Starting from this state, a frequency-selective electron spin resonance π-pulse with center frequency A is used to drive the system from $|00\rangle$ to $|10\rangle$. RF with variable duration drives the system between levels $|10\rangle$ and $|11\rangle$. Finally a selective ESR π-pulse at frequency A converts the population difference between nuclear spin sublevels into an optically detectable ESR signal. Transient nutation on a single nuclear spin shows no dephasing (Fig. 12.6C). Spin memory times of as long as 100 μs have been reported for ^{13}C nuclei in high-purity diamond [27]. Compared to the value obtained for single nuclei, it can be concluded that the hyperfine coupling to the electron spin of the NV center does not contribute as an additional source of decoherence to the single nuclear spin.

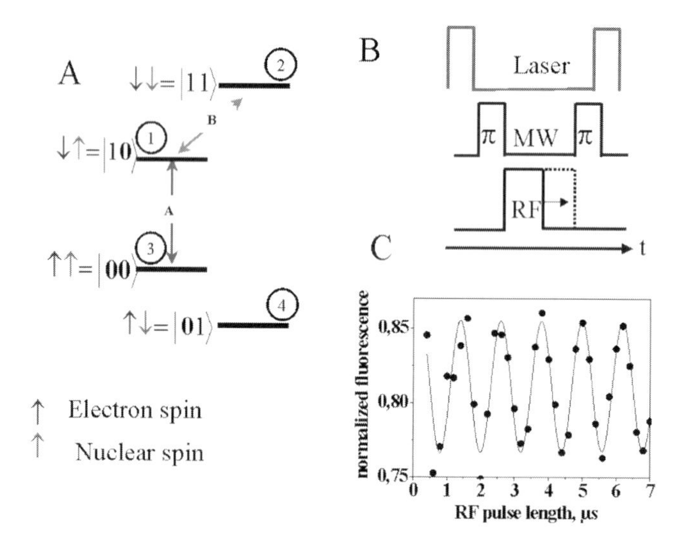

Figure 12.6: Spin energy level scheme (A) and pulse sequence (B) relevant for the ENDOR experiment. The energy levels describe the interaction of a single electron with a single ^{13}C nuclear spin in the ground state of the defect. The quantum numbers of states 3 and 4 are $mS = 0$, $mI = +1/2$ and $1/2$. The states 1 and 2 comprise two degenerate electron spin states $mS = 1$ with nuclear spin quantum numbers $mI = +1/2$ and $1/2$. The pulse sequence (B) used in the experiment comprises laser excitation, microwave (MW) and radio-frequency (RF) irradiation. (C) Transient nutation of a single ^{13}C nuclear spin.

12.6 Two-qubit Gate with Electron Spin and ^{13}C Nuclear Spin of Single NV Defect

The observation of Rabi nutations on ESR and NMR transitions provides the basis for a conditional two-qubit quantum gate. For a CROT gate one qubit is inverted depending on the state of the other qubit. A CROT gate is equivalent to a CNOT gate, except for a $\pi/2$ rotation of the nuclear spin around the z-axis [28]. In our experiments we have chosen the electron spin as control bit and the nuclear spin as the target bit. The CROT gate is then realized by a π-pulse on transition C (Fig. 12.6A). The result of the CROT gate is state $|11\rangle$ if the qubit was $|10\rangle$ before the application of the gate (see Fig. 12.6A).

In order to check the quality of the state prepared by the CROT gate in our experiment, density matrix tomography of the state after the gate has been carried out. To this end, a series of measurements on the diagonal as well as off-diagonal elements of the density matrix have been performed. For measurement of the diagonal elements, the signal strength of the transitions have been measured and normalized to the respective signal intensities of the initial state. The off-diagonal elements related to single quantum coherences have been reconstructed by first applying a $\pi/2$ pulse on the transition where the coherences should be measured. Subsequently, the amplitude of the Rabi nutations on the respective transition has been used to calculate the off-diagonal elements. Zero- or two-quantum coherences were first converted into measurable single-quantum coherences before their values were determined. An example of the density matrix reconstruction is shown in Fig. 12.7. The density matrix tomography shows the state of the system after a π-pulse on transition A and subsequent application to the CROT gate (π-pulse on transition B, Fig. 12.6A). In the reconstruction we assumed that the matrix is diagonal after the CROT gate and that it is real. In the ideal case, without decoherence and perfect pulse angles, the only non-zero matrix element of the density matrix after the CROT gate should be $\rho_{22} = 1$, provided that in the initial density matrix $\rho_{11} = 1$. However, in the present case the dephasing and finite linewidth need to be considered. This is why Fig. 12.4 also shows a numerical simulation of the density matrix after the gate.

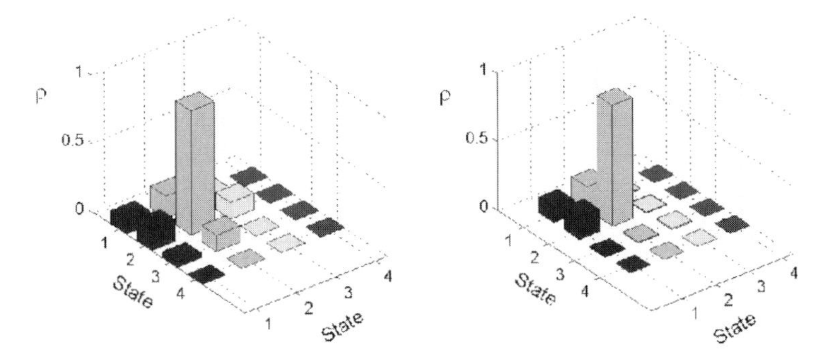

Figure 12.7: Density matrix of the state of the system after the CROT gate. The left part of the figure shows the experimentally determined values, and the right part shows the result of a simulation

The gate quality is measured by the fidelity F given by $F = \text{Tr}[\rho_P(t)\rho_I(t)]$ [29, 30], where $\rho_P(t)$ is the measured density matrix and $\rho_I(t)$ is the ideal one. The fidelity of a NOT gate on the nuclear spin is $F = 0.89$, less ideal than for the electron spin. This value limits the performance of the CROT gate. For a realistic comparison between experiment and theory, a simulation of the density matrix in Fig. 12.7 has been carried out by calculating $\rho(t) = S^{-1}\rho(t{=}0)S$, where S is a unitary matrix describing the action of the pulses on the spins in the rotating frame. By taking into account the linewidth of transitions A and C as well as the measured dephasing time and the pulse length used, the simulation reproduces the experiment well.

The NV center in diamond provides an opportunity to study the physics of single spins and small clusters of spins, or to create certain interesting quantum states with single spins. Entanglement between electron and nuclear spins had been using bulk ESR and NMR approaches. Since it is possible to precisely control the quantum state of single spins, a next logical step would be to create, e.g., Bell states and probe the quantum correlation among the two spins. This would be a test of Bell's inequality with spins in solids.

12.7 Outlook: Towards Scalable NV Based Quantum Processor

Well established single electron [23] and nuclear [31] spin measurement of the NV defect is a key step in solid state quantum information processing [32], as it allows for the assessment of the implemented quantum operations, such as the generation of spin entanglement. CNOT and other quantum operations between adjacent electron spins can be implemented using ESR analogues of well known NMR pulse sequences [33]. Switching on and off the interactions will be accomplished using refocussing in the time domain. Long decoherence time (100 μs at room temperature) and fast gates based on strong dipole–dipole coupling between closely spaced electron spins (50 MHz for 1 nm spacing) allow one to reach the quantum error correction fidelity threshold ($T_{\text{gate}}/T_2 = 10^{-5}$).

Further scaling up the quantum processor requires local addressing of spins in a long spin chain. The following methods can be explored for local addressing. First, optical selection in the frequency domain (requires low-temperature operation and Stark shift-based frequency tuning of NV centers). Second, ESR frequency selection by tuning the spin resonance frequency in a magnetic field gradient at room temperature. The difference in magnetic field between neighboring electron spins of the NV chain will produce the spectroscopic splitting of individual NV defects, which allow local addressing using selective ESR pulses.

Six years ago an innovative proposal of Kane [34] stimulated considerable interest in solid state quantum computing. The main advantage of the silicon approach is potential scalability with techniques developed by the semiconductor industry for producing integrated circuits. We believe that a quantum processor based on defects in diamond combine all the advantages of Kane's idea and a well developed optical readout technique. This might envision room-temperature quantum computing.

Acknowledgements

The authors are grateful for the financial support of the EU (QIPDD-ROSES) and the German Science foundation via the Schwerpunktprogramm Quanteninformationsverarbeitung and the graduate college Magnetic Resonance.

References

[1] Y. Mita, Phys. Rev. B 53, 11360–11364 (1996).

[2] G. Davies, Properties and Growth of Diamond (INSPEC, London, 1994).

[3] G. Davies, and M. F. Hamer, Proc. Roy. Soc. London A 384, 285–298 (1976).

[4] A. Zaitsev, Optical Properties of Diamond – A Data Handbook (Springer-Verlag, Berlin, 2001).

[5] A. T. Collins, M. F. Thomaz, and M. I. B. Jorge, J. Phys. C – Solid State Phys. 16, 2177–2181 (1983).
[6] E. van Oort, PhD Thesis (Amsterdam, 1990).
[7] N. R. S. Reddy, N. B. Manson, and E. R. Krausz, J. Lumin. 38, 46–47 (1987).
[8] J. P. D. Martin, J. Lumin. 81, 237–247 (1999).
[9] S. C. Rand, A. Lenef, and S. W. Brown, J. Lumin. 60-1, 739–741 (1994).
[10] A. Lenef, and S. C. Rand, Phys. Rev. B 53, 13441–13455 (1996).
[11] A. Lenef et al., Phys. Rev. B 53, 13427–13440 (1996).
[12] A. Dräbenstedt et al., Phys. Rev. B 60, 11503–11508 (1999).
[13] B. Kozankiewicz et al., J. Phys. Chem. A 104, 7464–7468 (2000).
[14] A. P. Nizovtsev et al., Optics Spectrosc. 94, 848–858 (2003).
[15] C. Kurtsiefer, S. Mayer, P. Zarda, and H. Weinfurter, Phys. Rev. Lett. 85, 290–293 (2000).
[16] A. Beveratos, R. Brouri, T. Gacoin, J. P. Poizat, and P. Grangier, Phys. Rev. A 6406 (2001).
[17] A. Beveratos et al., Eur. Phys. J. D 18, 191–196 (2002).
[18] S. C. Rand, in Properties and Growth of Diamond, G. Davies, Ed. (INSPEC, London, 1994), chap. 9.
[19] J. Wrachtrup, C. von Borczyskowski, J. Bernard, M. Orrit, and R. Brown, Nature 363, 244–245 (1993).
[20] J. Wrachtrup, C. von Borczyskowski, J. Bernard, R. Brown, and M. Orrit, Chem. Phys. Lett. 245, 262–267 (1995).
[21] J. Wrachtrup, C. von Borczyskowski, J. Bernard, M. Orrit, and R. Brown, Phys. Rev. Lett. 71, 3565–3568 (1993).
[22] A. P. Nizovtsev, S. Y. Kilin, C. Tietz, F. Jelezko, and J. Wrachtrup, Physica B 308, 608–611 (2001).
[23] F. Jelezko, T. Gaebel, I. Popa, A. Gruber, and J. Wrachtrup, Phys. Rev. Lett. 92, 076401 (2004).
[24] M. Luszczek, R. Laskowski, and P. Horodecki, Physica B – Condensed Matter 348, 292–298 (2004).
[25] X. F. He, N. B. Manson, and P. T. H. Fisk, Phys. Rev. B 47, 8816–8822 (1993).
[26] K. P. Dinse, and C. J. Winscom, in Triplet State ODMR Spectroscopy, R. H. Clarke, Ed. (Wiley, New York, 1982).
[27] K. Schaumburg, E. Shabanova, J. P. F. Sellschop, and T. Antony, Solid State Commun. 91, 735–739 (1994).
[28] P. C. Chen, C. Piermarocchi, and L. J. Sham, Phys. Rev. Lett. 8706, art-067401 (2001).
[29] J. F. Poyatos, J. I. Cirac, and P. Zoller, Phys. Rev. Lett. 78, 390–393 (1997).
[30] W. G. van der Wiel et al., Rev. Mod. Phys. 75, 1–22 (2003).
[31] F. Jelezko, T. Gaebel, I. Popa, M. G. A. Domhan, and J. Wrachtrup, Phys. Rev. Lett. 93, 130501 (2004).
[32] F. Jelezko, and J. Wrachtrup, J. Phys. – Condensed Matter R1089–R1104 (2004).
[33] J. Twamley, Phys. Rev. A 67 (2003).
[34] B. E. Kane, Nature 393, 133–137 (1998).

13 Quantum Dynamics of Vortices and Vortex Qubits

A. Wallraff[1], *A. Kemp, and A. V. Ustinov*

Physikalisches Institut III
Universität Erlangen-Nürnberg
Erlangen, Germany

13.1 Introduction

Vortices appear naturally in a wide range of gases and fluids, both on very large scales, e.g. when tornadoes form in the Earth's atmosphere, and on very small scales, e.g. in Bose–Einstein condensates of dilute atomic gases [1] or in superfluid helium [2], where their existence is a consequence of the quantum nature of the liquid. Collective nonlinear excitations such as the vortex considered here are ubiquitous in solid state systems (e.g. domain walls) and biological systems (e.g. waves on membranes), and have even been considered as model systems in particle physics. In superconductors, which we consider here, quantized vortices of the supercurrent [3], which are generated by magnetic flux penetrating into the material, play a key role in understanding the material properties [4] and the performance of superconductor-based devices [5,6]. At high temperatures the dynamics of vortices is essentially classical. At low temperatures, however, there are experiments suggesting the collective quantum dynamics of vortices [7,8]. Here we report on experiments in which we have probed for the first time the quantum dynamics of an *individual* vortex in a superconductor. By measuring the statistics of the vortex escape from a controllable pinning potential, we were able to demonstrate the quantization of the vortex energy within the trapping potential well and the quantum tunneling of the vortex through the pinning barrier.

The object that we have studied in our experiments is a vortex of electric current with a spatial extent of several tens of micrometers formed in a long superconducting tunnel junction. The electrodynamics of such a Josephson junction is governed by the phase difference φ between the macroscopic wave functions describing the superconducting condensate in the two electrodes [9]. It is a well established fact that the variable φ displays macroscopic quantum properties in point-like junctions at very low temperatures [10, 11]. Such *macroscopic* quantum phenomena are currently exploited for quantum information processing [12] using superconducting devices [10, 13–20]. In extended one- or two-dimensional Josephson junction systems, quantum tunneling in real space is to be expected for superconducting vortices, which are particle-like collective excitations of the phase difference φ. The small value of the expected mass of the vortex studied here suggests that quantum effects are likely to occur with vortices at low temperatures. Dissipative vortex tunneling has been given as a reason

[1]Current address: Department of Applied Physics, Yale University, New Haven, CT 06520, USA; email: andreas.wallraff@yale.edu

Quantum Information Processing, 2nd Edition. Edited by Thomas Beth and Gerd Leuchs

ISBN: 3-527-40541-0

for the non-vanishing relaxation rate of the magnetization in type-II superconductors as the temperature T is lowered toward zero. However, the quantum vortex creep model [8], which was suggested as an explanation for this behavior, faced orders of magnitude discrepancies with experimental data [21]. Alternative classical explanations have been suggested more recently to explain the low-temperature relaxation [22]. Nonetheless, macroscopic quantum tunneling was observed for states with many vortices in discrete arrays of small Josephson junctions [23, 24]. For typical arrays the calculated vortex mass is about 500 times smaller than the electron mass [7]. All previous research in this area has focussed on the collective behavior of a large number of vortices. Until now there had been no direct experimental observation of tunneling events of *individual* vortices.

In Sec. 13.2 we discuss the experimental observation of quantum tunneling and energy level quantization of an individual vortex. The formation and subsequent dissociation of vortex–antivortex pairs is covered in Sec. 13.3. In Sec. 13.4, the prospects of using quantum vortices in heart-shaped junctions as qubits for quantum information processing are evaluated and results on the manipulation of bistable vortex states are presented.

13.2 Macroscopic Quantum Effects with Single Vortices

Among the many different vortex structures in superconductors, there is a rather special type of vortices in long Josephson junctions. These vortices have a unique character of solitons – nonlinear waves that preserve their shape with time and propagate as ballistic particles [25]. These vortices are distinct from Abrikosov vortices in type-II superconductors, as they have no normal core and thus move with very low damping. In contrast to vortices in Josephson arrays [7], solitons in uniform long junctions do not generate any radiation during their motion and are well decoupled from other electromagnetic excitations in these systems. The quantum tunneling of Josephson vortices in long junctions has been predicted theoretically [26, 27]. In our experiments we have observed this effect for the first time [28].

13.2.1 Quantum Tunneling

We have probed the quantum properties of a single Josephson vortex in a current-biased annular junction subject to an in-plane magnetic field $\vec{H}$ (see inset of Fig. 13.1c). The junction of diameter $d = 100$ µm and width $w = 0.5$ µm is etched from a sputtered Nb/AlO_x/Nb thin film trilayer which is patterned using electron-beam lithography [29]. A photograph of the sample taken using an optical microscope is shown in Fig. 13.1a. Initially the vortex is topologically trapped in the junction by cooling the sample in a small perpendicular magnetic field $\vec{H}_{\mathrm{tr}}$ which is generated using a separate pair of coils, the axis of which is perpendicular to the junction plane (see Fig. 13.1b). A single vortex in an annular junction subject to an in-plane field behaves as a particle [30] in a tilted washboard potential [31, 32]. The component of the potential periodic in the vortex coordinate Θ is due to the interaction $\vec{\mu} \cdot \vec{H} \propto \cos\Theta$ of the vortex magnetic moment $\vec{\mu}$ with the external magnetic field $\vec{H}$ (see Fig. 13.1c). The tilt of the potential is proportional to the Lorentz force acting on the vortex, which is induced by the bias current I applied to the junction. The vortex may escape from a well in the tilted potential by a thermally activated process or by quantum tunneling. At low temperatures thermal

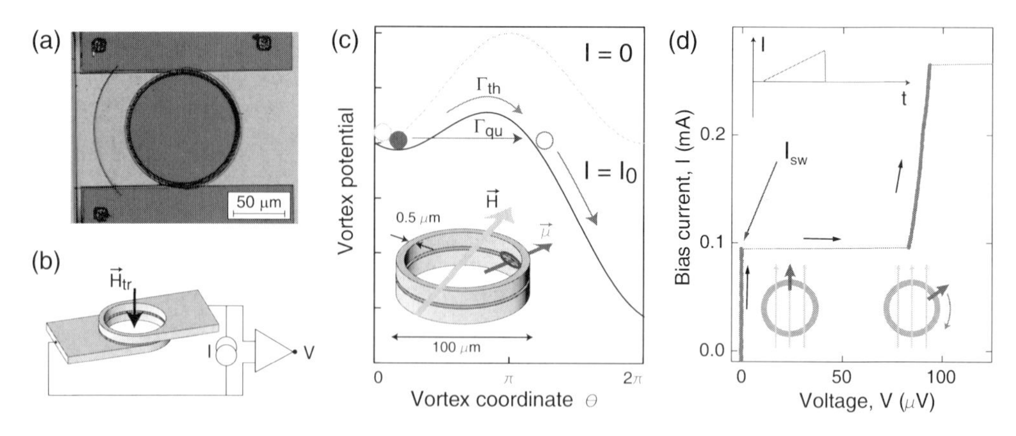

Figure 13.1: Sample, vortex potential, and switching current measurement. (a) Photograph of the sample taken with an optical microscope. (b) Sketch of bias lead configuration and direction of trapping field $\vec{H}_{\mathrm{tr}}$. We employ a bias lead geometry that minimizes self-field effects [36] and apply the magnetic field $\vec{H}$ generating the vortex potential in the plane of the junction but perpendicular to the bias leads. In this case the vortex is depinned in a location along the junction where self-fields are minimal. (c) A vortex with magnetic moment $\vec{\mu}$ trapped in an annular Josephson junction subject to an in-plane external magnetic field $\vec{H}$. The junction width and diameter are indicated (see inset). The resulting vortex potential at zero bias current I (dashed line) and at finite bias I_0 (solid line) is plotted versus the vortex coordinate Θ in the annulus. Different escape processes of the vortex are indicated. (d) Current–voltage characteristic showing the vortex depinning from the field induced potential at a random value of bias I_{sw} when ramping up the bias current at a constant rate in a sawtooth pattern (see upper inset). The transition of a pinned vortex state to a running vortex state is associated with a voltage appearing across the junction, which is proportional to the vortex velocity (see lower insets). In the experiment, the bias current is switched off immediately after a voltage is detected.

activation is exponentially suppressed and escape occurs by quantum tunneling. This process is identified by measuring the temperature dependence of the distribution P of depinning currents I_{sw} [33, 34] of the vortex trapped in the junction. The $P(I_{\mathrm{sw}})$ distributions are recorded by repeatedly ramping up the bias current at a constant rate $\dot{I}$ and recording the statistics of the current I_{sw} at which the vortex escapes from the well, which is associated with a switching of the junction from its zero-voltage state to a finite-voltage state (see Fig. 13.1d). Our measurement technique and setup have been tested and calibrated [34, 35] in experiments on macroscopic quantum tunneling of the phase in small Josephson junctions [11].

The bias current I_p at which vortex tunneling occurs with the highest probability (for a given bias current ramp rate) is found – as expected [31,32] – to be proportional to the applied magnetic field (data shown in Fig. 13.3c). This indicates that the shape of the potential is controlled in our experiment by both field and bias current. The bias-current-induced self-magnetic-field effect on the vortex potential has been minimized by using an appropriate bias lead configuration and field direction (see Fig. 13.1b).

To search for quantum tunneling of the vortex, the temperature dependence of the switching current distribution $P(I_{\mathrm{sw}})$ has been measured. In Fig. 13.2a such distributions are shown for a magnetic field of $H = 0.9$ Oe applied in the plane of the junction. It is clearly ob-

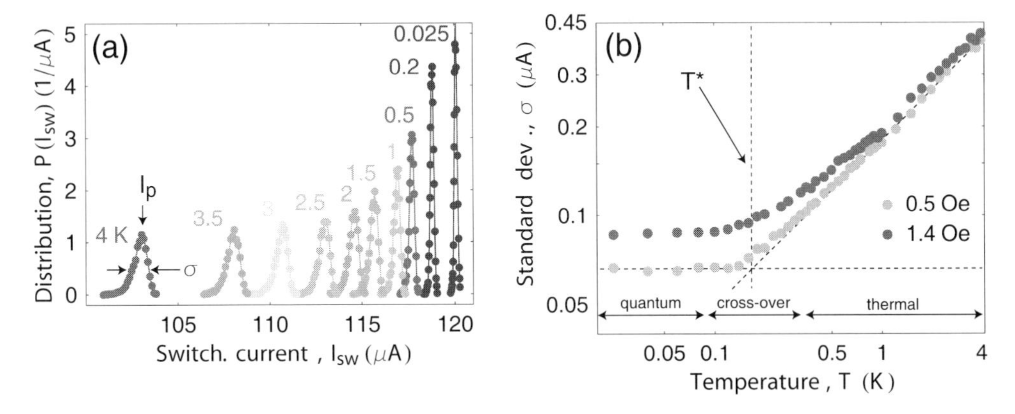

Figure 13.2: Thermal activation and quantum tunneling. (a) Switching current distributions $P(I_{\rm sw})$ at magnetic field $H = 0.9$ Oe for bath temperatures T between 4.0 and 25 mK. (b) Standard deviation σ of $P(I_{\rm sw})$ versus T for two values of field, indicating the cross-over in the vortex escape process from thermal activation to quantum tunneling. The cross-over temperature range around $T^\star$ is indicated. We have carefully verified experimentally that the saturation of the distribution width with temperature is not induced by excess noise or heating of the sample. The resolution of our measurement setup [34] is a factor of 3 to 4 larger than the most narrow distribution widths measured in these experiments.

served that the distribution width σ decreases with temperature and then *saturates* at low T. In Fig. 13.2b, σ is plotted versus temperature on a double logarithmic scale for two different values of the field. At high temperatures the distribution width is temperature-dependent, indicating the thermally activated escape of the vortex from the well. In the high-temperature limit σ is in good approximation proportional to $T^{2/3}$ as expected for a thermally activated escape of a particle from a washboard potential close to critical bias. The distribution width σ saturates at a temperature of about 100 mK. This behavior indicates the cross-over of the vortex escape process from thermal activation to quantum tunneling. At temperatures below 100 mK, σ is constant and the escape is dominated by quantum tunneling. As expected, the cross-over temperature $T^\star$ is dependent on magnetic field, but only rather weakly. We attribute this observation to the fact that the vortex may change its shape when traversing the barrier during the escape process. This aspect cannot be captured in the single particle model [30,31] but rather is a consequence of the fact that the vortex is a collective excitation.

13.2.2 Energy Level Quantization

To probe the energy levels of the vortex in the potential well, we have measured the vortex escape in the presence of microwave radiation using spectroscopic techniques that we have extensively tested on small Josephson junctions [35]. At low temperatures and in the absence of microwave radiation, the vortex tunnels out of the ground state of the potential well into the continuum. The occupation of excited states is exponentially small, if the level separation is larger than the temperature. By irradiating the sample with microwaves, the vortex can be excited resonantly from the ground state to the first excited state (see inset of Fig. 13.3b). In this case we observe a double-peaked structure in the switching current distribution (see

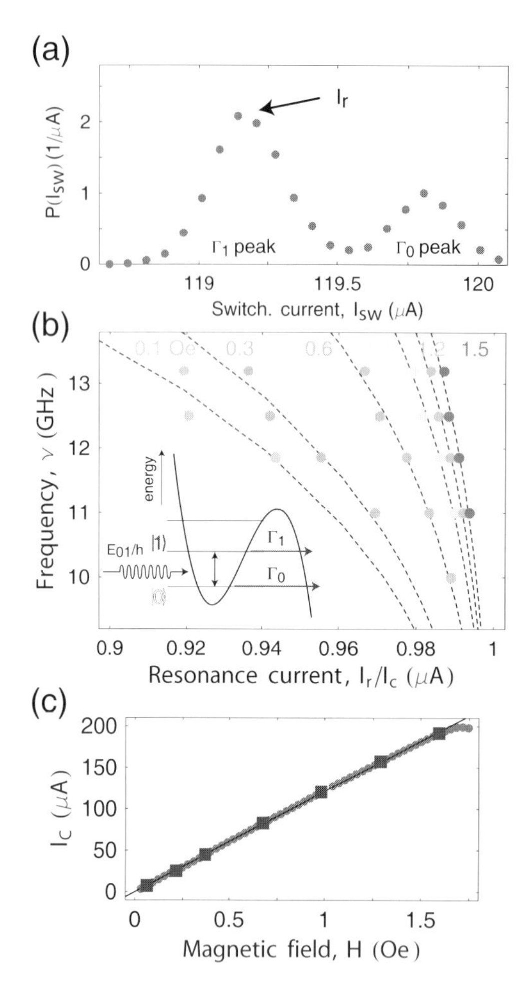

Figure 13.3: Vortex energy levels. (a) $P(I_{\mathrm{sw}})$ distribution at $H = 0.6$ Oe in the presence of microwave radiation of $\nu = 11$ GHz at $T = 25$ mK. The resonance current I_r at which tunneling from the first excited state is most probable is indicated. (b) Microwave frequency ν versus normalized resonance current $I_r/I_c(H)$ for magnetic fields H between 0.1 and 1.5 Oe. The dashed curves are fits to $\nu_{01}(H, I{=}0)\{1 - [I/I_c(H)]^2\}^{1/4}$. In the inset the tunneling from the ground state and from the first excited state populated by resonant microwave radiation is indicated. (c) Critical current I_c extracted from microwave spectroscopy (solid squares), most probable switching current I_p at $T = 25$ mK in the absence of microwaves (solid circles) and fit (solid line) versus magnetic field H.

Fig. 13.3a). The peak at higher bias current is due to the tunneling of the vortex from the ground state, whereas the peak at lower bias current corresponds to the tunneling out of the first excited state.

The energy level separation scales with both bias current and applied magnetic field. To investigate this property, we have spectroscopically determined the separation between the ground and the first excited state by varying the microwave frequency and the magnetic field. For each value of the magnetic field, we have determined the resonance current I_r (see Fig. 13.3a) for a few different microwave frequencies ν. In Fig. 13.3b, the applied microwave frequency ν is plotted versus the resonance current I_r normalized by the depinning current $I_c(H)$ at that field in the absence of microwaves and fluctuations. It is observed that all data points show the characteristic scaling of the energy level separation $\Delta E_{01}/h$ with the bias current as $\nu_{01}(H, I) = \nu_{01}(H, I{=}0)\{1 - [I/I_c(H)]^2\}^{1/4}$ as expected for the tilted washboard potential. The data at each field are fitted to this dependence using the characteristic frequency $\nu_{01}(H, I{=}0)$ of vortex oscillations at zero current and the depinning current $I_c(H)$ as fitting parameters (see dashed lines in Fig. 13.3b). The resonance frequency $\nu_{01}(H, I{=}0)$ and the depinning current $I_c(H)$ in the absence of fluctuations have been determined from the fit. As expected, the energy level separation ΔE_{01} increases with the field, and the depinning current I_c as determined from the spectroscopic data is linear in the field (see Fig. 13.3c), which is in excellent agreement with the current I_p measured directly in the absence of microwaves.

From the resonance frequency at a given bias current, we can estimate the cross-over temperature, which in the limit of small damping is given by $T^\star \simeq h\nu_{01}(H, I)/2\pi k_B$. Thus we can compare the cross-over temperature extracted from the temperature dependence of the switching current distributions to the predictions based on the data extracted from spectroscopic measurements. For $\nu_{01}(H, I)$ between 10 and 13 GHz, we find a value of $T^\star$ between approximately 75 and 100 mK, which is consistent with the measured saturation temperature in Fig. 13.2.

13.3 Vortex–Antivortex Pairs

The thermal and the quantum dissociation of a single vortex–antivortex (VAV) pair in an annular Josephson junction is experimentally observed and theoretically analyzed. In our experiments the VAV pair is confined in a pinning potential controlled by external magnetic field and bias current. The dissociation of the pinned VAV pair manifests itself in a switching of the Josephson junction from the superconducting to the resistive state. The observed temperature and field dependence of the switching current distribution is in agreement with the analysis. The cross-over from the thermal to the macroscopic quantum tunneling mechanism of dissociation occurs at a temperature of about 100 mK.

13.3.1 Thermal and Quantum Dissociation

In this section we report on the experimental observation of thermal and the quantum dissociation of a *single vortex–antivortex pair* [37]. States containing many VAV pairs are relevant to thin superconducting films or large two-dimensional Josephson arrays close to the Kosterlitz–Thouless transition [9]. A *single* VAV pair naturally appears in a long *annular* Josephson junction placed in an external magnetic field H parallel to the junction plane [31, 38] (Fig. 13.4a,b). For experiments we fabricated a junction of diameter $d = 100$ µm and width $w = 0.5$ µm, which was etched from a sputtered Nb/AlO_x/Nb thin-film trilayer and patterned using electron-beam lithography [29]. Its critical current density is 220 A/cm^2, the Josephson length is $\lambda_J \approx 30$ µm and the normalized junction length is $L \equiv \pi d/\lambda_J \approx 10.5$. The measured magnetic field dependence of the switching current is shown in Fig. 13.4c. In the field range $|H| < 1.5$ Oe (main central lobe), the switching of the Josephson junction from the superconducting state to the resistive one occurs through the *dissociation* of a single field-induced VAV pair confined in the potential well created by externally applied magnetic field and dc bias current. This process is confirmed by direct numerical simulations of the full sine–Gordon equation for an annular Josephson junction [31, 38] of length L. The numerically found magnetic field dependence of the switching current is in excellent agreement with the measurement (see solid line in Fig. 13.4c). Simulations of the magnetic field distribution in the junction clearly show the nucleation and subsequent dissociation of the VAV pair (see

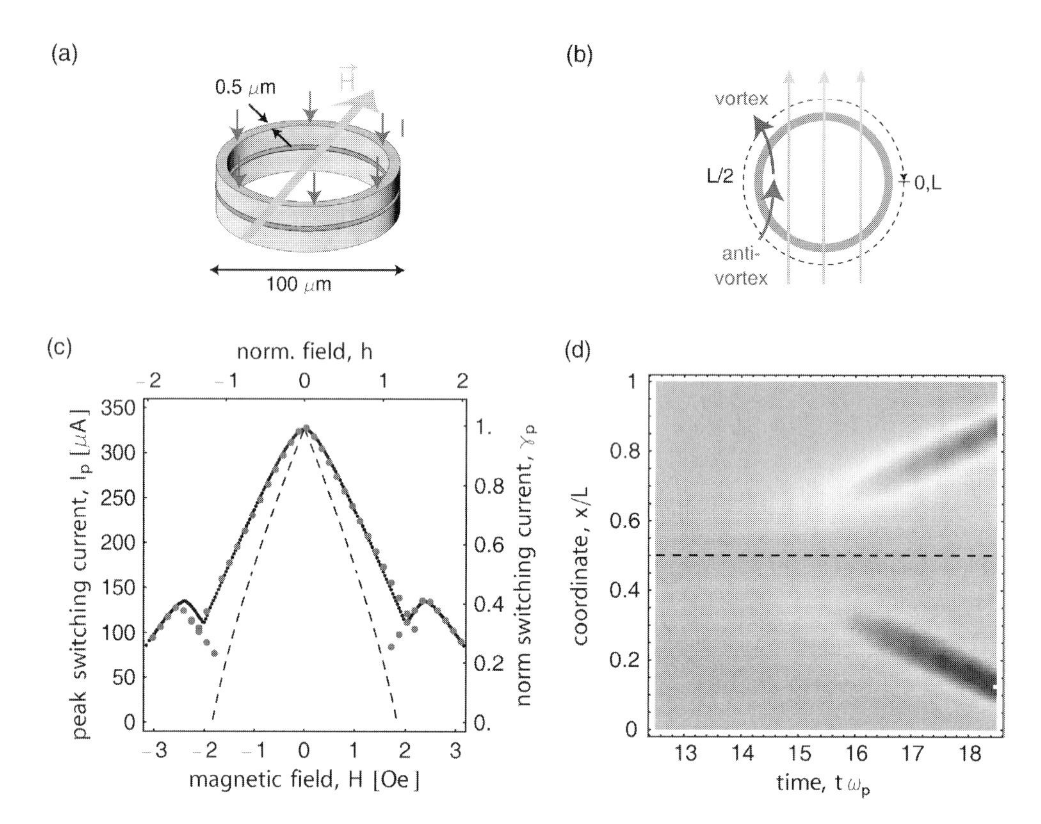

Figure 13.4: (a) Schematic view of a long annular Josephson junction *without trapped vortices* in an in-plane external magnetic field H with uniform bias current I. (b) Generation of a confined vortex–antivortex pair with the center coordinate at $x = L/2$. (c) Magnetic field dependence of switching currents: experimental data at $T = 100$ mK (circles), numerical calculation for $L = 10.5$ (solid line), and theoretical prediction of Eq. (13.6) (dashed line). (d) Numerically simulated evolution of the magnetic field distribution, as the Josephson junction switches into the resistive state. The emerging vortex and antivortex move in opposite directions.

Fig. 13.4d). Fluctuations, thermal and quantum, induce internal oscillations of the confined pair. At high temperature the dissociation then takes place by thermal activation over the barrier. At low temperature, *macroscopic quantum tunneling* through the barrier occurs. At fields $|H| > 1.5$ Oe the system becomes bistable as a well separated VAV pair penetrates in the junction. This state is perfectly reproduced by our numerical calculations (see the first sidelobes in Fig. 13.4c), and its quantum dynamics will be discussed elsewhere.

We have theoretically analyzed the penetration and the following dissociation of a VAV pair in the presence of a small magnetic field H and a large dc bias current I [37]. The bound vortex–antivortex pair is confined by a potential formed by the bias current and the magnetic field. Its dissociation can be mapped into the well known problem of particle escape from a cubic potential [37]. The probability of the dissociation depends on the height of the effective potential barrier

$$U_{\text{eff}}(\delta) = 2 \times 3^{5/4} h^{-1/4} [\delta - \delta_c(h)]^{3/2}, \tag{13.1}$$

where $\delta = (I_{c0} - I)/I_{c0} \ll 1$, with I_{c0} being the critical current of a long Josephson junction for $H = 0$, $h \propto H$ is the normalized external magnetic field, and $\delta_c(h) = 2h/3$ is the critical current in the absence of fluctuations.

At high temperatures, the dissociation is driven by thermal activation over this barrier. Using the known theory describing the escape from such a potential well [9, 11], we find the switching rate of a long Josephson junction from the superconducting state to the resistive state:

$$\Gamma_T(I) \propto \exp\left(-U_{\text{eff}}/k_B T\right). \tag{13.2}$$

Thus, at high temperature the standard deviation of the critical current σ increases with temperature and weakly depends on the magnetic field: $\sigma_T \propto T^{2/3} h^{1/6}$. Notice that σ_T *increases* with H, in contrast to the behavior of a small Josephson junction [9, 11], where $\sigma_T \propto [I_c(H)]^{1/3}$ decreases with H.

We have experimentally investigated the fluctuation-induced dissociation of the VAV pair by measuring the temperature and magnetic field dependence of the statistical distribution P of the switching currents $I < I_{c0}$ using techniques described in [34]. In Fig. 13.5a the temperature dependence of the switching current distribution measured at $H = 0$ is shown. At high temperatures the $P(I)$ distribution is temperature-dependent; at low temperatures a saturation is observed. In Fig. 13.5b, the standard deviation σ of $P(I)$ is plotted versus bath temperature T for two values of magnetic field; σ is well approximated by a $T^{2/3}$ dependence on the temperature, and the standard deviation is larger for the higher field as predicted in the above analysis. As clearly seen in Fig. 13.5b, σ decreases with temperature and saturates below a cross-over temperature of $T^\star \approx 100$ mK. At $T < T^\star$ the dissociation of the VAV pair occurs through a *macroscopic quantum tunneling* process. The cross-over temperature $T^\star \simeq \hbar\omega(\delta)/(2\pi k_B)$ is determined by the frequency $\omega(\delta)$ of small oscillations of the VAV pair. In the quantum regime the frequency $\omega(\delta) = 3^{3/8}/\sqrt{\chi}[\delta - \delta_c(h)]^{1/4}$ determines the oscillatory energy levels $E_n \simeq \hbar\omega(\delta)(n + 1/2)$ of the pinned VAV state. We examined these levels experimentally by performing microwave spectroscopy [39] as demonstrated earlier for the case of a single Josephson vortex trapped in a long Josephson junction [28] and summarize the results in the subsection below.

Neglecting dissipative effects, in the quantum regime the switching rate $\Gamma_Q(I)$ of the under-barrier dissociation can be estimated, as usual, in the WKB approximation, which yields

$$\Gamma_Q(I) \propto \exp\left(-\frac{36 U_{\text{eff}}(\delta)}{5\hbar\omega(\delta)}\right). \tag{13.3}$$

In this limit the standard deviation of the critical current σ is independent of temperature and (similar to the high-temperature case) it weakly increases with magnetic field.

The above analysis is based on an assumption that the pair's size is small with respect to the Josephson junction length. We find that the Josephson phase escape in the form of the dissociation of the pair occurs (in normalized units) as $h \geq \frac{3}{4}(L/2)^{-4}$. In the opposite limit of

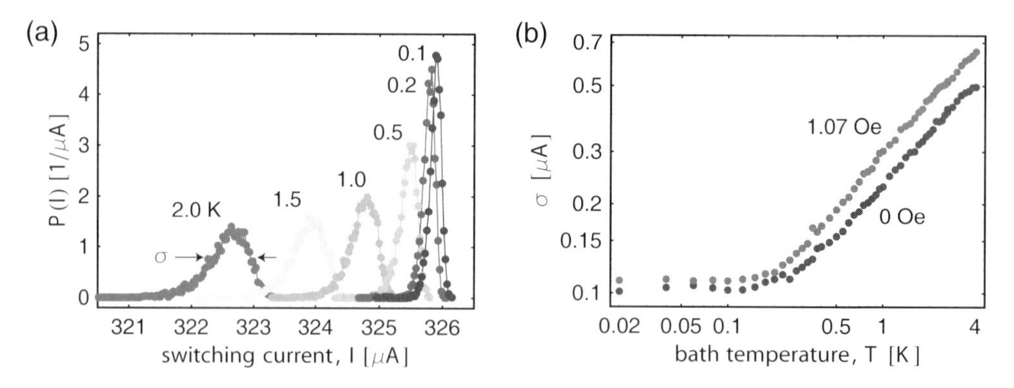

Figure 13.5: (a) Switching current distributions $P(I)$ at $H = 0$ shown at different bath temperatures T. (b) Standard deviation σ of $P(I)$ distributions versus bath temperature for the two indicated values of magnetic field.

very small magnetic field $h \lesssim (L/2)^{-4}$, the Josephson phase escape occurs homogeneously in the whole junction.[2]

The analysis presented above is valid for small magnetic fields, $h \ll 1$. As the magnetic field h increases, the critical current $I_c(h)$ is suppressed, and only a *qualitative* description of the VAV pair dissociation can be carried out. In the general case the pair size is $l_p \simeq [1-(I/I_{c0})^2]^{-1/4}$ (instead of $l_p \simeq \delta^{-1/4}$ that is valid for $h \ll 1$), and the amplitude of the state is $\xi_p \simeq \arccos(I/I_{c0})$ (instead of $\xi_p \simeq \sqrt{\delta}$). Following a similar procedure as in Ref. [37] we obtain the standard deviation dominated by the thermal fluctuations

$$\sigma_T \simeq \frac{T^{2/3} h^{2/3}}{\arccos[I_c(h)/I_{c0}]}, \tag{13.4}$$

and in the quantum regime

$$\sigma_Q \simeq \frac{h}{\{\arccos[I_c(h)/I_{c0}]\}^{8/5}}. \tag{13.5}$$

The standard deviation is determined by the magnetic field dependence of the critical current $I_c(h)$, which is given implicitly by the equation

$$h = \frac{3}{4}\sqrt{1-[I_c(h)/I_{c0}]^2}\,\arccos\left(\frac{I_c(h)}{I_{c0}}\right). \tag{13.6}$$

This dependence is shown in Fig. 13.4c by a dashed line. The discrepancy between analysis and numerical results in the values of I_c is a consequence of the fact that the analysis has been carried out for junctions with length $L/2\pi \gg 1$ but the experimentally investigated system only barely meets that limit ($L \approx 10.5$). However, the analytical predictions and numerical

[2] We do not consider here very long Josephson junctions where even in the absence of magnetic field fluctuation-induced vortex–antivortex pairs are generated in the Josephson junction [40–42].

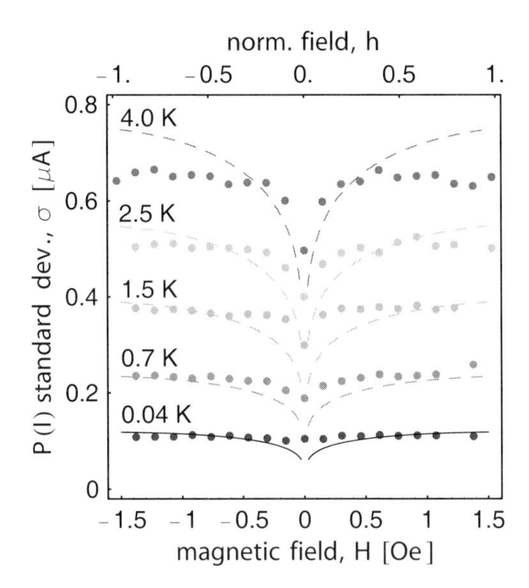

Figure 13.6: Standard deviation σ of $P(I)$ distribution versus magnetic field H in the temperature range between 40 mK and 4.0 K: experiment (dots), theory (dashed lines, Eqs. (13.4) and (13.5)). The region of magnetic field corresponds to the central lobe in the $I_c(H)$ dependence displayed in Fig. 13.4c.

results are in good accord as the length of the Josephson junction is increased to $L \geq 20$ (data not shown).

In Fig. 13.6, the measured field dependence of the switching current distribution width σ is shown for temperatures ranging from 40 mK to 4 K. The calculated dependences $\sigma_{T/Q}(h)$ are shown in the same figure by dashed lines. At each temperature the distribution width σ has a minimum, pronounced in the thermal regime, at zero magnetic field. In qualitative agreement with the analysis given by Eqs. (13.4) and (13.5), σ shows an *increase* and a following *saturation* with magnetic field in both thermal and quantum regimes. This behavior is characteristic for the fluctuation-induced dissociation of a VAV pair.

13.3.2 Energy Levels of a Bound Vortex–Antivortex Pair

As in the single vortex case, the energy levels can be probed spectroscopically. The model in [37] predicts the small oscillation frequency to be given by

$$\nu(I, H) = \nu_0(H) \left(1 - \frac{I}{I_c(H)}\right)^{1/4} . \tag{13.7}$$

Here the scaling of $\nu(I, H)$ with $I/I_c(H)$ is the same as for the vortex or small junction case at a constant value of H and $I/I_c(H)$ close to unity. The factor $\nu_0(H)$ is the magnetic-field-dependent internal oscillation frequency of the VAV pair at zero bias current. For a number of

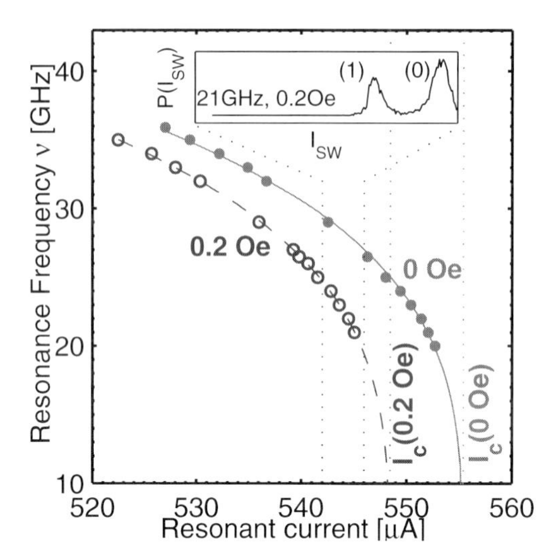

Figure 13.7: Resonance current versus frequency. Experimental data for zero magnetic field (filled circles) and 0.2 Oe (open circles) are compared to the model (solid line and dashed line, respectively). The respective critical currents are indicated by the dotted vertical lines. In the inset, the original switching-current probability distribution $P(I_{\rm sw})$ is indicated for a single data point using an enlarged current scale. The peak (1) corresponds to escape from the resonant level, and the peak (0) to escape from the ground state.

different values of magnetic field, we have measured the dependence of the resonance current on microwave frequency.

Experimental data for zero magnetic field and $H = 0.2$ Oe, which compare well to Eq. (13.7), are displayed in Fig. 13.7. Least-squares fits to the data for different magnetic fields yield both $I_c(H)$ and $\nu_0(H)$.

The theoretical dependence of $\nu_0(H)$ on $I_c(H)$ for VAV dissociation is modeled by [37]

$$\nu_0(H) = \frac{\omega_0}{2\pi}\frac{3^{3/8}}{\sqrt{\chi}}\left(\frac{I_c(H)}{I_{c0}}\right)^{1/4}, \tag{13.8}$$

where $\omega_0/2\pi$ is the plasma frequency of the junction at $H = 0$. This expression differs from that expected for phase escape in a small Josephson junction, where $\nu_0(H) \propto \sqrt{I_c(H)/I_{c0}}$. In Fig. 13.8 we fit the experimentally determined magnetic field dependence of the zero-bias oscillation frequency $\nu_0(H)$ to Eq. (13.8). For comparison, the expected oscillation frequency for homogeneous, small-junction-like escape of the phase is shown by the dotted line in the same plot. We note that the field dependence of the zero-bias oscillation frequency extracted from experimental data is in good agreement with the predictions based on our model for VAV dissociation, while the prediction for the homogeneous phase escape clearly disagrees with the experimental data.

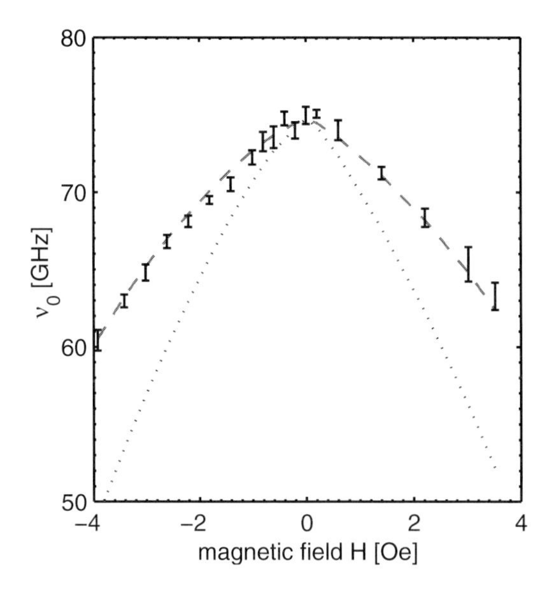

Figure 13.8: Field dependence of zero-bias oscillation frequency ν_0 extracted from measured data (solid error bars), and $\nu_0(H)$ calculated from $I_c(h)$ using the long-junction model (dashed line) and the small-junction model (dotted line).

13.4 The Josephson Vortex Qubit

In this section we discuss the design of a qubit based on a single Josephson vortex trapped in a shaped long Josephson junction. The vortex potential is formed due to its interaction with an in-plane magnetic field and a bias current applied to the junction. The profile of the potential is calculated using a standard perturbation approach. We examine the dependence of the potential properties on the junction shape and its electrical parameters and discuss the requirements for observing quantum effects in this system. We have developed and experimentally tested methods for the preparation and readout of vortex states of this qubit in the classical regime.

In the past years several different types of superconducting circuits [13–15, 43–45] based on small Josephson junctions in the phase or charge regime have been shown to achieve parameters that are favorable for quantum computation. In Ref. [46], a qubit based on the motion of a Josephson vortex in a long Josephson junction was proposed. A major difference from the small Josephson junction qubit proposals, where the effective potential is created by the Josephson or charging energy, is that in the vortex qubit the potential is formed by the magnetic interaction of the vortex magnetic moment with an external magnetic field, as described in Ref. [31].

In heart-shaped annular junctions two classically stable vortex states can be arranged, corresponding to two minima of the potential. While the external field is always applied in the plane of the long junction, its angle Θ and strength h can be varied. The bias current across the junction can be used to tilt the potential. These parameters allow one to manipulate and control the potential and to read out the qubit state using a depinning (zero-voltage) current measurement. The readout scheme and preparation of the state for this type of qubit was already demonstrated in the classical regime as briefly described in Ref. [47].

Here we describe how an effective double-well potential for the vortex can be constructed and determine the parameter range required to reach the quantum regime. A scheme to implement elementary single-bit quantum gates using two in-plane magnetic field components is presented. The single-vortex potential is calculated using a perturbation theory approach [30] and tunneling rates in the quantum regime are determined by numerical diagonalization of the Hamiltonian and compared to a calculation in the WKB approximation.

13.4.1 Principle of the Vortex Qubit

Josephson junctions that have a length significantly larger and a width w smaller than the Josephson penetration depth λ_J are called long junctions. These junctions are described by the phase difference $\varphi(q)$ of the two superconductors as the continuous degree of freedom along the spatial coordinate q (normalized to λ_J), where $0 < q < l$, where l is the length of the junction normalized to λ_J. A magnetic field threading the junction corresponds to a gradient in the phase difference along the junction. An electric field across the junction corresponds to a time derivative of φ. The dynamics of a long Josephson junction is governed by the perturbed sine–Gordon equation as discussed below.

Our experiments deal with long annular junctions. These consist of two stacked superconducting rings separated by a tunnel barrier. Since the flux in every ring is quantized, it is possible to realize a situation in which the difference between the fluxes in two rings is one flux quantum Φ_0. In this case a vortex of supercurrent carrying this flux quantum is formed along the junction. A resting vortex confines the magnetic flux to a characteristic size of λ_J. Since a moving vortex corresponds to moving magnetic flux, a voltage proportional to the speed of the vortex appears across the junction and can be detected. The electrical energy stored in the system is proportional to the square of the average voltage drop across the junction. It corresponds to the kinetic energy of a moving particle. It is therefore possible to consider a vortex as a quasiparticle moving in one dimension.

The junction can be biased by a current across it. A Lorentz-type driving force is exerted by the bias current on the vortex. The vortex magnetic moment also interacts with the external field by a magnetic dipole interaction. This makes it possible to create a potential for the vortex. The magnetic moment is always directed normal to the junction. By varying the angle of the junction centerline, it is possible to change the potential energy of the vortex as it moves along the junction.

Using the geometry shown in Fig. 13.9a, it is possible to generate a double-well potential for a vortex in the junction. We chose a shape that consists of a semicircle of radius R, and two connected arcs, which intersect each other at an angle of 2β. An external magnetic field h is applied, at an angle Θ. The field can be described by the components $h_x = h \sin \Theta$ and $h_y = h \cos \Theta$.

Assuming the vortex to carry a point-like magnetic moment, its stable positions of minimal magnetic energy can be easily found to be the regions where the junction is aligned perpendicular to the field. In reality, the vortex is distributed over a length on the order of λ_J. This changes the potential shape considerably, and may even change its qualitative features. In the

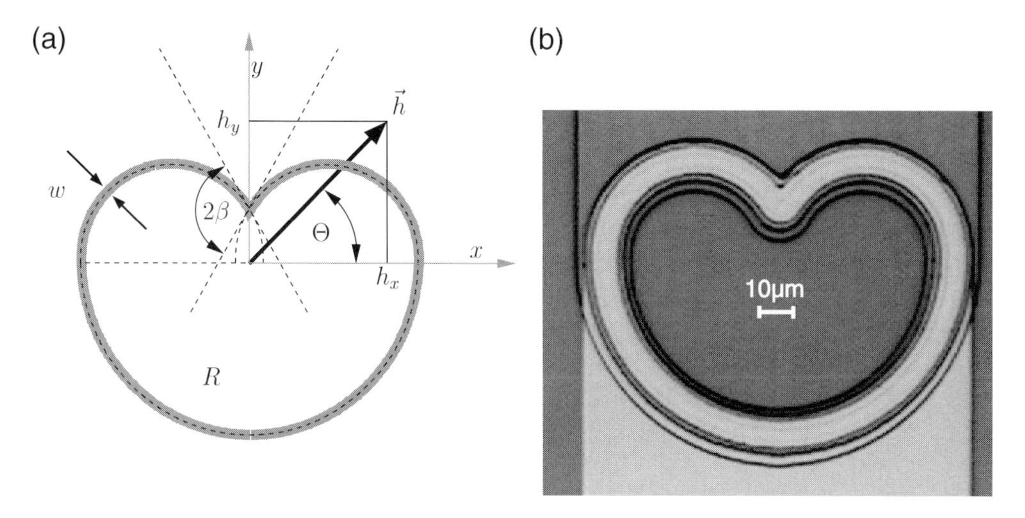

Figure 13.9: (a) The geometry of the heart-shaped junction, defined by the two geometrical parameters β and R. (b) Optical microscope picture of the junction used for the experimental test. Parameters are $R = 50$ μm, $\beta = 60°$, $w = 3$ μm, and $j_C = 796$ A/cm^2.

case of the double-well potential in the heart-shaped junction, it may cause the barrier that separates the two minima to vanish.

Here we derive the exact effective potential in order to determine the range of parameters and geometries suitable for experiments in the thermal and quantum regime. We show that the effective potential for a vortex inside a shaped junction is the sum of three terms, which depend on the external bias current, and the two in-plane field components. These are the three parameters that can be controlled during the experiment in order to realize degenerate bistable vortex states, to change the barrier between them, to lift the degeneracy in a controlled way, and, finally, to read out the state using a critical current measurement. Based on this approach we proceed to quantum mechanical calculations and the calculation of the depinning current.

13.4.2 Model

Classically, the evolution of the phase difference φ between the wave functions of the two superconducting electrodes forming a long Josephson junction is described by the sine–Gordon equation

$$\sin(\varphi) = \varphi_{qq} - \varphi_{tt}, \tag{13.9}$$

where the temporal coordinate is denoted by t, normalized to the plasma frequency ω_P^{-1}, discussed below. Subscripts denote partial derivatives.

Adding the terms for the inductive energy, the capacitive energy and the Josephson energy yields the corresponding Hamiltonian

$$H = \int_0^l \left(\tfrac{1}{2}\varphi_q^2 + \tfrac{1}{2}\varphi_t^2 + 1 - \cos\varphi\right) \mathrm{d}q, \tag{13.10}$$

with the characteristic energy scale E_0, discussed below. The temporal derivative φ_t corresponds to the normalized voltage across the junction; the spatial derivative φ_q corresponds to the normalized magnetic field in the junction.

Single-vortex solutions of Eq. (13.9) in a non-relativistic approximation for infinitely long junctions are given by

$$\varphi(q, q_0)^{\text{vortex}} = \pm \arctan[\exp(q - q_0 - vt)], \tag{13.11}$$

where q_0 denotes the vortex center of mass and v its velocity normalized to the Swihart velocity $\bar{c}$, which is the characteristic velocity for the electromagnetic waves in the junction.

In the case of weak magnetic field and small bias current, the interaction with field and current can be modeled using perturbation theory. In Ref. [30], a bias current γ (normalized to the critical current) was found to exert a driving force of $2\pi\gamma$ on the vortex. The influence of the external magnetic field on a shaped junction was studied theoretically [31] and experimentally [48] for a circular shape of the annular junction. The sine–Gordon equation was discussed previously for annular junctions of small ($2\pi r < \lambda_{\text{J}}$) and large radius ($2\pi r > \lambda_{\text{J}}$) in [38]. We use the overlap geometry, which is known to generate very little self-field of the bias current. In [49] we determined the self-field effect experimentally for circular geometries corresponding to the heart-shaped junctions by measuring the dependence of the critical current on the angle of the external magnetic field. Since no current flows during the operation of the qubit, we made no attempt to treat the self-field quantitatively. Before we investigate the more complex junction shapes further, we define the characteristic scales and normalizations.

In a long Josephson junction there are three important scales, which characterize the classical and quantum dynamics of the unperturbed system. The Josephson length

$$\lambda_{\text{J}} = \frac{\Phi_0}{2\pi}\sqrt{\frac{1}{L^* e_{\text{J}}}} \tag{13.12}$$

is determined by the inductance L^* and the Josephson coupling energy e_{J} per unit length in the junction. The length λ_{J} is the characteristic lateral dimension of the Josephson vortex at rest. The Josephson coupling energy is related to the critical current density j_{C} as $e_{\text{J}} = j_{\text{C}}\Phi_0/2\pi w$, where w denotes the width of the long Josephson junction.

Small-amplitude linear wave solutions of the phase are described by a dispersion relation. At the wave number $k = 0$ (a homogeneous oscillation over the whole junction), the corresponding frequency is the so-called Josephson plasma frequency ω_{P}, given by

$$\omega_{\text{P}} = \frac{2\pi}{\Phi_0}\sqrt{\frac{e_{\text{J}}}{C^*}}, \tag{13.13}$$

where C^* denotes the capacitance of the junction per unit length. The temporal coordinate of Eq. (13.9) is normalized by ω_{P}^{-1}. The product $\lambda_{\text{J}}\omega_{\text{P}} = \bar{c}$ is the Swihart velocity. All energies are normalized to the characteristic energy

$$E_0 = \frac{\Phi_0}{2\pi}\sqrt{\frac{e_{\text{J}}}{L^*}}\, w = e_{\text{J}}\lambda_{\text{J}}. \tag{13.14}$$

The energy unit E_0 is equal to the Josephson coupling energy of a small Josephson junction of area $\lambda_{\text{J}} w$. The rest energy of a vortex is equal to 8 in units of E_0, where half the energy is stored in the Josephson coupling and the other half is stored in the inductive energy. Since the speed of light equals unity in the normalized units, the rest mass of the vortex is $m_0 = 8$.

Applying these normalizations to Planck's constant $\hbar$, which has the unit of action, yields a normalized Planck's constant $\hbar_{\mathrm{norm}}$, given by

$$\hbar_{\mathrm{norm}} = \hbar \frac{\omega_{\mathrm{P}}}{E_0} = \hbar \left(\frac{2\pi}{\Phi_0} \right) \sqrt{\frac{L^*}{C^*}} \frac{1}{w}. \tag{13.15}$$

The normalized Planck's constant $\hbar_{\mathrm{norm}}$ does not depend directly on the Josephson coupling energy, but L^*, C^*, and e_{J} are related to each other through the barrier thickness.

13.4.3 Perturbative Calculation of Vortex Potential

We apply a perturbation theory approach similar to that of Refs. [30] and [31] and reduce the dynamics of the system to the center-of-mass motion of the vortex with its coordinate q_0 being the only degree of freedom. In the lowest order of perturbation theory, it is assumed that the phase gradient profile imposed by the external magnetic field and the phase gradient profile corresponding to a resting vortex do not influence each other, which requires that at least one of these is assumed to be small. The inductive energy term of Eq. (13.10) for $\varphi_q = \varphi_q^{\mathrm{vortex}} + \varphi_q^{\mathrm{ext}}$ yields

$$U^{\mathrm{ext}}(q_0) = \int_0^l \tfrac{1}{2} [\varphi_q^{\mathrm{vortex}}(q - q_0) + \varphi_q^{\mathrm{ext}}(q)]^2 \, \mathrm{d}q. \tag{13.16}$$

Expanding Eq. (13.16) yields

$$\begin{aligned} U^{\mathrm{ext}}(q_0) = &\int_0^l \tfrac{1}{2} \varphi_q^{\mathrm{vortex}}(q - q_0)^2 \mathrm{d}q + \int_0^l \tfrac{1}{2} \varphi_q^{\mathrm{ext}}(q)^2 \, \mathrm{d}q \\ &+ \int_0^l \varphi_q^{\mathrm{vortex}}(q - q_0) \varphi_q^{\mathrm{ext}}(q) \, \mathrm{d}q. \end{aligned} \tag{13.17}$$

Since we are interested only in the dependence of $U^{\mathrm{ext}}(q_0)$ on q_0, we can neglect the first two constant terms, which are the magnetic energy of a vortex at rest and the energy of the external magnetic field in the junction, respectively. The last term, which is a convolution of the externally introduced phase gradient with the phase gradient profile of the vortex, is the potential energy corresponding to the magnetic dipole interaction.

The influence of the bias current γ can be taken into account by adding a potential term corresponding to a constant driving force. This yields the total potential

$$U(q_0) = \varphi_q^{\mathrm{vortex}}(q - q_0) *_q \varphi_q^{\mathrm{ext}}(q) - 2\pi\gamma q_0 \tag{13.18}$$

for the vortex, where $*_q$ denotes the convolution in q. The effect of the convolution of the externally introduced phase gradient with the magnetic profile of the vortex is indicated in Fig. 13.10. The phase gradient (solid line) is induced by a field in the y-direction ($\Theta = \pi/2$) — see the heart-shaped junction depicted in Fig. 13.9a. While the derivative of the phase gradient is discontinuous at $q = 1/2l$, the resulting potential is smooth at this point. Furthermore, the

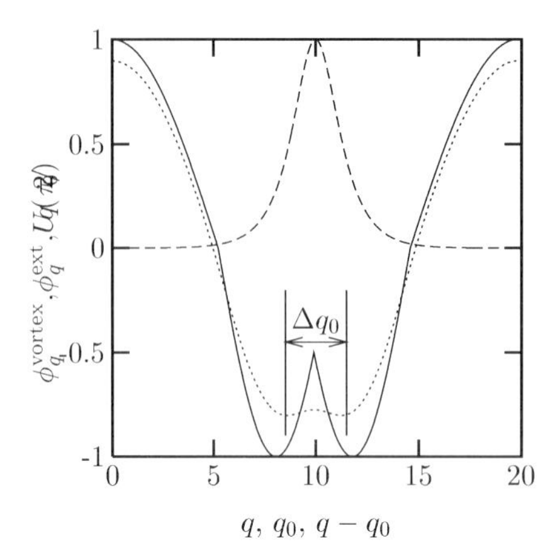

Figure 13.10: The phase gradient (solid line) $\varphi_q^{\text{ext}}(q)$ introduced by the external magnetic field, the phase gradient (long dashed line) $\phi_q^{\text{vortex}}(q-q_0)$ associated with the phase profile of a vortex (shifted to $q-q_0+10$ for clarity), and the effective potential (short dashed line) $U(q_0)$, given by Eq. (13.18) with $\gamma=0$.

separation Δq_0 between the minima and the height U_0 of the barrier is diminished by the convolution. In general, all local perturbations are smoothed to a length of the order of λ_J.

We now discuss how a specific geometry, like that of Fig. 13.9a, is related to the potential profile for an arbitrary field angle Θ. The phase difference gradients correspond to shielding currents, which flow in the superconducting electrodes of the junction. Since these shielding currents are orthogonal to the magnetic field, the phase gradient along the junction is proportional to the scalar product of the normal vector of the junction with the external magnetic field. This is equivalent to the physical interpretation of the vortex magnetic moment interacting with the external magnetic field. Expanding the scalar product into its (orthogonal) components

$$\varphi_q^{\text{ext}}(q) = \vec{n}(q)\cdot\vec{h} = n_x(q)h_x + n_y(q)h_y \tag{13.19}$$

yields an equation in which the phase gradient is linear in each component of the field. Since also the convolution of the vortex magnetic moment with the external magnetic field is a linear operation, it is possible to separate the convolution for the calculation of the potential into two components:

$$\begin{aligned}\varphi_q^{\text{vortex}}(q-q_0) *_q \varphi_q^{\text{ext}} = {}& \varphi_q^{\text{vortex}}(q-q_0) *_q n_x(q)h_x \\ &+ \varphi_q^{\text{vortex}}(q-q_0) *_q n_y(q)h_y.\end{aligned} \tag{13.20}$$

We now abbreviate $\varphi_q^{\text{vortex}}(q-q_0) *_q n_x(q)$ by U_x, and substitute Eq. (13.20) into Eq. (13.18). This yields

$$U(q_0) = U_x(q_0)h_x + U_y(q_0)h_y - 2\pi\gamma q_0 \tag{13.21}$$

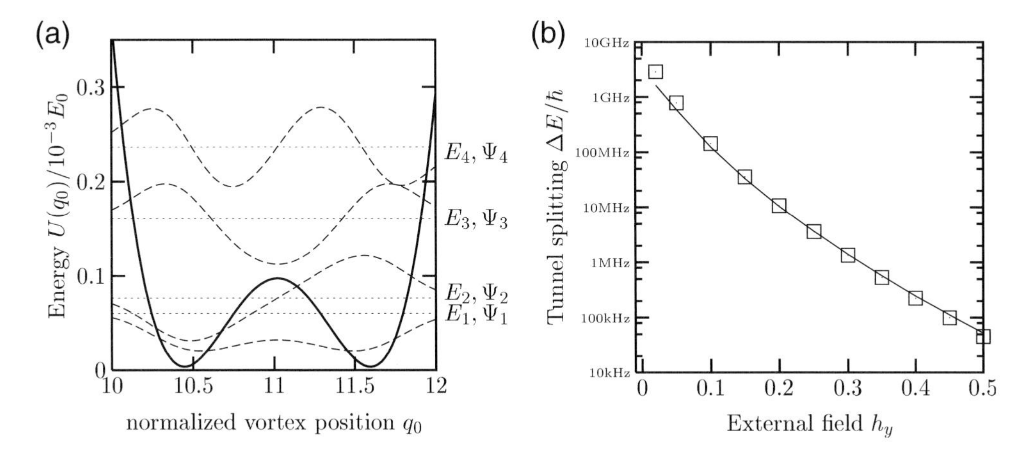

Figure 13.11: (a) Calculated lowest energy eigenstates for a junction of $w = 0.3$ μm. The energy levels are indicated by the dotted lines; the wave functions are indicated by the dashed lines. (b) Tunnel splitting between the states, according to the energy splitting of the lowest energy eigenstates.

as the total potential for the vortex motion.

We now return to the geometry in Fig. 13.9a. From the symmetry of the heart it can be seen immediately that $\vec{n}_y(q)$ is symmetric with respect to $q = 0$ and $q = l/2$, while $\vec{n}_x(q)$ is antisymmetric with respect to these points. Therefore also $U_y(q_0)$ and $U_x(q_0)$ are symmetric and antisymmetric, respectively. Of special interest is the region at $q = l/2$, since U_y forms a double-well potential there, if the radius R is large enough in relation to the Josephson length. The distance between the wells and the height of the barrier is strongly diminished with increasing Josephson length. At a certain critical value of the Josephson length, the barrier ceases to exist for this geometry.

13.4.4 Quantum Mechanics of a Vortex in a Double Well

The Schrödinger equation for the center-of-mass motion of a vortex in normalized units is given by

$$-i\hbar_{\mathrm{norm}}\Psi = \hat{H}\Psi = (\hat{T} + \hat{U})\Psi, \tag{13.22}$$

where $\hat{U}$ denotes the potential energy operator, and

$$\hat{T} - \frac{\partial}{\partial q^2}\frac{\hbar^2_{\mathrm{norm}}}{16} \tag{13.23}$$

corresponds to the kinetic energy operator for a particle of mass $m_0 = 8$. A numerical discretization and diagonalization of $\hat{H}$ yields the eigenstates, which are shown in Fig. 13.11a for $w = 0.3$ μm, $R = 50$ μm, and $j_{\mathrm{C}} = 1000$ A/cm^2.

The tunnel splitting between the two lowest energy eigenstates is proportional to the expected tunneling rate. The numerically calculated values for the tunnel splitting are shown as points in Fig. 13.11b, and tunneling rates calculated in WKB approximation are shown by the solid line. It can be seen that the tunneling rate can be tuned within the experimentally accessible field range by four orders of magnitude.

Applying a small x-component lifts the degeneracy of the potential, since an antisymmetric potential component will be added. If this component is small, perturbation theory yields a correction for the energy eigenvalues of the uncoupled states inside each well:

$$\Delta U = \langle \Psi | \hat{U}_x | \Psi \rangle h_x. \tag{13.24}$$

Since $U_x(q_0) = -U_x(-q_0)$, and the states localized in the left/right well are symmetric, one state is shifted up, while the other state is shifted down in energy.

At low temperatures, the vortex dynamics in the double-well potential is reduced to that of a two-state system, the Hamiltonian of which can be written as

$$H = \Delta E(h_y)\sigma_x + \Delta U \sigma_z, \tag{13.25}$$

where ΔE is the overlap of the ground states, found by the WKB method. It must be noted that $\Delta E(h_y)$ depends exponentially on h_y while ΔU depends only linearly on h_x. By controlling h_x and h_y it is possible to realize all single qubit operations.

13.4.5 Depinning Current and Qubit Readout

For the experimental test of the readout procedure in the classical regime [47], we are interested in the depinning current of the vortex, which is an experimentally accessible parameter. A vortex starts to move when the pinning force due to the external field is compensated by the force exerted by the bias current. Calculating the derivative of Eq. (13.21) yields the force acting on the vortex

$$F(q_0) = \frac{\partial U(q_0)}{\partial q_0} = \frac{\partial U_x(q_0)}{\partial q_0} h_x + \frac{\partial U_y(q_0)}{\partial q_0} h_y - 2\pi\gamma. \tag{13.26}$$

The equilibrium positions correspond to a zero net force. Using Eq. (13.21), we can write a condition for the equilibrium positions as

$$F_x(q_0)(h_x/h) + F_y(q_0)(h_y/h) = 2\pi(\gamma/h), \tag{13.27}$$

where F_x denotes $\partial U_x(q_0)/\partial q_0$.

Solutions of Eq. (13.27) vanish if local maxima/minima in the force are exceeded by $\gamma 2\pi$. A vortex trapped in one of these stable equilibrium positions will be depinned at these currents. The corresponding current is therefore called the depinning current $\gamma_{\mathrm{dep},i}$, where i is the index, if several stable positions exist. In Fig. 13.12, a potential without bias current is shown, together with the corresponding force. Each minimum of the potential, indicated by an open square, has one corresponding minimum (open circle) and maximum (solid circle) of the force. For a given angle Θ of the external field, (h_x/h) and (h_y/h) in Eq. (13.27) are constant. Therefore the ratio $\rho_{\mathrm{dep},i} = (\gamma_{\mathrm{dep},i}/h)$ is constant for a specific angle of the field.

The dependence of $\rho_{\mathrm{dep},i}$ on the angle of the external magnetic field can be found numerically. In Fig. 13.13, the dependence of $\rho_{\mathrm{dep},i}$ on the field angle Θ is plotted by two lines for

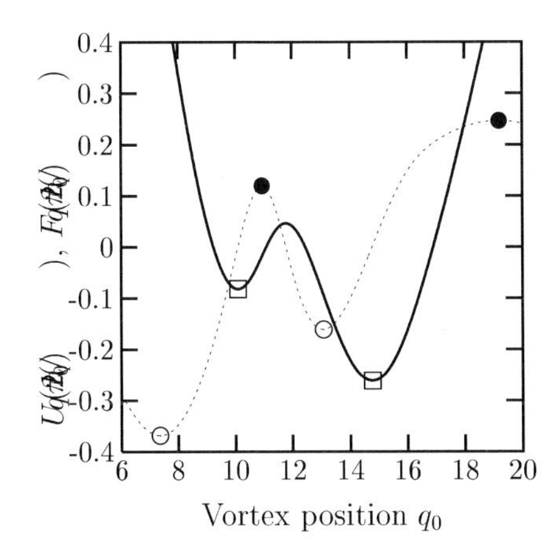

Figure 13.12: The minima of the potential $U(q_0)$ (solid line) are marked by open squares. The minima of the pinning force (dotted line) are marked by open circles, and the maxima by solid circles.

$i = 1$ and $i = 2$. We performed a test to find whether the vortex would be retrapped in the remaining stable positions. This happens if the spatial distance between the depinning position and the next maximum in the potential is so small that the vortex does not gain enough kinetic energy to overcome this maximum. If the current is increased, the vortex is depinned from this remaining position at the corresponding depinning current. In this case we plotted the latter value of the depinning current in Fig. 13.13.

We carried out an experimental test of the preparation and readout scheme proposed above using the junction shown in Fig. 13.9b. Figure 13.13 shows the dependence of the measured depinning current on the angle Θ of the external magnetic field. We measured the depinning current of the vortex after a clockwise and counterclockwise rotation as described in Ref. [47]. The two directions of rotation correspond to the circles and squares in Fig. 13.13, respectively.

From the dependence of the depinning current on the angle, the two ways of preparation can be associated with the numerically determined values of $\rho_{\mathrm{dep},i}$. At $\Theta = 270°$, we find numerically that $\rho_{\mathrm{dep},1} = \rho_{\mathrm{dep},2}$. This angle corresponds to an antisymmetric potential, which has a symmetric first derivative. Therefore the maximum values of the pinning force are identical. In Fig. 13.13 this theoretically predicted crossing is found in the experiment. The pinning for state (2) is slightly higher than expected. This small discrepancy of $\rho_{\mathrm{dep},i}$ indicates additional pinning possibly due to inhomogeneities, residual flux or geometrical reasons.

In the experiment, only at the readout angle $\Theta \approx 330°$ is the cross-over between retrapping in state (2) and depinning from state (1) observed as numerically predicted. The cross-over from retrapping in state (1) to detection of state (2) shows a large discrepancy from the numerical prediction. We attribute the difference between the experimental and numerical data in Fig. 13.13 in the range $\Theta \approx 45°$–$160°$ to damping, which has been neglected in the numerical calculation.

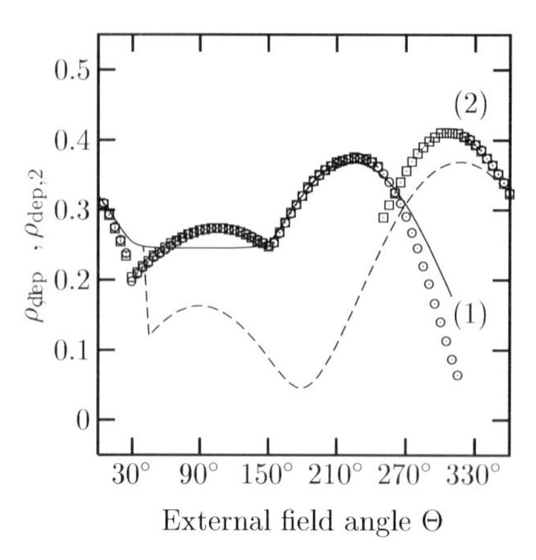

Figure 13.13: The normalized depinning current γ_{dep}/h plotted versus the angle Θ of the external field. Lines correspond to numerical calculation; squares (clockwise preparation of state, $i = 1$) and circles (counterclockwise preparation of state, $i = 2$) are the experimentally measured values.

13.5 Conclusions

In our experiments, we have observed for the first time the quantum tunneling of an individual vortex in a long Josephson junction [28]. We have also demonstrated the existence of well separated vortex energy levels in the potential well using microwave spectroscopy [28]. The dissociation of a bound vortex–antivortex pair, generated by magnetic field and bias current in a long junction, has been observed in the quantum regime [37]. These results are promising for use of Josephson vortices in long junctions as qubits for quantum information processing. The anharmonic oscillatory states of a vortex within a single well or, alternatively, the ground states of a vortex in a double-well potential can be used as basis states for a qubit [46, 50]. We have developed heart-shaped long junctions for use as qubits and have demonstrated a preparation and readout scheme for the vortex qubit in the classical regime [50] and thus verified the existence of bistable states. For the operation of the vortex qubit the width and height of the potential barrier and the symmetry of the double well are controlled by external fields. Macroscopic quantum coherence experiments using vortices in long submicrometer width junctions are currently in progress.

Acknowledgements

We gratefully acknowledge the contributions of C. Coqui, T. Duty, M. V. Fistul, Y. Koval, J. Lisenfeld, A. Lukashenko, and B. A. Malomed to the work described in this chapter. We thank A. Abdumalikov, G. Blatter, Takeo Kato, G. Schön, and S. Shnirman for fruitful discussions, A. Price for corrections to the manuscript, and IPHT Jena for the fabrication of one of the sets of samples used in this investigation. Partial financial support by the Deutsche Forschungsgemeinschaft (DFG) and D-Wave Systems Inc. is acknowledged.

References

[1] Abo-Shaeer, J. R., Raman, C., Vogels, J. M., & Ketterle, W. Observation of vortex lattices in Bose–Einstein condensates. *Science* **292**, 476–479 (2001).

[2] Blaauwgeers, R., Eltsov, V. B., Krusius, M., Ruohio, J. J., Schanen, R., & Volovik, G. E. Double-quantum vortex in superfluid ^{3}He-A. *Nature* **404**, 471–473 (2000).

[3] Huebener, R. P. *Magnetic Flux Structures in Superconductors*. Springer (2001).

[4] Bugoslavsky, Y., Perkins, G. K., Qi, X., Cohen, L. F., & Caplin, A. D. Vortex dynamics in superconducting MgB_2 and prospects for applications. *Nature* **410**, 563–565 (2001).

[5] Lee, C.-S., Janko, B., Derenyi, I., & Barabesi, A.-L. Reducing the vortex density in superconductors using the ratchet effect. *Nature* **400**, 337–340 (1999).

[6] Nori, F. and Savel'ev, S. Experimentally realizable devices for controlling the motion of magnetic flux quanta in anisotropic superconductors. *Nature Materials* **1**, 179–184 (2002).

[7] Fazio, R. & van der Zant, H. J. S. Quantum phase transitions and vortex dynamics in superconducting networks. *Phys. Rep.* **355**, 235–334 (2001).

[8] Blatter, G., Feigel'man, M. V., Geshkenbein, A. I., & Vinokur, V. M. Vortices in high-temperature superconductors. *Rev. Mod. Phys.* **66**, 1125–1388 (1994).

[9] Tinkham, M. *Introduction to Superconductivity*. McGraw-Hill (1996).

[10] Clarke, J., Cleland, A. N., Devoret, M. H., Esteve, D., & Martinis, J. M. Quantum mechanics of a macroscopic variable: the phase difference of a Josephson junction. *Science* **239**, 992–997 (1988).

[11] Devoret, M. H., Esteve, D., Urbina, C., Martinis, J., Clcland, A., & Clarkc, J. Macroscopic quantum effects in the current-biased Josephson junction. In *Quantum Tunneling in Condensed Media*. North-Holland (1992).

[12] Bennett, C. H. & DiVincenzo, D. P. Quantum information and computation. *Nature* **404**, 247–255 (2000).

[13] Nakamura, Y., Pashkin, Y. A., & Tsai, J. S. Coherent control of macroscopic quantum states in a single-Cooper-pair box. *Nature* **398**, 786–788 (1999).

[14] Friedman, J. R., Patel, V., Chen, W., Tolpygo, S. K., & Lukens, J. E. Quantum superposition of distinct macroscopic states. *Nature* **406**, 43–46 (2000).

[15] van der Wal, C. H., ter Haar, A. C. J., Wilhelm, F. K., Schouten, R. N., Harmans, C. J. P. M., Orlando, T. P., Lloyd, S., & Mooij, J. E. Quantum superposition of macroscopic persistent-current states. *Science* **290**, 773–777 (2000).

[16] Pashkin, Y. A., Yamamoto, T., Astafiev, O., Nakamura, Y., Averin, D., & Tsai, J. S. Quantum oscillations in two coupled charge qubits. *Nature* **421**, 823–826 (2003).

[17] Chiorescu, I., Nakamura, Y., Harmans, C. J. P. M., & Mooij, J. E. Coherent quantum dynamics of a superconducting flux qubit. *Science* **299**, 1869–1871 (2003).

[18] Yamamoto, T., Pashkin, Y. A., Astafiev, O., Nakamura, Y., & Tsai, J. S. Demonstration of conditional gate operation using superconducting charge qubits. *Nature* **425**, 941–944 (2003).

[19] Wallraff, A., Schuster, D. I., Blais, A., Frunzio, L., Huang, R.-S., Majer, J., Kumar, S., Girvin, S. M., & Schoelkopf, R. J. Circuit quantum electrodynamics: coherent coupling of a single photon to a Cooper pair box. *Nature* **431**, 162 (2004).

[20] Chiorescu, I., Bertet, P., Semba, K., Nakamura, Y., Harmans, C. J. P. M., & Mooij, J. E. Coherent dynamics of a flux qubit coupled to a harmonic oscillator. *Nature* **431**, 159–162 (2004).

[21] Hoekstra, H. F. T., Griessen, R., Testa, M. A., Brinkmann, M., Westerholt, K., Kwok, W. K., & Crabtree, G. W. General features of quantum creep in high-t_c superconductors. *Phys. Rev. Lett.* **80**, 4293–4296 (1998).

[22] Nicodemi, M. & Jensen, H. J. Creep of superconducting vortices in the limit of vanishing temperature: a fingerprint of off-equilibrium dynamics. *Phys. Rev. Lett.* **86**, 4378–4381 (2001).

[23] van der Zant, H. S. J., Fritschy, J. F. C., Orlando, T. P., & Mooij, J. E. Dynamics of vortices in underdamped Josephson-junction arrays. *Phys. Rev. Lett.* **66**, 2531–2534 (1991).

[24] Tighe, T. S., Johnson, A. T., & Tinkham, M. Vortex motion in two-dimensional arrays of small, underdamped Josephson junctions. *Phys. Rev. B* **44**, 10286–10290 (1991).

[25] Ustinov, A. V. Solitons in Josephson junctions. *Physica D* **123**, 315–329 (1998).

[26] Kato, T. & Imada, M. Macroscopic quantum tunneling of a fluxon in a long Josephson junction. *J. Phys. Soc. Jpn.* **65**(9), 2963–2975 (1996).

[27] Shnirman, A., Ben-Jacob, E., & Malomed, B. A. Tunneling and resonant tunneling of fluxons in a long Josephson junction. *Phys. Rev. B* **56**(22), 14677–14685 (1997).

[28] Wallraff, A., Lisenfeld, J., Lukashenko, A., Kemp, A., Fistul, M., Koval, Y., & Ustinov, A. V. Quantum dynamics of a single vortex. *Nature* **425**, 155 (2003).

[29] Koval, Y., Wallraff, A., Fistul, M., Thyssen, N., Kohlstedt, H., & Ustinov, A. V. Narrow long Josephson junctions. *IEEE Trans. Appl. Supercond.* **9**, 3957–3961 (1999).

[30] McLaughlin, D. W. & Scott, A. C. Perturbation analysis of fluxon dynamics. *Phys. Rev. A* **18**, 1652–1980 (1978).

[31] Grønbech-Jensen, N., Lomdahl, P., & Samuelsen, M. Phase locking of long annular Josephson junctions coupled to an external rf magnetic field. *Phys. Lett. A* **154**(1,2), 14–18 (1991).

[32] Ustinov, A. V. & Thyssen, N. Experimental study of fluxon dynamics in a harmonic potential well. *J. Low Temp. Phys.* **106**, 193–200 (1997).

[33] Fulton, T. A. & Dunkleberger, L. N. Lifetime of the zero-voltage state in Josephson tunnel junctions. *Phys. Rev. B* **9**, 4760–4768 (1974).

[34] Wallraff, A., Lukashenko, A., Coqui, C., Kemp, A., Duty, T., & Ustinov, A. V. Switching current measurements of large area Josephson tunnel junctions. *Rev. Sci. Instrum.* **74**, 3740 (2003).

[35] Wallraff, A., Duty, T., Lukashenko, A., & Ustinov, A. V. Multi-photon transitions between energy levels in a current-biased Josephson tunnel junction. *Phys. Rev. Lett.* **90**, 037003 (2003).

[36] Martucciello, N., Mygind, J., Koshelets, V. P., Shchukin, A., Filippenko, L., & Monaco, R. Fluxon dynamics in long annular Josephson tunnel junctions. *Phys. Rev. B* **57**(9), 5444–5449 (1998).

[37] Fistul, M. V., Wallraff, A., Koval, Y., Lukashenko, A., Malomed, B. A., & Ustinov, A. V. Quantum dissociation of a vortex–antivortex pair in a long Josephson junction. *Phys. Rev. Lett.* **91**, 257004 (2003).

[38] Martucciello, N. & Monaco, R. Static properties of annular Josephson tunnel junctions. *Phys. Rev. B* **54**(13), 9050 (1996).

[39] Kemp, A., Fistul, M. V., Wallraff, A., Koval, Y., Lukashenko, A., Malomed, B. A., & Ustinov, A. V. Energy level spectroscopy of a bound vortex–antivortex pair. In Delsing, P., Granata, C., Pashkin, Y., Ruggiero, B. & Silvestrini, P., Eds. *Quantum Computation: Solid State Systems*. Kluwer Academic / Plenum, 2004 (to appear).

[40] Maki, K. *Phys. Rev. Lett.* **39**, 46 (1977).

[41] Krive, I. V., Malomed, B. A., & Rozhavsky, A. S. *Phys. Rev. B* **42**, 273 (1990).

[42] Kato, T. Dimensional crossover by a local inhomogeneity in soliton-pair nucleation. *J. Phys. Soc. Jpn.* **69**, 2735 (200).

[43] Martinis, J. M., Nam, S., Aumentado, J., & Urbina, C. Rabi oscillations in a large Josephson-junction qubit. *Phys. Rev. Lett.* **89**, 117901 (2002).

[44] Vion, D., Aassime, A., Cottet, A., Joyez, P., Pothier, H., Urbina, C., Esteve, D., & Devoret, M. H. Manipulating the quantum state of an electrical circuit. *Science* **296**, 886–889 (2002).

[45] Yu, Y., Han, S., Chu, X., Chu, S.-I., & Wang, Y. Coherent temporal oscillations of macroscopic quantum states in a Josephson junction. *Science* **296**, 889–892 (2002).

[46] Wallraff, A., Koval, Y., Levitchev, M., Fistul, M. V., & Ustinov, A. V. Annular long Josephson junctions in a magnetic field: engineering and probing the fluxon potential. *J. Low Temp. Phys.* **118**(5/6), 543–553 (2000).

[47] Kemp, A., Wallraff, A., & Ustinov, A. V. Testing a state preparation and read-out protocol for the vortex qubit. *Physica C* **368**, 324 (2002).

[48] Ustinov, A. V., Malomed, B. A., & Thyssen, N. Soliton trapping in a harmonic potential: experiment. *Phys. Lett. A* **233**, 239 (1997).

[49] Kemp, A. Fluxon states in heart-shaped Josephson junction. Master's thesis, Physikalisches Institut III, Universität Erlangen-Nürnberg (2001).

[50] Kemp, A., Wallraff, A., & Ustinov, A. V. Josephson vortex qubit: design, preparation and read-out. *Phys. Stat. Sol. (b)* **233**(3), 472–481 (2002).

14 Decoherence in Resonantly Driven Bistable Systems

Sigmund Kohler and Peter Hänggi

Institut für Physik
Universität Augsburg
Augsburg, Germany

14.1 Introduction

A main obstacle for the experimental realization of a quantum computer is the unavoidable coupling of the qubits to external degrees of freedom and the decoherence caused in that way. A possible solution of this problem are error correcting codes. These, however, require redundant coding and, thus, a considerably higher algorithmic effort.

Yet another route to minimize decoherence is provided by the use of time-dependent control fields. Such external fields influence the coherent and the dissipative behavior of a quantum system and can extend coherence times significantly. One example is the stabilization of a coherent superposition in a bistable potential by coupling the system to an external dipole field [1, 2]. The fact that a driving field reduces the effective level splitting and therefore decelerates the coherent dynamics as well as the dissipative time evolution is here of cruical influence. A qubit is usually represented by two distinguished levels of a more complex quantum system and, thus, a driving field may also excite the system to levels outside the doublet that forms the qubit, i.e., cause so-called leakage. While a small leakage itself may be tolerable for the coherent dynamics, its influence on the quantum coherence of the system may be even more drastic. We demonstrate in this article that in a drivien qubit resonances with higher states, which are often ignored, may in fact enhance decoherence substantially.

A related phemomenon has been found in the context of dissipative chaotic tunneling near singlet-doublet crossings where the influence of so-called chaotic levels yields an enhanced loss of coherence [3, 4].

14.2 The Model and its Symmetries

We consider as a working model the quartic double well with a spatially homogeneous driving force, harmonic in time. It is defined by the Hamiltonian

$$H(t) = \frac{p^2}{2m} - \frac{1}{4}m\omega_0^2 x^2 + \frac{m^2\omega_0^4}{64E_\mathrm{B}}x^4 + Sx\cos(\Omega t). \tag{14.1}$$

The potential term of the static bistable Hamiltonian, H_DW, possesses two minima at $x = \pm x_0$, $x_0 = (8E_\mathrm{B}/m\omega_0^2)^{1/2}$, separated by a barrier of height E_B (cf. Fig. 14.1). The parameter ω_0 denotes the (angular) frequency of small oscillations near the bottom of each well. Thus,

Quantum Information Processing, 2nd Edition. Edited by Thomas Beth and Gerd Leuchs

ISBN: 3-527-40541-0

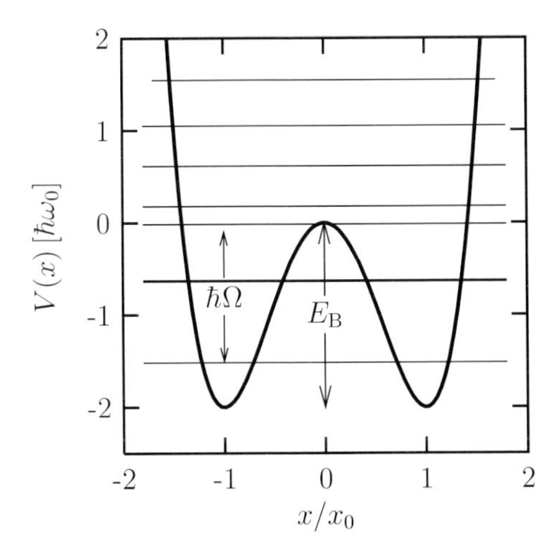

Figure 14.1: Sketch of the double well potential in Eq. (14.1) for $D = E_B/\hbar\omega_0 = 2$. The horizontal lines mark the eigenenergies in the absence of the driving; the levels below the barrier come in doublets.

the energy spectrum consists of approximately $D = E_B/\hbar\omega_0$ doublets below the barrier and singlets which lie above. As a dimensionless measure for the driving strength we use $F = S(8m\omega_0^2 E_B)^{-1/2}$.

The Hamiltonian (14.1) is T-periodic, with $T = 2\pi/\Omega$. As a consequence of this discrete time-translational invariance of $H(x, p; t)$, the relevant generator of the quantum dynamics is the one-period propagator [2, 5–8]

$$U(T, 0) = \mathcal{T} \exp\left(-\frac{\mathrm{i}}{\hbar}\int_0^T \mathrm{d}t\, H_{\mathrm{DW}}(t)\right), \tag{14.2}$$

where $\mathcal{T}$ denotes time ordering. According to the Floquet theorem, the Floquet states of the system are the eigenstates of $U(T, 0)$. They can be written in the form

$$|\psi_\alpha(t)\rangle = \mathrm{e}^{-\mathrm{i}\epsilon_\alpha t/\hbar}|\phi_\alpha(t)\rangle, \tag{14.3}$$

with

$$|\phi_\alpha(t + T)\rangle = |\phi_\alpha(t)\rangle.$$

Expanded in these Floquet states, the propagator of the driven system reads

$$U(t, t') = \sum_\alpha \mathrm{e}^{-\mathrm{i}\epsilon_\alpha (t-t')/\hbar}|\phi_\alpha(t)\rangle\langle\phi_\alpha(t')|. \tag{14.4}$$

The associated eigenphases ϵ_α, referred to as quasienergies, come in classes, $\epsilon_{\alpha,k} = \epsilon_\alpha + k\hbar\Omega$, $k = 0, \pm1, \pm2, \ldots$. This is suggested by a Fourier expansion of the $|\phi_\alpha(t)\rangle$,

$$\begin{aligned} |\phi_\alpha(t)\rangle &= \sum_k |\phi_{\alpha,k}\rangle\, \mathrm{e}^{-\mathrm{i}k\Omega t}, \\ |\phi_{\alpha,k}\rangle &= \frac{1}{T}\int_0^T \mathrm{d}t\, |\phi_\alpha(t)\rangle\, \mathrm{e}^{\mathrm{i}k\Omega t}. \end{aligned} \tag{14.5}$$

The index k counts the number of quanta in the driving field. Otherwise, the members of a class α are physically equivalent. Therefore, the quasienergy spectrum can be reduced to a single "Brillouin zone", $-\hbar\Omega/2 \leq \epsilon < \hbar\Omega/2$.

Since the quasienergies have the character of phases, they can be ordered only locally, not globally. A quantity that is defined on the full real axis and therefore does allow for a complete ordering, is the mean energy [2, 8]

$$E_\alpha = \frac{1}{T}\int_0^T \mathrm{d}t\, \langle\psi_\alpha(t)|\, H_{\mathrm{DW}}(t)\, |\psi_\alpha(t)\rangle \tag{14.6}$$

It is related to the corresponding quasienergy by

$$E_\alpha = \epsilon_\alpha + \frac{1}{T}\int_0^T \mathrm{d}t\, \langle\phi_\alpha(t)|\, \mathrm{i}\hbar\frac{\partial}{\partial t}\, |\phi_\alpha(t)\rangle . \tag{14.7}$$

Without the driving, $E_\alpha = \epsilon_\alpha$, as it should be. By inserting the Fourier expansion (14.5), the mean energy takes the form

$$E_\alpha = \sum_k (\epsilon_\alpha + k\hbar\Omega)\, \langle\phi_{\alpha,k}|\phi_{\alpha,k}\rangle . \tag{14.8}$$

This form reveals that the kth Floquet channel yields a contribution $\epsilon_\alpha + k\hbar\Omega$ to the mean energy, weighted by the Fourier coefficient $\langle\phi_{\alpha,k}|\phi_{\alpha,k}\rangle$. For the different methods to obtain the Floquet states, we refer the reader to the reviews [2, 8], and the references therein.

The invariance of the static Hamiltonian under parity $\mathsf{P} : (x, p, t) \to (-x, -p, t)$ is violated by the dipole driving force. With the above choice of the driving, however, a more general, dynamical symmetry remains. It is defined by the operation [2, 8]

$$\mathsf{P}_T : (x, p, t) \to (-x, -p, t + T/2) \tag{14.9}$$

and represents a generalized parity acting in the extended phase space spanned by x, p, and phase, i.e., time $t \bmod T$. While such a discrete symmetry is of minor importance in classical physics, its influence on the quantum mechanical quasispectrum $\{\epsilon_\alpha(S, \Omega)\}$ is profound: It devides the Hilbert space in an even and an odd sector, thus allowing for a classification of the Floquet states as even or odd. Quasienergies from different symmetry classes may intersect, while quasienergies with the same symmetry typically form avoided crossings. The fact that P_T acts in the phase space extended by time $t \bmod T$, results in a particularity: If, e.g., $|\phi(t)\rangle$ is an even Floquet state, then $\exp(\mathrm{i}\Omega t)|\phi(t)\rangle$ is odd, and vice versa. Thus, two equivalent Floquet states from neighboring Brillouin zones have opposite generalized parity. This means that a classification of the corresponding solutions of the Schrödinger equation, $|\psi(t)\rangle = \exp(-\mathrm{i}\epsilon t/\hbar)|\phi(t)\rangle$, as even or odd is meaningful only with respect to a given Brillouin zone.

14.3 Coherent Tunneling

With the driving switched off, $S = 0$, the classical phase space generated by H_{DW} exhibits the constituting features of a bistable Hamiltonian system: A separatrix at $E = 0$ forms the border between two sets of trajectories: One set, with $E < 0$, comes in symmetry-related pairs, each partner of which oscillates in either one of the two potential minima. The other set consists of unpaired, spatially symmetric trajectories, with $E > 0$, which encircle both wells.

Torus quantization of the integrable undriven double well implies a simple qualitative picture of its eigenstates: The unpaired tori correspond to singlets with positive energy, whereas the symmetry-related pairs below the top of the barrier correspond to degenerate pairs of eigenstates. Due to the almost harmonic shape of the potential near its minima, neighboring pairs are separated in energy approximately by $\hbar\omega_0$. Exact quantization, however, predicts that the partners of these pairs have small but finite overlap. Therefore, the true eigenstates come in doublets, each of which consists of an even and an odd state, $|\Phi_n^+\rangle$ and $|\Phi_n^-\rangle$, respectively. The energies of the nth doublet are separated by a finite tunnel splitting Δ_n. We can always choose the global relative phase such that the superpositions

$$|\Phi_n^{\mathrm{R,L}}\rangle = \frac{1}{\sqrt{2}}\left(|\Phi_n^+\rangle \pm |\Phi_n^-\rangle\right) \tag{14.10}$$

are localized in the right and the left well, respectively. As time evolves, the states $|\Phi_n^+\rangle$, $|\Phi_n^-\rangle$ acquire a relative phase $\exp(-\mathrm{i}\Delta_n t/\hbar)$ and $|\Phi_n^{\mathrm{R}}\rangle$, $|\Phi_n^{\mathrm{L}}\rangle$ are transformed into one another after a time $\pi\hbar/\Delta_n$. Thus, the particle tunnels forth and back between the wells with a frequency $\Delta_n/\hbar$. This introduces an additional, purely quantum-mechanical frequency scale, the tunneling rate $\Delta_0/\hbar$ of a particle residing in the ground-state doublet. Typically, tunneling rates are extremely small compared to the frequencies of the classical dynamics.

The driving in the Hamiltonian (14.1), even if its influence on the classical phase space is minor, can entail significant consequences for tunneling: It may enlarge the tunnel rate by orders of magnitude or even suppress tunneling altogether. For adiabatically slow driving, i.e. $\Omega \ll \Delta_0/\hbar$, tunneling is governed by the instantaneous tunnel splitting, which is always larger than its unperturbed value Δ_0 and results in an enhancement of the tunneling rate [9]. If the driving is faster, the opposite holds true: The relevant time scale is now given by the inverse of the quasienergy splitting of the ground-state doublet $\hbar/|\epsilon_1 - \epsilon_0|$. It has been found [9–11] that in this case, for finite driving amplitudes, $|\epsilon_1 - \epsilon_0| < \Delta_0$. Thus tunneling is always decelerated. When the quasienergies of the ground-state doublet (which are of different generalized parity) intersect as a function of F, the splitting vanishes and tunneling can be brought to a complete standstill by the purely coherent influence of the driving — not only stroboscopically, but also in continuous time [9–11].

So far, we have considered only driving frequencies much smaller than the frequency scale ω_0 of the relevant classical resonances. In this regime, coherent tunneling is well described within a two-state approximation [11]. Near an avoided crossing, level separations may deviate vastly, in both directions, from the typical tunnel splitting. This is reflected in time-domain phenomena ranging from the suppression of tunneling to a strong increase in its rate and to complicated quantum beats [12]. Singlet-doublet crossings, in turn, drastically change the quasienergy scales and replace the two-level by a three-level structure.

Three-level Crossings

A doublet which is driven close to resonance with a singlet can be adequately described in a three-state Floquet picture. For a quantitative account of such crossings and the associated

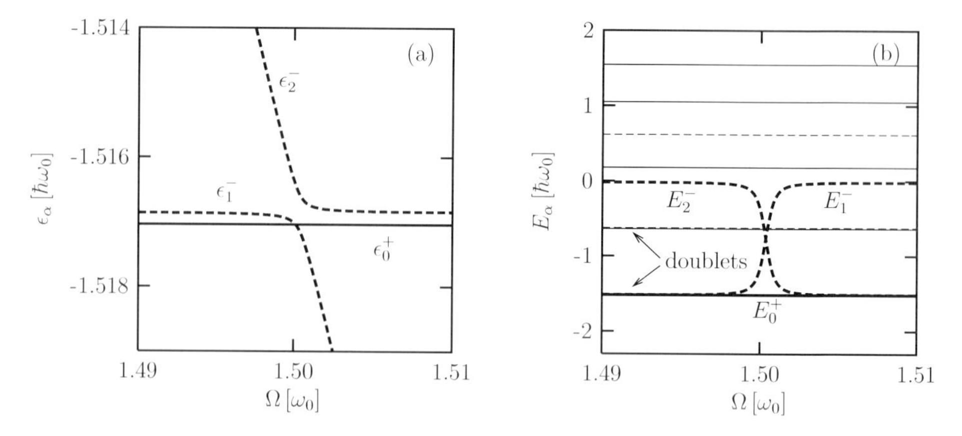

Figure 14.2: Quasienergies (a) and mean energies (b) found numerically for the driven double well potential with $D = E_\mathrm{B}/\hbar\omega_0 = 2$ and the dimensionless driving strength $F = 10^{-3}$. Energies of states with even (odd) generalized parity are marked by full (broken) lines; bold lines (full and broken) correspond to the states (14.16) which are formed from the singlet $|\phi_\mathrm{t}^-\rangle$ and the doublet $|\phi_\mathrm{d}^\pm\rangle$. A driving frequency $\Omega > 1.5\,\omega_0$ corresponds to a detuning $\delta = E_\mathrm{t}^- - E_\mathrm{d}^- - \hbar\Omega < 0$.

coherent dynamics, and for later reference in the context of the incoherent dynamics, we shall now discuss them in terms of a simple three-state model, which has been discussed in the context of chaotic tunneling [3, 13]. In order to illustrate the above three-state model and to demonstrate its adequacy, we have numerically studied a singlet-doublet crossing that occurs for the double-well potential, Eq. (14.1), with $D = 2$, at a driving frequency $\Omega \approx 1.5\,\omega_0$ and an amplitude $F = 0.001$ (Fig. 14.2).

Far outside the crossing, we expect the following situation: There is a doublet (subscript d) of Floquet states

$$\begin{aligned} |\psi_\mathrm{d}^+(t)\rangle &= \mathrm{e}^{-\mathrm{i}\epsilon_\mathrm{d}^+ t/\hbar}|\phi_\mathrm{d}^+(t)\rangle, \\ |\psi_\mathrm{d}^-(t)\rangle &= \mathrm{e}^{-\mathrm{i}(\epsilon_\mathrm{d}^+ +\Delta)t/\hbar}|\phi_\mathrm{d}^-(t)\rangle, \end{aligned} \tag{14.11}$$

with even (superscript $+$) and odd ($-$) generalized parity, respectively, residing on a pair of quantizing tori in one of the well regions. We have assumed the quasienergy splitting $\Delta = \epsilon_\mathrm{d}^- - \epsilon_\mathrm{d}^+$ (as opposed to the unperturbed splitting) to be positive. The global relative phase is chosen such that the superpositions

$$|\phi_\mathrm{R,L}(t)\rangle = \frac{1}{\sqrt{2}}\left(|\phi_\mathrm{d}^+(t)\rangle \pm |\phi_\mathrm{d}^-(t)\rangle\right) \tag{14.12}$$

are localized in the right and the left well, respectively, and tunnel back and forth with a frequency $\Delta/\hbar$.

As the third player, we introduce a Floquet state

$$|\psi_\mathrm{t}^-(t)\rangle = \mathrm{e}^{-\mathrm{i}(\epsilon_\mathrm{d}^+ +\Delta+\delta)t/\hbar}|\phi_\mathrm{t}^-(t)\rangle, \tag{14.13}$$

located mainly at the top of the barrier (subscript t), so that its time-periodic part $|\phi_{\mathrm{t}}^{-}(t)\rangle$ contains a large number of harmonics. Without loss of generality, its parity is fixed to be odd. Note that $|\phi_{\mathrm{d}}^{\pm}(t)\rangle$ are in general not eigenstates of the static part of the Hamiltonian, but exhibit for sufficiently strong driving already a non-trivial, T-periodic time-dependence. For the quasienergy, we assume that $\epsilon_{\mathrm{t}}^{-} = \epsilon_{\mathrm{d}}^{+} + \Delta + \delta = \epsilon_{\mathrm{d}}^{-} + \delta$, where the detuning $\delta = E_{\mathrm{t}}^{-} - E_{\mathrm{d}}^{-} - \hbar\Omega$ serves as a measure of the distance from the crossing. The mean energy of $|\psi_{\mathrm{t}}^{-}(t)\rangle$ lies approximately by $\hbar\Omega$ above the doublet such that $|E_{\mathrm{d}}^{-} - E_{\mathrm{d}}^{+}| \ll E_{\mathrm{t}}^{-} - E_{\mathrm{d}}^{\pm}$.

In order to model an avoided crossing between $|\phi_{\mathrm{d}}^{-}\rangle$ and $|\phi_{\mathrm{t}}^{-}\rangle$, we suppose that there is a non-vanishing fixed matrix element

$$b = \frac{1}{T}\int_0^T \mathrm{d}t\, \langle\phi_{\mathrm{d}}^{-}|H_{\mathrm{DW}}|\phi_{\mathrm{t}}^{-}\rangle > 0. \tag{14.14}$$

For the singlet-doublet crossings under study, we typically find that $\Delta \lesssim b \ll \hbar\Omega$. Neglecting the coupling with all other states, we model the system by the three-state (subscript 3s) Floquet Hamiltonian [3,4]

$$\mathcal{H}_{3\mathrm{s}} = \epsilon_{\mathrm{d}}^{+} + \begin{pmatrix} 0 & 0 & 0 \\ 0 & \Delta & b \\ 0 & b & \Delta+\delta \end{pmatrix} \tag{14.15}$$

in the three-dimensional Hilbert space spanned by $\{|\phi_{\mathrm{r}}^{+}(t)\rangle, |\phi_{\mathrm{d}}^{-}(t)\rangle, |\phi_{\mathrm{t}}^{-}(t)\rangle\}$. Its Floquet states are

$$\begin{aligned} |\phi_0^{+}(t)\rangle &= |\phi_{\mathrm{d}}^{+}(t)\rangle, \\ |\phi_1^{-}(t)\rangle &= \left(|\phi_{\mathrm{d}}^{-}(t)\rangle\cos\beta - |\phi_{\mathrm{t}}^{-}(t)\rangle\sin\beta\right), \\ |\phi_2^{-}(t)\rangle &= \left(|\phi_{\mathrm{d}}^{-}(t)\rangle\sin\beta + |\phi_{\mathrm{t}}^{-}(t)\rangle\cos\beta\right). \end{aligned} \tag{14.16}$$

with quasienergies

$$\epsilon_0^{+} = \epsilon_{\mathrm{d}}^{+}, \quad \epsilon_{1,2}^{-} = \epsilon_{\mathrm{d}}^{+} + \Delta + \frac{1}{2}\delta \mp \frac{1}{2}\sqrt{\delta^2 + 4b^2}, \tag{14.17}$$

and mean energies, neglecting contributions of the matrix element b,

$$\begin{aligned} E_0^{+} &= E_{\mathrm{d}}^{+}, \\ E_1^{-} &= E_{\mathrm{d}}^{-}\cos^2\beta + E_{\mathrm{t}}^{-}\sin^2\beta, \\ E_2^{-} &= E_{\mathrm{d}}^{-}\sin^2\beta + E_{\mathrm{t}}^{-}\cos^2\beta. \end{aligned} \tag{14.18}$$

The angle β describes the mixing between the Floquet states $|\phi_{\mathrm{d}}^{-}\rangle$ and $|\phi_{\mathrm{t}}^{-}\rangle$ and is an alternative measure of the distance to the avoided crossing. By diagonalizing the matrix (14.15), we obtain

$$2\beta = \arctan\left(\frac{2b}{\delta}\right), \quad 0 < \beta < \frac{\pi}{2}. \tag{14.19}$$

For $\beta \to \pi/2$, corresponding to $-\delta \gg b$, we retain the situation far right of the crossing, as outlined above, with $|\phi_1^{-}\rangle \approx -|\phi_{\mathrm{t}}^{-}\rangle$, $|\phi_2^{-}\rangle \approx |\phi_{\mathrm{d}}^{-}\rangle$. To the far left of the crossing, i.e. for

$\beta \to 0$ or $\delta \gg b$, the exact eigenstates $|\phi_1^-\rangle$ and $|\phi_2^-\rangle$ have interchanged their shape [3, 12]. Here, we have $|\phi_1^-\rangle \approx |\phi_d^-\rangle$ and $|\phi_2^-\rangle \approx |\phi_t^-\rangle$. The mean energy is essentially determined by this shape of the state, so that there is also an exchange of E_1^- and E_2^- in an exact crossing, cf. Eq. (14.18), while E_0^+ remains unaffected (Fig. 14.2b).

To study the dynamics of the tunneling process, we focus on the state

$$|\psi(t)\rangle = \frac{1}{\sqrt{2}}\left(\mathrm{e}^{-\mathrm{i}\epsilon_0^+ t/\hbar}|\phi_0^+(t)\rangle + \mathrm{e}^{-\mathrm{i}\epsilon_1^- t/\hbar}|\phi_1^-(t)\rangle\cos\beta + \mathrm{e}^{-\mathrm{i}\epsilon_2^- t/\hbar}|\phi_2^-(t)\rangle\sin\beta\right). \tag{14.20}$$

It is constructed such that at $t = 0$, it corresponds to the decomposition of $|\phi_{\mathrm{R}}\rangle$ in the basis (14.16) at finite distance from the crossing. Therefore, it is initially localized in the right well and follows the time evolution under the Hamiltonian (14.15). From Eqs. (14.12), (14.16), we find the probabilities for its evolving into $|\phi_{\mathrm{R}}\rangle$, $|\phi_{\mathrm{L}}\rangle$, or $|\phi_{\mathrm{t}}\rangle$, respectively, to be

$$\begin{aligned} P_{\mathrm{R,L}}(t) &= |\langle\phi_{\mathrm{R,L}}(t)|\psi(t)\rangle|^2 \\ &= \frac{1}{2}\left(1 \pm \left[\cos\frac{(\epsilon_1^- - \epsilon_0^+)t}{\hbar}\cos^2\beta + \cos\frac{(\epsilon_2^- - \epsilon_0^+)t}{\hbar}\sin^2\beta\right]\right. \\ &\quad \left. + \left[\cos\frac{(\epsilon_1^- - \epsilon_2^-)t}{\hbar} - 1\right]\cos^2\beta\sin^2\beta\right), \\ P_{\mathrm{t}}(t) &= |\langle\phi_{\mathrm{t}}(t)|\psi(t)\rangle|^2 = \left[1 - \cos\frac{(\epsilon_1^- - \epsilon_2^-)t}{\hbar}\right]\cos^2\beta\sin^2\beta. \end{aligned} \tag{14.21}$$

At sufficient distance from the crossing, there is only little mixing between the doublet and the resonant states, i.e., $\sin\beta \ll 1$ or $\cos\beta \ll 1$. The tunneling process then follows the familiar two-state dynamics involving only $|\phi_d^+\rangle$ and $|\phi_d^-\rangle$, with tunnel frequency $\Delta/\hbar$. Close to the avoided crossing, $\cos\beta$ and $\sin\beta$ are of the same order of magnitude, and $|\phi_1^-\rangle$, $|\phi_2^-\rangle$ become very similar to one another. Each of them has now support at the barrier top and in the well region, they are of a hybrid nature. Here, the tunneling involves all the three states and must be described at least by a three-level system. The exchange of probability between the two well regions proceeds via a "stop-over" at hte top of the barrier.

14.4 Dissipative Tunneling

The small energy scales associated with tunneling make it extremely sensitive to any loss of coherence. As a consequence, the symmetry underlying the formation of tunnel doublets is generally broken, and an additional energy scale is introduced, the effective finite width attained by each discrete level. As a consequence, the familiar way tunneling fades away in the presence of dissipation on a time scale t_{coh}. In general, this time scale gets shorter for higher temperatures, reflecting the growth of the transition rates. However, there exist counterintuitive effects: in the vicinity of an exact crossing of the ground-state doublet, coherence can be stabilized with higher temperatures [1] until levels outside the doublet start to play a role.

As a measure for the coherence of a quantum system we employ in this work the Renyi entropy [14]

$$S_\alpha = \frac{\ln \operatorname{tr} \rho^\alpha}{1-\alpha}. \tag{14.22}$$

In our numerical studies we will use S_2 which is related to the purity $\operatorname{tr}(\rho^2)$. It possesses a convenient physical interpretation: Suppose that ρ describes an incoherent mixture of n states with equal probability, then $\operatorname{tr}(\rho^2)$ reads $1/n$ and one accordingly finds $S_2 = \ln n$.

Floquet-Markov Master Equation

To achieve a microscopic model of dissipation, we couple the driven bistable system (14.1) bilinearly to a bath of non-interacting harmonic oscillators [8, 15, 16]. The total Hamiltonian of system and bath is then given by

$$H(t) = H_{\mathrm{DW}}(t) + \sum_{\nu=1}^{\infty} \left(\frac{p_\nu^2}{2m_\nu} + \frac{m_\nu}{2}\omega_\nu^2 \left(x_\nu - \frac{g_\nu}{m_\nu \omega_\nu^2} x \right)^2 \right). \tag{14.23}$$

Due to the bilinearity of the system-bath coupling, one can eliminate the bath variables to get an exact, closed integro-differential equation for the reduced density matrix $\rho(t) = \operatorname{tr}_{\mathrm{B}} \rho_{\mathrm{total}}(t)$. It describes the dynamics of the central system, subject to dissipation.

In the case of weak coupling, such that the dynamics is predominatly coherent, the reduced density operator obeys in good approximation a Markovian master equation. The Floquet states $|\phi_\alpha(t)\rangle$ form then a well-adapted basis set for a decomposition that allows for an efficient numerical treatment. If the spetral density of the bath influence is ohmic [8, 16], the resulting master equation reads [17, 18]

$$\dot{\rho}_{\alpha\beta}(t) = -\frac{\mathrm{i}}{\hbar}(\epsilon_\alpha - \epsilon_\beta)\rho_{\alpha\beta}(t) + \sum_{\alpha'\beta'} \mathcal{L}_{\alpha\beta,\alpha'\beta'} \rho_{\alpha'\beta'}. \tag{14.24}$$

The time-independent dissipative kernel

$$\begin{aligned} \mathcal{L}_{\alpha\beta,\alpha'\beta'} &= \sum_k (N_{\alpha\alpha',k} + N_{\beta\beta',k}) X_{\alpha\alpha',k} X_{\beta'\beta,-k} \\ &\quad -\delta_{\beta\beta'} \sum_{\beta'',k} N_{\beta''\alpha',k} X_{\alpha\beta'',-k} X_{\beta''\alpha',k} \\ &\quad -\delta_{\alpha\alpha'} \sum_{\alpha''k} N_{\alpha''\beta',k} X_{\beta'\alpha'',-k} X_{\alpha''\beta,k} \end{aligned} \tag{14.25}$$

is given by the Fourier coefficients of the position matrix elements,

$$X_{\alpha\beta,k} = \frac{1}{T}\int_0^T \mathrm{d}t\, \mathrm{e}^{-\mathrm{i}k\Omega t} \langle \phi_\alpha(t)|x|\phi_\beta(t)\rangle\rangle = X^*_{\beta\alpha,-k} \tag{14.26}$$

and the coefficients

$$N_{\alpha\beta,k} = N(\epsilon_\alpha - \epsilon_\beta + k\hbar\Omega), \quad N(\epsilon) = \frac{m\gamma\epsilon}{\hbar^2}\frac{1}{\mathrm{e}^{\epsilon/k_{\mathrm{B}}T} - 1} \tag{14.27}$$

which consist basically of the spectral density times the thermal occupation of the bath.

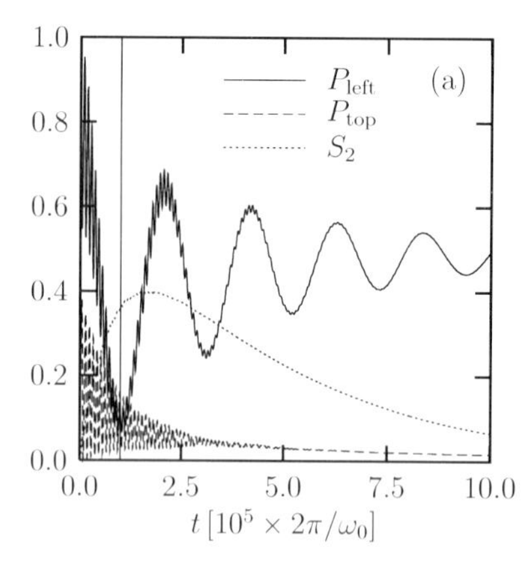

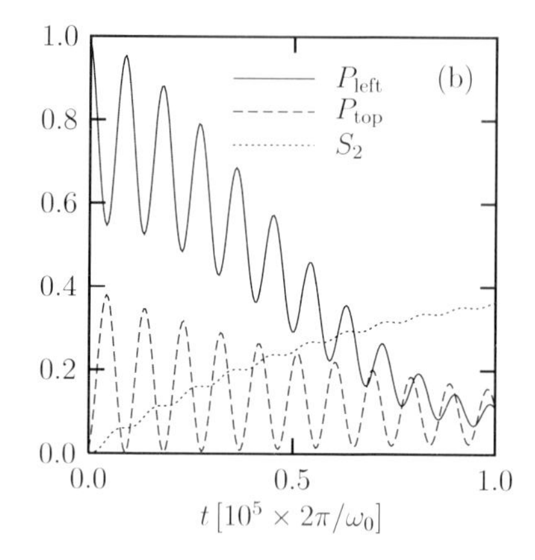

Figure 14.3: Time evolution of the state $|\phi_{\text{L}}\rangle$ at the center of the singlet-doublet crossing found for $D = 2$, $F = 10^{-3}$, and $\Omega = 1.5\,\omega_0$. The full line depicts the return probability and the broken line the occupation probability of the state at the top of the barrier. The dotted line marks the Renyi entropy S_2. Panel (b) is a blow-up of the marked region on the left of panel (a).

Dissipative Time Evolution

We have studied dissipative tunneling at the particular singlet-doublet crossing introduced in Sec. 14.3 (see Fig. 14.2). The time evolution has been computed numerically by integrating the master equation (14.24). As initial condition, we have chosen the density operator $\rho(0) = |\phi_{\text{L}}\rangle\langle\phi_{\text{L}}|$, i.e. a pure state located in the left well.

In the vicinity of a singlet-doublet crossing, the tunnel splitting increases and during the tunneling, the singlet $|\phi_{\text{t}}\rangle$ at the top of the barrier becomes populated periodically with frequency $|\epsilon_2^- - \epsilon_1^-|/\hbar$, cf. Eq. (14.21) and Fig. 14.3b. The large mean energy of this singlet results in an enhanced entropy production at times when it is well populated (dashed and dotted line in Fig. 14.3b). For the relaxation towards the asymptotic state, also the slower transitions within doublets are relevant. Therefore, the corresponding time scale can be much larger than t_{coh} (dotted line in Fig. 14.3a).

To obtain quantitative estimates for the dissipative time scales, we approximate t_{coh} by the growth of the Renyi entropy, averaged over a time t_{p},

$$\frac{1}{t_{\text{coh}}} = \frac{1}{t_{\text{p}}} \int_0^{t_{\text{p}}} \mathrm{d}t' \dot{S}_2(t') = \frac{1}{t_{\text{p}}} \Big(S_2(t_p) - S_2(0) \Big) \,. \tag{14.28}$$

Because of the stepwise growth of the Renyi entropy (Fig. 14.3b), we have chosen the propagation time t_{p} as an n-fold multiple of the duration $2\pi\hbar/|\epsilon_2^- - \epsilon_1^-|$ of a tunnel cycle. For this procedure to be meaningful, n should be so large that the Renyi entropy increases substantially during the time t_{p} (in our numerical studies from zero to a value of approximately 0.2). We find that at the center of the avoided crossing, the decay of coherence, respectively the entropy growth, becomes much faster and is essentially independent of temperature (Fig. 14.4a). At a temperature $k_{\text{B}}T = 10^{-4}\hbar\omega_0$ it is enhanced by three orders of magnitude. This indicates that transitions from states with mean energy far above the ground state play a crucial role.

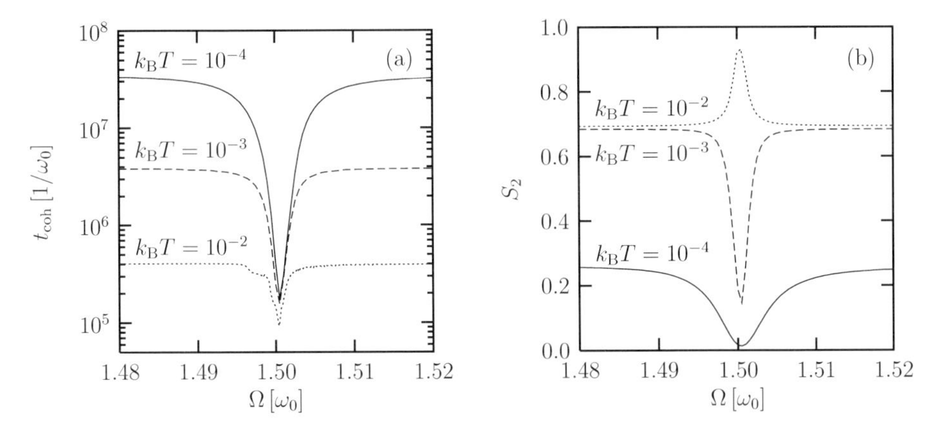

Figure 14.4: Decoherence time (a) and Renyi entropy S_2 of the asymptotic state (b) in the vicinity of the singlet-doublet crossing for $D = 2$, $F = 10^{-3}$, and $\Omega = 1.5\,\omega_0$. The temperature is given in units of $\hbar\omega_0$.

As the dynamics described by the master equation (14.24) is dissipative, it converges in the long-time limit to an asymptotic state $\rho_\infty(t)$. In general, this attractor remains time dependent but shares the symmetries of the central system, i.e. here, periodicity and generalized parity. However, the coefficients (14.25) of the master equation for the matrix elements $\rho_{\alpha\beta}$ are time independent and so the asymptotic solution also is. The explicit time dependence of the attractor has been effectively eliminated by the use of a Floquet basis.

To gain some qualitative insight into the asymptotic solution, we focus on the diagonal elements

$$\mathcal{L}_{\alpha\alpha,\alpha'\alpha'} = 2\sum_n N_{\alpha\alpha',n}|X_{\alpha\alpha',n}|^2, \quad \alpha \neq \alpha', \tag{14.29}$$

of the dissipative kernel. They give the rates of direct transitions from $|\phi_{\alpha'}\rangle$ to $|\phi_\alpha\rangle$. Within a golden rule description, these were the only non-vanishing contributions to the master equation to affect the diagonal elements $\rho_{\alpha\alpha}$ of the density matrix.

In the case of zero driving amplitude, the Floquet states $|\phi_\alpha\rangle$ reduce to the eigenstates of the undriven Hamiltonian H_{DW}. The only non-vanishing Fourier component is then $|\phi_{\alpha,0}\rangle$, and the quasienergies ϵ_α reduce to the corresponding eigenenergies E_α. Thus $\mathcal{L}_{\alpha\alpha,\alpha'\alpha'}$ only consists of a single term proportional to $N(E_\alpha - E_{\alpha'})$. It describes two kinds of thermal transitions: decay to states with lower energy and, if the energy difference is less than $k_{\mathrm{B}}T$, thermal activation to states with higher energy. The ratio of the direct transitions forth and back then reads

$$\frac{\mathcal{L}_{\alpha\alpha,\alpha'\alpha'}}{\mathcal{L}_{\alpha'\alpha',\alpha\alpha}} = \exp\left(-\frac{E_\alpha - E_{\alpha'}}{k_{\mathrm{B}}T}\right). \tag{14.30}$$

We have detailed balance and therefore the steady-state solution is

$$\rho_{\alpha\alpha'}(\infty) \sim \mathrm{e}^{-E_\alpha/k_{\mathrm{B}}T}\,\delta_{\alpha\alpha'}. \tag{14.31}$$

In particular, the occupation probability decays monotonically with the energy of the eigenstates. In the limit $k_B T \to 0$, the system tends to occupy mainly the ground state.

For a strong driving, each Floquet state $|\phi_\alpha\rangle$ contains a large number of Fourier components and $\mathcal{L}_{\alpha\alpha,\alpha'\alpha'}$ is given by a sum over contributions with quasienergies $\epsilon_\alpha - \epsilon_{\alpha'} + k\hbar\Omega$. Thus, a decay to states with "higher" quasienergy (recall that quasienergies do not allow for a global ordering) becomes possible due to terms with $k < 0$. Physically, it amounts to an incoherent transition under absorption of driving-field quanta. Correspondingly, the system tends to occupy Floquet states comprising many Fourier components with low index k. According to Eq. (14.8), these states have a low mean energy.

The effects under study are found for a driving with a frequency of the order ω_0. Thus, for a quasienergy doublet, not close to a crossing, we have $|\epsilon_\alpha - \epsilon_{\alpha'}| \ll \hbar\Omega$, and $\mathcal{L}_{\alpha'\alpha',\alpha\alpha}$ is dominated by contributions with $n < 0$, where the splitting has no significant influence. However, except for the tunnel splitting, the two partners in the quasienergy doublet are almost identical. Therefore, with respect to dissipation, both should behave similarly. In particular, one expects an equal population of the doublets even in the limit of zero temperature in contrast to the time-independent case.

In the vicinity of a singlet-doublet crossing the situation is more subtle. Here, the odd partner, say, of the doublet mixes with the singlet, cf. Eq. (14.16), and thus acquires components with higher energy. Due to the high mean energy E_t^- of the singlet, close to the top of the barrier, the decay back to the ground state can also proceed indirectly via other states with mean energy below E_t^-. Thus, $|\phi_1^-\rangle$ and $|\phi_2^-\rangle$ are depleted and mainly $|\phi_0^+\rangle$ will be populated. However, if the temperature is significantly above the splitting $2b$ at the avoided crossing, thermal activation from $|\phi_0^+\rangle$ to $|\phi_{1,2}^-\rangle$, accompanied by depletion via the states below E_t^-, becomes possible. Asymptotically, all these states become populated in a cyclic flow.

In order to characterize the coherence of the asymptotic state, we use again the Renyi entropy (14.22). According to the above scenario, we expect S_2 to assume the value $\ln 2$, in a regime with strong driving but preserved doublet structure, reflecting the incoherent population of the ground-state doublet. In the vicinity of the singlet-doublet crossing where the doublet structure is dissolved, its value should be of the order unity for temperatures $k_B T \ll 2b$ and much less than unity for $k_B T \gg 2b$ (Fig. 14.4b). This means that the crossing of the singlet with the doublet leads asymptotically to an improvement of coherence if the temperature is below the splitting of the avoided crossing. For temperatures above the splitting, the coherence becomes derogated. This phenomenon compares to chaos-induced coherence or incoherence, respectively, found in Ref. [3] for dissipative chaos-assisted tunneling.

14.5 Conclusions

For the generic situation of the dissipative quantum dynamics of a particle in a driven double-well potential, resonances play a significant role for the loss of coherence. The influence of states with higher energy alters the splittings of the doublets and thus the tunneling rates. We have studied decoherence in the vicinity of crossings of singlets with tunnel doublets under the influence of an environment. As a simple intuitive model to compare against, we have constructed a three-state system which in the case of vanishing dissipation, provides a faithful description of an isolated singlet-doublet crossing. The center of the crossing is characterized by a strong mixing of the singlet with one state of the tunnel doublet. The high mean energy of the singlet introduces additional decay channels to states outside the three-state system. Thus, decoherence becomes far more effective and, accordingly, coherent oscillations fade away on a much shorter time scale.

Acknowledgments

This work is supported by the Deutsche Forschungsgemeinschaft. Discussions with Thomas Dittrich are gratefully acknowledged.

References

[1] T. Dittrich, B. Oelschlägel, and P. Hänggi, Europhys. Lett. **22**, 5 (1993).

[2] M. Grifoni and P. Hänggi, Phys. Rep. **304**, 229 (1998).

[3] S. Kohler, R. Utermann, P. Hänggi, and T. Dittrich, Phys. Rev. E **58**, 7219 (1998).

[4] P. Hänggi, S. Kohler, and T. Dittrich, *Driven Tunneling: Chaos and Decoherence*, in Statistical and Dynamical Aspects of Mesoscopic Systems, Lecture Notes in Physics 547, p.125-157 (Springer, 2000)

[5] J. H. Shirley, Phys. Rev. **138**, B979 (1965).

[6] H. Sambe, Phys. Rev. A **7**, 2203 (1973).

[7] N. L. Manakov, V. D. Ovsiannikov, and L. P. Rapoport, Phys. Rep. **141**, 319 (1986).

[8] T. Dittrich, P. Hänggi, G.-L. Ingold, B. Kramer, G. Schön, and W. Zwerger, *Quantum Transport and Dissipation* (Wiley-VCH, Weinheim, 1998).

[9] F. Grossmann, T. Dittrich, P. Jung, and P. Hänggi, Phys. Rev. Lett. **67**, 516 (1991).

[10] F. Grossmann, P. Jung, T. Dittrich, and P. Hänggi, Z. Phys. B, **84**, 315 (1991).

[11] F. Grossmann and P. Hänggi, Europhys. Lett. **18**, 571 (1992).

[12] M. Latka, P. Grigolini, and B. J. West, Phys. Rev. A **50**, 1071 (1994).

[13] O. Bohigas, S. Tomsovic, and D. Ullmo, Phys. Rep. **223**, 43 (1993).

[14] A. Wehrl, Rep. Math. Phys **30**, 119 (1991).

[15] V. B. Magalinskiĭ, Zh. Eksp. Teor. Fiz. **36**, 1942 (1959), [Sov. Phys. JETP **9**, 1381 (1959)].

[16] A. O. Caldeira and A. L. Leggett, Ann. Phys. (N.Y.) **149**, 374 (1983); erratum: Ann. Phys. (N.Y.) **153**, 445 (1984).

[17] R. Blümel *et al.*, Phys. Rev. Lett. **62**, 341 (1989).

[18] S. Kohler, T. Dittrich, and P. Hänggi, Phys. Rev. E **55**, 300 (1997).

15 Entanglement and Decoherence in Cavity QED with a Trapped Ion

Werner Vogel and Christian Di Fidio

Arbeitsgruppe Quantenoptik, Fachbereich Physik
Universität Rostock
Rostock, Germany

15.1 Introduction

The recent experimental progress renders it possible to combine the experimental achievements in ion trapping and cavity QED. First experiments have demonstrated the possibility of trapping and manipulating a single ion inside a cavity by laser fields [1]. A setup of this type could be further developed to realize strong coupling of the internal and external degrees of freedom of the atom with a quantized mode of the cavity field. The exchange of photons between the ion and the cavity field entangles the quantum states of the trapped ion with those of the cavity field [2, 3]. Such new possibilities are of great interest since the role of entanglement and non-locality is nowadays widely discussed in all branches of the emerging fields of quantum information and quantum computation [4–6]. Moreover, quantum entanglement and non-locality are of great conceptual importance [7, 8] and constitute one of the most striking results of quantum theory [9, 10].

Trapped ions in leaky cavities have been regarded as good candidates to realize different schemes related to quantum information processing. While atomic states, including vibrational states of the quantized motion of a trapped ion, are good candidates for storing quantum information, photons are the preferred carriers for transferring the information from a given physical location to another one. For the transfer of information it is important that the photons can leak out of the cavity. There have been several proposals to utilize the interaction between the atom and the cavity field for physical realizations of quantum computing and quantum communication [11–14]. Atoms placed inside leaky optical cavities can be used to implement a distributed network [15] and for preparing and distributing motional-state entanglement [16]. Moreover, schemes have been proposed for teleporting atomic states between trapped atoms inside two distantly separated optical leaky cavities [17], and for teleporting motional quantum states [18].

In the present contribution we will deal with some aspects that are of basic interest for the field of quantum information. In sections 15.2 and 15.3 we will briefly review some of our results concerning the decoherence mechanism in Raman-driven trapped ions [19] and the preparation of particular entangled states [20] of Greenberger–Horne–Zeilinger (GHZ) type [21], respectively. Some new results are given in Sec. 15.4, where we consider the control of the photon number in a propagating light field. Moreover, in Sec. 15.5 will deal with the

Quantum Information Processing, 2nd Edition. Edited by Thomas Beth and Gerd Leuchs

ISBN: 3-527-40541-0

preparation of motional-state entanglement of spatially separated atoms. A brief summary is given in Sec. 15.6.

15.2 Decoherence Effects

Any type of application of trapped ions in high-Q cavities requires a good understanding of the decoherence effects acting in the system. Only a precise knowledge of the decoherence mechanisms allows one to optimize the desired quantum-state manipulations. The situations we are interested in typically consist in Raman-driven trapped ions, where a laser is used to control the interaction of atomic quantum states with the quantized cavity field. As a first step we will study decoherence in Raman-driven trapped ions. These effects are later on included in our studies of trapped ions in cavities.

Observations of Rabi oscillations of a Raman-driven trapped $^9\mathrm{Be}^+$ ion have led to pronounced damping effects [22]. Even under almost ideal conditions, when the atom was initially prepared in both the electronic and motional ground state, unexpectedly strong damping effects were found to occur. Attempts have been made to relate these damping effects to classical noise sources [23]. Later on, we have demonstrated that the observed decoherence effects, for an ion cooled to the ground state, can be understood from first principles without the need of introducing phenomenological noise sources [19]. In the experiment under consideration a dipole-forbidden transition $|1\rangle \leftrightarrow |2\rangle$ was driven in a Raman-scheme. The Raman coupling was enhanced by the use of off-resonant electric dipole transitions $|1\rangle \leftrightarrow |3\rangle$ and $|2\rangle \leftrightarrow |3\rangle$. It turns out that already a very small number of quantum jumps between the driven states and the auxiliary level $|3\rangle$ leads to the observed damping of Rabi oscillations.

For describing the decoherence effects caused by electronic transitions between the driven $|1\rangle \leftrightarrow |2\rangle$ transition and the auxiliary state $|3\rangle$, we apply quantum-trajectory methods [24]. The master equation for the density operator $\hat{\rho}$ for the states $|1\rangle$ and $|2\rangle$ and the motional subsystem can be given, after adiabatic elimination of the electronic level $|3\rangle$, in the form

$$\begin{aligned}\frac{\partial\hat{\rho}}{\partial t} &= \frac{1}{i\hbar}\left[\hat{H}'(t)\,\hat{\rho} - \hat{\rho}\,\hat{H}'^{\dagger}(t)\right] + \int_{-1}^{1} dq\,\hat{J}_q^{(1)}(t)\,\hat{\rho}\,\hat{J}_q^{(1)\dagger}(t) \\ &\quad + \int_{-1}^{1} dq\,\hat{J}_q^{(2)}(t)\,\hat{\rho}\,\hat{J}_q^{(2)\dagger}(t)\,. \qquad (15.1)\end{aligned}$$

The two (effective) jump operators $\hat{J}^{(1)}$ and $\hat{J}^{(2)}$ describe the jumps (via excitation of state $|3\rangle$) into the states $|1\rangle$ and $|2\rangle$, respectively. For notational simplicity in the following we will refer to the corresponding jumps as jumps 1 and jump 2. The jump operators also take into account the effects of the transitions on the motional states of the atom, for more details cf. [19]. The needed non-Hermitian Hamiltonian reads in rotating-wave approximation as

$$\hat{H}' = \left(1 + i\frac{\gamma+\gamma'}{2\Delta}\right)\hat{H}_{\mathrm{eff}}, \qquad (15.2)$$

herein Δ is the detuning of the lasers from the state $|3\rangle$ and γ (γ') is the decay rate of the state $|3\rangle$ to the state $|1\rangle$ ($|2\rangle$). The nonlinear Hamiltonian $\hat{H}_{\mathrm{eff}}$ is given by

$$\frac{\hat{H}_{\mathrm{eff}}}{\hbar} = -\left[\Omega i\eta\,\hat{a}^{\dagger}\hat{f}_1(\hat{a}^{\dagger}\hat{a};\eta)\hat{A}_{21} + \mathrm{H.c.}\right] - \Delta_{ac}\hat{A}_{11} - \Delta'_{ac}\hat{A}_{22}, \qquad (15.3)$$

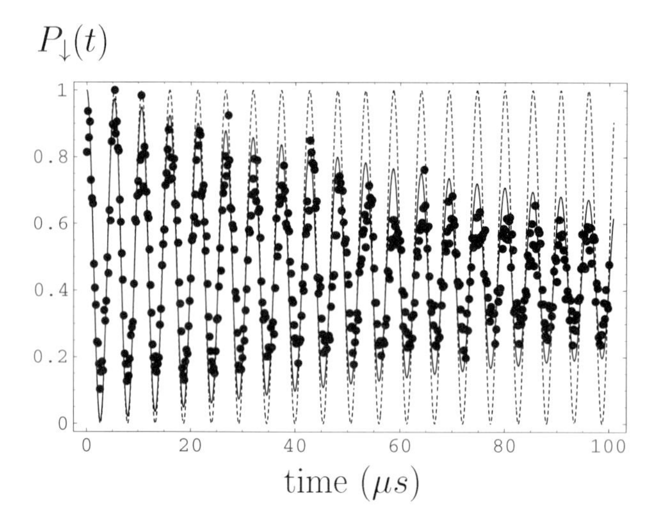

Figure 15.1: Occupation of the electronic state $|1\rangle$, $P_\downarrow(t)$, for an initial $|1,0\rangle$ state driven by the Raman interaction. The ensemble average (full line) over 20000 trajectories is calculated for the same parameters as in the experiment [22] and compared with the corresponding experimental data (dots).

the first term represents the well-known nonlinear Jaynes–Cummings Hamiltonian for the vibronic dynamics of a trapped ion [25]. The other terms are ac Stark shifts of size Δ_{ac} and Δ'_{ac} of the states $|1\rangle$ and $|2\rangle$ respectively. The Lamb–Dicke parameter η describes the localization of the motional wave-packet of the atom relative to the beat node of the driving lasers and Ω is the (electronic) Rabi frequency for the Raman transition. Moreover, $\hat{a}$ and $\hat{a}^\dagger$ are the annihilation and creation operators of vibrational quanta, respectively and the operators $\hat{A}_{ij}$ are defined as $\hat{A}_{ij} = |i\rangle\langle j|$ $(i, j = 1, 2)$. The operator function $\hat{f}_1(\hat{a}^\dagger\hat{a};\eta)$ in the motional Fock basis reads as

$$\hat{f}_m(\hat{a}^\dagger\hat{a};\eta) = \sum_{j=0}^{\infty} |j\rangle\langle j| \frac{j!}{(j+m)!} L_j^{(m)}(\eta^2)\, e^{-\eta^2/2}, \tag{15.4}$$

$L_j^{(m)}(x)$ being generalized Laguerre polynomials.

We have solved equation (15.1) by using the method of quantum-trajectories. Our simulation in Fig. 15.1 clearly shows a damping of the Rabi oscillations, in good agreement with the experimental data. Moreover, also in full agreement with the experiment, the oscillations are not centered around an average value of $1/2$, but the average decays as a function of time. The decay is caused by optical pumping into the state $|2,0\rangle$ that decouples from the Raman dynamics. This effect strongly supports the conclusion that electronic transitions play a substantial role for the decoherence of the trapped ions driven in Raman configurations.

It is noteworthy that the experiments have been performed for other initial preparations. From their data the authors [22] have inferred a dependence of the damping on the motional excitation of the form $\gamma_n \approx \gamma_0(n+1)^{0.7}$. This is not recovered by our simulations, since the decoherence due to quantum jumps is rather insensitive with respect to the motional excitation. Attempts have been made to explain the excitation dependence of the damping by combining our model with additional classical noise sources [26].

15.3 Greenberger–Horne–Zeilinger State

Let us consider a trapped ion interacting with a mode of a high-Q optical cavity of frequency ω_{C} as shown in Fig 15.2. The laser beam of frequency $\omega_{\mathrm{L}} + \delta\omega_{\mathrm{L}}$ is tuned to $\omega_{\mathrm{C}} - \omega_{\mathrm{L}} = \omega_{21} - \nu$, where $\delta\omega_{\mathrm{L}}$ is an adjustable laser-frequency shift, ω_{21} is the frequency of the electronic transition $|2\rangle \leftrightarrow |1\rangle$, and ν is the secular frequency of the RF–Paul trap along the cavity axis. The laser and the cavity mode are far detuned by Δ from the state $|3\rangle$ to provide the Raman coupling of states $|1\rangle$ and $|2\rangle$ on the first vibrational sideband. This yields a coupling between quantum states of the center-of-mass motion of the ion and the cavity filed. The cavity losses are described by the photon escape rate κ.

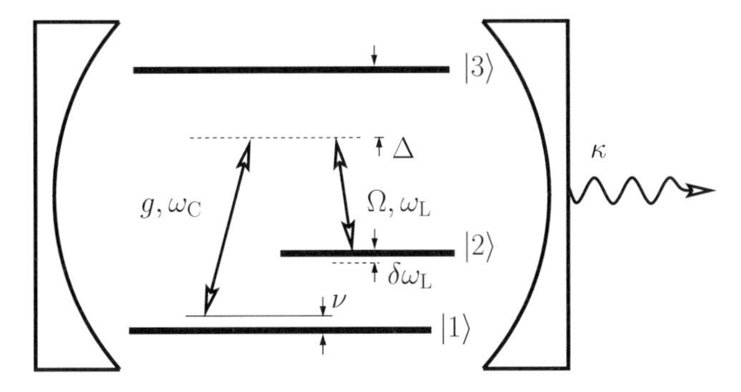

Figure 15.2: Raman-type excitation scheme for a trapped ion in an optical cavity.

The effective master equation for the density operator $\hat{\rho}$, including the electronic states $|1\rangle$ and $|2\rangle$ together with the motional subsystem and the cavity field, is of the form [20]

$$\frac{\partial \hat{\rho}}{\partial t} = \frac{1}{i\hbar}[\hat{H}_{\mathrm{eff}}, \hat{\rho}] - \frac{\kappa}{2}\left(\hat{b}^{\dagger}\hat{b}\,\hat{\rho} + \hat{\rho}\,\hat{b}^{\dagger}\hat{b} - 2\,\hat{b}\,\hat{\rho}\,\hat{b}^{\dagger}\right). \tag{15.5}$$

In vibrational rotating-wave approximation [25], the effective Hamiltonian $\hat{H}_{\mathrm{eff}}$ reads as

$$\hat{H}_{\mathrm{eff}} = \hbar\left[\Omega_{\mathrm{eff}}\,(\eta\,\hat{a}^{\dagger})\hat{f}_1(\hat{a}^{\dagger}\hat{a};\eta)\,\hat{b}^{\dagger}\hat{A}_{12} + \mathrm{H.c.}\right] + \hat{H}_{\mathrm{Stark}}\,, \tag{15.6}$$

where $\Omega_{\mathrm{eff}} = -\frac{1}{2}\,i\,g^{*}\,\Omega\,e^{i\varphi_{\mathrm{C}}}/\Delta$, and $\hat{H}_{\mathrm{Stark}}$ is the Stark shift term. The basic coupling consists in transitions $|2, m, n\rangle \leftrightarrow |1, m+1, n+1\rangle$, where in $|i, m, n\rangle$ the numbers i, m and n label the electronic state, the motional state and the number of cavity photons, respectively. The adjustable laser-frequency shift $\delta\omega_{\mathrm{L}}$ can be used to compensate Stark shifts, $\hat{b}$ ($\hat{b}^{\dagger}$) denotes the photon annihilation (creation) operator of the cavity field. Its phase φ_{C} describes the position of the center of the trap relative to cavity field. The Lamb–Dicke parameter η is now defined for the sum of wave numbers of laser and cavity field, $\eta \propto k_L + k_C$. The operator function $\hat{f}_m$ is given by Eq. (15.4).

For the absence of cavity losses and spontaneous emissions the system can be used to prepare GHZ state [27]. Starting from the initial state $|2,0,0\rangle$, one applies a resonant $\pi/2$-pulse interaction on the $|2,0,0\rangle \leftrightarrow |1,1,1\rangle$ transition to obtain the entangled three-particle-like state

$$|\Psi_{\rm GHZ}\rangle = \frac{1}{\sqrt{2}}\left(|2,\,0,\,0\rangle + |1,\,1,\,1\rangle\right). \tag{15.7}$$

In the following we will deal with the preparation of a GHZ state under more realistic conditions. The quality of the prepared state $\hat{\rho}$ is characterized by the fidelity

$$F(t) = \langle\Psi_{\rm GHZ}|\hat{\rho}(t)|\Psi_{\rm GHZ}\rangle. \tag{15.8}$$

Let us first consider the effects of the cavity losses when spontaneous emissions are negligibly small. For $\kappa \ll |\Omega_{0,0}|$, one obtains

$$F(t,\eta) = \frac{1}{2}e^{-\kappa t/2}\left[1 - \cos(\varphi)\sin(2\,|\Omega_{0,0}(\eta)|\,t)\right], \tag{15.9}$$

where $\varphi = \varphi_C + \Phi$, $g^*\Omega = |g^*\Omega|e^{i\Phi}$, and

$$|\Omega_{00}(\eta)| = |\Omega_{\rm eff}|\,\eta\, e^{-\eta^2/2} = \left|\frac{g\Omega}{2\,\Delta}\right|\,\eta\, e^{-\eta^2/2}. \tag{15.10}$$

The system is found close to the GHZ state (15.7) when a $\pi/2$-pulse is applied, that is $2|\Omega_{0,0}|\bar{t} = \pi/2$, and for $\varphi = \pi$. Inserting these parameters into Eq. (15.9), we obtain the fidelity as a function of η,

$$F(\bar{t},\eta) = \exp\left(-\frac{\kappa\pi}{8\,|\Omega_{00}(\eta)|}\right). \tag{15.11}$$

From Eq. (15.11) it is possible to obtain the optimum for the fidelity by maximizing $|\Omega_{00}(\eta)|$. This result has a simple explanation. For larger values of $|\Omega_{00}(\eta)|$ the generation time $\bar{t}$ is getting smaller. Thus, the role of cavity losses is reduced during the preparation time. From Eq. (15.10) we see that the fidelity is maximized for $\eta = 1$. As long as spontaneous emissions are disregarded, the choice of other η values, however, could be compensated for by increasing the laser field.

This situation changes drastically when spontaneous emissions play a significant role. In this case any increase of the laser field leads to increasing decoherence due to spontaneous emission. Our results in Figure 15.3 clearly show this effect. When the values of $|\Omega_{00}(\eta)|$ are set equal, the fidelity is much higher for $\eta = 1$ rather than for $\eta = 0.1$. Thus the optimum conditions for the generation of a GHZ state (15.7) are obtained for $\eta \approx 1$, that is far from the Lamb–Dicke regime.

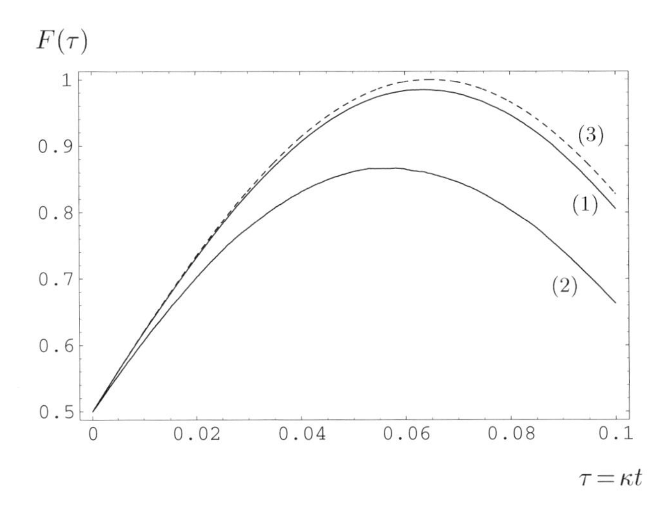

Figure 15.3: Fidelity $F(t)$ as a function of $\tau = \kappa t$, for $\kappa = 0.1$ MHz and $(\gamma + \gamma')/2\pi \simeq 19$ MHz. Curve (1) is for $\eta = 1$ and $|\Omega_{\text{eff}}| = 4$ MHz, curve (2) is for $\eta = 0.1$ and the same value of $|\Omega_{00}(\eta)|$ as in curve (1). The ensemble averages are performed over 10,000 trajectories. The dashed line (3) describes the case in the absence of cavity losses ($\kappa = 0$) and spontaneous emissions ($\gamma = \gamma' = 0$).

15.4 Photon-number Control

A precise control of the number of photons in a freely propagating light field has not been realized yet, but it would be of interest for various applications. For this purpose we consider an ion undergoing a Raman interaction with laser beam and cavity mode on the first motional sidebands: $\omega_{\text{L}} - \omega_{\text{C}} = \pm\nu$, cf. Fig. 15.4. In the master equation (15.5), we now use

$$\frac{\hat{H}_{\text{eff}}^{(+)}}{\hbar} = \left[-\frac{g^* \Omega}{2\Delta} e^{i\varphi_{\text{C}}} (i\eta \hat{a}^\dagger) \hat{f}_1(\hat{a}^\dagger \hat{a}; \eta) \hat{b}^\dagger + \text{H.c.} \right] + \hat{H}_{\text{Stark}} , \tag{15.12}$$

for $\omega_{\text{L}} - \omega_{\text{C}} = +\nu$. For $\omega_{\text{L}} - \omega_{\text{C}} = -\nu$ we replace $\hat{a}^\dagger \hat{f}_1 \hat{b}^\dagger$ with $\hat{f}_1 \hat{a} \hat{b}^\dagger$.

Now we proceed as follows. The initial state is $|\Psi(0)\rangle = |0, 0\rangle$, in $|m, n\rangle$ the numbers m and n are the number of motional quanta and the photon number in the cavity mode, respectively. The Raman interaction of frequency $\omega_{\text{L}} - \omega_{\text{C}} = \nu$ is switched on (controlled by the laser) for a time interval T_1. The Hamiltonian (15.12) leads to a chain of oscillations $|0, 0\rangle \leftrightarrow |1, 1\rangle \leftrightarrow \cdots \leftrightarrow |M, M\rangle \leftrightarrow \cdots$ in the ion-cavity system. Provided that no photon is emitted by the cavity during the time interval T_1, the system is in the entangled state

$$|\Psi(T_1)\rangle = \sum_{m=0} a_{m,m}(T_1) |m, m\rangle \tag{15.13}$$

for vibrational motion and cavity field, that resembles the state of photon pairs in optical parametric decay.

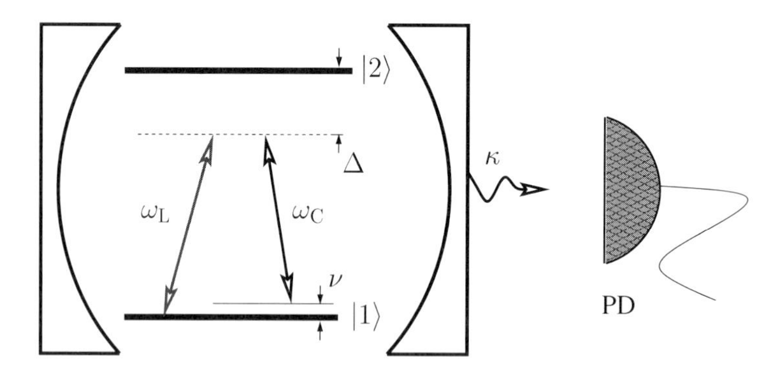

Figure 15.4: Trapped ion in a leaky optical cavity, Raman-driven on the first vibrational side-bands, $\omega_L - \omega_C = \pm\nu$. Photons leaking out of the cavity (at rate κ) are monitored by a photodetector PD.

For preparing a quantum state (15.13), it is important to reduce the probability of photon leakage from the cavity during the time interval T_1. This requires a good cavity, so that $\kappa T_1 \ll 1$. However, T_1 cannot be too small in order to get a sufficiently large weight factor $|a_{m,m}|$ for the desired Fock state. For example, for $\kappa T_1 = 0.6$ we may obtain a probability that no photon is emitted during T_1 of about 0.6. In this case we have tried to maximize the weighting of the state $|3,3\rangle$. If one detects a photon during the preparation interval T_1, one has to start the preparation from the beginning.

In the second step of our method, after preparation of the entangled state $|\Psi(T_1)\rangle$, the laser is turned off. During a time interval T_2, obeying the condition $\kappa T_2 \gg 1$ to ensure that all photons leak out of the cavity, one counts the number of photons in the cavity output port. Suppose that the detector has counted exactly N photons, the probability of which is $|a_{N,N}|^2$, see Eq. (15.13). In this case we end up in the state

$$|\Psi(T_1 + T_2)\rangle = |N, 0\rangle \,. \tag{15.14}$$

The ion is in the N-th motional Fock state and the cavity mode is in the vacuum state. Let us consider the situation for $N = 3$. The probability for preparing this state is given by $|a_{3,3}|^2$, for which we get a value of about 0.28. Altogether, including the value 0.6 of the probability of observing no photon in the first part of the preparation, the probability to end up in the state $|3,0\rangle$ is 0.17.

The state (15.14) can now be stored over some time interval. On demand the motional excitation can be converted into photons that escape from the cavity. For this purpose, in the third step of our procedure we turn on the laser again for a time interval T_3. Now the frequency is adjusted as $\omega_L - \omega_C = -\nu$ in order to realize the Hamiltonian $\hat{H}_{\text{eff}}^{(-)}$, containing the operator products $\hat{a}\hat{b}^\dagger$. They produce the couplings $|N,0\rangle \leftrightarrow |N-1,1\rangle \leftrightarrow \cdots \leftrightarrow |0,N\rangle$, that convert motional excitations of the ion into cavity photons. Due to the cavity losses the photons escape the cavity. The dynamics continues until the ion-cavity system ends in the ground state $|0,0\rangle$. Thus is ready to start again with the first step of our procedure. Exactly N photons have been emitted by the cavity, they are freely propagating and can be used for any desired application.

The time interval T_3 is chosen such that we can be practically certain that N photons have leaked out of the cavity. Note that one ma also try to externally prepare the motional Fock state of the ion and to obtain the state (15.14) by displacing the ion into the cavity, cf. the scheme used in [1]. Then one may proceed with the third step of our manipulations.

15.5 Entanglement of Separated Atoms

In order to achieve entanglement of two distant atoms, we consider now two cavities, each containing a trapped ion. The system arrangement of the cavities A and B is shown in Fig. 15.5. The photons leaking out from the two cavities are combined by a beam-splitter of transmissivity T and reflectivity R, where $|T| = |R|$. Two photodetectors, D_1 and D_2, are used to count the photons in the output ports of the beam-splitter. Note that a similar arrangement was considered for the teleportation of atomic (electronic) states [17], whereas here we will consider entanglement of the ion's motion.

The master equation for our combined system A plus B can be written in the Lindblad form as

$$\frac{\partial \hat{\rho}}{\partial t} = \frac{1}{i\hbar}\left[\hat{H}_{\mathrm{tot}}\,\hat{\rho} - \hat{\rho}\,\hat{H}_{\mathrm{tot}}^{\dagger}\right] + \hat{J}_1\,\hat{\rho}\,\hat{J}_1^{\dagger} + \hat{J}_2\,\hat{\rho}\,\hat{J}_2^{\dagger}\,. \tag{15.15}$$

The two jump-operators $\hat{J}_1$ and $\hat{J}_2$ describe a photon detection in D_1 and D_2, respectively, and are given by

$$\hat{J}_1 = \sqrt{\kappa}\,\hat{d}_1\,, \tag{15.16}$$

$$\hat{J}_2 = \sqrt{\kappa}\,\hat{d}_2\,. \tag{15.17}$$

where

$$\hat{d}_1 = (\hat{b}_A + i\hat{b}_B)/\sqrt{2}\,, \tag{15.18}$$

$$\hat{d}_2 = (i\hat{b}_A + \hat{b}_B)/\sqrt{2}\,. \tag{15.19}$$

The operators $\hat{b}_A$ and $\hat{b}_B$ are the photon annihilation operators of the cavity fields A and B, respectively. Moreover, the non-Hermitian Hamiltonian $\hat{H}_{\mathrm{tot}}$ is given by

$$\hat{H}_{\mathrm{tot}} = \hat{H}_{\mathrm{eff}}^{A} + \hat{H}_{\mathrm{eff}}^{B} - \frac{i}{2}\kappa\left(\hat{b}_A^{\dagger}\hat{b}_A + \hat{b}_B^{\dagger}\hat{b}_B\right), \tag{15.20}$$

where the two effective Hamiltonians $\hat{H}_{\mathrm{eff}}^{A}$ and $\hat{H}_{\mathrm{eff}}^{B}$ describe the ion-cavity systems A and B, respectively, each given by Eq. (15.12).

Now we will show how it is possible to exploit such setup to obtain an entangled state of the quantized motion of two well separated atoms. The evolution of the system, conditioned on the detections recorded by D_1 and D_2, can be appropriately treated by a quantum-trajectory treatment. Let us start from the initial state for the combined system A plus B

$$|\Psi^{\mathrm{tot}}(0)\rangle = |1,0,0\rangle_A|1,0,0\rangle_B\,, \tag{15.21}$$

where the first index in the brackets denotes the electronic level, the second index the number state for the vibrational motion, and the third index the number state for the cavity mode.

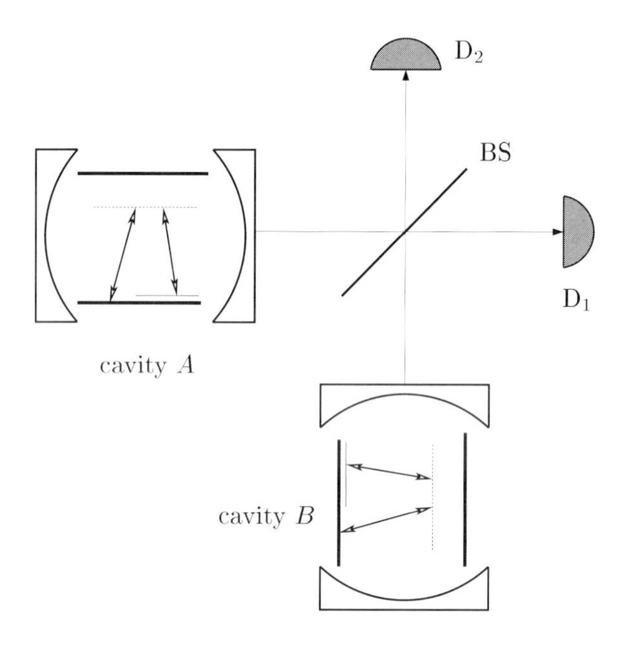

Figure 15.5: The photons leaking out from both the single-ended cavities A and B impinge on the 50:50 beam splitter BS and are detected by the detectors D_1 and D_2.

As in the previous section, we turn on the laser of frequency ω_L for a time interval T_1. We will be interested only in the case where in this time interval no detector registers any count. This suggest to choose $\kappa T_1 \ll 1$, so that in this time interval the probability to have a cavity decay is negligible. If a count in one of the detectors occurs in this time interval, this simply means that we have to start again. Let us now consider the evolution of the system under the influence of the laser of frequency ω_L in more detail. The Hamiltonians $\hat{H}^A_\mathrm{eff}$ and $\hat{H}^B_\mathrm{eff}$ contain elements proportional to $\hat{a}^\dagger_A \hat{b}^\dagger_A + \mathrm{H.c.}$ and $\hat{a}^\dagger_B \hat{b}^\dagger_B + \mathrm{H.c.}$, respectively. These produce a chain of oscillations $|1,0,0\rangle_A \leftrightarrow |1,1,1\rangle_A \leftrightarrow \cdots \leftrightarrow |1,N,N\rangle_A \leftrightarrow \cdots$ in system A and correspondingly in system B. Each oscillation $|1,m,m\rangle \leftrightarrow |1,m+1,m+1\rangle$ occurs with a Rabi frequency

$$\Omega_m = \left| \frac{g^* \Omega}{2\Delta} \right| \eta \, e^{-\eta^2/2} L^{(1)}_m(\eta^2) \, . \tag{15.22}$$

In this way the state of the combined system evolves from the state (15.21) to the state

$$|\Psi^\mathrm{tot}(T_1)\rangle = \Big(\sum_{m=0} a_{m,m}(T_1) |1,m,m\rangle_A \Big) \otimes \Big(\sum_{m=0} a_{m,m}(T_1) |1,m,m\rangle_B \Big) \, . \tag{15.23}$$

At this stage we turn off the lasers. It is noteworthy that one can choose η such that for a given $\bar{m}$ one has $L^{(1)}_{\bar{m}}(\eta^2) = 0$. In this way we reduce the possible states that can be actually reached by the evolution due to $\hat{H}^A_\mathrm{eff}$ and $\hat{H}^B_\mathrm{eff}$. In this case, from Eq. (15.22) one sees that the chain of oscillations ends at $|1,\bar{m},\bar{m}\rangle_A$ (the same in system B).

After the lasers are turned off, all we have to do is to wait for a time interval $\kappa T_2 \gg 1$ and look at the detectors. We are interested to consider only the following two cases. Either first a count in D_1 and then a count in D_2 was registered, or first a count in D_2 and then one in D_1, and no other counts. All the other cases are disregarded from further analysis. Of course, for each quantum-trajectory the time t_1 for the first detection and the time t_2 for the second detection are random. But independently of the particular registration times we end up in the state

$$|\Psi_{\rm c}^{\rm tot}(T_1+T_2)\rangle = \frac{1}{\sqrt{2}}\Big(|1,2,0\rangle_A|1,0,0\rangle_B + |1,0,0\rangle_A|1,2,0\rangle_B\Big). \tag{15.24}$$

The subscript c indicates that the realization of such state is "conditioned" on the registration of two counts (D_1, D_2) or (D_2, D_1). The probability of having the required counts is $2|a_{0,0}|^2|a_{2,2}|^2$. In this manner an entangled state of the vibrational Fock states 2 and 0 has been generated in two distantly separated trapped ions, the number of photons inside the two cavities being now zero.

15.6 Summary

We have studied some basic problems that are of importance for applications of trapped ions in quantum information processing. In particular this includes methods of the preparation of quantum states and the study of decoherence effects. We have identified the effect of spontaneous emissions to be an important source of decoherence in Raman-driven trapped ions. This effect must be taken into account in the study of quantum-state manipulations. For example, we have analyzed the preparation of GHZ states of a trapped ion in a cavity under fairly realistic conditions. We have also proposed a method for the control of the photon number in a freely propagating light field. Eventually, we have studied a scheme that is suited to entangle the quantized motion of two trapped ions positioned far apart from each other.

This work was supported by the Deutsche Forschungsgemeinschaft.

References

[1] G.R. Guthöhrlein, M. Keller, K. Hayasaka, W. Lange and H. Walther, Nature **414**, 49 (2001).

[2] H. Zeng, F. Lin, Phys. Rev. A **50**, R3589 (1994).

[3] A.C. Doherty, A.S. Parkins, S.M.Tan and D.F. Walls, Phys. Rev. A **57**, 4804 (1998).

[4] C.H. Bennett, Phys. Today **48**, 24 (October, 1995); I.L. Chuang and Y. Yamamoto, Phys. Rev. Lett. **76**, 4281 (1996).

[5] D.P. DiVincenzo, Science **270**, 255 (1995); I.L. Chuang, R.Laflamme, P.W. Shor, W.H. Zurek, *ibid.* **270**, 1633 (1995).

[6] A. Peres, *Quantum Theory: Concepts and Methods* (Kluwer Academic Publishers, Dordrecht, The Netherlands, 1987).

[7] E. Schrödinger, Naturwissenschaften **23**, 807, 823, 844 (1935); E. Schrödinger, Proc. Camb. Phil. Soc. **31**, 555 (1935); E. Schrödinger, Proc. Camb. Phil. Soc. **32**, 446 (1936),
[8] A. Einstein, B. Podolsky and N. Rosen, Phys. Rev. **47**, 777 (1935).
[9] J.S. Bell, *Speakable and unspeakable in quantum mechanics* (Cambridge University Press, Cambridge, England, 1998).
[10] B. d'Espagnat, *Veiled Reality* (Addison-Wesley Publishing Company, Reading, Massachusetts, 1995).
[11] T. Pellizzari, S.A. Gardiner, J.I. Cirac, and P. Zoller, Phys. Rev. Lett. **75**, 3788 (1995).
[12] J.I. Cirac, P. Zoller, H.J. Kimble, and H. Mabuchi, Phys. Rev. Lett. **78**, 3221 (1997).
[13] A. Beige, D. Braun, B. Tregenna, and P.L. Knight, Phys. Rev. Lett. **85**, 1762 (2000).
[14] S.J. van Enk, H.J. Kimble, J.I Cirac and P. Zoller, Phys. Rev. A **59**, 2659 (1999).
[15] A.S. Parkins and H.J.Kimble, J. Opt. B: Quantum Semiclass. Opt. **1**, 469 (1999).
[16] A.S. Parkins and E. Larsabal, Phys. Rev. A **63**, 012304 (2001).
[17] S. Bose, P.L. Knight, M.B. Plenio, and V. Vendral, Phys. Rev. Lett. **83**, 5158 (1999).
[18] A.S. Parkins and H.J. Kimble, in *Frontiers of Laser Physics and Quantum Optics, Proceedings of the International Conference on Laser Physics and Quantum Optics*, edited by Z. Xu, S. Xie, S.-Y. Zhu, and M.O. Scully (Springer, Berlin, 2000). also
[19] C. Di Fidio and W. Vogel, Phys. Rev. A **62**, 031802(R) (2000).
[20] C. Di Fidio, S. Maniscalco, W. Vogel, and A. Messina, Phys. Rev. A **65** (2002) 033825.
[21] D.M. Greenberger, M.A. Horne, A. Zeilinger, Phys. Today **46**(8) 22 (1993).
[22] D.M. Meekhof, C. Monroe, B.E. King, W.M. Itano and D.J. Wineland, Phys. Rev. Lett. **76**, 1796 (1996); D. Leibfried, D.M. Meekhof, C. Monroe, B.E. King, W.M. Itano, and D.J. Wineland, J. Mod. Opt. **44**, 2485 (1997);
[23] D.F. James, Phys. Rev. Lett. **81**, 317 (1998); S. Schneider and G.J. Milburn, Phys. Rev A. **57**, 3748 (1998); *ibid.* **59**, 3766 (1999); M. Murao and P.L. Knight, Phys. Rev. A **58**, 663 (1998).
[24] J. Dalibard, Y. Castin, and K. Mølmer, Phys. Rev. Lett. **68**, 580 (1992); K. Mølmer, Y. Castin, and J. Dalibard, J. Opt. Soc. Am. B **10**, 524 (1993); H.J. Carmichael, *An open systems approach to quantum optics*, Lecture Notes in Physics Vol. m18 (Springer, Berlin, 1993).
[25] W. Vogel and R.L. de Matos Filho, Phys. Rev. A **52** (1995) 4214;
[26] A.A. Budini, R.L. de Matos Filho, and N. Zagury, Phys. Rev. A **65**, 041402(R) (2002).
[27] V. Bužek, G. Drobný, M.S. Kim, G. Adam and P.L. Knight, Phys. Rev. A **56**, 2352 (1997).

16 Quantum Information Processing with Ions Deterministically Coupled to an Optical Cavity

Matthias Keller[1], *Birgit Lange*[1], *Kazuhiro Hayasaka*[2], *Wolfgang Lange*[1], *and Herbert Walther*[1]

[1] Max-Planck-Institut für Quantenoptik
Garching, Germany

[2] Communications Research Laboratory
Kobe, Japan

16.1 Introduction

Among the physical systems proposed for quantum information processing, single atoms and photons are of particular importance. In no other domain, individual quantum objects can be manipulated with comparable precision and at the same time kept well isolated from the decohering influence of the environment. Therefore, it is not surprising that the first demonstrations of conditional dynamics at the single quantum level, which is a precondition for the realization of quantum gates, were achieved with ions [1] and photons [2] as carriers of quantum information. In the case of photons, information may be stored using two orthogonal polarization states or the excitation of a single photon in a particular mode of the optical field. In the case of atoms, the information is stored in long-lived electronic states. These can be either hyperfine-levels of the ground state, or metastable states.

Thus far, atoms and photons have been employed in distinct areas of quantum information processing, exploiting their respective advantages. Photons, which can be transmitted easily to remote locations ("flying" qubits), have been applied mainly in quantum communication and quantum cryptography, where it is important to transfer entanglement over long distances [3]. On the other hand, when storage or local manipulation of quantum information is important, localized qubits are required, which may be stored long enough to perform complex quantum operations. For this purpose, strings of trapped ions are ideally suited, with coherence times of superpositions of hyperfine states on the order of minutes. The quantum states of the ions are manipulated by individually addressing them with lasers and read out with nearly perfect efficiency using the technique of electron shelving. Interactions between the qubits have been implemented by using the quantized vibrational excitation of the ion string [1,4].

A decisive advance in the development of the field is expected from the combination of local atomic quantum processing with photonic transmission of quantum states. Conceptually new possibilities would arise, enhancing quantum computational capabilities. For example, separate local quantum processors of moderate size could be linked in a quantum network, with quantum information being exchanged by means of photons [5,6]. Photons might also be employed in local quantum processing. There are several proposals to couple atoms or ions

Quantum Information Processing, 2nd Edition. Edited by Thomas Beth and Gerd Leuchs

ISBN: 3-527-40541-0

through their interaction with a quantized cavity field [7–11]. Here, photons are used instead of vibrational quanta to transfer quantum information between sites in an atomic quantum register. Even for exclusively photonic applications, a controlled atom–photon interface may be exploited as a source of triggered single-photon pulses. In quantum cryptography, for example, secure communication requires the use of single-photon states [12]. Recently, it was shown that linear optical elements are sufficient to realize quantum computation, if single-photon pulses are employed [13].

Consequently, the controlled coupling between atoms and photons is a goal of fundamental importance in quantum information processing. The coupling strength required to overcome damping in the system can only be reached if the interaction is enhanced by a cavity. The most important system parameter, the rate of coupling between atoms and field, is given by the dipole matrix element μ of the atomic transition at frequency ω and the mode-distribution $f(\mathbf{r})$ in the cavity:

$$g(\mathbf{r}) = \sqrt{\frac{\mu^2 \omega}{2\hbar\epsilon_0 V}}\, f(\mathbf{r}). \tag{16.1}$$

Here, $V = \int |f(\mathbf{r})|^2 d\mathbf{r}$ is the mode volume of the cavity. In the limit of very small V, reached in cavities with a length as low as 30 μm, coupling parameters g can surpass the rate Γ of atomic spontaneous decay and the damping rate κ of the cavity field. This is the strong-coupling regime of cavity-QED, in which atom and field are merged in a single combined quantum system with new eigenstates. It forms the physical basis for realizing an atom–photon interface. Among the recent achievements in the domain of cavity-QED are the observation of atomic trajectories via the cavity field [14, 15] as well as the stimulated Raman scattering of photons into the cavity mode [16].

A problem of previous cavity-QED experiments is the fact that the atoms interacting with the cavity field are moving through the cavity on a non-controlled path. Due to the spatial dependence of the coupling, which is apparent from Eq. (16.1), this leads to an atom–cavity interaction of random strength occurring at random times. If the atoms are slow enough, information on their position may be obtained through photons emitted from the cavity. In this way, a particular coupling may be realized by post-selection. However, deterministic interaction with the cavity requires stationary atoms.

16.2 Deterministic Coupling of Ions and Cavity Field

In order to realize a well-defined coupling with the optical field, the atom must be trapped inside the cavity. The scale on which localization is required is determined by the spatial variation of the electromagnetic field inside the cavity. In the direction of the cavity axis, a standing wave field builds up, implying that the atom must be confined to better than $\lambda/2$ to deterministically control the coupling. This condition is known as *Lamb–Dicke* localization.

Radio-frequency traps offer ideal conditions for localizing atomic particles in the Lamb–Dicke regime. They provide tight confinement, so that, in combination with laser Doppler-cooling, the required localization is easily achieved. At the same time, the trapping quadrupole field does not affect the internal levels of the ion, since its equilibrium position coincides with

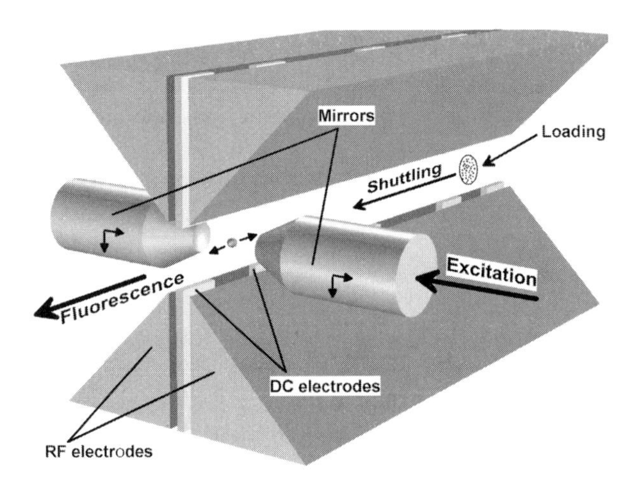

Figure 16.1: Drawing of the experimental setup used for coupling a single ion to the cavity field. The axial position of the ion is controlled by DC-electrodes, placed along the trap axis.

a node of the electric field. Typical trapping lifetimes are on the order of hours, making extended experiments with the same atom possible. Finally, due to the absence of material support structures from the interaction region, the optical field is free from perturbation.

The trap geometry chosen for our experiment [17] is that of a quadrupole mass filter with rf-electrodes of triangular cross section (Fig.16.1). In this way, lateral access to the trap axis is optimized, so that mirrors with tapered ends may be mounted close to the ion. By applying an alternating voltage with an amplitude $V_0 = 400$ Volt to diagonally opposite electrodes with a frequency $\Omega/2\pi = 12.7$ MHz, a rf-quadrupole potential is generated transversally to the trap axis. Averaged over a period of the rf-frequency, this potential results in harmonic confinement of the ion with an oscillation frequency $\omega_r/2\pi \approx 1.3$ MHz.

Particular care must be taken in the way ions are loaded in the setup. We use $^{40}Ca^+$-ions, which are created by electron impact ionization of a neutral calcium beam. In this process, charges are collected by the dielectric mirrors, resulting in stray fields that would make stable trapping impossible. Another problem is coating of the mirror surface by the atomic beam, degrading the reflectivity. We eliminated the detrimental effects of charging and coating of the mirrors by exploiting the linear geometry of our trap to load it in a region separated from the cavity by 25 mm (Fig. 16.1). Subsequently, the ion is transferred into the cavity with the help of five pairs of additional DC-electrodes in a time of 4 ms. With this novel loading method, stable trapping is possible even for small mirror separations.

After a single ion is moved to the cavity region, all residual dc-fields in the radial direction must be compensated with correctional voltages to place the ion precisely on the nodal line of the rf-field, which coincides with the trap axis. Otherwise, forced excitation of the radial oscillation of the ion by the trapping rf-field (rf-micromotion) would make it impossible to reach the Lamb–Dicke regime. The thermal motion of the ion is reduced by laser Doppler-cooling on the $^{40}Ca^+$resonance-transition at a wavelength of $\lambda = 396.8$ nm: By red-detuning the cooling laser and the cavity ($\Delta = \omega - \omega_0 \approx \Gamma/2$), a temperature of $T_D = \hbar\Gamma/2k_B \approx 500\,\mu K$

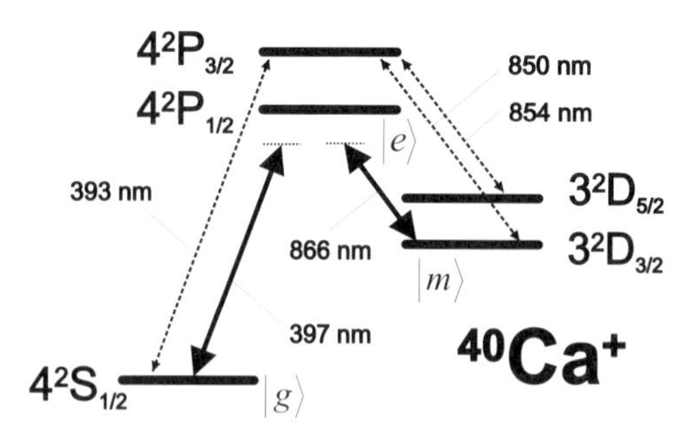

Figure 16.2: Level-scheme of the ^{40}Ca$^+$-ion with the relevant transitions for cavity QED.

is reached, determined by the natural linewidth $\Gamma = 2\pi \times 22.3$ MHz of the transition. The spread of the ion's wave packet then depends on the oscillation frequency ω_r in the potential well and the mass m of the ion [18]:

$$\Delta z = \sqrt{\frac{\hbar}{2m\omega_r} \coth\left(\frac{\hbar\omega_r}{2k_B T_D}\right)} \approx 50\ \mu\text{m}. \tag{16.2}$$

Figure 16.2 shows the level scheme of ^{40}Ca$^+$. Apart from the resonance transition $4^2S_{1/2} \to 4^2P_{1/2}$, there is a transition at $\lambda = 866$ nm from the excited $P_{1/2}$-level to the lower lying metastable $D_{3/2}$-state. The latter transition is particularly relevant for cavity-QED experiments, because in the infrared region ultra-low loss mirror coatings are available. However, if scattering and absorption losses are tolerable, a cavity may also be coupled to the ultraviolet transition. In this case, a repumping laser at 866 nm must be used to avoid population trapping in the metastable level.

16.3 Single-ion Mapping of Cavity-Modes

The spatial variation of the electromagnetic field at the center of a Fabry–Perot cavity is determined by the boundary conditions imposed by the mirrors with radius of curvature R as well as the length L of the cavity. In the initial experiments, the cavity was coupled to the resonance transition at $\lambda = 396.8$ nm. A mirror radius $R = 1$ cm and a cavity length $L = 6$ mm were chosen, resulting in a cavity waist $w_0 = 24$ µm.

In the direction transverse to the cavity axis (principal axes x and y), the field distribution is characterized by Hermite–Gauss modes. Together with the sinusoidal standing wave in the direction of the cavity axis (z-coordinate), the following mode-distribution is obtained for the TEM$_{nm}$ mode:

$$f_{nm}(x, y, z) = H_m(\sqrt{2}\,x/w_0)\; H_n(\sqrt{2}\,y/w_0)\; \exp\left(-\frac{x^2+y^2}{w_0^2}\right) \cos(kz), \tag{16.3}$$

where H_n is the Hermite polynomial of order n. A particular transverse mode TEM$_{nm}$ of the cavity field was excited by coupling light from a frequency doubled Ti:Sapphire laser with a

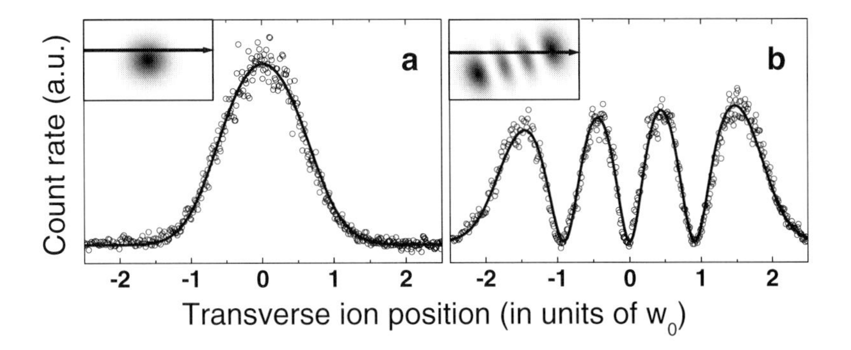

Figure 16.3: Transverse profiles of two Hermite–Gauss modes of the cavity, obtained by monitoring the ion's fluorescence count rate, while scanning over a range of 120 μm. a) TEM_{00}, b) TEM_{03}. The solid line is a fit using Eq. (16.3) and including saturation of the transition. The inset shows the calculated intensity distribution of the mode and indicates the scan path.

power of a few hundred nanowatts into the resonator. The cavity was tuned and locked by modulating its length with the help of a piezo-electric transducer on which one of the mirrors was mounted.

Equation (16.1) establishes a relation between the cavity-coupling $g(\mathbf{r})$ and the local intensity $I \sim |f_{nm}(\mathbf{r})|^2$ in the cavity. Therefore, the precision with which a certain value of g may be realized can be determined from the spatial resolution of the field structure $f_{nm}(\mathbf{r})$, obtained by using a single $^{40}\text{Ca}^+$-ion as a detector [17]. A measure of the local intensity, to which the ion is exposed, is the fluorescence intensity emitted by the ion upon excitation in the cavity field. For weak input intensity I_{in}, we obtain

$$I_{\text{fluo}}(\mathbf{r}) \sim \frac{\eta_m g(\mathbf{r})^2}{\kappa^2((\Gamma/2)^2 + \Delta^2)} I_{\text{in}}. \tag{16.4}$$

Here, κ is the cavity decay rate and η_m the mode matching efficiency. The fluorescent light emitted by the ion was collected with a lens ($f = 50$ mm, numerical aperture NA $= 0.17$), oriented in the direction of the trap axis and detected with a photomultiplier tube (overall detection efficiency 10^{-4}). While taking fluorescence data, the cooling laser at 397 nm was turned off.

Measuring I_{fluo} as a function of the position of the ion in the cavity yields an image of the mode distribution in the cavity, with a resolution given by the spread of the ion's wavefunction. The ion plays the role of a near-field probe, detecting the local field strength with minimum disturbance. One-dimensional cross sections of the transverse field distribution of the cavity mode were obtained by scanning the position of the ion along the trap axis with the help of the dc-electrodes providing axial confinement. By detecting the fluorescence rate at each point, a high-resolution map of the optical intensity distribution along the trap axis was obtained. Figure 16.3 shows scans of two transverse modes of the cavity obtained in this way. The fluorescence data is not entirely symmetric, because the principal axes of the cavity mode were slightly tilted with respect to the trap axis (angle $\alpha = 11°$).

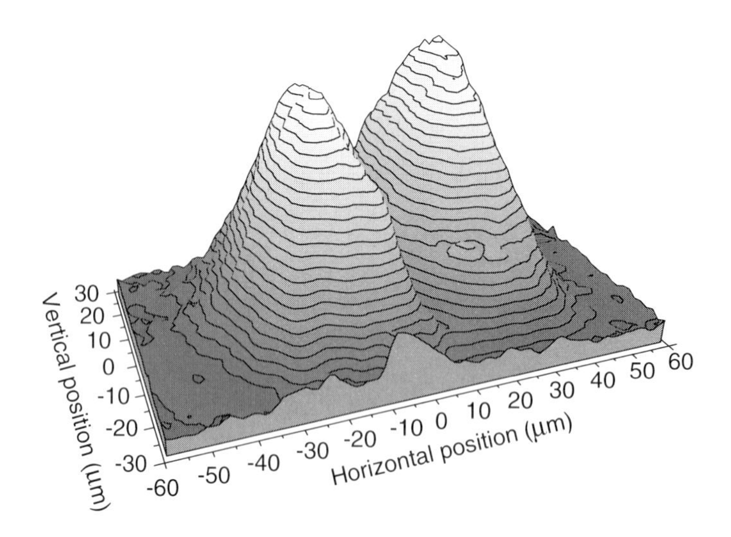

Figure 16.4: Two-dimensional intensity distribution of the TEM_{01}-mode of the cavity, measured with a single ion in an area of 120×60 µm^2. In the horizontal direction the ion was moved, while vertically the cavity position was changed relative to the ion.

In order to achieve optimum coupling between the ion and a transverse cavity mode, the vertical displacement must also be adjusted. Since moving the ion in this direction would lead to rf-micromotion, the vertical position of the cavity must be adapted instead. In the experiment this was achieved by mounting the cavity assembly on a piezoelectric translation stage with 100 µm travel. By combining the horizontal translation of the ion with vertical translation of the cavity, two-dimensional maps of the transverse cavity modes were obtained. An example of a two-dimensional intensity distribution recorded in this way is given in Fig. 16.4.

The scans of the transverse cavity modes with a single ion show that it is possible to perfectly localize the ion at an arbitrary position in the transverse field distribution, since the extension of the ion's wavepacket is much smaller than the transverse structures. The situation is different in the direction of the cavity axis. Here, the spread of the ion's position is only a factor of 5 smaller than the scale on which the cavity field varies. In this case, both the spatial structure of the field and the finite size of the ionic wave function in the trap determine the strength of the coupling between ion and electromagnetic field and hence the excitation of fluorescence by the cavity field.

The fluorescence intensity is obtained from the convolution of the wavepacket of the ion with spread $\Delta z = \sqrt{\langle z^2 \rangle}$ and the standing wave. In the TEM_{00}-mode at a transverse position $x = y = 0$ we have

$$I_{\text{fluo}}(z) \sim \frac{1}{2}\left[1 + \cos(2kz)\ \exp\bigl(-2(k\Delta z)^2\bigr)\right]. \tag{16.5}$$

Thus, from the z-dependence of the fluorescence intensity, the spread of the ion may be determined and hence the degree of control we have over the coupling between ion and field. To probe the fluorescence as a function of the ion's position on the cavity axis, the cavity was

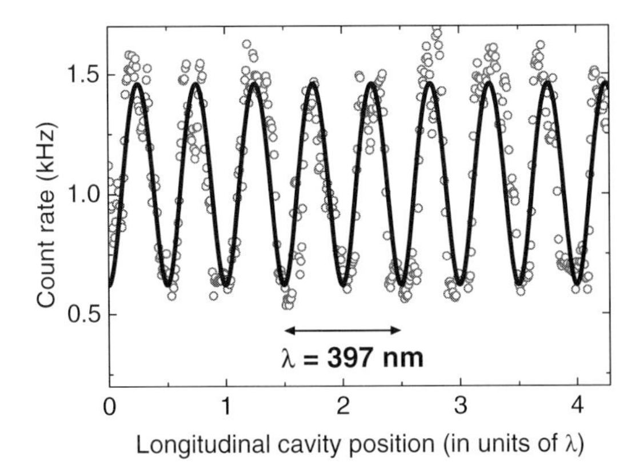

Figure 16.5: Single-ion mapping of the longitudinal structure of the cavity field. A visibility $V = 40\%$ is obtained, determined by the residual thermal motion of the Doppler-cooled ion. It corresponds to a resolution of $\Delta z = 42$ nm. The centroid of the ion's wave packet was determined to a precision of 16 nm.

translated piezo-electrically relative to the ion, which again was kept stationary to avoid the excitation of rf-micromotion.

Figure 16.5 shows the z-dependent fluorescence intensity. A pronounced sinusoidal pattern is observed, which maps the standing-wave cavity field. As predicted by Eq. (16.5), there is a finite visibility $V \approx 40\%$, which is related to the spread Δz of the ion's wave function:

$$V = \exp\left(-2(k\Delta z)^2\right) = \exp\left(-2\eta_{\mathrm{eff}}^2\right). \tag{16.6}$$

The visibility depends only on the effective Lamb–Dicke parameter η_{eff} of the ion. From the measurement, a value $\eta_{\mathrm{eff}} = 0.67$ is obtained, corresponding to a wave-packet size $\Delta z = 42$ nm.

The mapping of the cavity modes with a single-ion provides the first atomic-resolution images of the three-dimensional spatial structure of an optical field [17, 19–22]. At the same time, our results show that, by combining an ion-trap with an optical cavity, deterministic interaction of ion and field is possible. At present, thermal motion of the ion is the factor limiting the degree of control that can be achieved for the coupling. Still, the resolution is almost an order of magnitude better than the wavelength of the transition. Even higher resolution is within reach of standard ion-trapping technology. By using a steeper trap and cooling the ion to the ground state of motion [1, 23, 24], a resolution below 5 nm is feasible.

16.4 Atom–Photon Interface

The deterministic coupling of a single ion to the field in a high-finesse optical cavity is an important advance in the field of cavity-QED, where previously the lack of position control of the atom resulted in random interaction. With the Lamb–Dicke localization in the cavity field that we have demonstrated, the now well-defined atom–cavity coupling can be exploited for quantum information processing, where deterministic control of the system dynamics is of vital importance.

The stable Zeeman substate $|g\rangle = |\mathrm{S}_{1/2}, m{=}1/2\rangle$ and the metastable substate $|m\rangle = |\mathrm{D}_{3/2}, m{=}-3/2\rangle$ in the level scheme of the $^{40}\mathrm{Ca}^{+}$-ion (cf. Fig. 16.2) may be used as basis states for storing one bit of quantum information. Single qubit operations are performed by driving an off-resonant stimulated Raman transition involving the state $|e\rangle = |\mathrm{P}_{1/2}, m{=}-1/2\rangle$ [25]. As explained in the introduction, the realization of an interface between the quantum states of atoms and field is a goal of fundamental importance. It can be achieved by replacing one of the two Raman fields with the cavity. In the calcium system, the best choice for the cavity-coupling is the infrared transition $\mathrm{P}_{1/2} \leftrightarrow \mathrm{D}_{3/2}$, because in this frequency range ultra-low loss mirrors are available.

The interaction Hamiltonian contains a classical pulse $\Omega(t)$ driving the transition $|g\rangle \leftrightarrow |e\rangle$ and the cavity field coupled to the transition $|m\rangle \leftrightarrow |e\rangle$:

$$H_{\text{int}} = \hbar\Big(\Omega(t)\;|e\rangle\langle g| + g\;|e\rangle\langle m|\;a\Big) + \text{H.c.} \tag{16.7}$$

The detuning Δ of the UV pump-beam from resonance and the detuning δ of the cavity must be chosen large with respect to the linewidths and the Rabi-frequencies of the two transitions, to ensure that the intermediate $\mathrm{P}_{1/2}$-level is only virtually excited in the two-photon process, avoiding damping by spontaneous decay. The effective coupling obtained through the Raman process is determined by the atom-field coupling g, the Rabi-frequency Ω of the classical driving field and the detuning $\Delta \approx \delta$:

$$g_{\text{eff}} = \frac{g\,\Omega}{\Delta}. \tag{16.8}$$

In the case of Raman-resonance and neglecting atomic and cavity-damping, the system oscillates between the states $|g\rangle \otimes |0\rangle$ and $|m\rangle \otimes |1\rangle$, if it was in the state $|g\rangle \otimes |0\rangle$ originally. Here, the second ket designates the Fock-state of the cavity mode. The oscillation frequency is given by the effective coupling g_{eff}. If the system was in the state $|m\rangle \otimes |0\rangle$, no transition occurs.

In terms of quantum information processing, the system dynamics corresponds to a periodic exchange of information between the atoms and cavity photons. In particular, if the cavity is in the vacuum state $|0\rangle$ at $t=0$, then at the time $t=\pi/2g_{\text{eff}}$, the atom is in the state $|m\rangle$ unconditionally, and any initial superposition of the $|g\rangle$-level and the $|m\rangle$-level is mapped to a corresponding superposition of the states $|0\rangle$ and $|1\rangle$ of the cavity field:

$$(\alpha|g\rangle + \beta|m\rangle) \otimes |0\rangle \quad \xrightarrow[t=\pi/2g_{\text{eff}}]{} \quad |m\rangle \otimes (\alpha|1\rangle + \beta|0\rangle) \tag{16.9}$$

The complete transfer of population between the two basis states is known as a π-pulse. Note that Raman resonance, pulse area, as well as pulse length must be precisely controlled in order to achieve perfect population transfer.

For a complete description of the dynamics, damping due to spontaneous emission and cavity losses must be taken into account. While atomic decay may be avoided by choosing a sufficiently large detuning, cavity decay limits the time over which information may be stored in the cavity. Even for the best mirrors presently available, this will only be a couple of microseconds. Evidently, the cavity field is not well suited for long-term storage of quantum information. This task is better accomplished by atoms. The importance of the cavity-induced dynamics lies in the deterministic transfer of excitation from atoms to photons, providing a quantum mechanical link between the two systems. On completion of the transfer, the photons are emitted from the cavity, and thus are available for long-distance transmission of quantum information. The process may also be reversed, so that photonic quantum information may be transferred to an atom in a cavity. In both cases, maximum fidelity can only be achieved if the coupling parameters are known *a priori*, i.e. if the atom–cavity coupling is deterministic.

16.5 Single-Photon Source

A straightforward application of mapping an atomic state to a photon state is the deterministic generation of a pulse containing precisely one photon. Triggered single-photon emitters play an essential role in current quantum information processing schemes. In quantum cryptography, single photons are required for secure communication. Single photon sources allow a much higher transmission rate than attenuated coherent states, which have a large probability for containing no photon at all.

A number of experimental attempts have been made to realize a single-photon source. Among the systems that were used are PIN heterojunctions [26, 27], single quantum dots [28, 29] single molecules or single color centers in nanocrystals [30–32]. While producing antibunched light, these systems suffer from a low efficiency or a large spectral width of the output pulse. By contrast, a single ion in an optical cavity can be used to produce single photon pulses with almost 100% efficiency and Fourier-limited bandwidth, emitted into a well-defined Gaussian mode. Using an atom coupled to an optical cavity for generating single-photon pulses was first proposed by Law and Kimble [33].

The physical basis of single-photon emission in an ion-cavity system is the process of mapping the electronic state of a single ion to the cavity field (Sec. 16.4). In particular, for a single atom prepared in the state $|g\rangle$ ($|\alpha| = 1$), the dynamics described by Eq. (16.9) leads to precisely one photon in the cavity mode. The transfer of the photon to a mode of the free electromagnetic field is a consequence of the finite transmissivity of the cavity mirrors, leading to exponential damping of the cavity field. Since cavity decay occurs concurrently with the coherent atom–cavity evolution, the shape of the light pulse emitted from the cavity is determined by both processes. In particular, the peak of the single-photon pulse occurs at a time completely determined by the timing of the pump pulse (triggered emission). Since absorption and scattering in the dielectric coating of the mirrors are negligible (for the high quality coatings used in the experiment they amount to less than 5 ppm), the photon is delivered to the external mode with nearly 100% efficiency. It should be noted that in contrast to schemes using non-stationary atoms [16], in the case of a localized ion the scattering efficiency is equal to the overall efficiency, because the ion is always optimally coupled to the cavity field.

The determination of the pulse shape and the efficiency of single-photon emission requires a numerical solution of the master equation of the system. Using a Gaussian envelope for the pump pulse $\Omega(t)$, we obtain the single-photon pulse shape $P(t)$ and the overall emission

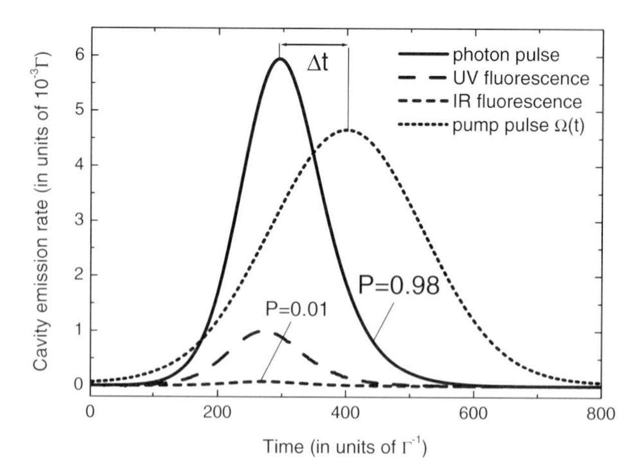

Figure 16.6: Single-photon pulse obtained with an ion resonantly coupled to the cavity ($g = 0.5\Gamma$) with ultra-low loss mirrors ($\kappa = 0.02\Gamma$). The peak Rabi-frequency of the resonant pump pulse, $\Omega_0 = 0.464\Gamma$, was chosen for optimum efficiency ($P_{\text{tot}} = 98\%$). There is a deterministic offset Δt between pump pulse and output pulse, allowing triggered emission of a single photon.

probability P_{tot} from the time-dependent solution $\rho(t)$ for the density matrix:

$$P(t) = 2\kappa\, Tr\{a^\dagger a \rho(t)\}; \qquad P_{\text{tot}} = \int_0^\infty P(t)dt. \tag{16.10}$$

The result for $g \gg \kappa$, obtained by using a short cavity ($L = 250$ µm) and the lowest transmissivity mirrors available (5 ppm), is shown in Fig. 16.6. A single photon is emitted from the cavity with $P_{\text{tot}} = 98\%$ efficiency, while in the remaining cases either no excitation occurs, or an infrared photon is scattered as fluorescent light to non-cavity modes. The single photon is generated in a completely deterministic way. From the figure it is apparent that the peak of the single photon pulse occurs at a fixed time before the pump pulse. The pump pulse $\Omega(t)$ not only triggers the single-photon pulse, but also determines its shape. A larger width of the pump pulse leads to a wider single photon pulse. The limiting values for the rise- and fall-time of the pulse are given by the inverse cavity damping κ^{-1}. Another important property of the output pulse is its spectral composition. A calculation of the pulse-spectrum shows that it is Fourier-limited, which is of great importance for applications in quantum information processing.

In order to emit more than one single photon from the cavity, the ion must be recycled to the ground state. This can be done by pumping on the $D_{3/2} \rightarrow P_{3/2}$-transition and the $D_{5/2} \rightarrow P_{3/2}$-transition. The repetition rate is limited by the decay rate of the $P_{3/2}$-level, which is 23.4 MHz.

16.6 Cavity-mediated Two-Ion Coupling

The linear trap described in Sec. 16.2 is ideally suited for storing linear crystals of multiple ions along the axis without exposing them to rf-micromotion. The Coulomb repulsion between the ions leads to equilibrium distances much larger than the transition wavelength, so that there is no direct dipole–dipole interaction between the ions. However, if two or more ions are individually coupled to the cavity, the optical field may be employed to mediate an interaction between their internal states. There have been a number of proposals to couple atoms or ions through the exchange of cavity photons [7, 9–11]. These are promising alternatives to schemes involving the ions' motional degrees of freedom, since there is no need for cooling the vibrational modes of the string below the Doppler temperature. Thus, using a cavity to perform quantum operations on adjacent pairs of ions in a long string is a viable route to a scalable quantum computer. Multi-ion dynamics in this system is implemented by translating the ion-chain after each operation, so that different pairs of ions are successively brought to interaction with the cavity field.

By storing an array of two ions in the trap and observing the total fluorescence signal as the crystal was translated along the trap axis, we could demonstrate that two ions could be positioned in the cavity mode with the same precision obtained in the one-ion case. We succeeded in matching the positions of the two ions to the two maxima of the TEM_{01}-mode (cf. Fig. 16.4). This is the configuration in which the interaction is least sensitive to position fluctuations of the ions. Each ion couples to the field with 74% of the single-ion coupling at the center of the TEM_{00}-mode. The deterministic coupling of two particles to a single cavity mode is a unique accomplishment. In cavity-QED experiments with atoms, the investigation of two-atom coupling is precluded by the random motion of free atoms and short storage times for single atoms in dipole traps.

When the size g of the coupling exceeds the decay rates of atoms and cavity, coherent exchange of excitation can occur without degradation due to damping or decoherence. In this regime, a single cavity mode may be used to mediate an interaction between two atoms individually coupled to it.

To investigate the exchange of quantum information between two ions through the cavity mode, we consider ion 1 initially in the ground state $|g_1\rangle$ and ion 2 in the metastable state $|m_2\rangle$. The Hamiltonian describing the cavity-mediated interaction is obtained as the two-ion generalization of Hamiltonian (16.7):

$$H_{\text{int}} = \hbar\Big(\Omega_1\,|e_1\rangle\langle g_1| + \Omega_2\,|e_2\rangle\langle g_2| + g\,|e_1\rangle\langle m_1|\,a + g\,|e_2\rangle\langle m_2|\,a\Big) + \text{H.c.} \quad (16.11)$$

In the absence of spontaneous emission and cavity damping, the interaction (16.11) causes the system to oscillate between the two-particle states $|g_1, m_2\rangle$ and $|m_1, g_2\rangle$. For a certain interaction time, the ions completely switch their excitation state.

Since even the best available mirrors have transmission losses of a few ppm, the effects of damping must be considered. Atomic spontaneous emission is avoided by driving the transition from the ground state to the metastable level with an off-resonant Raman pulse with a large detuning from resonance. Cavity decay affects the dynamics on a time-scale of κ^{-1}, since during the transfer the cavity mode is occupied. However, the cavity mode is occupied neither in the initial state $|g_1, m_2\rangle$ nor in the final state $|m_1, g_2\rangle$. Therefore, coupling may

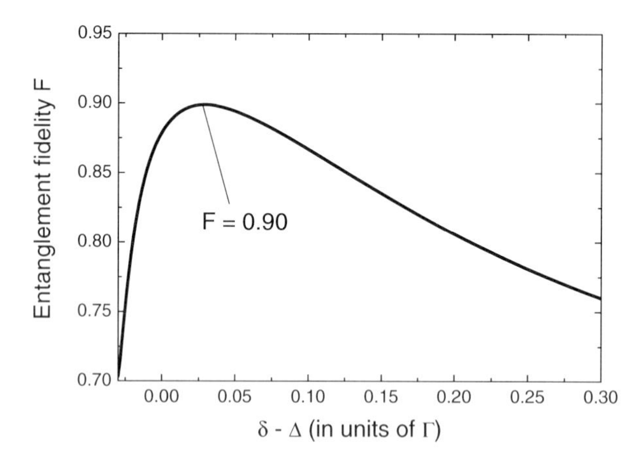

Figure 16.7: Fidelity of preparation of the state $|g_1, m_2\rangle + i|m_1, g_2\rangle$ as a function of detuning from Raman-resonance. The system parameters are $\Omega_1 = \Omega_2 = 0.02\Gamma$, $g = 0.5\Gamma$, $\kappa = 0.005\Gamma$, $\Delta = 7\Gamma$. At optimum detuning, a fidelity of 90% is reached.

also be achieved by virtual excitation of the cavity field. This is realized experimentally by detuning the cavity from the exact Raman-resonance. The transition from $|g_1, m_2\rangle$ to $|m_1, g_2\rangle$ corresponds to a fourth order process, involving virtual excitations of the $P_{1/2}$-levels in both ions as well as the cavity. The effective coupling scales with the detunings of the driving laser and the cavity as

$$g_{\text{eff}} = \frac{g^2 \Omega^2}{\Delta\, \delta\, (\delta - \Delta)}. \tag{16.12}$$

For off-resonant two-ion coupling, the effect of cavity damping is strongly reduced, at the cost of a longer time $\pi/2g_{\text{eff}}$ for a complete population transfer (π-pulse).

An application of the cavity-induced two-ion dynamics is the entanglement of the two particles. To this end, the population transfer is stopped after a time $\pi/4g_{\text{eff}}$ ($\pi/2$-pulse). At this time, the state of the system ideally should be entangled:

$$|\Psi\rangle = \frac{|g_1, m_2\rangle + i|m_1, g_2\rangle}{\sqrt{2}}. \tag{16.13}$$

The fidelity $F = \sqrt{\langle\Psi|\rho|\Psi\rangle}$ of the preparation of the two-atom entangled state $|\Psi\rangle$ is obtained from the density matrix ρ, calculated by numerically integrating the full master equation including atomic and cavity decay. Figure 16.7 shows that there is an optimum value for the detuning $\delta - \Delta$ from Raman resonance. If the detuning is chosen too large, the effective coupling according to Eq. (16.12) becomes too small, so that residual atomic decay can occur during the transfer. In the case of small detunings, there is an increased chance for excitation of the cavity mode, and hence cavity damping. For the parameters of Fig. 16.7, the maximum fidelity of a two-atom entangled state is 90%.

The controlled coupling of single ions or pairs of ions to a single mode of an optical cavity is a process of fundamental importance in the field of optical and atomic quantum information processing. It establishes a coupling between atomic qubits for local qubit-processing. It can also be employed to produce controlled single-photon pulses that are needed in quantum cryptography and photonic quantum information processing. Finally, it provides an interface to convert atomic quantum information to photonic qubits and vice versa and will therefore play an essential role in quantum information processing distributed over many local processors (quantum network).

References

[1] C. Monroe, D.M. Meekhof, B.E. King, W.M. Itano, and D.J. Wineland. *Demonstration of a fundamental quantum logic gate*. Phys. Rev. Lett. **75**, 4714 (1995).

[2] Q.A. Turchette, C.J. Hood, W. Lange, H. Mabuchi, and H.J. Kimble. *Measurement of conditional phase shifts for quantum logic*. Phys. Rev. Lett. **75**, 4710 (1995).

[3] G. Ribordy, J. Brendel, J.D. Gautier, N. Gisin, and H. Zbinden. *Long-distance entanglement-based quantum key distribution*. Phys. Rev. A **63**, 012309 (2001).

[4] C.A. Sackett, D. Kielpinski, B.E. King, C. Langer, V. Meyer, C.J. Myatt, M. Rowe, Q.A. Turchette, W.M. Itano, D.J. Wineland, and C. Monroe. *Experimental entanglement of four particles*. Nature **404**, 256 (2000).

[5] J. I. Cirac, P. Zoller, H. J. Kimble, and H. Mabuchi. *Quantum state transfer and entanglement distribution among distant nodes in a quantum network*. Phys. Rev. Lett. **78**, 3221 (1997).

[6] S. J. vanEnk, J. I. Cirac, and P. Zoller. *Ideal quantum communication over noisy channels: A quantum optical implementation*. Phys. Rev. Lett. **78**, 4293 (1997).

[7] T. Pellizzari, S.A. Gardiner, J.I. Cirac, and P. Zoller. *Decoherence, continuous observation, and quantum computing - a cavity QED model*. Phys. Rev. Lett. **75**, 3788 (1995).

[8] M.B. Plenio, S.F. Huelga, A. Beige, and P.L. Knight. *Cavity-loss-induced generation of entangled atoms*. Phys. Rev. A **59**, 2468 (1999).

[9] S. B. Zheng and G. C. Guo. *Efficient scheme for two-atom entanglement and quantum information processing in cavity QED*. Phys. Rev. Lett. **85**, 2392 (2000).

[10] M.S. Shahriar, J.A. Bowers, B. Demsky, P.S. Bhatia, S. Lloyd, P.R. Hemmer, and A.E. Craig. *Cavity dark states for quantum computing*. Opt. Commun. **195**, 411 (2001).

[11] E. Jane, M.B. Plenio, and D. Jonathan. *Quantum-information processing in strongly detuned optical cavities*. Phys. Rev. A **65**, 050302 (2002).

[12] C.H. Bennett and G. Brassard. *Quantum cryptography: Public-key distribution and coin tossing*. In Proc. IEEE Int. Conf. on Computers, Systems and Signal Processing, pages 175–179, New York. December (1984). IEEE.

[13] E. Knill, R. Laflamme, and G.J. Milburn. *A scheme for efficient quantum computation with linear optics*. Nature **409**, 46 (2001).

[14] C. J. Hood, T. W. Lynn, A. C. Doherty, A. S. Parkins, and H. J. Kimble. *The atom–cavity microscope: Single atoms bound in orbit by single photons*. Science **287**, 1447 (2000).

[15] P. W. H. Pinkse, T. Fischer, P. Maunz, and G. Rempe. *Trapping an atom with single photons*. Nature **404**, 365 (2000).

[16] M. Hennrich, T. Legero, A. Kuhn, and G. Rempe. *Vacuum-stimulated Raman scattering based on adiabatic passage in a high-finesse optical cavity*. Phys. Rev. Lett. **85**, 4872 (2000).

[17] G. R. Guthöhrlein, M. Keller, K. Hayasaka, W. Lange, and H. Walther. *A single ion as a nanoscopic probe of an optical field*. Nature **414**, 49 (2001).

[18] R.P. Feynman. *Statistical Mechanics*, pages 51–52. Frontiers in Physics. Benjamin/Cummings, Reading (1972).

[19] J. Michaelis, C. Hettich, A. Zayats, B. Eiermann, J. Mlynek, and V. Sandoghdar. *A single molecule as a probe of optical intensity distribution*. Opt. Lett. **24**, 581 (1999).

[20] B. Sick, B. Hecht, U. P. Wild, and L. Novotny. *Probing confined fields with single molecules and vice versa*. J. Microsc.-Oxf. **202**, 365 (2001).

[21] S. Gotzinger, S. Demmerer, O. Benson, and V. Sandoghdar. *Mapping and manipulating whispering gallery modes of a microsphere resonator with a near-field probe*. J. Microsc.-Oxf. **202**, 117 (2001).

[22] J. A. Veerman, M. F. Garcia-Parajo, L. Kuipers, and N. F. Van Hulst. *Single molecule mapping of the optical field distribution of probes for near-field microscopy*. J. Microsc.-Oxf. **194**, 477 (1999).

[23] E. Peik, J. Abel, T. Becker, J. von Zanthier, and H. Walther. *Sideband cooling of ions in radio-frequency traps*. Phys. Rev. A **60**, 439 (1999).

[24] C. F. Roos, D. Leibfried, A. Mundt, F. Schmidt-Kaler, J. Eschner, and R. Blatt. *Experimental demonstration of ground state laser cooling with electromagnetically induced transparency*. Phys. Rev. Lett. **85**, 5547 (2000).

[25] D.J. Heinzen and D.J. Wineland. *Quantum-limited cooling and detection of radio-frequency oscillations by laser-cooled ions*. Phys. Rev. A **42**, 2977 (1990).

[26] J. Kim, O. Benson, H. Kan, and Y. Yamamoto. *A single-photon turnstile device*. Nature **397**, 500 (1999).

[27] A. Imamoglu and Y. Yamamoto. *Turnstile device for heralded single photons - Coulomb-blockade of electron and hole tunneling in quantum-confined p-i-n heterojunctions*. Phys. Rev. Lett. **72**, 210 (1994).

[28] P. Michler, A. Kiraz, C. Becher, W.V. Schoenfeld, P.M. Petroff, L.D. Zhang, E. Hu, and A. Imamoglu. *A quantum dot single-photon turnstile device*. Science **290**, 2282 (2000).

[29] M. Pelton, C. Santori, G.S. Solomon, O. Benson, and Y. Yamamoto. *Triggered single photons and entangled photons from a quantum dot microcavity*. Eur. Phys. J. D **18**, 179 (2002).

[30] B. Lounis and W.E. Moerner. *Single photons on demand from a single molecule at room temperature*. Nature **407**, 491 (2000).

[31] C. Kurtsiefer, S. Mayer, P. Zarda, and H. Weinfurter. *Stable solid-state source of single photons*. Phys. Rev. Lett. **85**, 290 (2000).

[32] A. Beveratos, S. Kuhn, R. Brouri, T. Gacoin, J.P. Poizat, and P. Grangier. *Room temperature stable single-photon source*. Eur. Phys. J. D **18**, 191 (2002).

[33] C. K. Law and H. J. Kimble. *Deterministic generation of a bit-stream of single-photon pulses*. J. Mod. Opt. **44**, 2067 (1997).

17 Strongly Coupled Atom–Cavity Systems

Axel Kuhn, Markus Hennrich, and Gerhard Rempe

Max-Planck-Institut für Quantenoptik
Garching, Germany

17.1 Introduction

Worldwide, major efforts are made to realize decoherence-free systems for the storage of individual quantum bits (qubits) and to conditionally couple different qubits for the processing of quantum information. Ultra-cold trapped neutral atoms or ions are ideal to store quantum information in long-lived internal states. Photons are ideal to transmit quantum information between quantum memories. In this chapter, the coupling of internal atomic states to a quantized mode of the radiation field is discussed. We introduce an adiabatic coupling scheme between a single atom and an optical cavity, which is based on a unitary evolution of the coupled atom–cavity system and therefore is intrinsically reversible. It allows one to populate either Fock states on demand, or to emit single optical photons into a well-defined mode of the radiation field outside the cavity. Such a deterministic single-photon source is essential for optical quantum information processing with linear components [KLM01], and it makes possible quantum networking between different cavities, where the key requirement is the ability to interconvert stationary and flying qubits [CZKM97, DiV00]. These features distinguish the present scheme from other methods of Fock-state preparation in the microwave regime [MHN$^+$97, BVW01], where the photons remain trapped inside the cavity.

17.2 Atoms, Cavities and Light

The atom–cavity coupling scheme combines cavity quantum electrodynamics (CQED) with stimulated Raman scattering by adiabatic passage (STIRAP) [VFSB01]. To describe the scheme, we first introduce the relevant features of CQED and the Jaynes–Cummings model [JC63, SK93], then we focus on three-level atoms with two dipole transitions driven by two radiation fields. One of them comes from a laser, the other is that of a cavity strongly coupled to the atom. We show how to control the state vector of the system so that exactly one photon is emitted through one of the cavity mirrors, thereby forming a single-photon pulse [HLKR00, KHR02].

17.2.1 Field Quantization in a Fabry–Perot Cavity

A Fabry–Perot cavity with mirror separation l and reflectivity $\mathcal{R}$ has a free spectral range $FSR = \pi c/l$, and its finesse is defined as $\mathcal{F} = \pi\sqrt{\mathcal{R}}/(1-\mathcal{R})$. In the vicinity of a reso-

Quantum Information Processing, 2nd Edition. Edited by Thomas Beth and Gerd Leuchs

ISBN: 3-527-40541-0

nance, the transmission is a Lorentzian with a linewidth (FWHM) of $2\kappa = FSR/\mathcal{F}$, which is twice the decay rate of the cavity field, κ. Stable single-mode operation is achieved using curved mirrors, so that Hermite–Gaussian or Laguerre–Gaussian cavity eigenmodes are obtained. Within this chapter, we consider a single TEM_{00} mode with mode function $\psi(\mathbf{r})$ and resonance frequency ω_C. The electromagnetic field of the mode is quantized, so that its state vector is, in general, a superposition of photon-number states, the so-called Fock states, $|n\rangle$. Each contributing photon carries an energy of $\hbar\omega_C$, and for n photons in the mode the total energy is $\hbar\omega_C(n + \frac{1}{2})$, where $\hbar\omega_C/2$ is the zero-point energy. The equidistant energy spacing imposes an analogue treatment of the cavity to an harmonic oscillator. Consequently, creation and annihilation operators for a photon, $a^\dagger$ and a, respectively, are used to express the Hamiltonian of the cavity,

$$H_C = \hbar\omega_C \left(a^\dagger a + \frac{1}{2} \right) . \tag{17.1}$$

Note that this Hamiltonian does not include losses. In a real cavity, all photon number states decay until thermal equilibrium with the environment is reached. In the optical domain, this corresponds to the vacuum state, $|0\rangle$, with no photon in the cavity.

17.2.2 Two-Level Atom

We now analyze how the quantized field interacts with a two-level atom with ground state, $|g\rangle$, and excited state, $|e\rangle$, with energies $\hbar\omega_g$ and $\hbar\omega_e$, respectively, and transition dipole moment μ_{eg}. The Hamiltonian of the atom reads

$$H_A = \hbar\omega_g |g\rangle\langle g| + \hbar\omega_e |e\rangle\langle e| . \tag{17.2}$$

The coupling to the field mode of the cavity is expressed by the atom–cavity coupling constant,

$$g(\mathbf{r}) = g_0\, \psi(\mathbf{r}), \qquad \text{with} \qquad g_0 = \sqrt{(\mu_{eg}^2 \omega_C)/(2\hbar\epsilon_0 V)}, \tag{17.3}$$

where V is the mode volume of the cavity. In a closed system, any change of the atom's internal state must be reflected by an according change of the cavity's photon number, n. It follows that the interaction Hamiltonian of the atom–cavity system,

$$H_{int} = -\hbar g \left[|e\rangle\langle g| a + a^\dagger |g\rangle\langle e| \right] , \tag{17.4}$$

includes the creation and annihilation operators, $a^\dagger$ and a. For a given excitation number, n, only the pair of product states $|g, n\rangle$ and $|e, n-1\rangle$ is coupled, and for a cavity resonant with the atomic transition, the population oscillates with the Rabi frequency $\Omega_C = 2g\sqrt{n}$ between these states.

The eigenfrequencies of the total Hamiltonian, $H = H_C + H_A + H_{int}$, can be found easily. In the rotating wave approximation, they read

$$\omega_n^\pm = \omega_C \left(n + \frac{1}{2} \right) + \frac{1}{2} \left(\Delta_C \pm \sqrt{4ng^2 + \Delta_C^2} \right) , \tag{17.5}$$

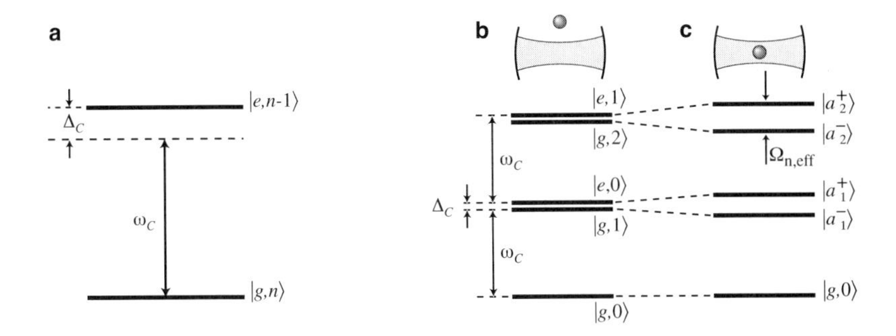

Figure 17.1: **(a)** A two-level atom with ground state $|g\rangle$ and excited state $|e\rangle$ coupled to a cavity containing n photons. In the dressed-level scheme of the combined atom–cavity system with the atom outside **(b)** or inside **(c)** the cavity, the state doublets are either split by Δ_C or by the effective Rabi frequency, $\Omega_{\mathrm{n,eff}}$, respectively.

where $\Delta_C = \omega_e - \omega_g - \omega_C$ is the detuning between the atom and the cavity. The level splitting between the two corresponding eigenstates, $\Omega_{\mathrm{n,eff}} = \sqrt{4ng^2 + \Delta^2}$, is the effective Rabi frequency of the population oscillation between states $|g, n\rangle$ and $|e, n-1\rangle$. This means that the cavity field stimulates the emission of an excited atom into the cavity, thus deexciting the atom and increasing the photon number by one. Subsequently, the atom is reexcited by absorbing a photon from the cavity field, and so forth. Since an excited atom and a cavity containing no photon initially are sufficient to start the oscillation between $|e, 0\rangle$ and $|g, 1\rangle$ at frequency $\sqrt{4g^2 + \Delta_C^2}$, this phenomenon is called vacuum-Rabi oscillation. For a resonant interaction, the oscillation frequency is $2g$, which is therefore called the vacuum-Rabi frequency.

So far, we have shown that the atom–cavity interaction splits the photon number states into doublets of non-degenerate dressed states, which are named after Jaynes and Cummings [JC63, SK93]. However, there is one exception to the rule: Without excitation, the atom and the cavity are in their ground states, $|g\rangle$ and $|0\rangle$, respectively. The only possible product state, $|g, 0\rangle$, is not coupled to any other state and therefore no splitting occurs.

17.2.3 Three-Level Atom

Now we consider an atom with a Λ−type three-level scheme providing transition frequencies $\omega_{eu} = \omega_e - \omega_u$ and $\omega_{eg} = \omega_e - \omega_g$ as depicted in Fig. 17.2. The $|u\rangle \leftrightarrow |e\rangle$ transition is driven by a classical light field of frequency ω_P with Rabi frequency Ω_P, the so-called pump field, and a cavity mode with frequency ω_C couples to the $|g\rangle \leftrightarrow |e\rangle$ transition. If we define the respective detunings as $\Delta_P = \omega_{eu} - \omega_P$ and $\Delta_C = \omega_{eg} - \omega_C$, and assume that the pump laser and the cavity only couple to their respective transitions, the behaviour of the atom–cavity system is described by the interaction Hamiltonian

$$H_{\mathrm{int}} = \hbar\Big[\Delta_P|u\rangle\langle u| + \Delta_C|g\rangle\langle g| - g(|e\rangle\langle g|a + a^\dagger|g\rangle\langle e|) - \frac{1}{2}\Omega_P(|e\rangle\langle u| + |u\rangle\langle e|)\Big]. \quad (17.6)$$

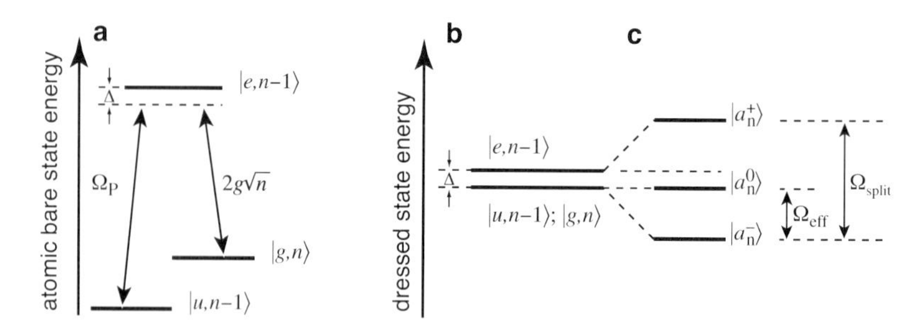

Figure 17.2: (a) A three-level atom driven by a classical laser field of Rabi frequency Ω_P, coupled to a cavity containing n photons. **(b)** Dressed-level scheme of the combined system for vanishing coupling strengths and for an atom interacting with laser and cavity **(c)**. The triplet is split by $\Omega_{\text{split}} = \sqrt{4ng^2 + \Omega_P^2 + \Delta^2}$. Note that for a detuning Δ which is large compared to all Rabi frequencies, the Raman transition $|u, n-1\rangle \leftrightarrow |g, n\rangle$ is driven at the effective Rabi frequency $\Omega_{\text{eff}} = \frac{1}{2}(\Omega_{\text{split}} - |\Delta|) \approx (4ng^2 + \Omega_P^2)/|4\Delta|$.

Given an arbitrary excitation number n, this Hamiltonian couples only the three states $|u, n-1\rangle$, $|e, n-1\rangle$, $|g, n\rangle$. For these triplets and a Raman-resonant interaction with $\Delta_P = \Delta_C \equiv \Delta$, the eigenfrequencies of the coupled system read

$$\begin{aligned} \omega_n^0 &= \omega_C\left(n + \frac{1}{2}\right) \quad \text{and} \\ \omega_n^{\pm} &= \omega_C\left(n + \frac{1}{2}\right) + \frac{1}{2}\left(\Delta \pm \sqrt{4ng^2 + \Omega_P^2 + \Delta^2}\right). \end{aligned} \tag{17.7}$$

Note that the Jaynes–Cummings state doublets of the two-level atom are now replaced by state triplets, with eigenstates

$$\begin{aligned} |\phi_n^0\rangle &= \cos\Theta|u, n-1\rangle - \sin\Theta|g, n\rangle, \\ |\phi_n^+\rangle &= \cos\Phi\sin\Theta|u, n-1\rangle - \sin\Phi|e, n-1\rangle + \cos\Phi\cos\Theta|g, n\rangle, \\ |\phi_n^-\rangle &= \sin\Phi\sin\Theta|u, n-1\rangle + \cos\Phi|e, n-1\rangle + \sin\Phi\cos\Theta|g, n\rangle, \end{aligned} \tag{17.8}$$

where the mixing angles Θ and Φ are given by

$$\tan\Theta = \frac{\Omega_P}{2g\sqrt{n}} \quad \text{and} \quad \tan\Phi = \frac{\sqrt{4ng^2 + \Omega_P^2}}{\sqrt{4ng^2 + \Omega_P^2 + \Delta^2} - \Delta}, \tag{17.9}$$

with Ω_P and g assumed to be real. We note that the interaction with the light lifts the degeneracy of the three eigenstates as soon as the Rabi-frequencies are non-zero. Furthermore, it must be emphasized that one of these states, namely $|\phi_n^0\rangle$, is neither subject to an energy shift, nor does the excited atomic state contribute to it. In the literature, $|\phi_n^0\rangle$ is therefore called a "dark state" since its population cannot be lost by spontaneous emission.

In the limit of vanishing Ω_P, the states $|\phi_n^{\pm}\rangle$ correspond to the Jaynes–Cummings doublet and the third eigenstate, $|\phi_n^0\rangle$, coincides with $|u, n-1\rangle$. Note that ω_n^0 is not affected by Ω_P or g. Therefore transitions between the dark states $|\phi_{n+1}^0\rangle$ and $|\phi_n^0\rangle$ are always in resonance with the cavity. This holds, in particular, for the transition from $|\phi_1^0\rangle$ to $|\phi_0^0\rangle \equiv |g, 0\rangle$ since the $n = 0$ state does not split (the corresponding states $|u, -1\rangle$ and $|e, -1\rangle$ do not exist).

17.2.4 Adiabatic Passage

Adiabatic passage in the optical domain is used for coherent population transfer in atoms or molecules for many years now. In two-level systems, it can be driven by a frequency chirp of a light field across the relevant resonance [Loy74, Loy78]. This technique has been successfully extended to three-level systems [YSHB99], and it is also used for velocity-selective excitation and Raman cooling of atoms [KPHS96, PKBS98]. Without chirp, but with a Raman transition driven by two distinct pulses of variable amplitudes, effects like electromagnetically induced transparency (EIT) [Har93, Har97], slow light [HHDB99, PFM$^+$01], and stimulated Raman scattering by adiabatic passage (STIRAP) [VFSB01] are observed. All these effects have been demonstrated with classical light fields.

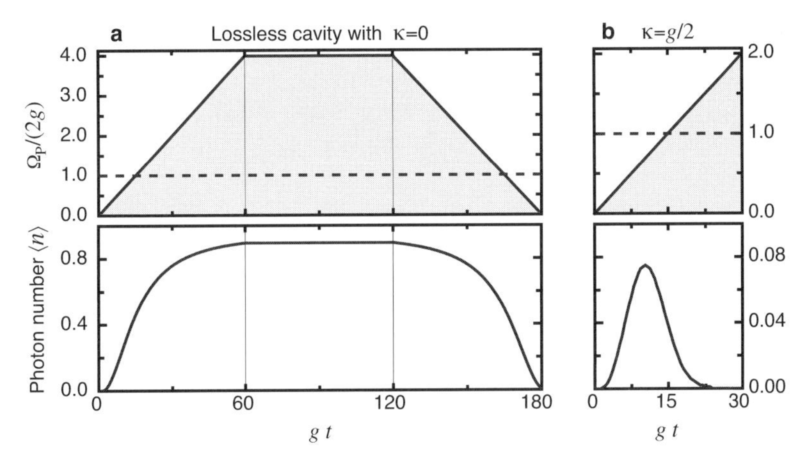

Figure 17.3: Relative Rabi frequency, $\Omega_P/(2g)$, and corresponding mean photon number, $\langle n\rangle$, as a function of the dimensionless time, gt, and an atom–cavity system initially prepared in state $|u, 0\rangle$. **(a)** For a lossless cavity, the mean photon number $\langle n\rangle$ is given by the population of state $|g, 1\rangle$, which obeys Eq. (17.10). When the pump pulse is turned off adiabatically, the photon is reabsorbed and the system returns to the initial state. **(b)** In case of a lossy cavity, the photon escapes during the rising edge of the pump pulse. For the chosen parameters, the area of the $\langle n\rangle$-pulse indicates a probability of $74\,\%$ for a single-photon emission.

These techniques have in common that the system's state vector, $|\Psi\rangle$, always coincides with a single eigenstate, e.g. $|\phi_n^0\rangle$, of the time-dependent interaction Hamiltonian. In principle, the time evolution of the system is completely controlled by the variation of this eigenstate. However, a more detailed analysis [Mes58, KR02] reveals that the eigenstates must change slowly with respect to the eigenfrequency differences. Adiabaticity is assured if the condition $|\omega_n^\pm - \omega_n^0| \gg |\langle\phi_n^\pm|\frac{d}{dt}|\phi_n^0\rangle|$ is met throughout the interaction, and as long as the system does not decay via some other channel. In this context, the non-decaying dark state, $|\phi_n^0\rangle$, is of enormous significance.

It follows that an atom–cavity system, once prepared in $|\phi_n^0\rangle$, should stay there, thus allowing one to control the relative population of the contributing product states, $|u, n-1\rangle$ and $|g, n\rangle$, simply by adjusting the pump Rabi frequency, Ω_P. To show this, let us start with a system initially prepared in state $|u, n-1\rangle$. As can be seen from (17.8), this state coincides with $|\phi_n^0\rangle$ if the condition $2g\sqrt{n} \gg \Omega_P$ is met in the beginning of the interaction. Once the system has been successfully prepared in the dark state, the ratio between the populations of the contributing states reads

$$\frac{|\langle u, n-1|\Psi\rangle|^2}{|\langle g, n|\Psi\rangle|^2} = \frac{4ng^2}{\Omega_P^2}. \tag{17.10}$$

As proposed in [KHBR99], we now assume that an atom in state $|u\rangle$ is placed in a cavity mode populated with $n-1$ photons driving the $|g, n\rangle \leftrightarrow |e, n-1\rangle$ transition with the effective Rabi frequency $2g\sqrt{n}$, so that the initial state coincides with $|\phi_n^0\rangle$. The atom is then exposed to a pump laser pulse coupling the $|u\rangle \leftrightarrow |e\rangle$ transition with a slowly rising amplitude that finally leads to $\Omega_P \gg 2g\sqrt{n}$. The system evolves from $|u, n-1\rangle$ to $|g, n\rangle$, and therefore the photon number increases by one. If decay is neglected, the successive application of this method would allow one to prepare arbitrary photon-number states [PMZK93]. To do so, the pump pulse must be turned off suddenly. Otherwise the system would adiabatically return to its initial state, as shown in Fig. 17.3(a).

The situation changes if the cavity-field decay time, κ^{-1}, is finite and comparable or shorter than the exciting laser pulse. Starting from the atom–cavity system prepared in state $|u, 0\rangle$, a single photon is generated and emitted from the cavity during the excitation process (see Fig. 17.3(b)). Once the photon escapes, the system reaches state $|g, 0\rangle$ and stays there, since it decouples from the cavity and the pump field.

17.3 Single-Photon Sources

So far, one of the most popular schemes used for the generation of single photons relies on parametric down-conversion. This scheme produces photons at more or less random times and, hence, requires post selection. Only during the last few years, different photon generation schemes have been demonstrated, like a single-photon turnstile device based on the Coulomb blockade mechanism in a quantum dot [KBKY99], the fluorescence of a single molecule [BLTO99, LM00], or a single colour centre (Nitrogen vacancy) in diamond [KMZW00, BBPG00], or the photon emission of a single quantum dot [MIM+00, YKS+02]. All these new schemes emit photons upon an external trigger event. However, the photons are spontaneously emitted into many modes of the radiation field and usually show a broad energy distribution. Only recently, cavity-enhanced spontaneous emission techniques for single-photon generation from a quantum dot have been proposed [BSPY00] and demonstrated [SPS+01, MKB+00, MRG+01]. However, all these single-photon emitters cannot transfer quantum information between atoms and photons in a coherent manner and, therefore, do not meet the essential requirement for quantum networking. The reason is that the emission process, namely an electronic excitation of the system followed by spontaneous emission, cannot be described by a Hamiltonian evolution and, hence, is irreversible.

17.3.1 Vacuum-Stimulated Raman Scattering

We now report on an experimental scheme [KHBR99, HLKR00] which employs the adiabatic passage technique discussed in section 17.2.4, and where the strong coupling of a single atom to a single cavity mode stimulates one branch of a Raman transition.

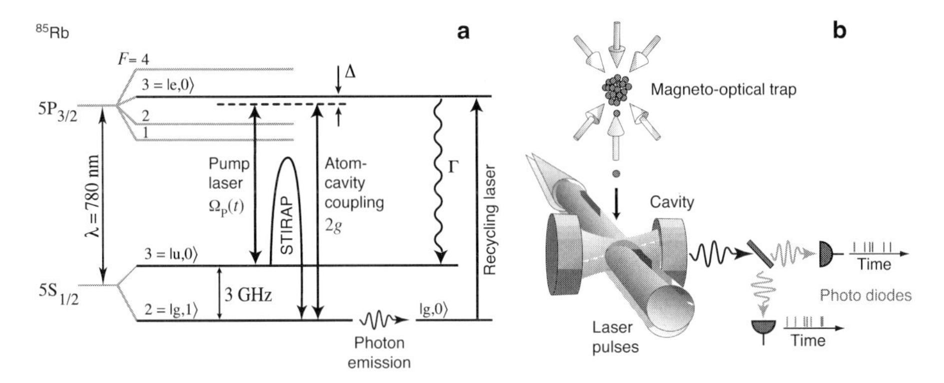

Figure 17.4: Scheme of the experiment. **(a)** Relevant energy levels and transitions in ^{85}Rb. The atomic states labeled $|u\rangle$, $|e\rangle$ and $|g\rangle$ are involved in the Raman process, and the states $|0\rangle$ and $|1\rangle$ denote the photon number in the cavity. **(b)** Setup: A cloud of atoms is released from a magneto–optical trap and falls through a cavity 20 cm below in about 8 ms with a velocity of 2 m/s. The interaction time of a single atom with the TEM_{00} mode of the cavity (waist $w_0 = 35\,\mu\text{m}$) amounts to about 17.5 µs. The pump and recycling lasers are collinear and overlap with the cavity mode. Photons emitted from the cavity are detected by a pair of photodiodes with a quantum efficiency of 50%.

Figure 17.4(a) shows the basic scheme of the process. A ^{85}Rb-atom is prepared in state $|u\rangle$ which is the $F = 3$ hyperfine state of the $5S_{1/2}$ electronic ground state. The atom is located in a high-finesse optical cavity which is near resonant with the transition between the $F = 2$ hyperfine state of the electronic ground state and the electronically excited $5P_{3/2}(F = 3)$ state, $|g\rangle$ and $|e\rangle$, respectively. Here, the average atom–cavity coupling constant, g, takes into account that neither the position of the atom in the cavity nor the magnetic quantum number of the atom is well defined in the experiment. We assume g to be constant while pump-laser pulses close to resonance with the $|u\rangle \leftrightarrow |e\rangle$ transition and with Rabi frequency $\Omega_P(t)$ are applied. As described in the preceding section, each pulse generates a single photon which leaves the cavity through that mirror which is designed as an output coupler. The pump pulse rises slowly so that the emission ends before $\Omega_P > 2g$ is reached. The dynamics of the simultaneous excitation and emission process determines the duration and, hence, the linewidth of the photon. When the photon is emitted, the final state of the coupled system, $|g, 0\rangle$, is reached. This state is not coupled to the one-photon manifold and the atom cannot be reexcited. This limits the number of photons per pump pulse and atom to one.

To emit a sequence of photons from one-and-the-same atom, the system must be transferred back to $|u, 0\rangle$ once an emission has taken place. To do so, we apply recycling laser pulses that hit the atom between consecutive pump pulses. The recycling pulses are resonant with the $|g\rangle \leftrightarrow |e\rangle$ transition and pump the atom to state $|e\rangle$. From there, it decays spontaneously to the initial state, $|u\rangle$. If an atom that resides in the cavity is exposed to a sequence of laser pulses, which alternate between triggering single-photon emissions and re-establishing the initial condition by optical pumping, a sequence of single-photon pulses is produced.

Figure 17.4(b) shows the apparatus. Atoms are released from a magneto–optical trap and pass through the TEM_{00} mode of the optical cavity. The flux of atoms[1] is freely adjusted between 0 and 1000 atoms/ms. In the cavity, they are exposed to the sequence of laser pulses. The cavity is 1 mm long and has a finesse of 60 000. One mirror has a larger transmission coefficient than the other so that photons leave the cavity through this output coupler with a probability of 90%. For each experimental cycle, all photon arrival times are recorded using two avalanche photodiodes which are placed at the output ports of a beam splitter.

17.3.2 Deterministic Single-Photon Sequences

Figure 17.5(a) displays an example of the photon stream recorded while single atoms fall through the cavity one after the other. Obviously, the photon sequence is different for each atom. In particular, not every pump pulse leads to a detected photon, since the efficiencies of photon generation and photon detection are limited.

In the experiment, the applied Rabi frequencies of the pump and recycling pulses have the shape of a saw-tooth and increase linearly, as displayed in Fig. 17.5(b). This leads to a constant rate of change of the dark state, $|\phi_1^0\rangle$, during the initial stage of the pump pulses and therefore optimal adiabaticity with minimal losses to the other eigenstates. The linear slope of the recycling pulses suppresses higher Fourier components and therefore reduces photon emission into the detuned cavity. Note that the recycling process is finished before the end of the pulse is reached, so that the final sudden drop in Rabi frequency has no effect.

Also shown in Fig. 17.5(c) are two measured arrival-time distributions of the photons. It is obvious that the pump-pulse duration of 2 μs is slightly too short, as the emitted photon pulse is not completely finished. The measured data agree well with the simulation (see Fig. 17.3(b)) if only photons from strongly coupled atoms are considered (solid line). For these we assume that several photons are detected within the atom–cavity interaction time. If solitary photons which we attribute to weakly coupled atoms are included in the analysis, the arrival-time distribution is given by the dotted line. Note that the envelope of the photon pulses is well explained by the expected shape of the single-photon wavepackets, and therefore cannot be attributed to an uncertainty in emission time which is not present for a unitary process. We emphasize that the pump-pulse duration was adjusted to maximize the number of photons per atom. Longer pump pulses would not truncate the photon pulses and, hence, would slightly increase the emission probability per pulse, but due to the limited atom–cavity interaction time the total number of photons per atom would be reduced.

The second-order intensity correlation of the photon stream is shown in Fig. 17.6. Displayed is the cross-correlation of the photon streams registered by the two photodiodes. It is defined as $g^{(2)}(\Delta t) = \langle P_{D1}(t)P_{D2}(t-\Delta t)\rangle \,/\, (\langle P_{D1}(t)\rangle \langle P_{D2}(t)\rangle)$, where $P_{D1}(t)$ and

[1]The flux is determined by a statistical analysis of the emitted light, with continuous pump- and recycling lasers exciting the atoms. As the cavity acts as an atom detector, atoms not interacting with the cavity mode are ignored, so that its spatial mode structure is automatically taken into account.

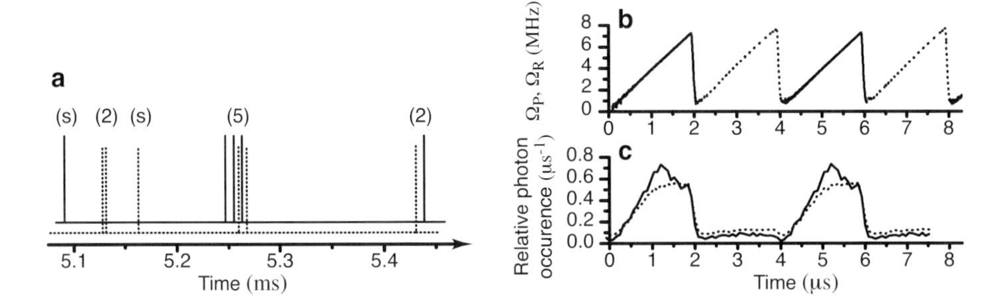

Figure 17.5: Photon emission. **(a)** Clip of the photon streams arriving at the photodiodes. Several sequences of two (2) and five (5) photon emissions are observed, with durations comparable to the atom–cavity interaction time. The solitary events (s) are either dark counts, or, more likely, photons coming from atoms that are only weakly coupled to the cavity. **(b)** Laser pulse shape: The atoms are periodically illuminated with 2 μs-long pulses from the pump (solid line) and the recycling laser (dotted line). **(c)** Arrival-time distribution of all photons emitted from the cavity (dotted line). The solid line shows the arrival-time distribution of photons emitted from strongly coupled atoms (see text), with $(g,\ \Omega^0_{P,R},\ \Delta,\ \Gamma,\ \kappa) = 2\pi \times (2.5,\ 8.0,\ -20.0,\ 6.0,\ 1.25)\,\text{MHz}$, where $\Omega^0_{P,R}$ are the peak Rabi frequencies of the pump- and recycling pulses, and Γ and κ are the atom and cavity-field decay rates, respectively.

$P_{D2}(t)$ are the probabilities to detect a photon at time t with photodiode $D1$ and $D2$, respectively. Note that all photon-arrival times are recorded to calculate the full correlation function, without the otherwise usual restriction of a simple start/stop measurement which would consider only neighbouring events.

All measured correlation functions oscillate with the same periodicity as the sequence of pump pulses, which indicates that photons are only emitted during the pump pulses, and no emissions occur when recycling pulses are applied. The nearly Gaussian envelope of the comb-like function is obviously a consequence of the limited atom–cavity interaction time. The most remarkable feature, however, is the (nearly) missing correlation peak at $\Delta t = 0$. To discuss this, note that for an atomic flux of 10 atoms/ms the probability to meet an atom in the cavity is 15 %, while the probability to have more than one atom is 1.4 %, which is not negligible. It is therefore clear that additional atoms also coupled to the cavity give rise to excess photons. In this case, the central correlation peak does not vanish completely (Fig. 17.6a). For a smaller atomic flux of 3.4 atoms/ms, the probability to find a single atom inside the cavity is 5.7 %, while the probability of having more than one atom is only 0.18 %. This is negligible, and indeed, the correlation peak at $\Delta t = 0$ vanishes (Fig. 17.6b). Even in case of a high atom flux, a straightforward approach allows one to isolate the single-atom contribution to the signal. Photons coming from different atoms give rise to a periodic contribution to the correlation function, which extends well beyond the atom–cavity interaction time and continues to oscillate with constant amplitude. In the regime $|\Delta t| \to \infty$, a periodic function is fitted to the correlation function and then continued into the relevant time regime around $\Delta t \approx 0$. This fit function, which is shown as a hatched area in Fig. 17.6a, is then subtracted from the measured correlation function. It follows that only the correlations between photons emitted from one-

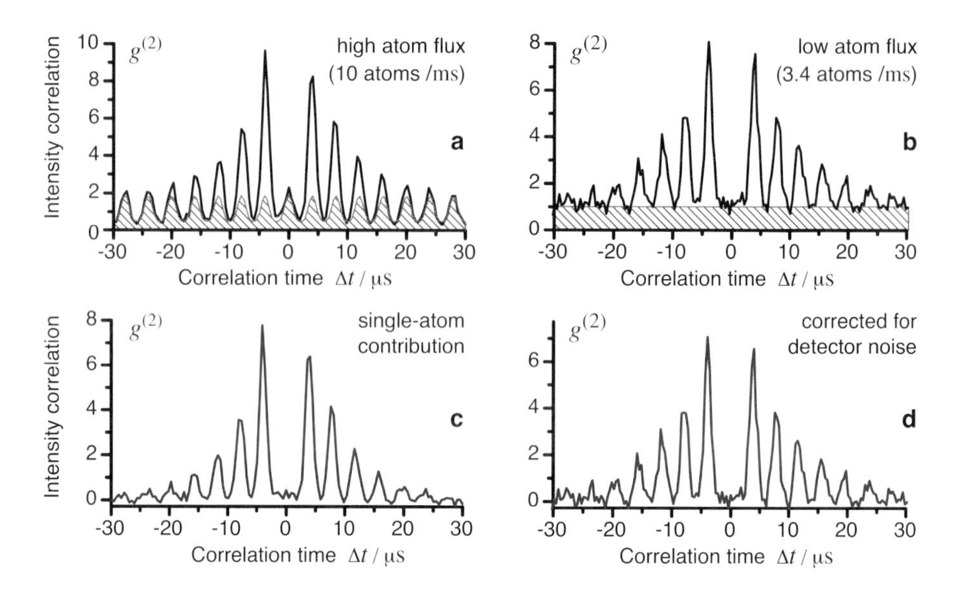

Figure 17.6: Second-order intensity correlation of the emitted photon stream. **(a)** High atom flux, averaged over 4997 experimental cycles (loading and releasing of the atom cloud) with a total number of 151089 photon counts. The hatched area represents correlations mainly caused by excess photons emitted from different atoms. For correlation times larger than the atom–cavity interaction time, only this different-atoms contribution persists. **(b)** Low atom flux, averaged over 15000 experimental cycles with a total number of 184868 photon counts. The probability to have more than one atom inside the cavity can be neglected here. The hatched area represents mainly correlations between photons and detector-noise counts. **(c,d)** Single-atom contribution to the correlation function, obtained after subtraction of the different-atom contribution from (a) or the detector-noise contribution from (b), respectively.

and-the-same atom remain (Fig. 17.6c). In fact, the central correlation peak disappears. Note also that in case of the smaller atom flux (Fig. 17.6b), the minimum of $g^{(2)}(\Delta t)$ does not reach zero, but equals one, a value solely caused by correlations between photons and detector noise counts. Subtraction of this noise contribution, shown as a hatched area in Fig. 17.6b, from the correlation function leads again to the single-atom contribution to $g^{(2)}(\Delta t)$ (Fig. 17.6d). The two data sets displayed in Fig. 17.6(c and d) demonstrate the nonclassical character of the emitted light. Strong antibunching together with a vanishing correlation around $\Delta t = 0$ is observed. This proves that the number of photons per pump pulse and atom is limited to one, and no second emission occurs before the atom is recycled to its initial state.

We emphasize that the detection of a first photon signals the presence of an atom in the cavity and fixes the atom number to one. The photons emitted from this atom during subsequent pump pulses dominate the photon statistics and give rise to antibunching. Such an antibunching would not be observed for faint laser pulses, since a random photon statistics applies to each pulse.

The areas of the different peaks of the correlation function in Fig. 17.6 reflect the probability for the emission of further photons from one-and-the-same atom. They are determined from a lengthy but straightforward calculation, which relates the number of correlations per pulse with the total number of photons. Using the data displayed in Fig. 17.6(c and d), the result for the conditional emission of another photon during the (next, 3rd, 4th, 5th, 6th, 7th) pump pulse is (8.8, 5.1, 2.8, 1.4, 0.8, 0.5) %. Note that the probabilities for subsequent emissions decrease, since the photon emission probability, P_{emit}, depends on the location of the moving atom. It is highest for an atom in an antinode, and decreases if the atom moves away from this point. It is not possible to control the atom's location in the present experiment, but it is possible to calculate $P_{\mathrm{emit}}(z)$ from the experimental data. Here, z is the atom's vertical position relative to the cavity axis, and $P_{\mathrm{emit}}(z)$ is averaged over all possible atomic trajectories in the horizontal xy-plane. Assuming a Gaussian z-dependence, the deconvolution of $g^{(2)}(\Delta t)$ gives $P_{\mathrm{emit}}(z) = 0.17 \exp\left[-(z/15.7\,\mu\mathrm{m})^2\right]$. For $z = 0$, the average photon-emission probability of 17% is smaller than the calculated value of 67% for an atom in an antinode of the cavity. It follows that a system combining a cavity and a single atom at rest in a dipole trap [SRPG01, KAS$^+$01], or a single ion at rest in a rf-trap [GKH$^+$01, MKB$^+$], should allow one to generate a continuous bit-stream of single photons with a large and time-independent efficiency [LK97, KHBR99]. The photon repetition rate is limited by the atom–cavity coupling constant, g, which one could push into the GHz regime by using smaller cavities of wavelength-limited dimensions in, e.g., a photonic bandgap material.

17.4 Summary and Outlook

Vacuum-stimulated Raman scattering driven by an adiabatic passage in a strongly coupled atom–cavity system can be used to generate single photons on demand. It is possible to generate a sequence of several photons on demand from one-and-the-same atom while it interacts with the cavity. These photons are generated in a well-defined radiation mode. They should have the same frequency, defined by the Raman-resonance condition, and a Fourier-transform limited linewidth. It follows that one can expect the photons to be indistinguishable and, therefore, ideal for all-optical quantum computation schemes [KLM01]. Finally, we state that the photon generation process depends on the initial state of the atom interacting with the cavity. If the atom is prepared in a superposition of states $|g, 0\rangle$ and $|u, 0\rangle$ prior to the interaction, this state will be mapped onto the emitted photon. The photon-generation process is unitary, so that it can be reversed in time and a second atom, placed in another cavity, could act as a photon receiver. This should allow one to transfer quantum states between different atoms, which is the key to quantum communication in a distributed network of optical cavities [CZKM97].

This work was partially supported by the European Union through the IST(QUBITS) and IHP(QUEST) programs.

References

[BBPG00] R. Brouri, A. Beveratos, J.-P. Poizat, and P. Grangier. Photon Antibunching in the Fluorescence of Individual Color Centers in Diamond. *Opt. Lett.*, 25:1294–1296, 2000.

[BLTO99] C. Brunel, B. Lounis, P. Tamarat, and M. Orrit. Triggered Source of Single Photons Based on Controlled Single Molecule Fluorescence. *Phys. Rev. Lett.*, 83:2722–2725, 1999.

[BSPY00] O. Benson, C. Santori, M. Pelton, and Y. Yamamoto. Regulated and entangled photons from a single quantum dot. *Phys. Rev. Lett.*, 84:2513–2516, 2000.

[BVW01] S. Brattke, B. T. H. Varcoe, and H. Walther. Generation of Photon Number States on Demand via Cavity Quantum Electrodynamics. *Phys. Rev. Lett.*, 86:3534–3537, 2001.

[CZKM97] J. I. Cirac, P. Zoller, H. J. Kimble, and H. Mabuchi. Quantum State Transfer and Entanglement Distribution Among Distant Nodes in a Quantum Network. *Phys. Rev. Lett.*, 78:3221–3224, 1997.

[DiV00] D. P. DiVincenzo. The physical implementation of quantum computation. *Fortschr. Phys.*, 48:771, 2000.

[GKH+01] G. R. Guthörlein, M. Keller, K. Hayasaka, W. Lange, and H. Walther. A single ion as a nanoscopic probe of an optical field. *Nature*, 414:49–51, 2001.

[Har93] S. E. Harris. Electromagnetically Induced Transparency with Matched Pulses. *Phys. Rev. Lett.*, 70:552–555, 1993.

[Har97] S. E. Harris. Electromagnetically Induced Transparency. *Phys. Today*, 50(7):36, 1997.

[HHDB99] L. V. Hau, S. E. Harris, Z. Dutton, and C. H. Behroozi. Light speed reduction to 17 metres per second in an ultracold atomic gas. *Nature*, 397:594–598, 1999.

[HLKR00] M. Hennrich, T. Legero, A. Kuhn, and G. Rempe. Vacuum-Stimulated Raman Scattering Based on Adiabatic Passage in a High-Finesse Optical Cavity. *Phys. Rev. Lett.*, 85:4872–4875, 2000.

[JC63] E. T. Jaynes and F. W. Cummings. Comparison of Quantum and Semiclassical Radiation Theories with Application to the Beam Maser. *Proc. IEEE*, 51:89–109, 1963.

[KAS+01] S. Kuhr, W. Alt, D. Schrader, M. Müller, V. Gomer, and D. Meschede. Deterministic Delivery of a Single Atom. *Science*, 293:278–280, 2001.

[KBKY99] J. Kim, O. Benson, H. Kan, and Y. Yamamoto. A Single Photon Turnstile Device. *Nature*, 397:500–503, 1999.

[KHBR99] A. Kuhn, M. Hennrich, T. Bondo, and G. Rempe. Controlled Generation of Single Photons from a Strongly Coupled Atom–Cavity System. *Appl. Phys. B*, 69:373–377, 1999.

[KHR02] A. Kuhn, M. Hennrich, and G. Rempe. Deterministc single-photon source for distributed quantum networking. *Phys. Rev. Lett.*, 89:067901, 2002.

[KLM01] E. Knill, R. Laflamme, and G. J. Milburn. A scheme for efficient quantum computing with linear optics. *Nature*, 409:46–52, 2001.

[KMZW00] C. Kurtsiefer, S. Mayer, P. Zarda, and H. Weinfurter. Stable Solid-State Source of Single Photons. *Phys. Rev. Lett.*, 85:290–293, 2000.

[KPHS96] A. Kuhn, H. Perrin, W. Hänsel, and C. Salomon. Three Dimensional Raman Cooling using Velocity Selective Rapid Adiabatic Passage. In K. Burnett, editor, *OSA TOPS on Ultracold Atoms and BEC*, volume 7, pages 58–65. OSA, 1996.

[KR02] A. Kuhn and G. Rempe. Optical cavity qed: Fundamentals and application as a single-photon light source. In *Varenna Lecture Notes 2001*, in print. Int. School of Physics Enrico Fermi, IOS, 2002.

[LK97] C. K. Law and H. J. Kimble. Deterministic Generation of a Bit-Stream of Single-Photon Pulses. *J. Mod. Opt.*, 44:2067–2074, 1997.

[LM00] B. Lounis and W. E. Moerner. Single Photons on Demand from a Single Molecule at Room Temperature. *Nature*, 407:491–493, 2000.

[Loy74] M. M. T. Loy. Self-induced rapid adiabatic passage. *Phys. Rev. Lett.*, 32:814–817, 1974.

[Loy78] M. M. T. Loy. Two-photon adiabatic inversion. *Phys. Rev. Lett.*, 41:473–475, 1978.

[Mes58] A. Messiah. *Quantum Mechanics*, volume 2, chapter 17. J. Wiley & Sons, NY, 1958.

[MHN+97] X. Maître, E. Hagley, G. Nogues, C. Wunderlich, P. Goy, M. Brune, J.-M. Raimond, and S. Haroche. Quantum Memory with a Single Photon in a Cavity. *Phys. Rev. Lett.*, 79:769–772, 1997.

[MIM+00] P. Michler, A. Imamoglu, M. D. Mason, P. J. Carson, G. F. Strouse, and S. K. Buratto. Quantum Correlation Among Photons from a Single Quantum Dot at Room Temperature. *Nature*, 406:968–970, 2000.

[MKB+] A. B. Mundt, A. Kreuter, C. Becher, D. Leibfried, J. Eschner, F. Schmidt-Kaler, and R. Blatt. Coupling a single atomic quantum bit to a high finesse optical cavity. e-print quant-ph/0202112. *Phys. Rev. Lett.*(to be published)

[MKB+00] P. Michler, A. Kiraz, C. Becher, W. V. Schoenfeld, P. M. Petroff, L. Zhang, E. Hu, and A. Imamoglu. A Quantum Dot Single Photon Turnstile Device. *Science*, 290:2282–2285, 2000.

[MRG+01] E. Moreau, I. Robert, J. M. Gérard, I. Abram, L. Maniv, and V. Thierry-Mieg. Single-mode solid-state single photon source based on isolated quantum dots in pillar microcavities. *Appl. Phys. Lett.*, 79:2865–2867, 2001.

[PFM+01] D. F. Phillips, A. Fleischhauer, A. Mair, R. L. Walsworth, and M. D. Lukin. Storage of Light in Atomic Vapor. *Phys. Rev. Lett.*, 86:783–786, 2001.

[PKBS98] H. Perrin, A. Kuhn, I. Bouchoule, and C. Salomon. Sideband cooling of neutral atoms in a far-detuned optical lattice. *Europhys. Lett.*, 42:395–400, 1998.

[PMZK93] A. S. Parkins, P. Marte, P. Zoller, and H. J. Kimble. Synthesis of Arbitrary Quantum States Via Adiabatic Transfer of Zeeman Coherence. *Phys. Rev. Lett.*, 71:3095, 1993.

[SK93] B. W. Shore and P. L. Knight. The Jaynes–Cummings Model. *J. Mod. Opt.*, 40:1195, 1993.

[SPS+01] C. Santori, M. Pelton, G. Solomon, Y. Dale, and Y. Yamamoto. Triggered Single Photons from a Quantum Dot. *Phys. Rev. Lett.*, 86:1502–1505, 2001.

[SRPG01] N. Schlosser, G. Reymond, I. Protsenko, and P. Grangier. Sub-poissonian loading of single atoms in a microscopic dipole trap. *Nature*, 411:1024–1027, 2001.

[VFSB01] N. V. Vitanov, M. Fleischhauer, B. W. Shore, and K. Bergmann. Coherent manipulation of atoms and molecules by sequential laser pulses. *Adv. At. Mol. Opt. Phys.*, 46:55–190, 2001.

[YKS+02] Z. Yuan, B. E. Kardynal, R. M. Stevenson, A. J. Shields, C. J. Lobo, K. Cooper, N. S. Beattie, D. A. Ritchie, and M. Pepper. Electrically Driven Single-Photon Source. *Science*, 295:102–105, 2002.

[YSHB99] L. P. Yatsenko, B. W. Shore, T. Halfmann, and K. Bergmann. Source of metastable H(2s) atoms using the Stark chirped rapid-adiabatic-passage technique. *Phys. Rev. A*, 60:R4237–R4240, 1999.

18 A Relaxation-free Verification of the Quantum Zeno Paradox on an Individual Atom

Ch. Balzer, Th. Hannemann, D. Reiß, Ch. Wunderlich, W. Neuhauser, and P. E. Toschek

Universität Hamburg
Institut für Laser-Physik
Hamburg, Germany

Abstract: The temporal evolution of a quantum system is frustrated by observing the system, even when there is no back-action on the system. This much-disputed Quantum Zeno Paradox – a clue to which is entanglement – is verified on an individual atomic ion: The evolution of the ion's spin, microwave-driven on the ground-state hyperfine resonance, alternates with probing the ion's quantum state by attempts of laser-excited resonance scattering. Enhanced chance of survival marks even the lower "dark" state correlated with detection of a null signal. – A previous conclusive purely optical experiment and related work is summarized.

18.1 Introduction

It is well known that entanglement of quantum states is the principal resource for quantum information processing. Entangled quantum states are characterized by two (or more) degrees of freedom whose eigenstates are correlated. The eigenstate of one degree of freedom in which the system may be found predetermines the occupied eigenstate of the other degree of freedom. The canonical example of this particular quantum feature is the pair of correlated particles of an EPR experiment [1]. With a bound particle, an internal and a motional degree of freedom may be made entangled by excitation of a vibrational sideband of the internal resonance [2]. The vibration is capable of coupling the particle to neighbours for the processing of quantum information [3]. Moreover, individual particles are usually manipulated by radiatively addressing one particular resonance, and probing the result by irradiation of light resonant on another one. Here, the particle's state on the addressed resonance ($|0\rangle$ or $|1\rangle$), and of the scattered light ($|\text{on}\rangle$ or $|\text{off}\rangle$) turn entangled. In this way, a kind of indirect quantum measurement based on correlation of state and detected signal becomes feasible [4].

The reiterated observation of an observable of a quantum system involved in the temporal evolution of that system makes the system again and again projected into an eigenstate, preferentially into its initial state, and slows down or even impedes the evolution [5, 6]. Is this "quantum Zeno" effect (QZE) [7] perhaps induced by physical reaction of the detected quantity on the system, or rather by the gain of information in the measurement? It has been shown that these alternatives are indistinguishable *in principle*, when the system under scrutiny consists of an ensemble, and the result of a measurement is an expectation value. The alternatives

Quantum Information Processing, 2nd Edition. Edited by Thomas Beth and Gerd Leuchs

ISBN: 3-527-40541-0

may be discriminated, however, on an *individual* quantum system, when each measurement yields an eigenvalue as its result [8–10].

At this point, entanglement of the radiatively driven quantum system and the detected signal comes into play. Since this feature can allow for back-action-free determination of the system's state, measurements based on entanglement that indeed proved the evolution of the system impeded would qualify as the *origin* of impediment. This particular kind of inhibition, having been labelled "quantum Zeno paradox" (QZP) [11], would exclude, as its source, physical back action by the meter upon the quantum system. The demonstration of this inhibition would prove the feasibility of such a back-action-free measurement that is essential for applications to quantum information processing, and in particular for evading certain kinds of decoherence.

Recently, an experiment on the quantum evolution of an individual atomic ion under reiterated probing the ion's state has been reported that satisfies the above preconditions [12]. The inhibition of the ion's laser-driven evolution on an electronic, dipole-forbidden line by repeated attempts of making the ion scatter resonance light was demonstrated. This evidence was derived from the statistics of uninterrupted sequences of *equal* results, that is, all results in any sequence of measurements signal either "scattered light on", or "off". In particular, the "off" results that were correlated with the ion being excited in a metastable electronic state represent the desired demonstration, since the corresponding measurements satisfy the condition of quantum non-demolition [8, 13]. On the other hand, although the decay of that "dark" state via an E2 line is rather weak, the data nevertheless require the consideration of the minute relaxation with their quantitative evaluation, and with the modelling of the dynamics of the system. Thus, another verification seems desirable that relies on a quantum system in its ground state. Such an experiment has been performed, making use of two alternative schemes of sequential microwave-optical double resonance on an individual ion confined in an electrodynamic ion trap.

18.2 The Hardware and Basic Procedure

The structure of the experiment is shown in Fig. 18.1. A single $^{171}Yb^{+}$ ion was driven on its 12.6 GHz ground-state $F = 0 \rightarrow 1$, $m_F = 0$ hyperfine resonance by a resonant microwave pulse, such that a certain probability of excitation prevailed. If excitation to the $F = 1$ level had happened, the subsequent laser pulse was scattered off the ion's resonance line $S_{1/2}(F = 1, m_F = 0) - P_{1/2}$ and signaled the preceding microwave excitation. Whereas spurious optical pumping to the $m_F = \pm\, 1$ levels leaves the cyclic laser excitation leaky and decouples the ion from the microwave driving [14], the "dark" state $F = 0$, characterized by absent light scattering, lacks any relaxation. Consequently, observations whose results are "off" are well-suited for proving the inhibition of the evolution as generated by the measurement.

The frequency-doubled output of a Ti:sapphire laser (369 nm, 100 kHz bandwidth) served as the probe light. It was tuned some 10 MHz down from the ion's $S_{1/2} - P_{1/2}$ resonance in order to provide laser cooling for the ion. The light scattered off the ion was photon-counted. Data acquisition as well as the control of the laser parameters was performed in real time. Spuriously pumping the ion into the $D_{3/2}$ metastable level was undone by the 935-nm light of

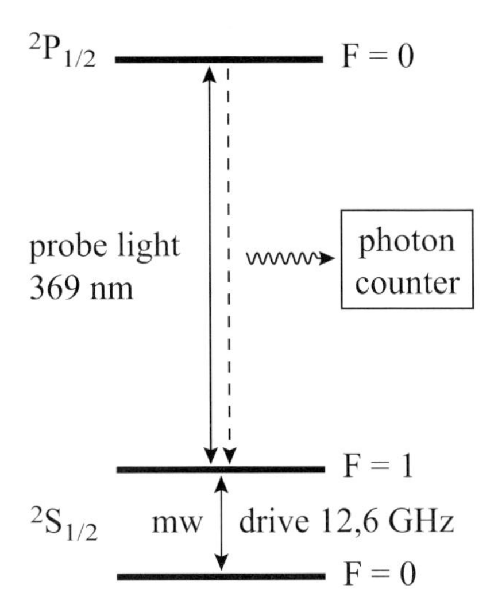

Figure 18.1: Relevant energy levels of $^{171}Yb^+$ion, microwave driving radiation, resonant probe light, and detection of resonance fluorescence.

a diode laser that immediately repumped the ion into its ground state. – The spatial position of the ion was kept at the node of the rf trapping field. Any deviation from this position is accompanied by residual micro-motion that was monitored, by phase-sensitive detection of spurious correlation with the trapping voltage, and eliminated.

The driving radiation was made up of microwave pulses of duration τ and Rabi frequency Ω, whose pulse area $\Omega\tau$ equals the phase angle of nutation, θ. The microwave frequency ω must be set on resonance ω_0 very precisely. Even minute detuning was identified when driving the ion by microwave double pulses and subsequently probing the ion's spin state by a laser pulse (Fig. 18.2). This probing randomly yields an "on" result (ion in $F = 1$ level) or an "off" result (ion in $F = 0$ level). Some 500 of these results, recorded upon stepwise incremented temporal separation of the two driving pulses, make a trajectory of measurements. When many of these trajectories – in fact 50 – were superimposed, this ensemble of results showed a Ramsey pattern of interference fringes [15]. Fitting this pattern by the standard sine-square transition probability yields the off-resonance detuning of the preset microwave frequency, and the initial phase.

The pulse area θ of the driving pulses was set by stepwise incrementing the length of a single driving pulse, each step followed by a pulse of probe light. Superimposed trajectories of results reveal nutational oscillation of the ion's spin, such that the desired pulse length was precisely preset to a well-defined fraction or multiple of π.

The actual measurements, following a first scheme, were arranged according to Fig. 18.3. Gating the laser light by an acousto-optical deflector generated 1.5 ms-long probe pulses applied to the ion between the driving pulses, whose pulse area ($\theta = 2\pi$ at $\tau = 4.9$ ms) was set to particular values in subsequent runs. The scattered probe light was recorded by a photon counter gated open in synchronism with the probe pulses. When a laser pulse made appear a

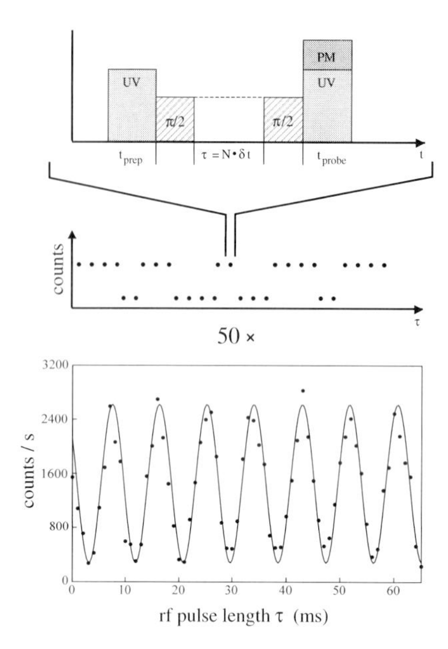

Figure 18.2: Setting the driving microwave radiation on resonance: A test measurement on the single $^{171}Yb^+$ ion starts with a laser pulse of length $t_{prep} = 50$ms that pumps the ion, via the wing of the $S_{1/2}$ $(F = 1) - P_{1/2}(1)$ resonance, into the $F = 0$ level of the ground state. Then, two microwave $\pi/2$ driving pulses with temporal separation τ were applied. Their frequency was $\omega = 12{,}642815750$ GHz $\times$ 2π, and their duration 1.225 ms. Finally, a laser pulse was applied to the ion, the photon counter(PC) gated open, and scattered light registered. The separation τ was incremented, by $\delta\tau$, with every next measurement. A series of 500 measurements make a trajectory. After each trajectory, the microwave and UV laser light were made to simultaneously irradiate the ion for 100 ms in order to provide laser cooling to the ion. This procedure was repeated 50 times, and the results of the 50 trajectories were accumulated; they show Ramsey-type interference fringes. Fitting the data with the standard transition probability for the $\pi/2$ double pulse yielded the detuning 112 Hz, and the initial phase $3/2\pi$.

burst of scattered light, the ion was considered to have been successfully excited to the $F = 1$ hyperfine level of its ground state by the preceding driving pulse. No fluorescence showing up was considered the signature of the ion having survived on the "dark" level, $F = 0$. A driving pulse of a particular length makes the ion coherently prepared in the pertaining superposition ground state $|0\rangle$ cos θ+$|1\rangle$ sin θ, and the subsequent pulse of the probe light makes the ion "read out" to be found in one of its corresponding eigenstates. The results of some 10^4 of these measurements form a trajectory that is segmented in a string of "waiting" intervals, every second one made up by a random number of equal results "on", alternating with "off" intervals, and so forth. The statistical distributions of all sequences of equal results that are present in these waiting intervals is analysed for the signature of the inhibiting effect of measurement upon the quantum evolution of the ion.

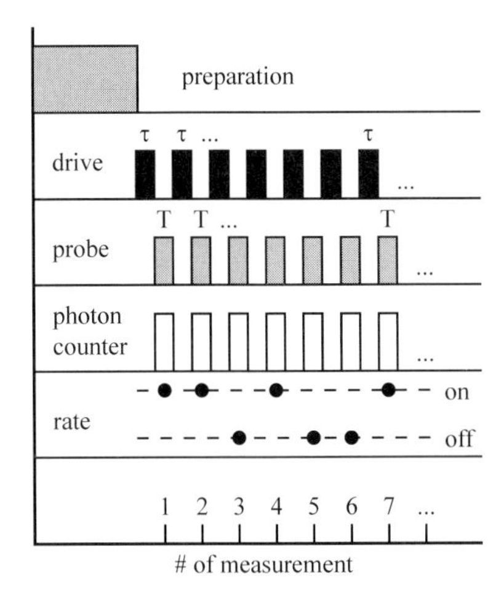

Figure 18.3: Temporal schedule of preparation (first and second row) and measurements (third and fourth row), according to the first scheme. Trajectory of results (fifth and sixth row) made up of "waiting" intervals of a random number of "on" and "off" results alternating.

18.3 First Scheme: Statistics of the Sequences of Equal Results

We want to derive, from the recorded data, an empirical measure for the probability of the ion to survive, during q sequential measurements, in its same state, $|1\rangle$, correlated with "on", or $|0\rangle$, correlated with "off". This probability is related to the frequency of occurence $U(q)$ of a sequence of q equal results in the trajectory. If this trajectory is long enough, $U(q)$ may be identified with the corresponding conditional probability of the ion to be found in one particular state, and to survive the next $q-1$ pulse pairs:

$$U(q) \;=\; U(1)V(q-1). \tag{18.1}$$

This equality is the signature of QZE, if the probability of survival, V, is made to include the frustrating effect of the measurements. Let us assume, that the driven coherent evolution of the ion is in fact impeded by the probing, leaving the ion in an *eigenstate:* A sequence of equal results means the ion having been set back to the *same* state as the one it has been in before the application of the driving pulse. Under this condition, the conditional probability of surviving the application of q drive and probe pulse pairs is simply

$$V(q) \;=\; p^q\,, \tag{18.2}$$

where $p = \cos^2(\Omega\tau/2)$ is the probability of the ion remaining in its initial state of a relaxationless two-level system upon the irradiation with a resonant coherent pulse of area $\theta = \Omega\tau$ [16].

The two-level system is represented by the driven hyperfine resonance of the ion. Note that we need not consider any complications related to relaxation, since there is no intrinsic loss with the ground-state hyperfine resonance, and phase fluctuations of the microwave are negligible. Thus, the probability of the ion staying in the same state as before under the attempt of acting by the driving pulse should not differ in the states 0 and 1: $p_1 = p_0 \equiv p$. However, as was mentioned above, optical pumping among the Zeeman levels of state 1 by the probe light provides the ion with some energy relaxation, as soon as "on" results are involved [14]. Thus, sequences of "on" results would require a more sophisticated evaluation, taking into account both transversal and longitudinal relaxation. Normalised distributions of the frequencies of

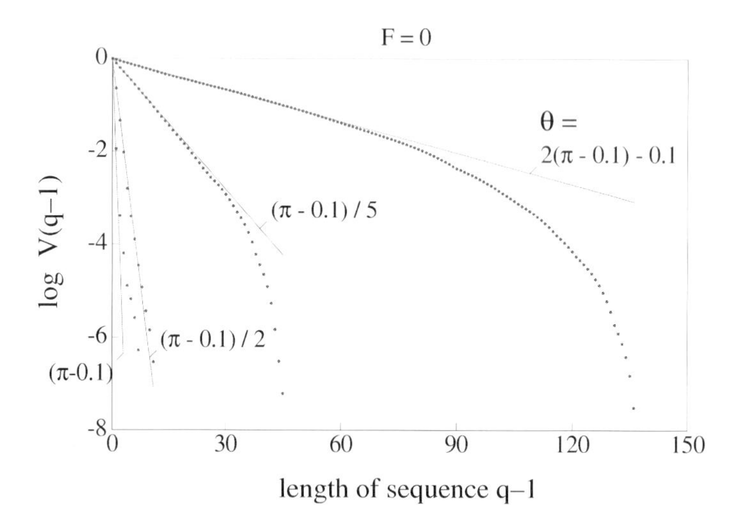

Figure 18.4: Probability $U(q)/U(1)$ of uninterrupted sequences of q results *all* of them "off", when ion was initially prepared in the "off" state ($F = 0$). The lines show the distribution of probability $V(q-1)$ for the ion *not* undergoing a flip of its spin to the "on" state during the entire sequence. Length of trajectories: 2000 measurements of 4.9 ms driving time and 2 ms probing time. See text.

"off" sequences vs q, that is $U(q)/U(1)$, are shown in Fig. 18.4, on a log scale, for nominally $\theta = \pi$, $\pi/2$, $\pi/5$, and $2\pi - 0.1$. Also shown are lines representing calculated distributions of $V(q-1)$ that have been made to fit the data by varying θ. The fitting procedure is very sensitive: deviations from the best fit as small as $\delta\theta = 10^{-5}$ are recognizable [13]. It turns out, that the area values of the microwave pulses actually preset deviated from their nominal values by 3%. At large q, the data show mild deficiency of long sequences, marking slightly excessive excitation to the $F = 1$ level. This feature indicates some dephasing of the driven spin dynamics to happen on the time scale of seconds, as it was verified by model calculations.

The recorded normalized distributions $U(q)/U(1)$ have been found to agree with the largest possible probabilities of survival $V(q-1)$,. With small pulse area θ, these probabilities are compatible with having *undone* the effect of almost *each* driving pulse by the subsequent probe pulse. This agreement is the signature of QZE, and, in particular with the "off" sequences, the signature of QZP.

18.4 Second Scheme: Driving the Ion by Fractionated π-Pulses

A previous proposal for the verification of QZE included the interaction of the quantum system by a π-pulse of driving radiation resonant with a dipole-forbidden line. During this pulse the system was supposed to be irradiated by n laser pulses resonant with a neighbouring resonance line [17]. An experiment according to this scheme was performed on 5000 beryllium ions in an electromagnetic trap [18]. Although the results of this experiment completely agreed with the predictions of quantum mechanics, their relevance for the verification of QZE was questioned on various counts two of which seem indeed appropriate:

1. The irradiation by the n light pulses was considered n "measurements". However, only their cumulative result was recorded, at the end of the π-pulse. Thus, back and forth transitions in an individual ion, or in pairs of ions, go unnoticed in this type of "non-selective" measurement [19] with the corresponding probability of *net* survival

$$P_{00}^{(n)} = \frac{1}{2}\left(1 + \cos^n(\pi/n)\right). \tag{18.3}$$

   The distribution of results observed in such an experiment and found to obey Eq. 18.3 cannot prove QZE, let alone QZP.

2. Observations on an ensemble cannot discriminate the effects of potential dynamic interactions of quantum system and meter from the effects of the measurement, that is the updated knowledge about the system. However, discrimination is indeed feasible with an *individual* quantum system to be observed [8–10]. This is so since the result of a measurement on an individual system is one of its *eigenvalues*, whereas the measurement on an ensemble yields an *expectation* value [13].

The experiment on a single ion reported in the previous sections is easily moulded in a shape similar to an implementation of Cook's proposal, however, avoiding the drawbacks mentioned above: Here, the quantum system is a single ion, and the n intermittant probe interventions are accompanied by simultaneously recording the signal of light scattering ("on"), or of its absence ("off"). Consequently, the adopted improved strategy amounts to the performance of "selective" measurements qualified for the verification of the QZE, and even of the QZP.

After a preparatory laser pulse that pumps the ion into the $F = 0$ ground state, via the wing of the resonance line $S_{1/2}(F = 1) \rightarrow P_{1/2}(F = 0)$, the ion was irradiated by a series of n driving pulses of area $\theta = \pi/n$. This fractionated attempt of excitation would result in a complete π-flop of the ion into state $F = 1$, provided that dephasing is safely negligible over the full driving time (Fig. 18.5a and b). Now, pulses of probe light were made to illuminate the ion during the $n - 1$ intermissions between the fractional driving pulses, and the photon counter was synchronously gated open in order to register potential light scattering as the signature of the ion's excitation after any one of the driving pulses. The final result – scattered light on or off – generated by the last (n-th) fractional probe pulse of the series was separately registered (Fig. 18.5c). This series of measurements, complete with preparations, probe interventions, and observations, was reiterated 2000/n times.

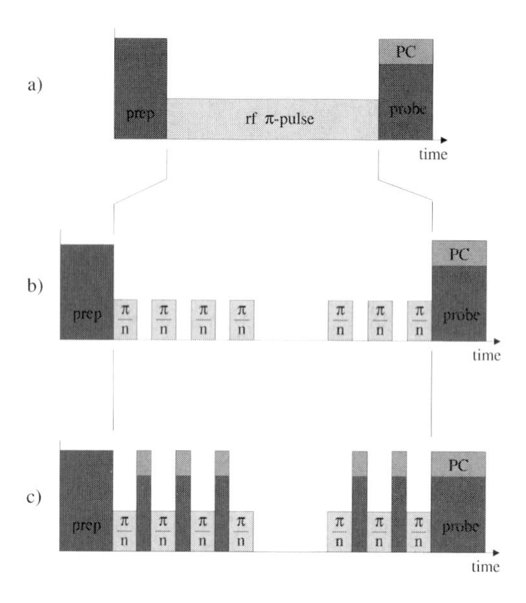

Figure 18.5: Scheme of measurement with one driving π-pulse (a). Same with fractionated π-pulse (n-times π/n), no intermediate probing (b). Same, but π/n-pulses alternating with probe pulses (c). PC: photon counting.

Only those series of results are considered to represent *survival* of the ion in its initial state $|0\rangle$ that include no "on" result with any one of the n observations in the series. The number of such series found, normalized by the total number of series, is supposed to approach the probability of n-times survival,

$$P_{00}(n) = \cos^{2n}(\pi/2n) . \tag{18.4}$$

The probability $P_{ij}(n)$ indicates selective measurements, where $i = 0$ means equal results "off" in the first $n-1$ measurements, and j the result of the final measurement. This evaluation is equivalent to discarding all series from the ensemble of survival histories as soon as an "on" result shows up in the $n-1$ intermediate observations, and it warrants the measurement being selective [10].

The numbers of series that indicate complete (n-fold) survival, i.e. the "frequencies" of complete survival, have been determined for $n = 0, 1, 2, 3$, and 9; they are shown as the dark grey histogram in Fig. 18.6. Black bars mark the probability of n-fold survival as calculated

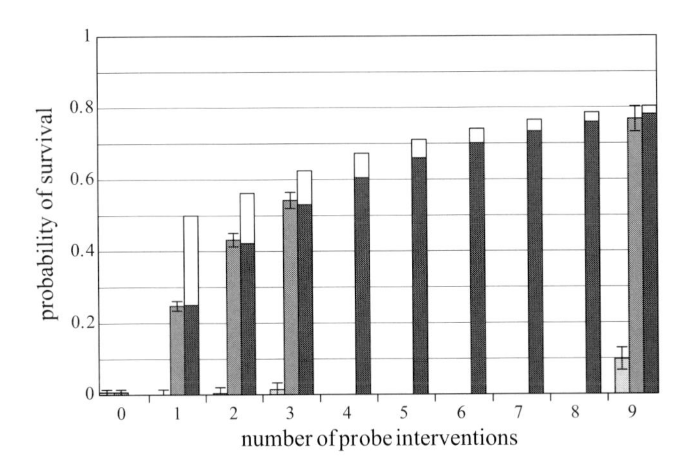

Figure 18.6: Probability of survival in the "off" state with intermediate probing (dark grey bars), and with no probing (light grey bars), vs number n of π-pulse partition. The probability is evaluated by counting as "favourable" those measurements that show only "off" results in each of the n observations. Driving time 2.9 ms, probing time 3 ms. Probability of survival in a "selective" measurement, calculated after Eq. 18.4 (black bars). If evaluated from the *entire* ensemble of results, including intermediate "on" results, the probability of survival would represent a non-selective measurement. This probability is calculated after Eq. 18.3 (white bars).

from Eq. 18.4. These probabilities agree with the recorded frequencies within the variances of the latter data. Light grey bars show frequencies of survival from data that have been recorded when the intervention by the probe laser pulses was lacking. Here, the probability of survival is supposed to vanish, since an effective π-pulse drives the ion into state $|1\rangle$ with certainty, the excitation being a deterministic act. Spurious decoherence leaves some finite chance of survival at long times, for example, at $n = 9$.

The white histogram that underlies the other ones shows, for comparison, the net probabilities of survival from Eq. 18.3 expected to show up with the results of the final probing, when all the results of the intermediate probing were *ignored*. Such a strategy of non-selective measurements on an ensemble of measurements would be unable to verify QZE, like that on an ensemble of quantum objects [18]. In principle, experimental data for this type of measurement could be generated by ignoring, in the recorded set of data, the results of the $n - 1$ intermediate observations. Unfortunately, these data include "on" results, and thus are subject to the systematics from spurious optical pumping of the ion into the $m_F = \pm 1$ Zeeman sub-levels.

The observed probabilities of survival are based on detected frequencies of occurrence of series indentified as "favourable", that is obeying the required conditions, in particular n successive "off" detections per series. These frequencies have been corrected for initial faulty ion preparation (18% mean value) and false detection of one of the n results of photon counting in a given series of π/n pulses. The latter error was determined as follows: The probability distributions of the counting rates of "on" (1) and "off" (0), vs the counted "number of photons", are approximately Poissonian and overlap each other. Their distinction is optimum when pulses that contain less than two photon counts are identified as "off" results (0). In fact, the "off" distribution overlaps this threshold by 2% (mean value). This overlap is identified as the risk of an individual false detection, from which the rate of misinterpreting a series of n measurements was determined. Multiple false detection within a series was ignored. – The error bars of the recorded data represent the variances of the on-off binomial distributions.

The measured and corrected frequencies of survival of the ion's state upon the action of the n-times fractionated π-pulse and intertwined probe interventions vanish at $n = 0$ and increase to 77% at $n = 9$. They agree, within the statistical errors as indicated by the variances of the on-off binomial distributions, with the values of the probability for selective observations, calculated from Eq. 18.4 (black bars). This finding on an individual quantum system verifies the QZE, and even the QZP.

18.5 Conclusions

Entangled states are a resource for reaction-free measurement on individual quantum systems. The exploitation of this resource requires a degree of freedom of the system whose state may be deliberately and coherently set by the observer, i.e., for example, the radiatively driven resonance of a two-level system. In terms of quantum information processing, this is a "one-qubit quantum gate". In addition, it requires a part of the observing device, a scattered light field, for example, whose observed states are correlated with the states of the driven resonance. An experiment along this line has been designed and performed on an individual particle in order to verify the QZE by measurements that yield null results and disprove physical interaction of the quantum object – the driven resonance of the particle – with the probe radiation, that is part of the "meter". Such an arrangement suits even the conditions for verification of QZP [11].

A single $^{171}Yb^{+}$ ion was driven, on its ground-state $F = 0 \rightarrow 1$ resonance, by resonant microwave pulses of predetermined area. In a first approach, intermittant probing by laser light resonant on the resonance transition of the ion did or did not elicit resonance scattering to become recorded. From the statstics of sequences of reiterated driving and probing whose results were "off", and that were not interrupted by results yielding the scattered light "on", the probability of survival of the ion in the ground state $F = 0$ was derived and found to agree with the ion being set back, by each probing, to its initial eigenstate. This probability may be compared with its hypothetical values expected when the act of measurement would lack the effect of setting back the system. Any such value is exceeded by the actual probability the more, the longer the sequence of the "off" results continues, that is, the longer the ion survives in its initial "dark" ground-state.

In the second approach, fractionated π pulses were used for driving the individual ion on its hyperfine resonance. The fractional driving pulses alternated with pulses of probe light *and* simultaneous recording of the scattered light, or of the absence of scattering. The probabilities of surviving probe interventions were found to increase with the number n.

Both approaches make known the micro-state of the *ensemble* of *observations*, unlike with an ensemble of particles. In contrast with a previous experiment on such an ensemble, these data represent *selective* measurements, as it is indispensable for the verification of QZE [11].

Also, they allow one to discriminate the effect of the *measurement* against that of physical intervention of the meter upon the quantum system [8–10], and to verify even QZP.

The present experiment differs from a previous one [12] by complete absence of relaxation on the quantum object with the preparations and measurements used for the evaluation: Here, the dark state is not metastable, but rather it is the $F = 0$ ground state of the ion.

18.6 Survey of Related Work

In the frame of the "Schwerpunktprogramm Quanteninformationsverarbeitung der DFG", the work outlined above and a body of related work have been performed. The results have been published on various occasions and in varying context:

1. "The Quantum Zeno Effect – Evolution of an Atom Impeded by Measurement", Ch. Balzer, R. Huesmann, W. Neuhauser, P.E. Toschek – Opt.Communic. **180**, 115-120 (2000). [12]

   Original publication of the first and purely optical variant of a conclusive experiment on the verification of the QZE/QZP.

2. "What does an Observed Quantum System Reveal to its Observer?", P.E. Toschek, Ch. Wunderlich. – Eur.Phys. J.D. **14**, 387-396 (2001). [13]

   Interpretation of (1.) in a broad context, and discussion of extensions.

3. "Evolution of an Atom Impeded by Measurement: The Quantum Zeno Effect", Ch. Wunderlich, Ch. Balzer, P.E. Toschek. – Z. Naturforschung **56**a, 106-164 (2001).

   Invited Report at the International Workshop on *Mysteries, Puzzles, and Paradoxes in Quantum Mechanics*, 17.-23.09.2000, Gargnano, Lake Garda, Italy.

4. "Quantum Measurement and Nonclassical Vibration of an Ion in a Trap", R. Huesmann, Ch. Balzer, B. Appasamy, Y. Stalgies, P.E. Toschek. – AIP Conference Proceedings 457, Woodbury, New York, p. 252-259.

   Invited Conference Report on the experimental and fundamental preconditions of (1.) at the *International Conference on Trapped Charged Particles and Fundamental Physics*, 31.08.-04.09.1998, Pacific Grove / Monterey, Cal., USA.

5. "Evolution of an Atom Impeded by Measurement", Ch. Balzer, T. Hannemann, Ch. Wunderlich, W. Neuhauser, P.E. Toschek. – Conference Digest QThD 84, 177.

   Conference Report on (1.) at the *CLEO/Europe-IQEC*, 10.-15.09.2000, Nice, France.

6. "Der Quanten-Zeno Effekt", Ch. Balzer, R. Huesmann, W. Neuhauser, P.E. Toschek. – Verhandl. DPG (VI) **34**, 380, Q3.5 (1999).

   National Conference Report on the topic of (1.).

7. "Light-Induced Decoherence in the Driven Evolution of an Atom", Ch. Balzer, Th. Hannemann, D. Reiss, W. Neuhauser, Ch. Wunderlich, P.E. Toschek. – Laser Physics **12**, 729-735 (2002). [14]

Conference Report at the *Tenth International Laser Physics Workshop (LPHYS '01)*, 03.-07.07.2001, Moscow, Russia, on the design of decoherence on individually prepared and addressed trapped ions.

8. "From Spectral Relaxation to Quantified Decoherence", Ch. Balzer, Th. Hannemann, D. Reiss, W. Neuhauser, P.E. Toschek, Ch. Wunderlich. – *Laser Physics at the Limit*, Springer-Verlag, Berlin, Heidelberg, New York 2001, p. 233-241.

   Festschrift contribution.

9. "A Relaxationless Demonstration of the Quantum Zeno Paradox on an Individual Atom", Ch. Balzer, Th. Hannemann, Ch. Wunderlich, W. Neuhauser, P.E. Toschek – Submitted to Optics Communications.

   Original publication of the second and microwave-optical variant of a conclusive experiment on the verification of the QZE/QZP.

10. "What does an Observed Atom Reveal to its Observer?", Ch. Balzer, P.E. Toschek. – Laser Physics **12**, 253-261 (2002).

    Plenary Lecture at the *Tenth International Laser Physics Workshop (LPHYS '01)*, 03.-07.07.2001, Moscow, Russia.

11. "What does an Observed Atom Reveal to its Observer?", P.E. Toschek, Ch. Balzer. – J.Opt. B: Quantum Semiclass. Opt. **4**, S450 (2002).

    Invited Report at the International Workshop on *Mysteries, Puzzles, and Paradoxes in Quantum Mechanics*, 27.08.-01.09.2001, Gargnano, Lake Garda, Italy.

12. "Light-Induced Decoherence in the Driven Evolution of an Atom", Ch. Balzer, T. Hannemann, W. Neuhauser, Ch. Wunderlich, P.E. Toschek. – Verhandl. DPG (VI) **36**, 193, Q 9.19 (2001).

    Report on the topic of (7.) at the *Seventh European Conference on Atomic and Molecular Physics* (ECAMP VII), Berlin, 02.-06.04. 2001.

13. "Entanglement and Inhibited Quantum Evolution", P.E. Toschek.

    Invited Report at the *Third International Conference on Trapped Charged Paticles and Fundamental Interactions*, 25.-30.08.2002, Wildbad Kreuth.

14. "A Relaxation-Free Verification of the Quantum Zeno Paradox on an Individual Atom", Ch. Balzer, Th. Hannemann, D. Reiß, Ch. Wunderlich, W. Neuhauser, P.E. Toschek, (this paper).

    Report on (9.) at the *DFG Workshop Quanten-Informationsverarbeitung*, 28.-30.01.2002, Bad Honnef.

15. "Raman Cooling and Heating of two Trapped Ba^+ Ions", D. Reiss, K. Abich, W. Neuhauser, Ch. Wunderlich, P.E. Toschek. – Phys. Rev. A **65**, 053401 (2002).

    Original publication on the vibrational and spectral dynamics of two ions in an electrodynamic trap.

16. "Austausch der elektronischen Anregung in zwei kalten gespeicherten Ionen", K. Abich, Ch. Wunderlich, P.E. Toschek. – Verhandl. DPG (VI) **35**, 1084, Q 34.3 (2000).

    National Conference Report on the topic of (15.).

17. "Laser Cooling and Heating of Two Trapped $^{138}Ba^{+}$ Ions", D. Reiss, K. Abich, W. Neuhauser, Ch. Wunderlich, P.E. Toschek. – Verhandl. DPG (VI) **36**, 197, Q 11.3 (2001).

    Report on the topic of (15.) at the *Seventh European Conference on Atomic and Molecular Physics* (ECAMP VII), Berlin, 02.-06.04.2001.

18. "Thermisch aktivierte Übergänge zweier gespeicherter Ba^{+}-Ionen in einem bistabilen Fallenpotential", K. Abich, D. Reiss, A. Keil, P.E. Toschek, W. Neuhauser, Ch. Wunderlich. – Verhandl. DPG (VI) **37**, 150, Q 511.2 (2002).

    National Conference Report extending (15.).

19. "Spektroskopie des E2-Übergangs $S_{1/2}$-$D_{5/2}$ in Ba^{+}," A. Keil, K. Abich, D. Reiss, P.E. Toschek, Ch. Wunderlich, W. Neuhauser. – Verhandl. DPG (VI) **37**, 142, Q 433.14 (2002).

    Poster at a National Conference.

Acknowledgements

This work was also supported by the Hamburgische Wissenschaftliche Stiftung.

References

[1] A. Einstein, B. Podolsky, N. Rosen, Phys. Rev. **45**, 777 (1935).

[2] B. Appasamy, Y. Stalgies, P. E. Toschek, Phys. Rev. Lett. **80**, 2805 (1998).

[3] J. I. Cirac, P. Zoller, Phys. Rev. Lett. **74**, 4091 (1995).

[4] B. Appasamy, I. Siemers, Y. Stalgies, J. Eschner, R. Blatt, W. Neuhauser, P. E. Toschek, Appl. Phys. B **60**, 473 (1995).

[5] L.A. Khalfin, Pis'ma Zh. Eksp. Teor. Fiz. **8**, 106 (1968), [JETP Lett. **8**, 65 (1968)].

[6] L. Fonda, G.C. Ghirardi, A. Rimini, R. Weber, Nuovo Cimento A **15**, 689 (1973).

[7] B. Misra, E. C. G. Sudarshan, J. Math. Phys. (N.Y.) **18**, 756 (1977).

[8] V. B. Braginsky, F. Ya. Khalili, *Quantum Measurement*, Cambridge University Press, Cambridge, 1992.

[9] T. P. Spiller, Phys. Lett. A **192**, 163 (1994).

[10] O. Alter, Y. Yamamoto, Phys. Rev. A **55**, 2499 (1997).

[11] D. Home, M. A. B. Whitaker, Annals of Physics **258**, 237 (1997).

[12] Ch. Balzer, R. Huesmann, W. Neuhauser, P. E. Toschek, Optics Communic. **180**, 115 (2000).

[13] P. E. Toschek, Ch. Wunderlich, Eur. Phys. J. D **14**, 387 (2001).

[14] Ch. Balzer, Th. Hannemann, D. Reiss, W. Neuhauser, Ch. Wunderlich, P. E. Toschek, Laser Physics **12**, 729 (2002).

[15] R. Huesmann, Ch. Balzer, Ph. Courteille, W. Neuhauser, P. E. Toschek, Phys. Rev. Lett. **82**, 1611 (1999).

[16] See, e.g., U. & L. Fano, *Physics of Atoms and Molecules*, The University of Chicago Press, Chicago, 1970.

[17] R. Cook, Phys. Scripta T **21**, 49 (1988).

[18] W. M. Itano, D. J. Heinzen, J. J. Bollinger, D. J. Wineland, Phys. Rev. A **41**, 2295 (1990); *ibid.* **43**, 5168 (1991).

[19] H. Nakazato, M. Namiki, S. Pascazio, H. Rauch, Phys. Lett. A **217**, 203 (1996).

19 Spin Resonance with Trapped Ions: Experiments and New Concepts

K. Abich, Ch. Balzer, T. Hannemann, F. Mintert, W. Neuhauser, D. Reiß, P. E. Toschek, and Ch. Wunderlich

Universität Hamburg
Institut für Laser-Physik
Hamburg, Germany

19.1 Introduction

Confined in the field free center of an electrodynamic trap in ultra high vacuum, trapped ions provide us with individual localized quantum systems well isolated from the environment. The first preparation and detection of an individual ion was reported in [1] whereas storage and detection of a collection of ions was demonstrated even earlier in Paul and Penning traps [2]. Since then a large variety of intriguing experiments were carried out, for instance, the demonstration of optical cooling [3] and experiments related to a variety of fundamental physical questions (for instance [4].) Also, for precision measurements and frequency standards the use of trapped ions is well established. The great potential that trapped ions have as a physical system for quantum information processing (QIP) was first recognized in [5], and important experimental steps have been undertaken towards the realization of an elementary quantum computer with this system (for instance, [6–9].) At the same time, the advanced state of experiments with trapped ions reveals the difficulties that still have to be overcome.

Yb^+ and Ba^+ ions are well suited for experiments on QIP. Investigations on these ion species complement each other since they represent different types of qubits ($^{171}Yb^+$: Hyperfine transition; Ba^+ : Electric quadrupole transition). Sections 19.2 and 19.3 describe electron spin resonance experiments with individual trapped $^{171}Yb^+$ ions on basic issues of quantum mechanics that are relevant for QIP. These experiments also demonstrate the ability to prepare arbitrary states of the hyperfine qubit of individual $^{171}Yb^+$ ions with very high precision – a prerequisite for QIP. New theoretical concepts for ion trap experiments taking advantage of spin resonance techniques are outlined in Section 19.4. This new approach may lead to experiments combining advantageous features of two experimental 'worlds': spin resonance techniques (e.g. nuclear magnetic resonance), usually carried out with ensembles of molecules, can be applied to individual qubits in ion traps where spin–spin coupling is an adjustable parameter.

Cooling of the collective motion of several particles is another basic requirement for implementing conditional quantum dynamics with trapped ions. An experimental study of the collective vibrational motion of two trapped $^{138}Ba^+$ ions cooled by two light fields is described in Section 19.5.

Quantum Information Processing, 2nd Edition. Edited by Thomas Beth and Gerd Leuchs

ISBN: 3-527-40541-0

19.2 Self-learning Estimation of Quantum States

The concept of a quantum state is a central ingredient of quantum theory. Thus, the determination of an arbitrary state of a quantum system is an important task in quantum physics, and is particularly relevant in QIP and communication where quantum mechanical 2-state systems (qubits) are elementary constituents. The determination of the set of expectation values of observables associated with a specific quantum state is complicated by the fact that after a measurement of one observable, information on the complementary observable is no longer available. Only if infinitely many identical copies of a given state were available could this quantum state be characterized accurately. This requirement forbids the exact determination of an unknown arbitrary quantum state, and it is of interest to investigate ways to gain optimal knowledge of a given quantum state using finite resources.

Quantum states of various physical systems such as light fields, molecular wave packets, motional states of trapped ions and atomic beams have been determined experimentally (For a review of recent work see, for instance, [10].) With qubits, a first indication of the appropriate operations to be carried out (when two identical qubits are available) to gain maximal information was given in [11]. It was strongly suggested that optimal information gain is achieved when a suitable measurement on both particles together is performed. The suggestion in [11] that the optimal measurement for determining a quantum state of two identically prepared qubits needs to be carried out on both particles together, was proven in [12]. In more technical terms this means that the operator characterizing the measurement does not factorize into components that act in the Hilbert spaces of individual particles only. In [12] it was also shown that the same is true when $N = 1, 2, 3, \ldots$ identically prepared qubits are available, and the optimal fidelity that can be reached is $(N + 1)/(N + 2)$. Further theoretical work paved the way towards a possible experimental realization of optimal quantum state estimation by showing that finite positive operator valued measurements are sufficient for optimal state estimation [13]. However, the proposed strategies to read out information encoded in the quantum state of a given number N of identical qubits require intricate measurements using a basis of entangled states. In addition, all N qubits have to be available simultaneously. It is desirable to have a measurement strategy at hand that gives an estimate of a quantum state with high fidelity even if N measurements are performed sequentially on each individual qubit, that is, in a factorizing basis. The authors of [14] consider the problem of quantum state reconstruction of large ensembles of identically prepared quantum states using separate measurements on each particle.

Estimating a quantum state can also be viewed as the decoding procedure at the receiver end of a quantum channel necessary to recover elements of an alphabet that have been encoded in quantum states by a sender (see, for instance, [15].) Debugging of a quantum algorithm is another possible application for quantum state estimation [16]. Once a quantum algorithm has been implemented, it has to be tested, for instance by checking the state of a certain qubit in the course of the computation. In such a case the qubits are only available sequentially and efficient estimation is desirable, that is, a large overlap of the estimated state with the true one while keeping the number of repetitions of the algorithm as small as possible.

Quantum state estimation of qubits with fidelity close to the optimum (even with a moderate number of measurements) is possible when a self-learning algorithm is used [16]. When using this algorithm, N members of an ensemble of quantum systems identically prepared in a pure state,

$$|\psi\rangle = |\psi\rangle_1 \otimes |\psi\rangle_2 \otimes \ldots \otimes |\psi\rangle_N \,, \tag{19.1}$$

are measured individually, that is, they do not have to be available simultaneously. In other words, the measurement operator $\hat{M}$ employed to estimate the state can be written as a tensor product, and we have

$$\hat{M}|\psi\rangle = \hat{m}_1|\psi\rangle_1 \otimes \hat{m}_2|\psi\rangle_2 \otimes \ldots \otimes \hat{m}_N|\psi\rangle_N \,. \tag{19.2}$$

The operators $\hat{m}_n$ project onto the orthonormal basis states

$$|\theta_m^{(n)}, \phi_m^{(n)}\rangle = \cos\frac{\theta_m^{(n)}}{2}|0\rangle + \sin\frac{\theta_m^{(n)}}{2}e^{i\phi_m^{(n)}}|1\rangle \text{ and } |\pi - \theta_m^{(n)}, \pi + \phi_m^{(n)}\rangle \tag{19.3}$$

The first experimental realization of a self-learning measurement on an individual quantum system in order to estimate its state is reported in [9]. The projector $\hat{m}_n$ of measurement n is varied in real time during a sequence of N measurements conditioned on the results of all previous measurements $\hat{m}_l$ ($l < n \leq N$) in this sequence. The cost function that is optimized when proceeding from measurement n to $n+1$ is the fidelity of the estimated state after measurement $n+1$.

The $S_{1/2}$ ground-state hyperfine doublet of a single $^{171}Yb^+$ ion confined in a Paul trap represents the qubit (Figure 19.1). The

$$|0\rangle \equiv |S_{1/2}, F=0\rangle \leftrightarrow |S_{1/2}, F=1, m_F=0\rangle \equiv |1\rangle \tag{19.4}$$

transition is driven by a quasiresonant microwave (mw) field at 12.6 GHz. Since the system is virtually free of decoherence, arbitrary $SU(2)$ transformations can be performed by appropriate mw pulses. Resonance fluorescence scattered on the $S_{1/2}$(F=1) $\leftrightarrow$ $P_{1/2}$ transition driven by a laser at 369 nm serves for state-selective detection with efficiency $\eta > 97\%$. Optical pumping into the $D_{3/2}$ state is prevented by illuminating the ions with laser light near 935 nm driving the transition $|D_{3/2}, F{=}1\rangle \rightarrow |[3/2]_{1/2}\rangle$.

A prerequisite for self-learning estimation is the ability to prepare arbitrary states of this $SU(2)$ system with very high precision. The coherence time of the hyperfine qubit in $^{171}Yb^+$ is very long on the time scale of qubit operations and is essentially limited by the coherence time of microwave radiation used to drive the qubit transition. Figure 19.2a) shows the experimentally determined excitation probability (Rabi oscillations) of state $|1\rangle$ (single $^{171}Yb^+$ ion) as a function of the mw pulse length applied to the ion. The observed Rabi oscillations are free of decoherence over experimentally relevant time scales. However, the contrast of the oscillations is below unity, since in these experiments the initial state $|0\rangle$ was prepared with probability 0.9. This limitation will be addressed in future experiments. Figure 19.2b) displays data from a Ramsey-type experiment [17] where the ion precesses freely for a prescribed time between two subsequent mw $\pi/2$ pulses. This experimental signal, too, is essentially free of decoherence and the contrast of the Ramsey fringes is only limited by the finite preparation efficiency. The data in Figure 19.2 show that single-qubit operations are carried out with high precision, an important prerequisite for scalable quantum computing.

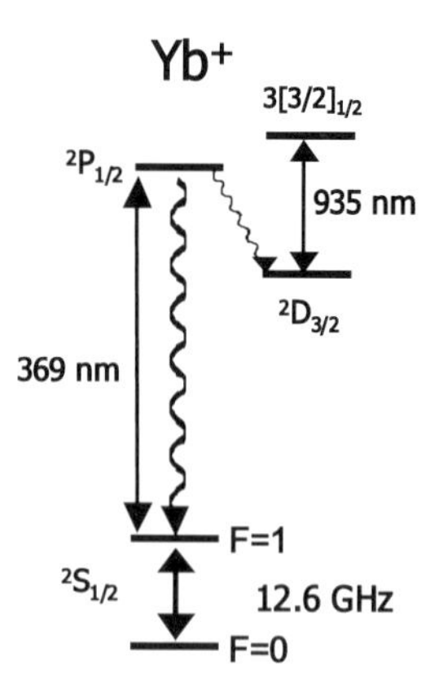

Figure 19.1: The hyperfine transition of individual $^{171}Yb^+$ ions $|S_{1/2}F = 1, m_F = 0\rangle \leftrightarrow |S_{1/2}F = 0,\rangle$ is coherently driven, while the optical resonance $|S_{1/2}F = 1\rangle \leftrightarrow |P_{1/2}F = 0\rangle$ serves for state selective detection by resonantly scattered light.

19.3 Experimental Realization of Quantum Channels

The well controlled generation of fragile quantum superposition states, their storage, and their transfer between different quantum systems is a prerequisite for any type of quantum information processing. The transfer of quantum information from atoms to electromagnetic fields is also an important issue, when trapped ions held in spatially separated traps act as qubits. Then the transfer of quantum information between different locations may be achieved by using an electromagnetic field as a mediator.

A change of a qubit state arises when it propagates in time and in space, that is, when it is transmitted through a quantum channel. A physical realization of a quantum channel could be, for instance, the transmission of quantum information encoded in the polarization of individual photons propagating through an optical fiber. The optimal reconstruction of quantum states has been the topic of experiments described in the previous section. Now we consider explicitly the influence of the environment on a quantum state once it has been prepared, that is, we investigate the influence of the quantum channel on the transmission of quantum information.

The state of a qubit is completely determined by the expectation values $\langle\sigma_x\rangle$, $\langle\sigma_y\rangle$, and $\langle\sigma_z\rangle$, and the density matrix describing its state can be written as

$$\rho = \frac{1}{2}(I + \vec{s} \cdot \vec{\sigma}) \tag{19.5}$$

where $\vec{s} \cdot \vec{\sigma} = \langle\sigma_x\rangle\sigma_x + \langle\sigma_y\rangle\sigma_y + \langle\sigma_z\rangle\sigma_z$, and $\sigma_{x,y,z}$ are the Pauli matrices.

In [18] it was shown that any quantum channel, that is, in our context, arbitrary dynamics of a two-level system can be cast in the form

$$\vec{s}^{\,\prime} = \hat{M}\vec{s} + \vec{v} \,. \tag{19.6}$$

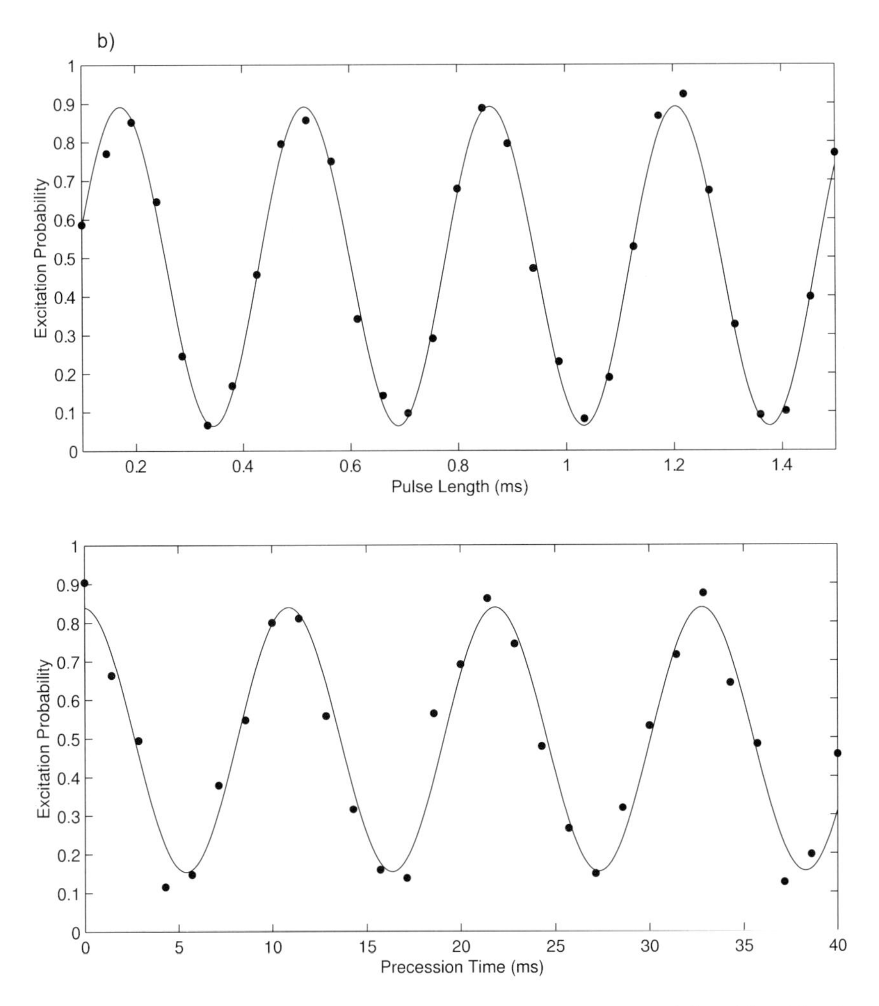

Figure 19.2: a) Rabi oscillations: Excitation probability of state $|1\rangle \equiv |S_{1/2}F = 0, m_F = 0\rangle$ of a single $^{171}\mathrm{Yb}^+$ ion averaged over 200 preparation-detection cycles as a function of mw pulse length t_{mw} (solid circles). The solid line results from a fit giving the Rabi frequency $\Omega_R = 2.9165 \times 2\pi$ kHz. The contrast of Rabi oscillations is below unity, since in these experiments the initial state $|0\rangle$ was prepared with probability 0.9. b) Data from a Ramsey-type experiment [17] where the ion undergoes free precession between two subsequent mw pulses (detuning $\delta = 91.3 \times 2\pi$Hz, averaged over 100 realizations.) This experimental signal, too, is essentially free of decoherence and the contrast of the so-called Ramsey fringes is limited only by the finite preparation efficiency. The data displayed in this figure show that single-qubit operations are carried out with high precision, an important prerequisite for scalable quantum computing.

where $\hat{M} \in \mathbb{R}^{3\times 3}$ and $\vec{v} \in \mathbb{R}^3$. Equation 19.6 yields $\vec{s}\,'$, the Bloch vector of the qubit after it has traversed the quantum channel characterized by $\hat{M}$ and $\vec{v}$. Various quantum channels have been realized experimentally and the coefficients

$$M_{ab} = 2P_{ab} - P_{az} - P_{a(-z)} \,, \qquad v_a = P_{az} + P_{a(-z)} - 1 \qquad (19.7)$$

have been determined by measuring the probabilities (or rather relative frequencies) $P_{ab} = \langle a| \rho' |b\rangle$, where ρ' is the density matrix describing the qubit state after the quantum channel, and $a, b \in \{x, y, z\}$. An explicit expression for the parameters determining $\hat{M}$ and $\vec{v}$ is given in [19].

Applying coherent and incoherent operations to the hyperfine qubit of $^{171}Yb^+$ we realized and completely characterized a polarization rotating quantum channel, a phase damping quantum channel acting in the xy−plane, and a phase damping quantum channel acting in an arbitrary plane [20]. Combinations of the aforementioned channels can also be realized. Incoherent disturbances to a quantum channel are realized by applying to the qubit small amounts of light close to 369 nm, thus inducing well-defined quantities of longitudinal and/or transversal relaxation during coherent microwave operations [21]. This light-induced decoherence is readily applicable to individually addressed quantum systems, it may be switched on and off immediately, and it is reproducible. Another way to produce a desired quantum channel is to apply a noisy magnetic field with well-defined spectral properties in conjunction with coherent microwave operations [20].

The properties of a quantum channel, that is, propagation of quantum information in time or space, has been investigated under the influence of well controlled disturbances, and the parameters characterizing the quantum channel can be adjusted at will. Thus a model system is realized to investigate, for example, the reconstruction of quantum information after transmission through a noisy quantum channel.

Transfer of quantum states becomes important when quantum information is distributed among different quantum processors, as is envisaged, for instance, for ion trap quantum information processing. Furthermore, codes for quantum information processing, and in particular error correction codes may be tested for their applicability under well defined, non-ideal conditions.

19.4 New Concepts for QIP with Trapped Ions

The extension of the experimental techniques described above to a collection of ions confined in a linear electrodynamic trap will be dealt with in this section. The distance between neighboring ions, δz is determined by the mutual Coulomb repulsion of the ions and the trapping potential characterized by the angular frequency, ν_1 of the center-of-mass (COM) vibrational mode of the ion string in the axial direction. Typically, δz is of the order of a few μm; for example, $\delta z \approx 7$ μm for $N = 10$ $^{171}Yb^+$ ions with $\nu_1 = 2\pi$ 100 kHz. In order to prepare qubits individually (single qubit operations), electromagnetic radiation is aimed at one ion at a time, that is, it must be focused to a spot size much smaller than δz. Therefore, *optical* radiation is usually required for individual addressing of qubits in ion traps [22].

In addition to the ability to perform arbitrary single-qubit operations, a second ingredient is required for QIP: conditional quantum dynamics with, at least, two qubits. Any quantum algorithm can then be synthesized using these elementary building blocks [23]. While two internal states of each trapped ion serve as a qubit, communication between these qubits necessary for conditional dynamics is achieved via the vibrational motion of the ion string as a whole [5, 6, 24, 25]. Therefore, it is necessary to couple external (motional) and internal degrees of freedom.

Common to all experiments – related either to QIP or other research fields – that require some kind of coupling between internal and external degrees of freedom of atoms is the use of optical radiation for this purpose. The absorption or emission of a photon by an ion is accompanied by the recoil of the ion with momentum $\hbar k$, where $k = 2\pi/\lambda$, and λ is the wavelength of the applied radiation field. This recoil may set in motion the ion confined in a harmonic trapping potential. The parameter determining the coupling strength between internal and motional dynamics is the Lamb–Dicke parameter (LDP)

$$\eta \equiv \sqrt{\frac{(\hbar k)^2}{2m}/\hbar\nu_1} = \frac{\Delta z_1\, 2\pi}{\lambda} \tag{19.8}$$

the square of which gives the ratio between the change in kinetic energy of the atom due to the absorption or emission of a photon and the quantized energy spacing of the harmonic trapping potential. The mass of the ion is denoted by m, and $(\Delta z_1)^2 = \hbar/2m\nu_1$ is the mean square deviation of the COM vibrational mode's ground state wave function in position space. Only if η is nonvanishing will the absorption or emission of photons be possibly accompanied by a change of the motional state of the atom. Trapping a $^{171}Yb^+$ ion, for example, with $\nu_1 = 2\pi \times 100$ kHz gives $\Delta z_1 \approx 17$nm and it is clear that driving radiation in the *optical* regime is necessary to couple internal and external dynamics of trapped ions.

19.4.1 Spin Resonance with Trapped Ions

It would be desirable for ion trap experiments to take advantage of the highly developed technological resources used in spin resonanc experiments (e.g. nuclear magnetic resonance.) In particular, employing microwave radiation with extremely long coherence time compared to optical radiation allows for precise and, on the time scale of typical experiments, virtually decoherence free manipulation of qubits. In what follows it is outlined how in a linear ion trap with an additional axial magnetic field gradient, $\partial_z B$, i) ions can be individually addressed in frequency space using *microwave* radiation, and, ii) the Hamiltonian governing the interaction between microwave radiation and ions in such a modified ion trap is formally identical with the Hamiltonian describing atom-light interaction in usual harmonic traps [26]. However, the usual LDP η is replaced by a new effective LDP $|\eta'| = \sqrt{\eta^2 + \epsilon^2}$. The coupling constant ϵ is proportional to $\partial_z B/\nu_1^{3/2}$ while η is negligibly small in the microwave domain [27].

In the last part of this section it is shown how the ion string can be treated like a molecule in nuclear magnetic resonance (NMR) experiments: conditional dynamics with ions can be carried out taking advantage of pairwise spin–spin coupling between individual qubits [28].

Individual addressing of qubits in a modified ion trap Applying a magnetic field gradient $\vec{B} = bz \cdot \hat{z} + B_0$ along the axial direction of a linear ion trap causes a z-dependent Zeeman shift of the internal ionic states $|0\rangle$ and $|1\rangle$. Thus the transition frequency $\omega_{01}^{(j)}$, $j = 1 \ldots N$, of each ion is individually shifted and the qubits can be addressed in frequency space. When separating the qubit resonance frequencies through the application of a magnetic field gradient, overlap between the motional sidebands of the qubit transitions has to be avoided. Therefore, the gradient has to be chosen such that

$$\delta\omega \geq 2\nu_N + \nu_1 \tag{19.9}$$

where $\delta\omega = \partial_z\omega_{01}\,\delta z$ is the frequency shift between two neighboring ions (compare Figure 19.3), and ν_N is the angular frequency of the highest axial vibrational mode.

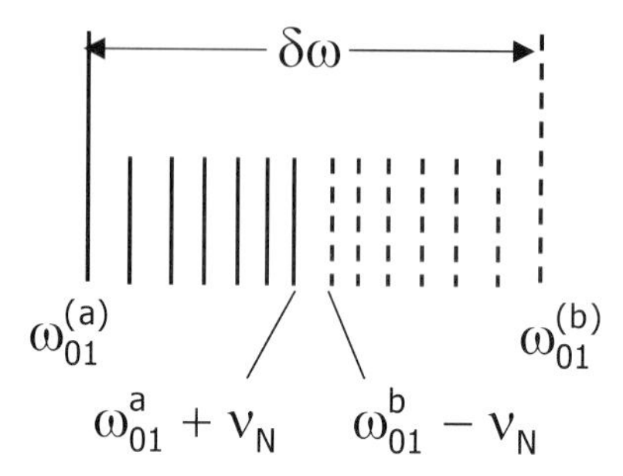

Figure 19.3: Schematic drawing of the resonances of qubits a and b with N accompanying sideband resonances. The angular frequency ν_N corresponds to the highest axial vibrational mode.

For example, $N = 10$ $^{171}Yb^+$ ions with $\nu_1 = 2\pi \times 100$ kHz require $\partial_z B \approx 10$ T/m [27]. Choosing the field gradient carefully ensures that even coincidences of second-order sideband resonances can be suppressed efficiently. A gradient having the necessary magnitude can be conveniently generated using a coil of 1 mm diameter (approximately the size of the ion traps employed in our experiments) with 3 windings and running a current of 3.3 A through them. This configuration produces a field gradient up to 20 T/m over the required distance. With additional coils the gradient can be modelled to have a desired spatial dependence. Small permanent magnets or microfabricated structures allow for the generation of gradients of a few hundred T/m.

Coupling internal and external dynamics In the previous paragraphs it was shown that a magnetic field gradient applied to a linear ion trap allows for individual addressing of ions in frequency space. In order to create entangled states between internal and motional degrees of freedom of one particular ion it is obviously necessary to couple internal and external dynamics. If microwave or radio-frequency radiation is used, the recoil on the ion upon absorption or emission of a photon is not sufficient to excite motional states of the ion (the LDP is vanishingly small.) However, in the presence of a magnetic field gradient, motional quanta can nevertheless be created or annihilated in conjunction with changing the internal state of an ion. The physical origin of this effect will be discussed in what follows.

The position dependent Zeeman shift gives rise to a force acting on the ion in addition to the electrodynamic and Coulomb potentials such that its equilibrium position is slightly different, depending on whether the ion resides in state $|0\rangle$ or $|1\rangle$. Consequently, if an electromagnetic field is applied to drive this qubit resonance, a transition between the two states $|0\rangle$

and $|1\rangle$ will be accompanied by a change of position of the ion,

$$z_0 = -\hbar \frac{\partial_z \omega_{01}}{m\nu^2} . \tag{19.10}$$

In phase space of the harmonic oscillator of the ionic vibration, this corresponds to a shift along the position coordinate occurring together with a shift along the momentum coordinate due to photon recoil. (The latter, however, is negligibly small in the microwave regime.) Thus, the oscillator will be excited. In [27] it is shown that the formal description of this coupling between internal and external dynamics is identical to the one used for the coupling induced by optical radiation. The usual LDP is replaced by an effective new parameter $\eta' \approx \epsilon_{nl}$ where

$$\epsilon_{nl} \equiv S_{nl} \frac{z_0^{(nl)}}{\Delta z_n} = S_{nl} \frac{\Delta z_n \partial_z \omega_{01}^{(l)}}{\nu_n} . \tag{19.11}$$

The strength of the coupling between an ion's internal dynamics and the motion of the ion string is different for each ion l and depends on the vibrational mode n: S_{nl} is the dimensionless coefficient of the unitary matrix that diagonalizes the dynamical matrix. It is a measure for how much ion l participates in the motion of vibrational mode n. The usual LDP η, too, contains this mode and ion dependent factor. The numerator on the rhs of Equation 19.11 contains the spatial derivative of the resonance frequency of qubit l times the extension Δz_n of the ground state wave function of mode n, that is, the variation of the internal transition frequency of qubit l when it is moved by a distance Δz_n. Thus, the coupling constant ϵ_{nl} is proportional to the ratio between this frequency variation and the vibrational frequency of mode n.

All optical schemes devised for conditional quantum dynamics with trapped ions can also be applied in the microwave regime, despite the negligible recoil associated with this type of radiation. This includes, for instance, the proposal presented in [5] that requires cooling to the motional ground state, and the proposals reported in [24] and [25] (the latter two work also with ions in thermal motion.)

Trapped ions as an N-qubit molecule The Hamiltonian describing a string of trapped two-level ions in a trap with axial magnetic field gradient (without additional radiation used to drive internal transitions) has been shown to read [28]

$$H = \frac{\hbar}{2} \sum_{n=1}^{N} \omega_n(z_{0,n}) \sigma_{z,n} + \sum_{n=1}^{N} \hbar \nu_n (a_n^\dagger a_n) - \frac{\hbar}{2} \sum_{n<l}^{N} J_{nl} \sigma_{z,n} \sigma_{z,l} . \tag{19.12}$$

The first sum on the rhs of 19.12 represents the internal energy of the collection of N ions where the qubit angular resonance frequency of ion n at its equilibrium position $z_{0,n}$ is denoted by ω_n. The second term sums the energy of N axial vibrational modes. These first two terms represent the usual Hamiltonian for a string of two-level ions confined in a harmonic potential. The new spin–spin coupling term (last sum in 19.12) arises due to the presence of the magnetic field gradient. Here,

$$J_{ab} \equiv \sum_{n=1}^{N} \nu_n \epsilon_{na} \epsilon_{nb} . \tag{19.13}$$

The pairwise coupling 19.13 between qubits a and b is mediated by the vibrational motion. Therefore, it contains terms quadratic in ϵ, and the coupling of qubit a and b to the vibrational motion has to be summed over all modes.

In a "real" molecule different nuclear spins share binding electrons that generate a magnetic field at the location of the nuclei, and the energy of a nuclear spin exposed to the electrons' magnetic field depends on the charge distribution of the binding electrons. If a particular nuclear spin is flipped, the interaction with the surrounding electrons will slightly change the electrons' charge distribution which in turn may affect the energy of other nuclear spins. This indirect spin–spin coupling is realized here in a different way: the role of the binding electrons is replaced by the vibrational motion of the ions.

Usual ion trap schemes take advantage of motional sidebands that accompany qubit transitions. Instead, the spin–spin coupling that arises in a suitably modified trap may be directly used to implement conditional dynamics using NMR methods. The collection of trapped ions can thus be viewed as an N-qubit molecule with adjustable coupling constants [28]. Making use of this spin–spin coupling does not involve real excitation of vibrational motion. In this sense it is similar to a scheme for conditional quantum dynamics that uses optical 2-photon transitions detuned from vibrational resonances [24], and, thus should be tolerant against thermal motion of the ions.

Many phenomena that were only recently studied in the optical domain form the basis for techniques belonging to the standard repertoire of coherent manipulation of nuclear and electronic magnetic moments associated with their spins. NMR experiments have been tremendously successful in the field of QIP taking advantage of highly sophisticated experimental techniques [29]. NMR experiments usually work with macroscopic ensembles of spins and considerable effort has to be devoted to the preparation of pseudo-pure states of spins with initial thermal population distribution. This preparation leads to exponentially growing cost (with the number N of qubits) either in signal strength or the number of experiments involved, since the fraction of spins in their ground state is proportional to $N/2^N$. Trapped ions, on the other hand, provide individual qubits – for example hyperfine states as described in this work – well isolated from their environment. It would be desirable to combine the advantages of trapped ions and NMR techniques in future experiments using either "conventional" ion trap methods, but now with microwave radiation as outlined above, or, as described in 19.4.1, treating the ion string as a N-qubit molecule with adjustable coupling constants.

19.4.2 Simultaneous Cooling of Axial Vibrational Modes

Vibrational modes of a collection of ions serve as bus-qubits and allow for the implementation of conditional quantum dynamics. Thus, cooling of the ions' motional degrees of freedom is necessary for QIP, and it is important to have cooling techniques that are suitable to reach low vibrational excitation.

For usual ion trap quantum computing it is desirable to cool not only the vibrational mode used as a bus-qubit, but also other modes that take no active role as bus-qubits ("spectator modes"), since the Rabi frequency for transitions between internal states of the ions depends on the motional state of all modes. Therefore, the increasing number of vibrational modes that have to be cooled represents a serious obstacle on the way towards using a larger number of ions for QIP.

Excitation on the so-called red sideband transition together with subsequent spontaneous emission at the transition frequency ω_{01} will reduce the kinetic energy of a trapped atom (sideband cooling). Applying sideband cooling sequentially to each of the vibrational modes will leave little time for quantum logic operations between cooling cycles. With many ions such sequential cooling might not work at all, since after having cooled the last one of N axial modes, the first one may already be considerably affected by heating. Therefore it is desirable to find new methods that allow for simultaneous and efficient cooling of multiple vibrational modes of many ions.

A new method for simultaneous sideband cooling of axial vibrational modes of ions confined in a linear trap is described in [30]: A magnetic field gradient applied to an electrodynamic ion trap can be designed such that all first order red sidebands of all axial vibrational modes nearly coincide, and thus can all be sideband cooled by irradiating the ion string with electromagnetic radiation only at a single frequency. Either laser light or microwave radiation can be used for sideband excitation to implement this method for simultaneous cooling of many vibrational modes. This cooling method is compatible with the new concept for ion trap QIP outlined above [30].

19.5 Raman Cooling of two Trapped Ions

Another avenue towards quantum computation with trapped ions is the use of an electric quadrupole transition (E2-transition) as a qubit [7, 8, 31]. We carry out experiments with Ba^+ ions where the E2-transition between the ground state $S_{1/2}$ and the metastable excited $D_{5/2}$ state is investigated.

Cooling of the collective motion of several particles is prerequisite for implementing conditional quantum dynamics on trapped ions. A study of the collective vibrational motion of two trapped $^{138}Ba^+$ ions cooled by two light fields close to the resonances corresponding to the $S_{1/2}$-$P_{1/2}$ (493 nm, green light) and $P_{1/2}$-$D_{3/2}$ (649 nm, red light) transitions is described in [33]. These light fields serve at the same time for state selective scattering of resonance fluorescence.

When two ions are confined in a nearly spherically symmetric Paul trap, and if they are sufficiently laser cooled, then we always observe them crystallizing [32] at the same locations (Figure 19.4.) The crystallization at preferred locations is explained by the slight asymmetry of the effective trapping potential, that is, $\nu_x \neq \nu_y \neq \nu_z$, where $\nu_{x,y,z}$ are the angular frequencies of the center-of-mass-mode of the secular motion in different spatial directions. The Coulomb potential makes the ions repel each other, and the ion crystal tends to align along the axis of weakest confinement by the electrodynamic potential.

The potential along the $z-$direction is steeper than in the xy-plane. Consequently, if cooled well enough, the ions stay in this plane. Since $\nu_y > \nu_x$ they are not free to rotate in the xy-plane either. Instead they have to surmount a potential barrier along the azimuthal coordinate in order to exchange places (figure 1 in [33]). This is only possible, if the vibrational energy of the relative motion of the two ions in the $\tilde{y}$-mode [33] exceeds the azimuthal barrier height. Depending on the parameter settings (detuning and intensity) of the cooling lasers, the motional states 'fixed' and 'rotating' corresponding to different temperatures are indeed observed experimentally (Figure 19.4.)

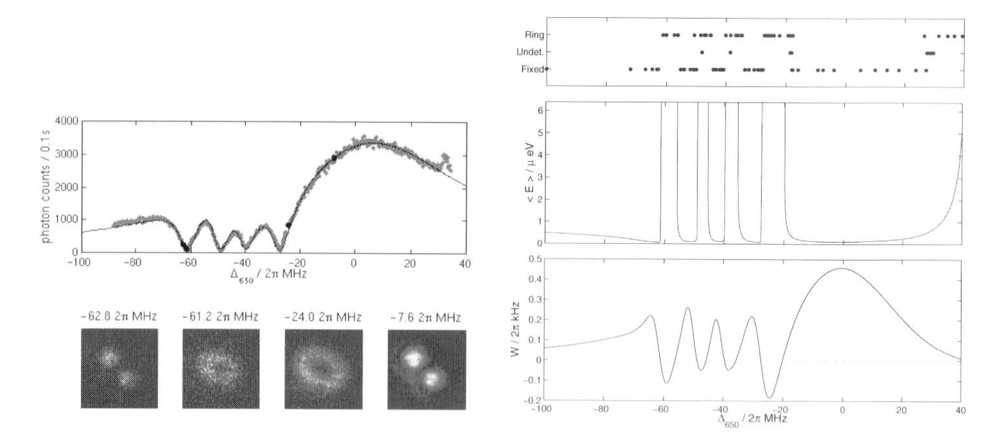

Figure 19.4: Left top: Fluorescence of two trapped ions as a function of laser detuning. Left bottom: Two trapped Ba^+ ions show different motional states depending on laser parameters. Spatial distribution of the two ions at the detunings indicated above. Right top: Observed motional states for different detunings of the 650 nm light. The dots correspond to individual observations. Right middle: Mean motional energy in the $\tilde{y}$-mode calculated from theory. Right bottom: Cooling rate for the $\tilde{y}$-mode calculated from theory. Figure taken from [33].

It turns out that the transition from an ion crystal to the ring structure occurs at that detunings of the red laser where theory predicts laser cooling to turn into heating. The ions gain enough energy from scattering photons to surmount the azimuthal potential barrier and appear as a ring on the spatially resolving photo detector. The transition from cooling to heating occurs when the red laser is scanned across a dark resonance $S_{1/2}(m)$-$D_{3/2}(m')$ with frequency increasing: as soon as it becomes blue detuned with respect to the closest dark resonance, the cooling rate vanishes and even turns negative (i.e, heating occurs.) Increasing the laser's frequency even more means that the red laser is further blue detuned with respect to the dark resonance that was just passed. At the same time, however, the next resonance is approached relative to which the laser is red detuned and the cooling rate increases again. It should be noted that Raman scattering responsible for these processes occurs when both lasers are *red* detuned relative to the main resonance.

Very good agreement is found between the theoretical prediction of the transition of the ions' motional state and experimental observations. In addition, parameter regimes of the laser light irradiating the ions are identified that imply most efficient laser cooling and are least susceptible to drifts, fluctuations, and uncertainties in laser parameters. Cooling of all vibrational modes of two ions is achieved with suitable parameters as, for instance, in figure 5 in [33]. When applied to cooling of a string of ions in a linear trap, the multidimensional parameter space allows one to identify regions where cooling is most efficient for all modes.

We gratefully acknowledge support by the Deutsche Forschungsgemeinschaft, the Bundesministerium für Bildung und Forschung, and the Quantum Information Processing Network of Excellence.

References

[1] W. Neuhauser, M. Hohenstatt, P. E. Toschek, and H. G. Dehmelt, Phys. Rev. A **22**, 1137 (1980).

[2] E. Fischer, Z. Physik **156**, 1 (1959); D. Church and H. Dehmelt, J. Appl. Phys. **40**, 3421 (1969); R. Ifflander and G. Werth, Opt. Comm. **21**, 411 (1977).

[3] W. Neuhauser, M. Hohenstatt, P. E. Toschek, and H. G. Dehmelt, Phys. Rev. Lett. **41**, 233 (1978); D. J. Wineland, R. E. Drullinger, and F. L. Walls, Phys. Rev. Lett. **40**, 1639 (1978).

[4] F. Diedrich and H. Walther, Phys. Rev. Lett. **58**, 203 (1987); M. Schubert, I. Siemers, R. Blatt, W. Neuhauser, and P. Toschek, Phys. Rev. Lett. **68**, 3016 (1992).

[5] J. I. Cirac and P. Zoller, Phys. Rev. Lett. **74**, 4091 (1995).

[6] C. Monroe, D. M. Meekhof, B. E. King, W. M. Itano, and D. J. Wineland, Phys. Rev. Lett. **75**, 4714 (1995).

[7] B. Appasamy, Y. Stalgies, and P. E. Toschek, Phys. Rev. Lett. **80**, 2805 (1998).

[8] Ch. Roos *et al.*, Phys. Rev. Lett. **83**, 4713 (1999).

[9] T. Hannemann, D. Reiß, C. Balzer, W. Neuhauser, P. E. Toschek, and Ch. Wunderlich, Phys. Rev. A **65**, 050303(R)/1 (2002).

[10] M. Freyberger, P. Bardroff, C. Leichtle, G. Schrade, and W. Schleich, Phys. World **10**, 41 (1997); V. Bužek, R. Derka, G. Adam, and P. L. Knight, Ann. Phys. **266**, 454 (1998); I. A. Walmsley and L. Waxer, J. Phys. B **31**, 1825 (1998).

[11] A. Peres and W. Wootters, Phys. Rev. Lett. **66**, 1119 (1991).

[12] S. Massar and S. Popescu, Phys. Rev. Lett. **74**, 1259 (1995).

[13] R. Derka, V. Bužek, and A. K. Ekert, Phys. Rev. Lett. **80**, 1571 (1998); J. I. Latorre, P. Pascual, and R. Tarrach, Phys. Rev. Lett. **81**, 1351 (1998).

[14] R. D. Gill and S. Massar, Phys. Rev. A **61**, 042312 (2000).

[15] K. R. W. Jones, Phys. Rev. A **50**, 3682 (1994); E. Bagan, M. Baig, A. Brey, and R. Munoz-Tapia, Phys. Rev. Lett. **85**, 5230 (2000); A. Peres and P. F. Scudo, Phys. Rev. Lett. **86**, 4160 (2001).

[16] D. G. Fischer, S. H. Kienle, and M. Freyberger, Phys. Rev. A **61**, 032306 (2000).

[17] N. F. Ramsey, *Molecular Beams* (Oxford University Press, Oxford, 1956).

[18] A. Fujiwara and P. Algoet, Phys. Rev. A **59**, 3290 (1999).

[19] M. A. Cirone *et al.*, arXiv:quant-ph/0108037 (2001).

[20] T. Hannemann, D. Reiß, Ch. Balzer, P. E. Toschek, W. Neuhauser, and Ch. Wunderlich, in preparation .

[21] Ch. Balzer, T. Hannemann, D. Reiß, W. Neuhauser, P. E. Toschek, and Ch. Wunderlich, in *Laser Physics at the Limit* (Springer Verlag, Heidelberg-Berlin-New York, 2001), pp. 233–241; Ch. Balzer, T. Hannemann, D. Reiß, W. Neuhauser, P. E. Toschek, and Ch. Wunderlich, Laser Physics **12**, 729 (2002).

[22] H. C. Nägerl *et al.*, Phys. Rev. A **60**, 145 (1999).

[23] D. P. DiVincenzo, Phys. Rev. A **51**, 1015 (1995); A. Barenco *et al.*, Phys. Rev. A **52**, 3457 (1995).

[24] A. Sorensen and K. Molmer, Phys. Rev. A **62**, 022311/1 (2000).

[25] D. Jonathan, M. B. Plenio, and P. L. Knight, Phys. Rev. A **62**, 042307 (2000).

[26] W. Vogel, D.G. Welsch, S. Wallentowitz, *Quantum Optics. An Introduction* (Wiley-VCH, Berlin, 2001), Chap. 12.

[27] F. Mintert and Ch. Wunderlich, Phys. Rev. Lett. **87**, 257904 (2001).

[28] Ch. Wunderlich, in *Laser Physics at the Limit* (Springer Verlag, Heidelberg-Berlin-New York, 2001), pp. 261–271.

[29] L. M. K. Vandersypen *et al.*, Nature **414**, 883 (2001).

[30] D. Reiß, G. Morigi, and Ch. Wunderlich, in preparation.

[31] P. A. Barton *et al.* Phys. Rev. A **62**, 032503 (2000).

[32] Th. Sauter, H. Gilhaus, I. Siemers, R. Blatt, W. Neuhauser, and P. E. Toschek, Z. Phys. D **10**, 153 (1988).

[33] D. Reiß, K. Abich, W. Neuhauser, Ch. Wunderlich, and P. E. Toschek, Phys. Rev. A **65**, 053401/1 (2002).

20 Controlled Single Neutral Atoms as Qubits

V. Gomer, W. Alt, S. Kuhr, D. Schrader, and D. Meschede

Institut für Angewandte Physik
Universität Bonn
Bonn, Germany

Recent experimental progress in trapping, observation and manipulation of single neutral atoms for purposes of quantum information processing is presented.

20.1 Introduction

The hardware requirements for a quantum information processor (QIP) [1] are clearly contradictory: On the one hand, nearly perfect isolation from environment is demanded to insure pure initial quantum states and preservation of their coherent character. On the other hand, to perform a quantum logical operation a strong interaction with environment has to be switched on and off at will. These stringent requirements rule out most known physical systems. The most natural candidates come from quantum optics where well-isolated individual atoms and photons are manipulated in a controlled environment with well-understood couplings [2, 3].

The experimentally most advanced implementation of a QIP using pure quantum states of single atomic particles is based on trapped ions [4–9]. There have been several recent proposals using trapped *neutral* atoms as quantum memory elements. In most of these proposals the quibits are stored in the long-lived hyperfine states of distinguishable trapped atoms. As an alternative to ions, neutral atoms enjoy even weaker coupling to the environment facilitating long relaxation and coherence times. The price for this attractive property is practically zero intrinsic interaction between neutral atoms.

To perform two-qubit gates with neutral atoms different physical mechanisms have been proposed: controlled cold collisions [10–13, 46, 52], induced dipole–dipole interaction [14–17], magnetic spin–spin interaction [18] and cavity quantum electrodynamics (CQED) [9, 19–30].

Most schemes proposed are still far beyond the present experimental possibilities. One common crucial issue of all proposals is the fact that controlled coherent interaction assumes experimentally controlled precise positioning of individual atoms. To illustrate this point let us consider the case of CQED-based QIP.

20.2 Cavity QED for QIP

The atom–cavity system seems to be a "natural" candidate for QIP implementation. This strongly interacting system intrinsically entangles atomic degrees of freedom with the quan-

Quantum Information Processing, 2nd Edition. Edited by Thomas Beth and Gerd Leuchs

ISBN: 3-527-40541-0

tum field of the cavity in the absence of dissipation. The cavity field serves as a quantum information bus between the two atoms, and the atoms interact by "sharing" a single photon field of the cavity. In microwave regime quantum engineering of entangled states and elementary quantum gates with highly-excited Rydberg atoms have been already demonstrated [31–33].

The advantage of the *optical* regime is the possibility to detect and further utilize the light leaving the cavity at the level of single photons. This promises not only an implementation of single-photon sources for quantum cryptography [30, 34, 35] but also the use of single photons for the exchange of quantum information between distant nodes of a quantum network consisting of small-scale quantum computers [36–38]. Actually this is the only proposed physical system with the ability to interconvert "stationary and flying qubits" [1]. Another advantage of near infrared or optical frequencies is the possibility to selectively address individual qubits. It is also important to note that the optical cavity field is typically in the vacuum state and experiences non-zero excitation only during the gate operation itself – no coherence of the cavity field is required from gate to gate.

All theoretical proposals of quantum gates with atoms and cavities require strong atom–cavity coupling. It means that the one-photon Rabi frequency must be larger than all other characteristic frequencies in the system: the damping rate of the cavity, the spontaneous decay rate of the atom and the reciprocal interaction time between the atom and the cavity. These requirements leads to the necessity of cavity mirrors with the highest available reflectivity placed at unpleasantly small separations of 100 μm or less and the use of slow laser-cooled atoms [39–42].

Because of the spatial structure of the light field inside the cavity, the coupling parameter is strongly position dependent. Since theoretical proposals for quantum gates assume good control of the atom–cavity coupling, storing an atom in an antinode of the cavity mode, where this parameter is maximum, would be the ideal scheme in order to allow a full-control over the coupling. However, the state-of-the-art for putting atoms into such a tiny cavity is to use atoms released from a magneto–optical trap (MOT) falling down [39, 41] or launched towards the cavity [40, 42]. In this situation atoms enter the cavity at random times and, even worse, at random positions. Obviously, the present experimental challenge of optical CQED is to get a good control of the atomic motion within or nearby optical resonators. This is why many experimentalists have been dreaming about the ability at the very beginning to take exactly, let us say 2 atoms and to put them into the cavity in a controlled way.

20.3 Single Atom Controlled Manipulation

One possible scenario is the preparation of an exact known number of atoms in a certain quantum state and their controlled transportation into the cavity to perform a quantum gate. Such a "qubit conveyor belt" on which qubits can be initialized and then brought to the region where active computation is taking place is not restricted to CQED systems only but is also of more general interest for any QIP [1] and is similar to the current trend in architecture of ion-based large-scale QIP with separated memory and interaction regions [43].

Note that motional control of individual atomic particles has been already perfected in the field of ion trapping, see [9, 44, 45]. However, many experimental difficulties arise, because ion traps do not easily fit within the required very small cavity volumes, and strong optical transitions in the near infrared, which are convenient for optical CQED, are not readily available with ions.

Neutral laser-cooled atoms can be trapped and manipulated using magnetic [46] or optical dipole forces [47]. The latest can show up in various forms from a single focused laser beam [48,49], standing waves [50], optical lattices [51] or dipole microtrap arrays [52].

By using far-red-detuned optical dipole traps we show here how to position neutral atoms in space fast and with high precision. The main properties of the stored atoms and possibilities to manipulate them are characterized. Our efforts here focus on the development of experimental techniques for trapping, observation and manipulation of *single* neutral atoms using laser fields.

20.4 How to Prepare Exactly 2 Atoms in a Dipole Trap?

We start with an exactly known atom number in a MOT which is then transferred into a dipole trap with 100% efficiency. Observation of single atoms in a MOT has been pioneered in [53]. The MOT is loaded from the background Cs vapour with a very low pressure. The high magnetic field gradient of 400 G/cm localizes the trapped atoms to a region of diameter 30 μm and also provides a low loading rate of 2 atoms/min only. This ensures that accidental loading events during the measurement procedure are negligible. We speed up the loading process by temporarily lowering the magnetic field gradient to 40 G/cm. This results in a larger capture cross section which significantly increases the loading rate [54]. Finally the field gradient is returned to its initial value, effectively freezing the number of the atoms in the trap for many seconds. Varying the time during which the field gradient is low enables us to select a specific *mean* atom number. This system could easily deliver one atom within 100 ms. A home-designed imaging optics [55] allows us to detect $8 \cdot 10^4$ fluorescence photons per second from a single atom stored in a MOT and to determine the exact number of atoms within a millisecond. Now this exactly known number of atoms can be easily transferred between the MOT and the dipole trap many times without loosing the atoms [48, 50]. A prerequisite for 100% transfer is a thorough alignment of the dipole trap laser onto the MOT. As a sensitive alignment criterion we use the fact that the dipole trapping laser shifts the atomic transition out of resonance, which lowers the fluorescence rate of the MOT. The dipole trap laser is therefore superposed with the MOT by minimizing the fluorescence rate of a single trapped atom [56]. To transfer cold atoms from the MOT into the dipole trap, both traps are simultaneously operated for several milliseconds before we switch off the MOT [48].

Another way to load one single atom into a dipole trap has been recently demonstrated in [49] by overlapping a standard MOT with thousands of atoms with an optical dipole trap of submicrometre size. Because of the extremely small trapping volume, only one atom can be loaded at a time.

20.5 Optical Dipole Trap

Our dipole trap consists of two counter-propagating Gaussian laser beams with equal intensities and frequencies [50, 56]. Both dipole trap laser beams are derived from a single Nd:YAG laser ($\lambda = 1064$ nm) which is far red detuned from the $6S_{1/2} \rightarrow 6P_{1/2,3/2}$ transitions of cesium ($\lambda_{D1} = 894$ nm, $\lambda_{D2} = 852$ nm). The laser beams have parallel linear polarization and thus produce a standing wave interference pattern. For a total power of 4 W and a beam waist of 30 μm, the potential depth corresponds to 1.3 mK.

The maximum photon scattering rate amounts to 15 photons/s for our parameters. This scattering rate yields a recoil heating rate of about 2 μK/s which is negligible in our experiment. Neither is technical heating due to intensity fluctuations and pointing instabilities of the trapping laser beams as discussed in detail in [57] observable in this experiment [58]. The measured trap lifetime of 25 s is rather limited by background collisions [50, 56].

The temperature of the atoms in the dipole trap is similar to the temperatures in our high-gradient MOT [59] as measured by observation atom losses during adiabatic reduction of the trap depth [58]. We therefore conclude that the MOT effectively cools the atoms into the dipole trap to about Doppler temperature, in contrast to the traps used in our previous experiments [48], which had a 10 times higher trap depth.

The axial oscillation frequency of 330 kHz has been measured by resonant and parametric excitation of the oscillatory motion of a single atom in the dipole trap [58]. The measured temperature of 0.2 mK and oscillation frequency indicate a mean oscillatory quantum number of 13.

20.6 Relaxation and Decoherence

To perform single-qubit operations in this system is relatively simple: the setting of a qubit state, in our case one of both hyperfine sublevels $F = 3, 4$ of the cesium ground state, is realized by the method of optical pumping. Here, only a few photons need to be scattered before we can be confident the system has relaxed. When we wish to couple resonantly to the hyperfine splitting we can use both microwave radiation directly or two-photon Raman transitions. The advantage of an optical Raman transition is the possibility to selectively address individual atoms and, additionally to use the same Raman laser pulses for cooling.

As an example we show in Fig. 20.1a Raman spectrum recorded in copropagating geometry at a magnetic field of 0.4 G. We do Raman spectroscopy by preparing the atoms in the $F = 4$ hyperfine state, by applying a Raman pulse, and by counting the number of atoms in $F = 3$. For this purpose, a strong and short laser pulse selectively pushes the atoms in $F = 4$ out of the trap. Finally, we use the MOT as a 100% efficiency single atom detector. Note that this simple but robust technique represents a nearly 100% efficiency state-selective detection of a single atom. For copropagating Raman laser beams one does not expected to observe vibrational sidebands in the spectrum. Note, however, that the Zeeman structure of the lowest qubit state $F = 3$ is clearly resolved. We currently use such Raman spectroscopy to learn more about relaxation mechanisms in our system.

The main characteristics of an quantum memory element are relaxation and coherence times of its qubit states. The hyperfine state relaxation rate due to spontaneous Raman scattering from the trapping laser is two orders of magnitude smaller than the total photon scattering rate [60]. Indeed, we measured hyperfine state lifetimes on the order of several seconds [48].

However, the most interesting question in respect to possible applications of trapped neutral atoms for QIP is how long (or how short) is the lifetime of a coherent superposition of

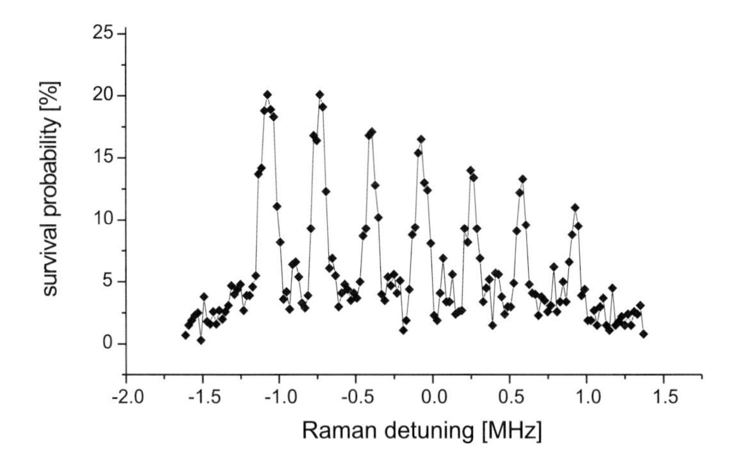

Figure 20.1: Raman spectroscopy in copropagating geometry shows the resolved Zeeman structure of the $F = 3$ state at a magnetic field 0.4 G. Every point is an average over about 150 atoms.

qubit states in this system? The destructive interference of Raman scattering amplitudes [60] leading to long relaxation times promises similarly long intrinsic coherence times in this system. As expected, first measurements clearly show a strong dependency of decoherence on the dipole trap depth and indicate that the coherence time can easily exceed 100 ms.

The main *inhomogeneous* broadening mechanism here originates from the dephasing due to differential light shift in both hyperfine states [61] and thermal motion of the atom in the dipole potential. Note that all results presented here were obtained without any additional cooling of atoms inside the dipole trap. We just get for free the Doppler temperature from the MOT only and the need for an additional cooling is obvious. High measured oscillation frequency of the trapped atoms is a perfect prerequisite for the resolved sideband Raman cooling methods, that can control the motion of trapped atoms extremely well. Several groups have laser-cooled atoms to the lowest bound state in lattice wells [62–64] and controlled their localized quantum wavepacket states [65–68].

Together with state selective detection this will allow Raman cooling to the oscillatory ground state on the level of a single atom. In this case the localization of an atom in the "softest" radial direction of the dipole trap (corresponding, however, to the most critical axial direction of the cavity) will be about 100 nm. The corresponding variation of the strength of the cavity coupling will not exceed 10%, an already acceptable value if one uses some kind of the rapid adiabatic passage technique during atom–cavity interaction.

Any additional cooling of stored atoms will also improve the performance of our optical conveyor belt.

20.7 Qubit Conveyor Belt

An atom initially trapped in the stationary standing wave dipole trap can be moved along the optical axis by changing the frequency difference of the two counter-propagating laser beams contributing to the standing-wave dipole trap. This causes the potential wells to move at a certain velocity. To control the frequency difference, both dipole trap laser beams pass through acousto-optical modulators (AOMs) [56]. While both AOMs are driven with the same frequency the standing wave pattern is at rest and atoms can be loaded into the dipole trap. To accelerate them along the dipole trap axis one of the AOMs is driven by a phase-continuous frequency ramp. In a similar fashion they can be decelerated and brought to a stop at a predetermined position along the standing wave [56].

Using this optical conveyor belt we have realized controlled transport of a single or any desired small number of neutral atoms over a distance of a centimeter with sub-micrometer precision [50]. The maximum achieved acceleration allows us to move an atom by 6 mm in 2 ms only [56]. To comparison, the distance between the MOT and the cavity at current Caltech experiments is 5 mm [39,41].

20.8 Outlook

In summary, we have realized a system compatible with most of the requirements for a CQED-based QIP: individual neutral atoms stored in a dipole trap with long relaxation and coherence times. One-qubit gate operations and a reliable read-out seem to be feasible and the optical conveyor belt promises controlled positioning of the atoms inside the cavity.

Future work will include the demonstration of strong state-dependent conditional atom–cavity coupling, keeping the atom inside the cavity. Then one needs to learn how to transfer photons back and forth between two atoms in the same cavity. One simple possible method to realize it is to multiply load the conveyor belt with single atoms with well defined separations. Using more complicated field configurations giving full 3D control of trapped atoms is also conceivable [69]. A more advanced approach is to use two conveyor belts (don't tell this to our phd students!) allowing to independently position and entangle any two atoms inside the cavity. The main challenge will be to combine all these different techniques in a single experiment.

Acknowledgement

We would like to thank Bernd Ueberholz, Daniel Frese and Wade Smith for their early contributions to this work, and Igor Dotsenko and Yevhen Miroshnychenko for their experimental assistance to the Raman spectroscopy and useful discussions in preparing the manuscript. We have received support from the *Schwerpunktprogramm Quanten-Informationsverarbeitung*, SPP 1078 of the Deutsche Forschungsgemeinschaft and the state of Nordrhein-Westfalen.

References

[1] D. DiVincenzo, The physical implementation of quantum computation, Fortschr. Phys. **48**, 771-783 (2000)

[2] Special issue on experimental proposals for quantum computation, Fortschr. Phys. **48**, (2000)

[3] C. Monroe, Quantum information processing with atoms and photons, Nature **416**, 238 (2002)

[4] J.I. Cirac and P. Zoller, Quantum Computations with Cold Trapped Ions, Phys. Rev. Lett. **74**, 4091 (1995)

[5] C. Monroe, D.M. Meekhof, B.E. King, W.M. Itano, and D.J. Wineland, Demonstration of a fundamental logic gate, Phys. Rev. Lett. **75**, 4714 (1995)

[6] A. Steane, The ion trap quantum information processor, Appl. Phys. **B64**, 623-642, (1997)

[7] D.J. Wineland, C. Monroe, W.M. Itano, D. Leibfried, B.E. King, D.M. Meekhof, Experimental issues in coherent quantum-state manipulation of trapped atomic ions, J. Res. Natl. Inst. Stand. Tech. **103**, 259-328 (1998)

[8] C.A. Sackett, D. Kielpinski, B.E. King, C. Langer, V. Meyer, C.J. Myatt, M. Rowe, Q.A. Turchette, W.M. Itano, D.J. Wineland, and C. Monroe, Experimental entanglement of four particles, Nature **404**, 256 (2000)

[9] M. Keller, B. Lange, W. Lange, and H. Walther; Quantum Information Processing with Ions Deterministically Coupled to an Optical Cavity, in "Quantum Information Processing", G. Leuchs and T. Beth (eds.), Wiley-VCH, Weinheim 2003.

[10] D. Jaksch, H.-J. Briegel, J.I. Cirac, C.W. Gardiner, and P. Zoller, Entanglement of atoms via cold controlled collisions, Phys. Rev. Lett. **82**, 1975 (1999)

[11] T. Calarco, E.A. Hinds, D. Jaksch, J. Schmiedmayer, J.I. Cirac, and P. Zoller, Quantum gates with neutral atoms: Controlling collisional interactions in time-dependent traps, Phys. Rev. **A61**, 022304 (2000)

[12] K. Eckert, J. Mompart, X.X. Yi, J. Schliemann, D. Bruß, G. Birkl, and M. Lewenstein, Quantum computing in optical microtraps based on the motional states of neutral atoms, quant-ph/0206096

[13] G. Cennini, G. Ritt, C. Geckeler, R. Scheunemann, and M. Weitz; Towards Quantum Logic with cold atoms in a CO2-Laser Optical Lattice, in "Quantum Information Processing", G. Leuchs and T. Beth (eds.), Wiley-VCH, Weinheim 2003.

[14] G.K. Brennen, C.M. Caves, P.S. Jessen, and I.H. Deutsch, Quantum Logic Gates in Optical Lattices, Phys. Rev. Lett. **82**, 1060 (1999)

[15] D. Jaksch, J.I. Cirac, and P. Zoller, S.L. Rolston, R. Cote and M.D. Lukin, Fast Quantum Gates for Neutral Atoms, Phys. Rev. Lett. **85**,2208 (2000)

[16] I.V. Bargatin, B.A. Grishanin, and V.N. Zadkov, Analysis of radiatively stable entanglement in a system of two dipole-interacting three-level atoms, Phys. Rev. **A61** 052305 (2000); I.V. Bargatin, B.A. Grishanin, V.N. Zadkov, Generation of entanglement in a system of two dipole-interacting atoms by means of laser pulses, Fortschr. Phys. Vol. **48**, 637 (2000); quant-ph/9903056

[17] I.E. Protsenko, G. Reymond, N. Schlosser, and P. Grangier, Operation of a quantum phase gate using neutral atoms in microscopic dipole traps, Phys. Rev. **A65**, 052301 (2002)

[18] L. You and M.S. Chapman, Quantum entanglement using trapped atomic spins, Phys. Rev. **A62**, 052302 (2000)

[19] A. Barenco, D. Deutsch, A. Ekert, and R. Jozsa, Conditional Quantum Dynamics and Logic Gates, Phys. Rev. Lett. **74**, 4083 (1995)

[20] T. Sleator and H. Weinfurter, Realizable Universal Quantum Logic Gates, Phys. Rev. Lett. **74**, 4087 (1995)

[21] Q.A. Turchette, C.J. Hood, W. Lange, H. Mabuchi, and H.J. Kimble, Measurement of Conditional Phase Shifts for Quantum Logic, Phys. Rev. Lett. **75**, 4710 (1995)

[22] P. Domokos, J.M. Raimond, M. Brune, and S. Haroche, Simple cavity-QED two-bit universal quantum logic gate: The principle and expected performances, Phys. Rev. **A52**, 3554 (1995)

[23] T. Pellizzari, S.A. Gardiner, J.I. Cirac, P. Zoller, Decoherence, Continuous Observation, and Quantum Computing: A Cavity QED Model, Phys. Rev. Lett. **75**, 3788 (1995).

[24] L.G. Lutterbach and L. Davidovich, Non-classical states of the electromagnetic field in cavity QED, Opt. Expr. **3**, 147 (1998)

[25] A. Hemmerich, Quantum entanglement in dilute optical lattices, Phys. Rev. **A60**, 943 (1999)

[26] V. Giovannetti, D. Vitali, P. Tombesi, and A. Ekert, Scalable quantum computation with cavity QED systems, Phys. Rev. **A62**, 032306 (2000)

[27] A. Beige, D. Braun, B. Tregenna, P. L. Knight, Quantum Computing Using Dissipation to Remain in a Decoherence-Free Subspace, Phys. Rev. Lett. **85**, 1762 (2000)

[28] S.B. Zheng and G.-C. Guo, Efficient Scheme for two-atom entanglement and quantum information processing in cavity QED, Phys. Rev. Lett. **85**, 2392 (2000).

[29] A. Recati, T. Calarco, P. Zanardi, J.I. Cirac, and P. Zoller, Holonomic quantum computation with neutral atoms, quant-ph/0204030 (2002)

[30] A. Kuhn, M. Hennrich, and G. Rempe; Strongly Coupled Atom–Cavity Systems, in "Quantum Information Processing", G. Leuchs and T. Beth (eds.), Wiley-VCH, Weinheim 2003.

[31] E. Hagley, X. Maître, G. Nogues, C. Wunderlich, M. Brune, J. M. Raimond, and S. Haroche, Generation of Enstein-Podolsky-Rosen Pairs of Atoms, Phys. Rev. Lett. **79**, 1 (1997)

[32] A. Rauschenbeutel, G. Nogues, S. Osnaghi, P. Bertet, M. Brune, J.-M. Raimond, and S. Haroche, Step-by-Step engineered multiparticle entanglement, Science **288**, 2024 (2000)

[33] J.M. Raimond, M. Brune, and S. Haroche, Manipulating quantum entanglement with atoms and photons in a cavity, Rev. Mod. Phys. **73**, 565 (2001)

[34] C.K. Law and H.J. Kimble, Deterministic generation of a bit-stream of single-photon pulses, J. Mod. Opt. **44**, 2067 (1997)

[35] M. Hennrich, T. Legero, A. Kuhn, and G. Rempe, Vacuum-Stimulated Raman Scattering Based on Adiabatic Passage in a High-Finesse Optical Cavity, Phys. Rev. Lett. **85**, 4872 (2000)

[36] J.I. Cirac, P. Zoller, H.J. Kimble, and H. Mabuchi, Quantum State Transfer and Entanglement Distribution among Distant Nodes in a Quantum Network, Phys. Rev. Lett. **78**, 3221 (1997)

[37] S.J. van Enk, J.I. Cirac, and P. Zoller, Ideal Quantum Communication over Noisy Channels: A Quantum Optical Implementation, Phys. Rev. Lett. **78**, 4293 (1997)

[38] S.J. van Enk, J.I. Cirac, and P. Zoller, Photonic Channels for Quantum Communication, Science **279**, 205 (1998)

[39] J. Ye, D.W. Vernooy, and H.J. Kimble, Trapping of Single Atoms in Cavity QED, Phys. Rev. Lett. **83**, 4987 (1999)

[40] P. Münstermann, T. Fischer, P.W.H. Pinkse, G. Rempe, Single slow atoms from an atomic fountain observed in a high-finesse optical cavity, Optics Comm. **159**, 63 (1999)

[41] C.J. Hood, T.W. Lynn, A.C. Doherty, A.S. Parkins, H.J. Kimble, The atom–cavity microscope: single atoms bound in orbit by single photons, Science **287**, 1447 (2000).

[42] P.W.H. Pinkse, T. Fischer, P. Maunz and G. Rempe, Trapping an atom with single photons, Nature **404**, 365 (2000).

[43] D. Klepinski, C. Monroe, and D.J. Wineland, Architecture for a large-scale ion-trap quantum computer, Nature **417**, 709 (2002)

[44] G.R. Guthörlein, M. Keller, K. Hayasaka, W. Lange, and H. Walther, A single ion as a nanoscopic probe of an optical field, Nature **414**, 49 (2001)

[45] K. Abich, Ch. Balzer, T. Hannemann, F. Mintert, W. Neuhauser, D. Reiß, P. E. Toschek, and Ch. Wunderlich; Spin Resonance with Trapped Ions: Experiments and New Concepts, in "Quantum Information Technology", G. Leuchs and T. Beth (eds.), Wiley-VCH, Berlin, Weinheim, New York 2003.

[46] P. Krüger, A. Haase, R. Folman, and J. Schmiedmayer; Quantum information processing with neutral atoms on atom chips, in "Quantum Information Processing", G. Leuchs and T. Beth (eds.), Wiley-VCH, Weinheim 2003.

[47] R. Grimm, M. Weidemüller, Y.B. Ovchinnikov, Optical dipole traps for neutral atoms, Adv. At. Mol. Opt. Phys. **42**, 95 (2000).

[48] D. Frese, B. Ueberholz, S. Kuhr, W. Alt, D. Schrader, V. Gomer and D. Meschede, Single Atoms in an Optical Dipole Trap: Towards a Deterministic Source of Cold Atoms, Phys. Rev. Lett. **85**, 3777 (2000).

[49] N. Schlosser, G. Reymond, I. Protsenko, and P. Grangier, Sub-poissonian loading of single atoms in a microscopic dipole trap, Nature **411**, 1024 (2001)

[50] S. Kuhr, W. Alt, D. Schrader, M. Müller, V. Gomer, D. Meschede, Deterministic Delivery of a Single Atom, Science **293**, 278 (2001), published online 14 June 2001; 10.1126/science.1062725

[51] see I.H. Deutsch and P.S. Jessen, Optical lattices, Adv. At. Mol. Opt. Phys. **37**, 95-138 (1996); G. Grynberg and C. Robilliard, Cold atoms in dissipative optical lattices, Phys. Rep. **355**, 335-451 (2001) and references therein

[52] R. Dumke, M. Volk, T. Müther, F. B. J. Buchkremer, W. Ertmer, and G. Birkl; Quantum Information Processing with Atoms in Optical Micro-Structures, in "Quantum Information Processing", G. Leuchs and T. Beth (eds.), Wiley-VCH, Weinheim 2003.

[53] Z. Hu, and H. J. Kimble, Observation of a single atom in a magneto–optical trap, Opt. Lett. **19**, 1888 (1994); F. Ruschewitz, D. Bettermann, J.L. Peng and W. Ertmer, Statistical investigations on single trapped neutral atoms, Europhys. Lett. **34**, 651 (1996); D. Haubrich, H. Schadwinkel, F. Strauch, B. Ueberholz, R. Wynands, and D. Meschede, Observation of individual neutral atoms in magnetic and magneto–optical traps, Europhys. Lett. **34**, 663 (1996)

[54] D. Haubrich, A. Höpe and D. Meschede, A simple model for optical capture of atoms in strong magnetic quadrupole fields, Optics Comm. **102**, 225 (1993)
[55] W. Alt, An objective lens for efficient fluorescence detection of single atoms, Optik **113**, 142 (2992); physics/0108058
[56] D. Schrader, S. Kuhr, W. Alt, M. Müller, V. Gomer, D. Meschede, An optical conveyor belt for single neutral atoms, Appl. Phys. B **73**, 819 (2001)
[57] T.A. Savard, K.M. O'Hara, and J.E. Thomas, Laser-noise-induced heating in far-off resonance optical traps, Phys. Rev. A **56**, R1095 (1997); M.E. Gehm, K.M. O'Hara, T.A. Savard, and J.E. Thomas, Dynamics of noise-induced heating in atom traps, *ibid.* **58** 3914 (1998)
[58] W. Alt, D. Schrader, S. Kuhr, M. Müller, V. Gomer, and D. Meschede, Single atoms in a standing-wave dipole trap, to be published
[59] A. Höpe, D. Haubrich, G. Müller, W. G. Kaenders and D. Meschede, Neutral Cesium Atoms in Strong Magnetic-Quadrupole Fields at Sub-Doppler Temperatures, Europhys. Lett. **22**, 669 (1993)
[60] R.A. Cline, J.D. Miller, M.R. Matthews, and D.J. Heinzen, Spin relaxation of optically trapped atoms by light scattering, Optics Lett. **19**, 207 (1994)
[61] A. Kaplan, M.F. Andersen, and N. Davidson, Suppression of inhomogenious broadening in rf spectroscopy of optically trapped atoms, arXov:physics/0204082 (2002)
[62] H. Perrin, A. Kuhn, I. Bouchoule and C. Salomon, Sideband cooling of neutral atoms in a far-detuned optical lattice, Europhys. Lett. **42**, 395-400 (1998)
[63] S. E. Hamann, D. L. Haycock, G. Klose, P. H. Pax, I. H. Deutsch, P. S. Jessen, Resolved-sideband Raman cooling to the ground state of an optical lattice, Phys. Rev. Lett. **80**, 4149 (1998)
[64] V. Vuletic, C. Chin, A.J. Kerman, and S. Chu, Degenerate Raman Sideband Cooling of Trapped Cesium Atoms at Very High Atomic Densities, Phys. Rev. Lett. **81**, 5768 (1998)
[65] G. Raithel, W.D. Phillips, and S.L. Rolston, Collapse and Revivals of Wave Packets in Optical Lattices, Phys. Rev. Lett. **81**, 3615 (1998)
[66] I. Bouchoule, H. Perrin, A. Kuhn, M. Morinaga, and C. Salomon, Neutral atoms prepared in Fock states of a one-dimensional harmonic potential, Phys. Rev. **A59**, R8 (1999);
[67] M. Morinaga, I. Bouchoule, J.-C.Karam, and C. Salomon, Manipulation of Motional Quantum States of Nuetral Atoms, Phys. Rev. Lett. **83**, 4037 (1999)
[68] D.L. Haycock, P.M. Alsing, I.H. Deutsch, J. Grondalski, and P.S. Jessen, Mesoscopic Quantum Coherence in an Optical Lattice, Phys. Rev. Lett. **85**, 3365 (2000)
[69] M.P. MacDonald, L. Paterson, K. Volke-Sepelveda, J. Alt, W. Sibbert, and K. Dholakia, Creation and Manipulation of Three-Dimensional Optically Trapped Structures, Science **296**, 1101 (2002)
[70] H. J. Metcalf, P. van der Straten, *Laser Cooling and Trapping*, Springer, New York (1999)
[71] V.I. Balykin, V.G. Minogin, V.S. Letokhov, Electromagnetic Trapping of cold atoms, Rep. Prog. Phys. **63**, 1429 (2000)
[72] N. Davidson, H.J. Lee, C.S. Adams, M. Kasevich, and S. Chu, Long Atomic Coherence Times in an Optical Dipole Trap, Phys. Rev. Lett. **74**, 1311 (1995)
[73] H. Mabuchi, J. Ye, H.J. Kimble, Full observation of single-atom dynamics in cavity QED, Appl. Phys. **B 68** 1095 (1995); quant-ph/9805076

21 Towards Quantum Logic with Cold Atoms in a CO_2 Laser Optical Lattice

G. Cennini, G. Ritt, C. Geckeler, R. Scheunemann, and M. Weitz

Physikalisches Institut der Universität Tübingen
Tübingen, Germany

21.1 Introduction

Recent progress in the field of quantum optics has opened the door to an increasing number of fascinating applications, ranging from the generation of entangled quantum states [1] to the production and manipulation of Bose–Einstein condensates of atomic gases [2]. Current information technology is based on classical physics, and we expect interesting perspectives if the laws of quantum physics are applied to this branch of science. The conceptual link between quantum mechanics and information processing was proposed by Feynman [3] and Deutsch [4] in the early 1980s. A decade later, an astonishing example of the power of quantum information processing was given by Shor in an algorithm for factorizing large numbers into primes [5]. Shor proved that the computational time can be polynomial with the input size for a quantum computer, while it grows exponentially in the case of conventional (classical) information processing. As a further motivation towards the interest in research on quantum information processing, let us refer to Moore's law, which states that about every one and a half years the microprocessors of our personal computers double their speed while simultaneously reducing their size by a factor of two. At the time of writing, it is not clear for how much longer the present silicon technology can withstand such a terrific pace. However, it seems foreseeable that in the not-too-distant future the logic gates of computers' microprocessors will be so small that they consist of only a few atoms each. One then expects that the laws of quantum mechanics will significantly affect the processing of information.

This chapter is organized as follows. We first briefly recall the basic conditions for implementing a quantum algorithm, pointing out the present state of the art achieved in several branches of experimental physics. We here focus on the prospects of cold atoms trapped in mesoscopic optical lattices for scalable quantum computing. Subsequently, we review experimental work concerning the resolving and addressing of atoms in individual sites of a CO_2 laser optical lattice and forced evaporative cooling of rubidium atoms to quantum degeneracy in both running wave and lattice geometries. We finally discuss future prospects.

Quantum Information Processing, 2nd Edition. Edited by Thomas Beth and Gerd Leuchs

ISBN: 3-527-40541-0

21.2 Entanglement and Beyond

The possibility of operating with coherent superpositions and entangled states is essential for a quantum computer [6]. There are several conditions that a set of real quantum objects must fulfill in order to allow for the application of quantum computing. First of all, we require a quantum register, as, e.g., several two-level quantum systems that are isolated from the environment for sufficiently long to maintain coherence throughout the computation process. We then should envision a mechanism for coupling the two-level systems by performing unitary gate operations — in other words we have to implement a quantum bus channel connecting our quantum objects, the so-called quantum bits (qubits). And last, but not least, we have to employ a technique for reading out the final result of our computation. This last requirement is not only needed to work out the outcomes of the whole process, but it is also necessary to set the quantum register to a well-known initial state [7].

So far, an increasing number of theoretical proposals towards the physical implementation of quantum gates has been published. Cirac and Zoller [8] suggested implementing quantum logic in coherent superpositions of two hyperfine ground states of cold ions trapped in a Paul trap. Wineland and coworkers experimentally demonstrated a controlled-NOT gate with a single trapped ion [9]. This and other early experiments have successfully demonstrated the operational principle of quantum logic gates. In ion traps, the implementation of real quantum algorithms, which require many entangled particles, is ultimately limited by decoherence originated by the Johnson noise of the trap endcaps. Other experimental techniques have been implemented for the realization of quantum logic gates: cold atoms in high-finesse cavities [10] and nuclear spins with NMR techniques [11]. In 'flying qubits' cavity-QED experiments the quantum register consists of photons of an optical cavity. These modes are coupled to others via an atom placed inside the cavity. In NMR experiments, the quantum register is formed by nuclear spins of molecules in a strong magnetic field, and the quantum bus channel is provided by spin–spin interactions. In both cases the scalability poses tough limitations on the construction of a quantum processing machine. For example, NMR techniques suffer from an exponential decrease of the signal-to-noise ratio after only a few quantum gates. Several interesting proposals on quantum logic gates in solid state devices have been published. In principle, solid state devices are easy to scale up, but nonetheless their main drawback is still represented by decoherence caused by the strong interaction with a complex environment.

A recently proposed technique for implementing and testing quantum logic gates with neutral atoms uses far-detuned optical lattices. In optical lattices, atoms are trapped in the nodes or antinodes of optical standing waves [12]. Several researchers suggested that these systems are attractive and effective tools for quantum information processing [13–16]. In optical lattices, the quantum bits can be encoded in an internal hyperfine state of atoms, with a single trapped atom per lattice site. The lattice can then be regarded as a very dense and periodic quantum register, in which it is possible to perform multiple-particle entanglement operations in parallel. The quantum bus channel can, e.g., be implemented by using a state-dependent lattice potential and controlled elastic collisions between atoms in adjacent lattice sites, as proposed by Jaksch *et al.* [15] (see also [14]). It has been pointed out that in such systems efficient schemes for quantum error corrections and fault-tolerant computing can be straightforwardly implemented due to the inherent possibility of parallel operation [15, 17]. The addressing of single qubits is facilitated when using a trapping wavelength far above that of the atomic absorption wavelength. Optical lattices based on the focussed and retro-reflected radiation derived from a CO_2 laser operating near 10.6 μm seem to be promising candidates for such experiments.

21.3 Quantum Logic and Far-detuned Optical Lattices

In optical lattices, atoms are trapped in light-shift potentials created by the interference of two or more laser beams. The nature of the trapping force arises from the coherent interaction between the induced atomic dipole moment and the laser electric field [18–20]. When the laser frequency is tuned to the red side of an atomic resonance, the dipole force attracts atoms towards the interference maxima of the light field. For optical fields with frequency far below all electric dipole resonances of ground-state atoms (i.e., 'quasi-static' fields), the trapping potential is given by

$$U = -\tfrac{1}{2}\alpha_s|\vec{E}|^2, \tag{21.1}$$

where α_s denotes the static atomic polarizability of the ground state and $\vec{E}$ is the electric field associated with the optical field. Due to the large detuning, the photon scattering rate is very small. In particular, the Rayleigh photon scattering rate in the quasi-static regime is given by

$$\Gamma_S = \frac{16r_0^2 P}{3\hbar w_0^2}\left(\frac{m_e\alpha_s}{e^2}\right)\omega^3, \tag{21.2}$$

where w_0 is the laser beam waist, r_0 is the classical electron radius, m_e is the electron mass, and ω denotes the optical frequency of the laser light. The ω^3 term is a phase space factor, which, in addition to the large detuning from resonance, reduces the photon scattering for laser fields with small optical frequency. Spontaneous Raman scattering is suppressed by a further factor $(\Delta_{\mathrm{FS}}\omega/\omega_{\mathrm{atom}})^2$, where ω_{atom} is the frequency of the atomic resonance. In our experiments we use CO_2 laser light whose wavelength is near 10.6 μm, and typical values of the Rayleigh scattering rate on trapped rubidium atoms are of the order of 600 s. This results in a small coupling of the atoms to the environment, so that the effect of decoherence can be kept small.

The addressing of an individual qubit in the lattice by means of focussed, resonant laser beams requires the optical resolving of the lattice sites. Optical resolving is possible when the spatial distance between adjacent lattice sites $\lambda_{\mathrm{laser}}/2$ is much larger than the wavelength of illumination, i.e., the atomic absorption wavelength λ_{atom}. This criterion is well fulfilled in our experiments, where the lattice spacing of 5.3 μm is almost an order of magnitude above the absorption wavelength of the rubidium D2 line (780 nm).

When a lattice potential that depends on the internal atomic state is used, local interactions (such as elastic collisions) can produce an entanglement of atoms. We here focus on such local techniques, as they may ultimately lead to highly parallel quantum gates. The internal atomic structure plays an important role in the following discussion. For alkali atoms trapped in an extremely far-detuned laser field, the detuning from resonance is much larger than both the hyperfine and fine structure splitting. This condition is met when CO_2 laser radiation is used. The trapping potential for ground-state alkali atoms here corresponds to that of a $J = 0$ to

$J' = 1$ transition, as the ac Stark shift for all ground-state sublevels is identical and does not depend on the internal atomic state. However, a state-dependent lattice potential can be obtained when introducing additional laser beams with detuning comparable to the upper-state fine structure splitting, but still large compared to the hyperfine structure. In this case, the electron spin can no longer be neglected and one can show that the ac Stark shift depends on the laser polarization.

Thus, the idea is to transfer the atoms trapped in a CO_2 laser field into a closer resonant manipulation lattice during gate operations. The manipulation lattice with laser detuning comparable to the atomic fine structure splitting, i.e., tuned between the $P_{1/2}$ and $P_{3/2}$ levels, produces an ac Stark shift depending on the polarization of the light field and, for a particular detuning, some of the ground-state magnetic sublevels are trapped in σ^+ polarized light, while others are trapped in σ^- polarized light. It is then possible to distinguish two atomic "species", which can be manipulated individually with two oppositely circularly polarized standing waves. When the qubits are stored in a coherent superposition of two ground-state sublevels $|g+\rangle$ and $|g-\rangle$, which are trapped in σ^+ and σ^- polarized light, respectively, a quantum logic phase gate can be realized with a transient spatial shift operation of, e.g., the $|g+\rangle$ components by one lattice separation, leading to a cold collision with the component in $|g-\rangle$ of the atom in the next lattice site (see Figs. 21.1 and 21.2).

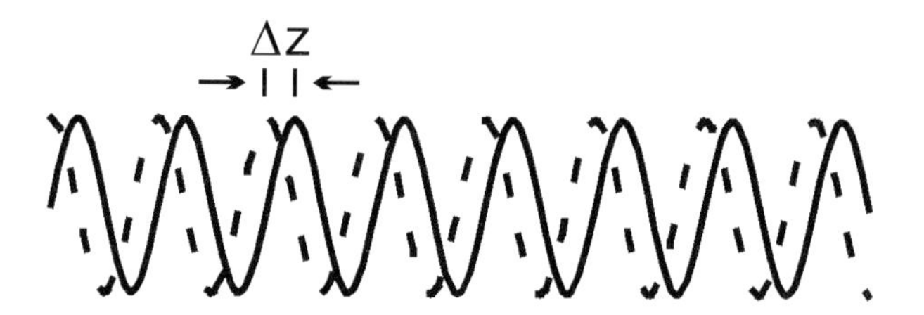

Figure 21.1: Spatial intensity dependence of two 1D standing waves with σ^+ (solid) and σ^- (dashed), respectively, circular polarization. The two standing waves are spatially shifted relative to each other by Δz. For a suitable laser detuning, atoms in the ground-state level $|g+\rangle$ are pulled into the intensity maxima of the σ^+ polarized standing wave, whereas atoms in the $|g-\rangle$ state are trapped in the maxima of the σ^- polarized standing wave.

The interaction between two cold atomic wavepackets is determined by the s-wave scattering length and can be mathematically expressed in terms of a contact potential. The resulting interaction leads to a conditional phase shift in the atomic wavefunction [15]. Let us assume that a standing wave of the manipulation light is oriented along the same axis as the infrared wave generated by the CO_2 laser, and that the CO_2 laser wavelength is chosen such that it equals an integer multiple of the near-resonant light, e.g., $\lambda_{CO_2} = 13\lambda_{res}$. After preparing the qubits in the CO_2 laser lattice, the retro-reflected 10.6 μm beam is extinguished and the closer resonant light is activated. The atoms are then transferred into the near-resonant standing wave and they occupy precisely every 13th lattice site. In this near-resonant lattice, conditional dynamics can be implemented. The described scheme should enable the realization of entangled two- and more particle quantum states with the possibility of addressing individual qubits.

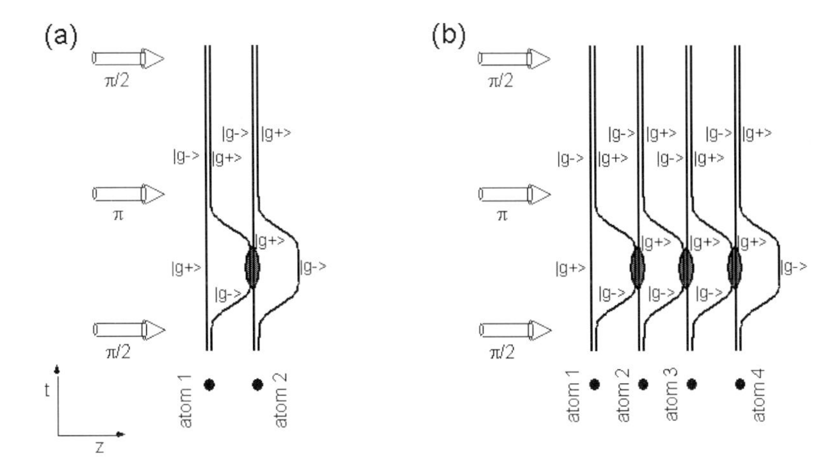

Figure 21.2: Scheme of a quantum logic gate for (a) two and (b) four atoms. The quantum bits are stored in each atom as a coherent superposition of the ground-state levels $|g+\rangle$ and $|g-\rangle$. In the shaded regions, matter waves are phase shifted by cold coherent collisions.

21.4 Resolving and Addressing Cold Atoms in Single Lattice Sites

In this section we review the realization of a mesoscopic optical lattice, implemented with cold atoms trapped in a CO_2 laser standing wave [21]. Figure 21.3 shows a scheme of the experimental setup, as used by two of the authors while at the Max-Planck-Institut für Quantenoptik (MPQ) in Garching near Munich. A single-mode CO_2 laser generated up to 50 W mid-infrared radiation near 10.6 μm. The light passed through an acousto-optic modulator (AOM) in order to provide optical isolation and allow for control of the beam intensity. The transmitted beam entered a vacuum chamber and was focussed by an adjustable ZnSe lens placed inside the vacuum chamber to a beam waist of typically 35 μm. By means of a further lens and an external retro-reflecting mirror, an intense standing wave was formed. Atoms of the isotope ^{85}Rb were collected and pre-cooled in a magneto–optical trap (MOT), which was loaded from the thermal gas emitted by heated rubidium dispensers. The cooling light, tuned a few linewidths to the red of the $5S_{1/2}, F = 3$ to $5P_{3/2}, F = 4$ cycling transition, was emitted by a free-running diode laser injection-locked to a second, grating-stabilized diode. In order to repump rubidium atoms into the cooling transition, a grating-stabilized third laser locked to the $5S_{1/2}, F = 2$ to $5P_{3/2}, F = 3$ transition was used.

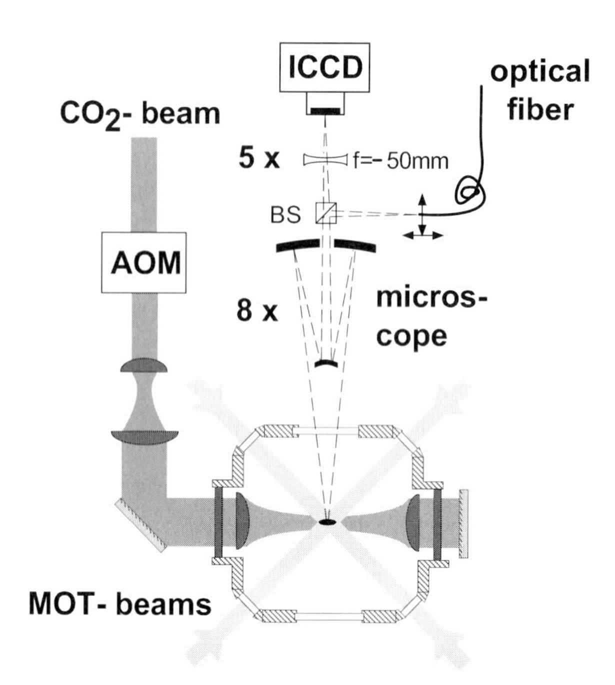

Figure 21.3: Schematic of the experimental apparatus.

During a typical experimental run, roughly 5×10^7 atoms were collected in the MOT during a 3 s loading phase. Then they were compressed for 20 ms in a temporal dark MOT [22] realized by further detuning the cooling laser and by simultaneously reducing the repumping laser intensity by a factor of 10. During this phase, the CO_2 laser beam was switched on, allowing for rubidium atoms to be loaded into its focal region. At the end of the dark MOT phase, all resonant beams were extinguished and the magnetic field was switched off. By this time, rubidium atoms were only trapped by the lattice potential. In order to detect the trapped atoms, the MOT beams were pulsed on after a variable amount of time, and the fluorescence scattered by the trapped atoms was imaged onto both a calibrated photodiode and an intensified CCD camera (ICCD). The typical numbers of atoms trapped in the lattice was near 1×10^6. These atoms were distributed over more than 100 pancake-shaped lattice sites. The trap lifetime was 3.4 s, being limited by collisions with the thermal background gas (10^{-9} mbar background pressure). With a mid-infrared input power of 14 W, the trap depth [13] was $U_0/\hbar = 43.5$ MHz, corresponding to a temperature of 1.4 mK.

The vibrational frequencies of the CO_2 laser lattice were measured by parametrically exciting the trapped atoms [13]. For this measurement, we periodically modulate the CO_2 laser beam intensity with an AOM. We observed a significant trap loss, caused by parametric vibrational excitation, when the modulation frequency was close to twice a trap vibrational frequency. With typical parameters of 14 W optical power and 35 μm waist radius, it was possible to measure oscillation frequencies in the central trap region of $\nu_z = 54(5)$ kHz in the axial and $\nu_r = 4.2(5)$ kHz in the radial direction. For these parameters, the Lamb–Dicke limit, corresponding to an oscillation frequency above the photon recoil energy in frequency units $\nu_{\text{rec}} = h/2m\lambda^2$ (3.8 kHz for ^{85}Rb), is therefore fulfilled in all three spatial dimensions.

For a more complete characterization of the trap, the atomic temperature was measured with a time-of-flight technique. The observed values were strongly dependent on the trapping laser power. This power dependence is attributed to the difference in the ac Stark shifts of ground and excited states of the rubidium atoms, which for high laser powers becomes comparable to the upper-state hyperfine splitting, and then significantly reduces the efficiency of sub-Doppler cooling mechanisms. For maximum laser power, the measured temperatures approached the Doppler limit of the rubidium atom (140 μK), while for small CO_2 laser power (near 4 W power in the trapping beam), we observed temperatures as low as 10 μK. At those low values of the laser power we measured the maximum atomic phase space density $n\lambda_{dB}^3$ of about 1/300 in the optical lattice (see [23]). The tight confinement in the steep microtraps gave an atomic phase space density roughly three orders of magnitude above that of a conventional MOT. The atomic density reached values of several 10^{13} cm^{-3}, leading to a high collisional rate.

Finally, it was possible to spatially resolve atoms trapped in single lattice sites (see Fig. 21.4). For this imaging of the lattice, the power of the CO_2 laser trapping field was lowered to 4 W, which reduces the atomic velocity distribution spread. The trapped atoms were irradiated with resonant light by pulsing on the MOT beams for a period of up to 100 μs. The frequency of this light was chosen to be resonant with the $5S_{1/2}, F = 3$ to $5P_{3/2}, F = 4$ transition at the bottom of the central potential wells. During the exposure, the MOT repumping light was also activated. The scattered atomic fluorescence was imaged onto an ICCD camera by a long-distance microscope (1.9 μm spatial resolution) placed 10 cm away from the trap center outside the vacuum system. Figure 21.4a shows rubidium atoms localized in the periodically spaced pancake-like microtraps of 5.3 μm period. Each of these sites contained up to 4×10^3 atoms. The CO_2 laser trapping beam was left on during the entire cycle. Figure 21.4b depicts an image taken by only illuminating a single trapping site for a period of 100 μs. Here, around 10 μW of additional resonant light was sent through the core of an optical fiber and then imaged via a beamsplitter through the microscope onto the sample. Repumping light was again provided by the MOT repumping beams. The exposure shows atoms localized in one distinct potential well of the standing wave, with the neighboring lattice sites suppressed by a factor of approximately 2.3. The other lattice sites were filled, but are not visible here.

Finally, for the image shown in Fig. 21.4c, a single lattice site was removed before the exposure with an intense resonant laser beam. Here, after loading the atoms into the trap, a 10 μs long pulse of light through the fiber with the same frequency, but with 20 times higher intensity, was applied. Again the MOT repumping beams were used to provide the necessary repumping light. The population of a single lattice site was almost completely removed, while atoms in the neighboring sites were affected much less by the short pulse. By varying the position of the optical fiber along the axial direction of the lattice, it was possible to address a different lattice site within the optical field of view, which comprised around 50 lattice sites. These pictures demonstrate that it is in principle possible to address single qubits, as is necessary to read in and out quantum information in our optical lattice.

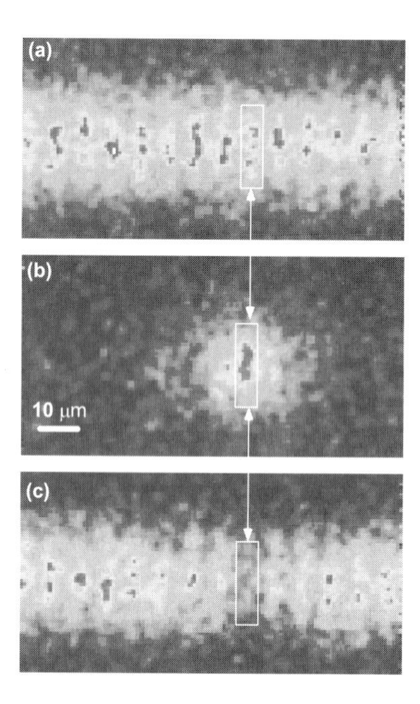

Figure 21.4: Images of the lattice after the following manipulations: (a) No manipulation of the atoms in the microtraps. (b) Only atoms from a single lattice site are illuminated during the exposure with a focussed laser beam. (c) Image of the lattice after removing atoms in one lattice site before the exposure.

21.5 Recent Work

To prevent decoherence during gate operations, it is mandatory to cool the atoms into the vibrational ground state of the lattice potential [15]. Such ground-state cooling can be achieved either by using Raman sideband cooling technique [24] or by forced evaporation of trapped atoms [25, 26]. Given the high atomic densities of several 10^{13} cm^{-3} possible in CO_2 laser traps with laser cooling alone (discussed in the previous section) [23], the initial conditions seem especially good for evaporative cooling techniques. Other recent work has impressively shown that forced evaporation in CO_2 laser traps can achieve rapid cooling to quantum degeneracy [27–29]. In our Tübingen lab, we have successfully implemented evaporative cooling of ^{87}Rb atoms in a single running beam dipole trap configuration to quantum degeneracy (see Fig. 21.5). We initially load the dipole trap with 6×10^7 atoms from a magneto–optical trap. Further cooling is then achieved by lowering the dipole trapping potential in a controlled way.

In initial experiments, we investigated a crossed dipole trapping geometry to reach Bose–Einstein condensation following Chapman's work [27]. Subsequently, we moved to a less alignment-sensitive single beam trapping geometry. Here, we use a very small waist size (about 27 μm) to enhance the collision rate. Bose–Einstein condensation is observed after 7 s of evaporative cooling time, during which the single trapping laser beam is ramped from an initial power of 28 W to a final power of 200 mW. In this way we produce a spinor condensate with 12 000 atoms distributed over the $m_F = -1, 0$, and $+1$ components of the $F = 1$ ground state. The critical temperature of the condensate is 200 nK. Note that all previous experiments achieving quantum degeneracy directly in dipole traps have required either a crossed dipole trapping geometry or Feshbach resonances to enhance the collisional rate [27–30].

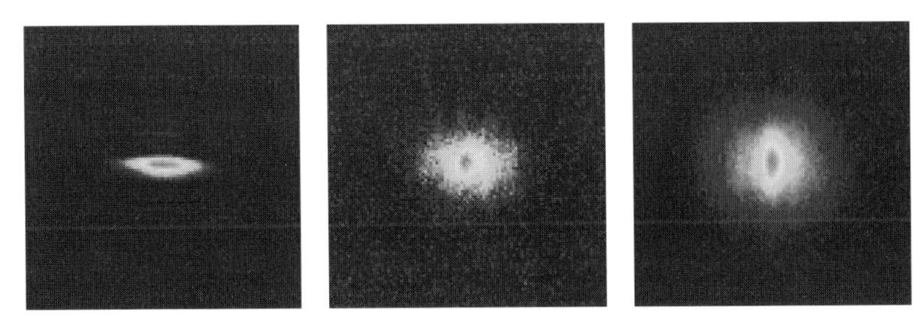

Figure 21.5: Free expansion of a Bose–Einstein condensate generated in a single-beam CO_2 laser dipole trap. Shown is a series of shadow images of the atomic cloud recorded after allowing for different free expansion times. The left image was taken with nearly no free expansion, thus showing the cigar-shaped spatial distribution of the trapped cloud. The middle and right images were recorded after 8 ms and 15 ms of free expansion. At the latter time, the asymmetry of the cloud is inverted due to the anisotropic expansion of the mean field energy.

In subsequent experiments we have attempted to obtain stable trapping throughout the evaporative cooling phase only for atoms in the field-insensitive ($m_F = 0$) Zeeman state. In the single dipole trap geometry, the confinement along the beam axis is relatively weak, so that already with a moderate field gradient atoms in the field-sensitive spin projections are removed from the trap by magnetic forces (see Fig. 21.6). As an additional effect, the $m_F = 0$ population is enhanced in the condensate, which we attribute to sympathetic cooling with atoms in $m_F = \pm 1$ states during the cooling phase. In the trap, we produce a condensate with 7000 atoms in the $m_F = 0$ spin projection. The chemical potential here is first-order insensitive to magnetic field fluctuations [31, 32].

By lowering the optical trapping potential we further succeeded in coupling out a monoenergetic atom laser beam from this magnetic insensitive condensate with a length of 1 mm [32]. For the atom laser we estimated a brightness of approximately 7×10^{27} atoms $s^2\ m^{-5}$.

Recently, we have moved to a CO_2 laser lattice geometry. At present, we enhance the loading of atoms into the central lattice sites by during the dark MOT phase switching from a perfect standing wave to a geometry with two spatially separated optical beams. This is achieved by misaligning the retro-reflection mirror with a piezo. For the evaporative cooling phase, we then switch back to a perfect lattice geometry. In this way, we produce a periodic array of typically 15 micro-condensates spaced by 5.3 m. We have experimentally observed the multiple-order interference pattern when releasing the condensed atoms from the lattice. At present, the number of atoms per site is still relatively large (of order 200), which we are planning to reduce in future. The individual sites should be addressable by means of focussed laser beams.

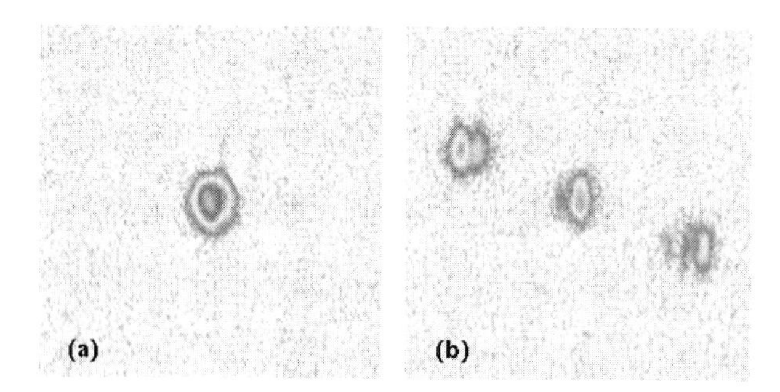

Figure 21.6: False-color shadow images of the atomic cloud after 15 ms of free expansion (field of view: 0.33 mm square). (a) Stern–Gerlach magnetic field gradient applied throughout the experiment, so that a pure $m_F = 0$ condensate is produced. (b) Field gradient activated only during free expansion phase. The three spin projections $m_F = -1$, 0, and 1 of a spinor condensate are visible as separate clouds.

In this system, we expect novel possibilities in the direct microscopic study of quantum tunneling and quantum phase transitions. Unity occupation of lattice sites (as is required for quantum logic) can be achieved by making use of a phase transition from a superfluid Bose–Einstein phase to a Mott insulator phase induced when increasing the ratio between the on-site interaction potential to the tunneling matrix element. Such a phase transition is predicted by the Bose–Hubbard model [33] and has recently been demonstrated experimentally in a near-resonant optical lattice [34]. To conclude, optical lattices are of interest in a wide spectrum of physical fields, ranging from model systems for solid state theory to its fascinating prospects in studies of entangled states and scalable quantum logic.

Acknowledgements

We acknowledge the contributions of F. S. Cataliotti, S. Friebel, T. W. Hänsch, and J. Walz during the Munich phase of this project, and we thank H. Briegel and P. Zoller for useful and stimulating discussions.

References

[1] K. Mattle, H. Weinfurter, P. G. Kwait, and A. Zeilinger, Phys. Rev. Lett. **76**, 4656 (1996); A. Furusawa, J. L. Sørensen, S. L. Braunstein, C. A. Fuchs, H. J. Kimble, and E. S. Polzik, Science, Oct. 23, 706–709 (1998);
D. Boschi, S. Branca, F. De Martini, L. Hardy, and S. Popescu, Phys. Rev. Lett. **80**, 1121 (1998).

[2] M. Inguscio, S. Stringari, and C. Wieman (Eds.), Bose–Einstein Condensation in Atomic Gases, Amsterdam: IOS Press (1999).

[3] R. P. Feynman, Simulating physics with computers, Int. J. Theor. Phys. **21**, 467 (1982).

[4] D. Deutsch, Quantum theory, the Church–Turing principle and the universal quantum computer, Proc. Roy. Soc. A **400**, 97 (1985).

[5] P. W. Shor, Algorithms for quantum computation: discrete logarithms and factoring, Proc. 35th IEEE Symp. on Foundations of Computer Science, S. Goldwasser, Ed., Los Alamos, CA: IEEE Computer Soc. Press (1994), p. 124.

[6] C. H. Bennett, Quantum information and computing, Physics Today, Oct. 24 (1995).

[7] D. P. DiVincenzo, The physical implementation of quantum computation, quant-ph/0002077.

[8] J. I. Cirac, and P. Zoller, Phys. Rev. Lett. **74**, 4091 (1995).

[9] C. Monroe, D. M. Meekhof, B. E. King, W. M. Itano, and D. J. Wineland, Phys. Rev. Lett. **75**, 4714 (1995).

[10] Q. A. Turchette, C. J. Hood, W. Lange, H. Mabushi, and H. J. Kimble, Phys. Rev. Lett. **75**, 4710 (1995).

[11] N. A. Gershenfeld, and I. L. Chuang, Bulk spin-resonance quantum computing, Science **274**, 350 (1997).

[12] P. S. Jessen, and I. H. Deutsch, Adv. At. Mol. Opt. Phys. **37**, 95 (1996), and references therein.

[13] S. Friebel, C. D'Andrea, J. Walz, M. Weitz, and T. W. Hänsch, Phys. Rev. A **57**, R20 (1998).

[14] G. Brennen, C. Caves, P. Jessen, and I. Deutsch, Phys. Rev. Lett. **82**, 1060 (1999).

[15] D. Jaksch, H.-J. Briegel, J. I. Cirac, C. W. Gardiner, and P. Zoller, Phys. Rev. Lett. **82**, 1975 (1999).

[16] A. Hemmerich, Phys. Rev. A **60**, 943 (1999).

[17] H.-J. Briegel, T. Calarco, D. Jaksch, J. I. Cirac, and P. Zoller, J. Mod. Opt. **47**(2/3), 415 (2000).

[18] S. Chu, J. E. Bjorkholm, A. Ashkin, and A. Cable, Phys. Rev. Lett. **57**, 314 (1985);
see also: R. Grimm, M. Weidemüller, and Yu. B. Ovchinnikov, Adv. At. Mol. Opt. Phys. **42**, 95 (2000).

[19] C. Cohen-Tannoudji, Atomic Motion in Laser Light, J. Dalibard, J. M. Raimond, and J. Zinn-Justin, Eds., Les Houches, Session LIII, 1990, Elsevier Science (1992).

[20] J. P. Gordon, and A. Ashkin, Motion of atoms in a radiation trap, Phys. Rev. A **21**, 1606 (1980).

[21] R. Scheunemann, F. S. Cataliotti, T. W. Hänsch, and M. Weitz, Phys. Rev. A **62**, R51801 (2000).

[22] W. Ketterle, K. B. Davis, M. A. Joffe, A. Martin, and D. E. Pritchard, Phys. Rev. Lett. **70**, 2253 (1997).

[23] S. Friebel, R. Scheunemann, J. Walz, T. W. Hänsch, and M. Weitz, Appl. Phys. B **67**, 699 (1998).

[24] S. Hamann, D. Haycock, G. Klose, D. Pax, I. Deutsch, and P. Jessen, Phys. Rev. Lett. **80**, 4149 (1998);
H. Perrin, A. Kuhn, I. Bouchoule, and C. Salomon, Europhys. Lett. **42**, 395 (1998);
V. Vuletic, C. Chin, A. Kerman, and S. Chu, Phys. Rev. Lett. **81**, 5768 (1999).

[25] K. B. Davis, M. O. Mewes, and W. Ketterle, Appl. Phys. B **60**, 155 (1995).

[26] C. S. Adam, H. Jin Lee, N. Davidson, M. Kasevich, and S. Chu, Phys. Rev. Lett. **74**, 3579, (1995).

[27] M. D. Bennett, J. A. Sauer, and M. S. Chapman, Phys. Rev. Lett. **87**, 10404 (2001).

[28] S. R. Granade, M. E. Gehm, K. M. O'Hara, and J. E. Thomas, Phys. Rev. Lett. **88**, 120405, (2002).

[29] T. Weber, J. Herbig, M. Mark, H.-C. Nägerl, and R. Grimm, Science **299**, 232 (2003).

[30] Y. Takasu, K. Maki, K. Komori, T. Takano, K. Honda, M. Kumakura, T. Yabuzaki, and Y. Takahashi, Phys. Rev. Lett. **91**, 040404 (2003).

[31] G. Cennini, G. Ritt, C. Geckeler, and M. Weitz, Appl. Phys. B **77**, 773 (2003).

[32] G. Cennini, G. Ritt, C. Geckeler, and M. Weitz, Phys. Rev. Lett. **91**, 240408 (2003).

[33] D. Jaksch, C. Bruder, J. I. Cirac, C. W. Gardiner, and P. Zoller, Phys. Rev. Lett. **81**, 3108 (1998).

[34] M. Greiner, O. Mandel, T. Esslinger, T. W. Hänsch, and I. Bloch, Nature **415**, 39 (2002).

22 Quantum Information Processing with Atoms in Optical Micro-Structures

R. Dumke, M. Volk, T. Müther, F. B. J. Buchkremer, W. Ertmer, and G. Birkl

Institut für Quantenoptik
Universität Hannover
Hannover, Germany

22.1 Introduction

Following the spectacular theoretical results in the field of quantum information processing of recent years [1], there is now also a growing number of experimental groups working in this area. Among the many currently investigated approaches, which range from schemes in quantum optics to superconducting electronics [1, 2], the field of atomic physics seems to be particularly promising due to the remarkable experimentally achieved control of single qubit systems and the understanding of the relevant coherent and incoherent processes. While there have been successful implementations of quantum logic with charged atomic particles in ion traps [3], quantum information schemes based on neutral atoms [4–10] are an attractive alternative due to the weak coupling of neutral atoms to their environment. A further attraction of neutral atoms lies in the fact that many of the requirements for the implementation of quantum computation [11] are potentially met by the newly emerging miniaturized and integrated atom optical setups.

These miniaturized setups can be obtained by using different types of micro-fabricated structures: The trapping and guiding of neutral atoms in micro-fabricated **charged and current carrying** structures have been pursued by a number of groups in recent years [12–19]. A new approach to generate miniaturized and integrated atom optical systems has been recently introduced by our group [20]: We proposed the application of micro-fabricated **optical** elements (microoptical elements) for the manipulation of atoms and atomic matter waves with laser light. This enables one to exploit the vast industrial and research interest in the field of applied optics directed towards the development of micro-optical elements, which has already led to a wide range of state-of-the-art optical system applications [21, 22] in this field. Applying these elements to the field of quantum information processing, however, constitutes a novel approach [20,23,24]. Together with systems based on miniaturized and microfabricated mechanical as well as electrostatic and magnetic devices, the application of microoptical systems will launch a new field in atom optics which we call **ATOMICS** for **AT**om **O**ptics with **MIC**ro-**S**tructures. This field will combine the unique features of devices based on the quantum mechanical behavior of atomic matter waves with the tremendous potential of micro- and nanofabrication technology and will lead to setups that are very attractive for quantum information processing.

Quantum Information Processing, 2nd Edition. Edited by Thomas Beth and Gerd Leuchs

ISBN: 3-527-40541-0

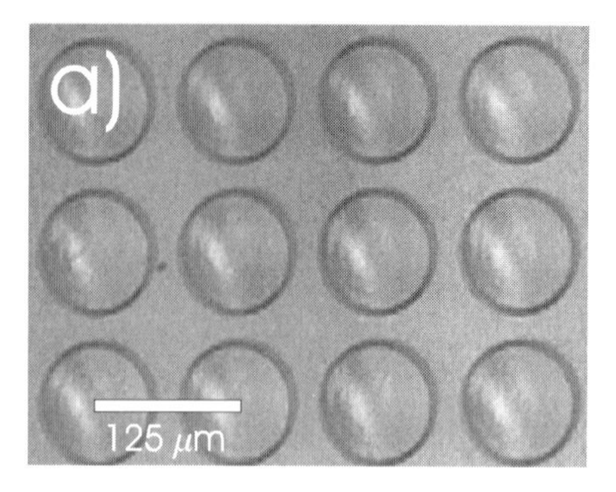

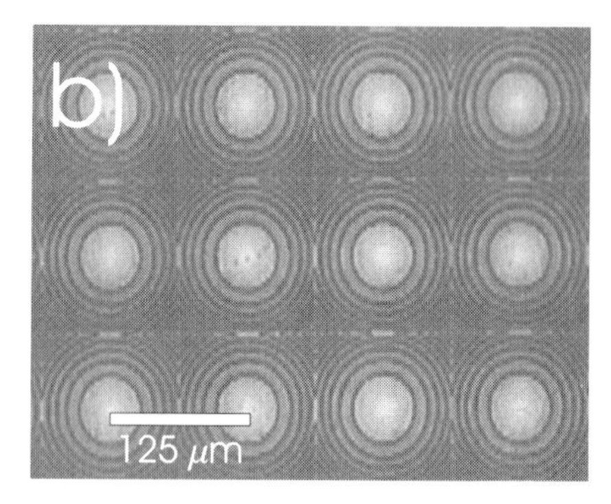

Figure 22.1: Refractive (a) and diffractive (b) array of spherical microlenses.

22.2 Microoptical Elements for Quantum Information Processing

A special attraction of using microoptical elements lies in the fact, that most of the currently used techniques in atom manipulation are based on the interaction of light with atoms. The use of micro-fabricated optical elements is therefore in many ways the canonical extension of the conventional optical methods into the micro-regime, so that much of the knowledge and experience that has been acquired in atom optics can be applied to this new regime in a very straightforward way. There are however, as we will show in the following, a number of additional inherent advantages in using microoptics which significantly enhance the possibilities of atom manipulation and will lead to a range of new developments that were not achievable until now: The use of state-of-the-art lithographic manufacturing techniques adapted from semiconductor processing enables the optical engineer to fabricate structures with dimensions in the micrometer range and submicrometer features with a large amount of flexibility and in a large variety of materials (glass, quartz, semiconductor materials, plastics, etc.). The flexibility of the manufacturing process allows the realization of complex optical elements which create light fields not achievable with standard optical components. Another advantage lies in the fact, that microoptics is often produced with many identical elements fabricated in parallel on the same substrate, so that multiple realizations of a single conventional setup can be created in a straightforward way (scalability). A further attraction of the flexibility in the design and manufacturing process of microoptical components results from the huge potential for integration of different elements on a single substrate, or, by using bonding techniques, for the integration of differently manufactured parts into one system. No additional restrictions arise from the small size of microoptical components since for most applications in atom optics, the defining parameter of an optical system is its numerical aperture, which for microoptical components can be as high as NA=0.5, due to the small focal lengths achievable.

Among the plethora of microoptical elements that can be used for quantum information processing applications are refractive or diffractive microoptics, computer generated holograms, microprisms and micromirrors, integrated waveguide optics, near-field optics, and integrated techniques such as planar optics or micro-opto-electro-mechanical systems (MOEMS). Excellent overviews of microoptics can be found in [21, 22].

In this paper we give an overview of the novel possibilities that arise for quantum information processing with neutral atoms if one employs micro-fabricated optical elements. We show how crucial components for miniaturized systems for quantum information processing with neutral atoms can be realized with this approach [20, 23, 24].

22.3 Experimental Setup

For the experiments presented here, we employ a two-dimensional array of spherical, diffractive microlenses with a focal length of 625 μm and a lens diameter and separation of 125 μm (Fig. 22.1). The microlens array is made of fused silica and contains 50×50 diffractive lenslets. By focusing a laser beam with this array, one obtains an array of foci, with a separation of 125 μm. (Fig. 22.2a). For a red-detuned laser beam this results in an array of dipole traps, each analogous to a trap obtained by a single focused laser beam [25–27]. Figure 22.3 shows the first realization of this type of multiple dipole trap based on the application of microoptics [24].

The trapping light is derived from a 500 mW amplified diode laser system and is sent through a rubidium gas cell heated to a temperature of 110°C serving as a narrowband absorption filter. This reduces the strong background of amplified spontaneous emission by at least two orders of magnitude otherwise preventing the operation of a dipole trap due to scattering of resonant photons. The light is then sent through an acoustooptical switching device and through a polarizer, which ensures a high degree of linear polarization. The remaining light (typical power P = 100–200 mW, typical detuning $\Delta\lambda$ = 0.2 to 2 nm below the $5S_{1/2}(F = 3) \rightarrow 5P_{3/2}(F' = 4)$ transition at 780 nm ('red-detuning')) is focused by the microlens array. In order to have full optical access for atom preparation and detection, we image the focal plane of the microlens array onto a magnetooptical trap (MOT) with the help of two achromats (focal length of 300 mm, diameter of 50 mm, i.e. no significant reduction of numerical aperture), so that we can collect atoms in the MOT without any limitations imposed by the presence of the microstructure (Fig 22.2 (b)). Thus, we obtain an array of foci with a separation of 125 μm and a spot size of (7 ± 2) μm ($1/e^2$-radius of intensity). The optical transfer of the trapping light has the additional advantage that we can place the microoptical

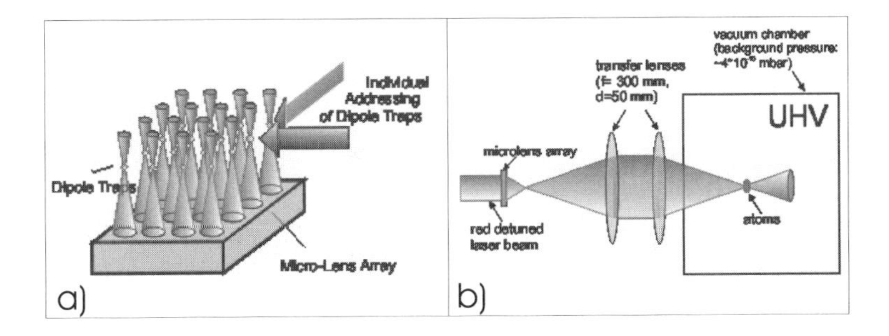

Figure 22.2: (a) A two-dimensional array of laser foci is created by focusing a single laser beam with an array of microlenses. (b) Optical setup for transfering the focal plane into the vacuum chamber.

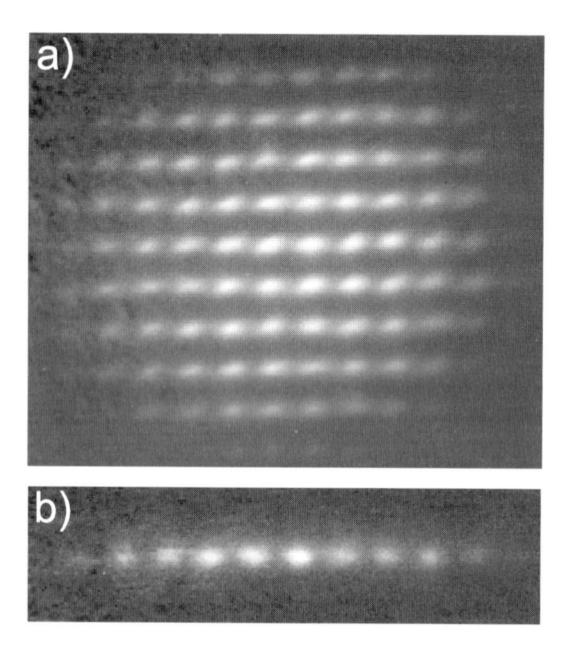

Figure 22.3: (a) Two- and (b) one-dimensional arrays of rubidium atoms trapped in arrays of dipole traps. The traps are created using a microoptical lens array and are separated by 125 μm. The brightest traps contain about 10^3 atoms.

system outside the vacuum chamber and thus can switch between a variety of microoptical elements easily.

We load the array of dipole traps and detect the trapped atoms similar to [28, 29]: We start with a MOT of 10^7 to 10^8 ^{85}Rb atoms which we overlap for several hundreds of ms with the dipole trap array and optimize the loading process for highest atom number. The MOT is then switched off and the atoms are held in the dipole traps for a variable storage time (typically 25 to 60 ms). This time is long enough for untrapped atoms to leave the detection region. The primary MOT light and the repumper are switched on again for a period of approximately 1 ms to detect the trapped atoms via spatially resolved detection of fluorescence with a spatial resolution of 17 μm (rms-spread of smallest observed structures).

22.4 Scalable Qubit Registers Based on Arrays of Dipole Traps

The structure obtained by the microlens array is suitable for a scalable qubit register [24]. The trapping potential is effectively independent of the computational state, realized as the hyperfine ground states of ^{85}Rb. So, quantum information inscribed into atomic states can be stored in such a register. In the experiment we obtain a two-dimensional array of approximately 80 well separated dipole traps with a potential depth of about 1 mK containing up to

10^3 atoms (Fig. 22.3a). The number of filled traps is limited by the size of the laser beam illuminating the microlens array and by the initial MOT size. The apparent larger extent of the individual traps in the horizontal direction in all images presented in this paper is caused by the detection optics being horizontally tilted relative to the beam axis of the trap light, necessary to avoid trapping light entering the camera aperture. The detection efficiency of our setup is already high enough to be able to detect atom samples of fewer than 100 atoms per trap. We are currently optimizing the detection efficiency to allow the observation of single atoms as well [27, 30, 31]. Illumination of only one row of the microlens array leads to a one-dimensional array of dipole traps (Fig. 22.3b). For the traps of this array (power per trap P = 3 mW, $\Delta\lambda$ = 0.4 nm) the calculated potential depth is $U_0/k_B = 2.5$ mK, which agrees within a factor of 2 with the one inferred from the measured radial oscillation frequency of 7.5 kHz [32]. The discrepancy can be fully explained by the known uncertainties in determining the laser power per trap, the focal waist, and the oscillation frequencies. The lifetime of the atoms in the traps is 35 ms, which is most probably limited by heating due to scattering of residual near resonant light not completely absorbed from the trapping beam [33]. Using a time-of-flight technique, we determined the atom temperature to be below 20 μK, which suggests the presence of an additional cooling mechanism during the loading phase as also observed in [34].

22.5 Initialization, Manipulation and Readout

In addition to its scalability, our approach is especially suited to fulfill another requirement for the physical implementation of quantum information processing, namely the ability to selectively address, initialize, and read out individual qubits [24]. The large lateral separation between the dipole traps enables us to selectively address the individual traps in a straightforward fashion. We demonstrate this by addressing a near-resonant laser beam onto one of the dipole traps for a few ms after the loading process is completed (see also [35]). This heats the atoms out of the addressed dipole trap. As can be seen in Figure 22.4 (a), no atoms are left at the site of the addressed dipole trap, while the atoms at the adjacent sites remain unaffected. Also only a single trap can be read out by addressing the dipole trap with a focused detection beam (Fig. 22.4b). So the other traps remain unaffected and can contribute to addition

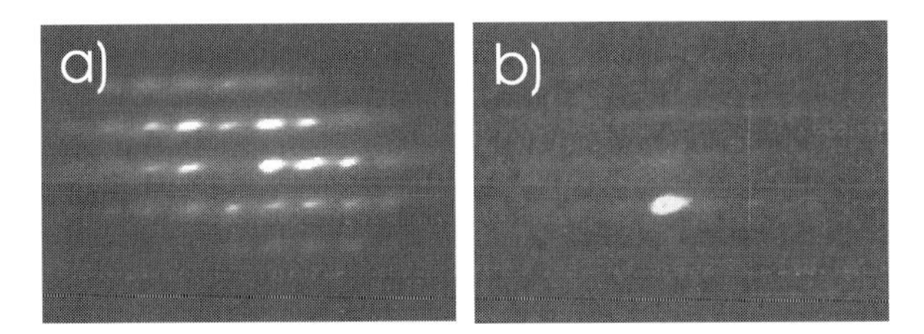

Figure 22.4: Demonstration of the selective addressability of individual trap sites: (a) By focusing a near-resonant laser beam onto one of the dipole traps during the storage period, the atoms in this trap are removed, while the other dipole traps remain unaffected. (b) One trap can be illuminated and the qubit state read without disturbing the atoms in the other traps.

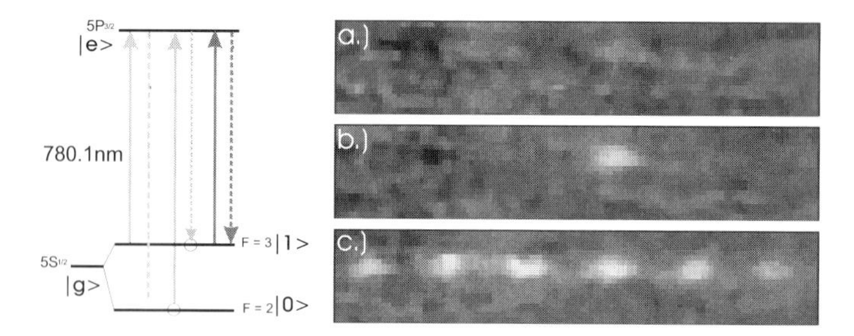

Figure 22.5: Site-specific and state-selective initialization and readout of quantum states. All sites of a one-dimensional array of dipole traps are filled with atoms. For detection only light resonant with the $5S_{1/2}(F = 3) \rightarrow 5P_{3/2}(F' = 4)$ transition is applied. (a) All atoms are in the F=2 state, they do not scatter the detection light. (b) The atoms at one site have been transfered to the F=3 state and can be detected. (c) The atoms at all sites have been transfered to the F=3 state.

calculations. By two-dimensional scanning of the addressing beam or by illuminating each lenslet individually with spatially modulated addressing light, every site can be addressed individually. This opens the possibility to selectively prepare and manipulate the qubits in the individual traps.

As a next step, we demonstrated the site-specific and state selective initialization and readout of atomic quantum states (Fig. 22.5) [24]. Here, we illuminate a one-dimensional array (analogous to Fig. 22.3b) only with light resonant with the $5S_{1/2}(F = 3) \rightarrow 5P_{3/2}(F' = 4)$ transition (i.e. the repump light switched off) during detection. Since the atoms are almost exclusively in the lower hyperfine groundstate $5S_{1/2}(F = 2)$ after the loading phase, they do not scatter the detection light (Fig. 22.5a) unless we actively pump them into the upper hyperfine groundstate $5S_{1/2}(F = 3)$ during the time the atoms are stored in the dipole traps, i.e. prior to the detection phase. This could be demonstrated for one (Fig. 22.5b) or, alternatively, all (Fig. 22.5c) of the trap sites. This demonstrates the site-specific and state-selective initialization and detection capability of our approach.

22.6 Variation of Trap Separation

While many of the advantages of our system result from the large lateral separation of the individual trap sites, it is also possible to actively control the distance between individual traps if smaller or adjustable distances are required, e.g. for quantum gate operations and the entanglement of atoms via atom-atom interactions. In our setup, this can be accomplished by illuminating one microlens array with two beams under slightly different angles (Fig. 22.6a) [24] (see also [27]), which results in two interleaved sets of arrays of trapped atoms. Interference effects between the two laser beams are prevented by using orthogonal linear polarizations. Figure 22.6b shows two vertically displaced sets of arrays of trapped atoms with a mutual separation of 45 μm. The separation only depends on the angle between the two laser beams and can easily be changed, especially to smaller values. By reducing the relative angle to zero, overlapping traps can be created. Figure 22.6c shows vertical cross-sections through one pair

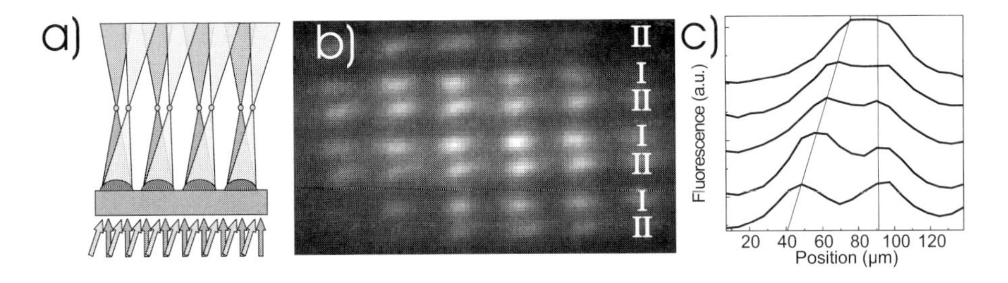

Figure 22.6: (a) and (b): Two interleaved sets of arrays of trapped atoms (Arrays I and II in (b)) are obtained by illuminating one microlens array with two beams under slightly different angles. The distance between traps of the two arrays is approximately 45 μm. (c) Cross-sections through pairs of dipole traps as in (b) (rotated by 90°) for different separations between the traps ranging from 18 μm to 45 μm. Image sections showing two pairs of traps with decreasing separation from bottom to top corresponding to the cross-sections at left.

of dipole traps for different angles between the two laser beams and thus different site separations. We could also change the separation of the traps and thus move the atoms in one set of traps in real time by deflecting one beam with the help of a fast acoustooptical beam deflector.

With this technique it should become possible to selectively prepare qubits at large distances, then decrease the distance for gate operations and entanglement via atom-atom interactions and then increase the distance again for the read out of the qubit states.

22.7 Implementation of Qubit Gates

In order to demonstrate the full potential of our approach for quantum information processing, we evaluate the criteria given in [11]. Scalability, site-specific and state-selective initialization and readout, as well as the ability to change the separation of trapping sites at will are the most important advantages of our approach, and have been demonstrated in the previous sections. In addition, the concept presented here and the underlying technology fulfill the remaining two criteria, namely long relevant decoherence times and the availability of a universal set of quantum gates as well.

Single qubit gates can easily be implemented relying on the existing concepts and technology for the coherent manipulation of internal atomic states. In addition, a large majority of the proposed two-qubit gates can be realized with our approach as well. A suitable implementation of a two-qubit phase gate, based on controlled cold collisions can be achieved with state selective switching of the trapping potentials as proposed for magnetic microtraps [5, 6]. Atoms are prepared in the microtraps provided by the microoptical lens array and the qubits are initialized in the previously discussed way. The hyperfine ground states $5S_{1/2}(F = 2)$ can be denoted as the computational state $|0\rangle$ and the $5S_{1/2}(F = 3)$ as the computational state $|1\rangle$. The gate operation will be induced by overlapping two microtraps (trapping potential) with an additional potential (gate potential). This potential connects the two trapping potentials with a shallow well along the axis connecting the two microtraps but strongly confines the atoms in the other directions. The resulting potential is shown in Figure 22.7 (a). The next step is

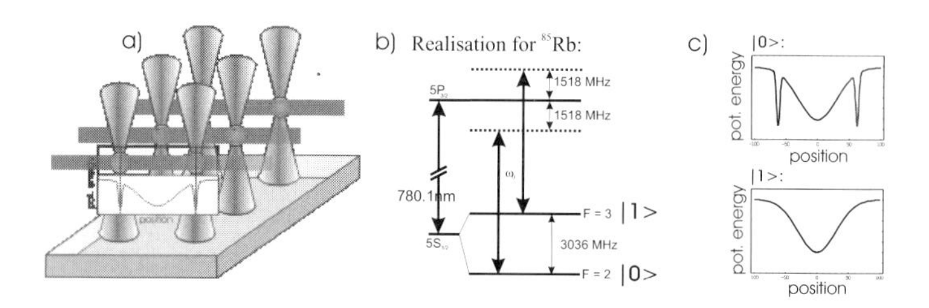

Figure 22.7: Possible implementation of a phase gate: (a) Two microtraps are connected by a gate potential. (b) Realization of state-selective potentials. (c) Resulting trapping potentials for the atoms in the different qubit states.

to overlap the trapping potential with a state selective potential, which can be realized by illuminating the microlens array in addition to the trapping light with a state selective light field. The state-selective potential for ^{85}Rb can be realized with a laser field with a detuning $\Delta\nu$ of 1.5 GHz (250 linewidths) below (red-detuning) the $5S_{1/2}(F = 2) \rightarrow 5P_{3/2}$ transition. This same laser field is blue-detuned for atoms in the F=3 state with a detuning of $\Delta_{HFS} - \Delta\nu =$ 1.5 GHz (with Δ_{HFS} the hyperfine splitting). (Fig. 22.7b)

Choosing the right intensities during the gate operation for the trapping and the state selective light fields, the potential for atoms in state $5S_{1/2}(F = 2)$ will not change. Atoms in the $5S_{1/2}(F = 3)$ state are confined only by the gate potential with the trapping and the state-selective potentials compensating each other (Fig. 22.7c). During the gate operation the trapping potentials have to be laterally shifted a small distance apart from each other to avoid uncontrolled interaction of the atoms.

If the two atoms are in state $|0\rangle$ there is no change in their potential (Fig. 22.8a). If one of the atoms is in state $|1\rangle$, this atom starts to oscillate in the gate potential without interacting with the other atom (Figs. 22.8b, c). If both atoms are in the state $|1\rangle$, they will oscillate in the gate potential and interact due to cold collisions (Fig. 22.8d). After a certain interaction time, the atoms will accumulate a phase equal to π and the gate and state-selective potentials are switched off when the atoms are back at their initial position, so they can be trapped again in the microtraps.

With the specific lens array we used here (beam waist 7 μm) and a standard set of laser parameters (Ti:Sapphire laser, wavelength 825 nm, power per trap 50 mW) several tens of traps with radial vibrational frequencies of 10 kHz and a decoherence time (assumed to be limited by spontaneous scattering) of 50 ms can be obtained [24]. These parameters allow the implementation of collisional gates [5, 6] with a decoherence time being at least 50 times longer than the gate time.

The full potential of our concept can be exploited by using available lens arrays optimized for large numerical aperture: it has been shown experimentally [36] that beam waists smaller than 1 μm can be achieved with microlens arrays similar to the one used in this work. With standard laser parameters (wavelength 850 nm, power per trap 1 mW) radial vibrational frequencies of 50 kHz and decoherence times of 150 ms can be obtained. This allows the implementation of various types of quantum gates [5, 8–10].

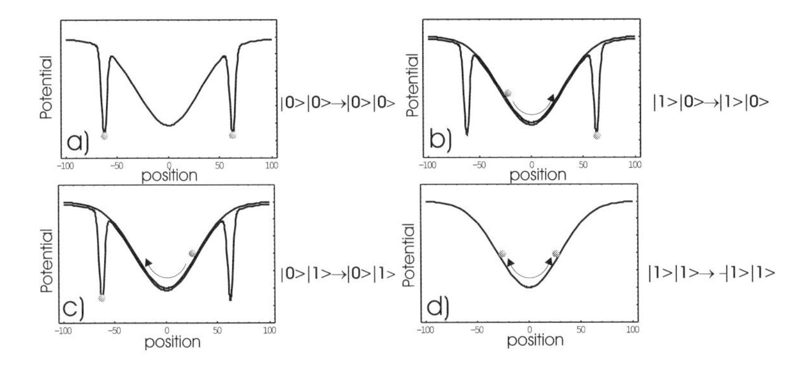

Figure 22.8: Possible implementation of a phase gate: (a) Both atoms in state $|0\rangle$ remain separated. (b) and (c) If one of the atoms is in state $|1\rangle$, this atom starts to oscillate in the gate potential, without interacting with the other atom. (d) For both atoms in the state $|1\rangle$, they will oscillate in the gate potential and interact due to cold collisions.

One class are motional gates, where the qubits are encoded in the two lowest vibrational states of each trap. These quantum gates are based on tunneling and cold collisions via adiabatically approaching neighboring traps. With this mechanism one can realize a phase gate [9] or a $\sqrt{SWAP}$ gate [10]. For a $\sqrt{SWAP}$ gate, the gate time is on the order of a few tens of milliseconds, assuming the previously discussed parameters. The spontaneous scattering time is $\sim$ 150 ms. However for strongly confined atoms, only a small fraction of the spontaneous scattering processes leads to a change in the vibrational state [37], giving significantly larger coherence times.

Due to its potentially short gate times and its insensitivity to the temperature of the atoms and to the variations in atom-atom separation, the Rydberg gate [8] deserves specific attention: Using the parameters and scaling laws discussed in [8], for atoms in microlens dipole traps with a waist and a respective minimum trap separation of 1 µm, a gate time of about 1 µs can be achieved. This gate time is 20 times shorter than the oscillation period and 10^4 times shorter than the decoherence time, giving favorable experimental conditions. Further evidence for the potential of this approach can be drawn from the fact that this array of 1 µm-waist dipole traps represents a 2D extension of the dipole trap demonstrated in [27]. For this trap a detailed investigation [38] gives proof for the feasibility of Rydberg gates and explicitly shows that for trap separations of 1 to 5 µm, gate times of 1 to 10 µs are possible.

We conclude this discussion by pointing out that since our dipole trap arrays can give the same trap parameters as single dipole traps or standing wave dipole potentials, the single atom loading schemes experimentally demonstrated in [27] and [31] can be extended to our configurations in a straightforward fashion. In addition, with the achievable vibrational frequencies being larger than the recoil frequency, sideband-cooling of atoms to the vibrational ground state of the trap arrays should be possible. Finally, the demonstration of the loading of periodic dipole potentials with single atoms from a Bose-Einstein condensate [39] might allow an extremely efficient means of single atom loading into vibrational ground states of our trap arrays.

Acknowledgements

This work is also supported by the project ACQUIRE (IST-1999-11055) of the European Commission. We thank D. Bruß, K. Eckert, J. Mompart, A. Sanpera, and M. Lewenstein for stimulating discussions and a productive collaboration.

References

[1] for a summary see: A. M. Steane, Rep. Prog. Phys. **61**, 117 (1998); J. Gruska, *Quantum Computing*, (McGraw-Hill, London, 1999); D. Bouwmeester, A. Ekert, and A. Zeilinger (ed.), *The Physics of Quantum Information*, (Springer, Berlin, 2000), and references therein.

[2] Special Issue of Fortschritte der Physik **48**, issue 9-11 (2000) and references therein.

[3] C. Monroe, D. M. Meekhof, B. E. King, S. R. Jeffers, W. M. Itano, D. J. Wineland, and P. Gould, Phys. Rev. Lett. **74**, 4011 (1995); C.A. Sackett, D. Kielpinski, B. E. King, C. Langer, V. Meyer, C. J. Myatt, M. Rowe, Q. A. Turchette, W. M. Itano, D. J. Wineland, and C. Monroe, Nature **404**, 256 (2000); M. A. Rowe, D. Kieplinski, V. Meyer, C. A. Sackett, W. M. Itano, C. Monroe, and D. J. Wineland, Nature **409**, 791-794 (2001).

[4] E. Hagley, X. Maître, G. Nogues, C. Wunderlich, M. Brune, J. M. Raimond, and S. Haroche, Phys. Rev. Lett. **79**, 1 (1997).

[5] D. Jaksch, H. J. Briegel, J. I. Cirac, C. W. Gardiner und P. Zoller, Phys. Rev. Lett. **82**, 1975 (1999).

[6] T. Calarco, E. A. Hinds, D. Jaksch, J. Schmiedmayer, J. I. Cirac, and P. Zoller, Phys. Rev. A **61**, 022304 (2000).

[7] G. K. Brennen, C. M. Caves, P. S. Jessen, and I. H. Deutsch, Phys. Rev. Lett. **82**, 1060 (1999).

[8] D. Jaksch, J. I. Cirac, P. Zoller, S. L. Rolston, R. Côte, and M. D. Lukin, Phys. Rev. Lett. **85**, 2208 (2000).

[9] E. Charron, E. Tiesinga, F. Mies, and C. Williams, Phys. Rev. A **88**, 077901 (2002).

[10] K. Eckert, J. Mompart, X. X. Yi, J. Schliemann, D. Bruss, G. Birkl, and M. Lewenstein, Phys. Rev. A 66, 042317 (2002); J. Mompart, K. Eckert, W. Ertmer, G. Birkl, and M. Lewenstein, Phys. Rev. Lett. **90**, 147901 (2003).

[11] D. P. DiVincenzo, Fortschr. Phys. **48**, 9 (2000).

[12] J. D. Weinstein and K. G. Libbrecht, Phys. Rev. A **52**, 4004 (1995).

[13] For an overview of early experimental work see: E. A. Hinds and I. G. Hughes, J. Phys. D: Appl. Phys. **32**, R119 (1999).

[14] J. Schmiedmayer, Eur. Phys. J. D **4**, 57 (1998); R. Folman, P. Krüger, D. Cassettari, B. Hessmo, T. Maier, and J. Schmiedmayer, Phys. Rev. Lett. **84**, 4749 (2000), D. Cassettari, B. Hessmo, R. Folman, T. Maier, and J. Schmiedmayer, Phys. Rev. Lett. **85**, 5483 (2000).

[15] J. Reichel, W. Hänsel, and T. W. Hänsch, Phys. Rev. Lett. **83**, 3398 (1999); W. Hänsel, J. Reichel, P. Hommelhoff, and T. W. Hänsch, Phys. Rev. Lett. **86**, 608 (2001).

[16] D. Müller, D. Z. Anderson, R. J. Grow, P. D. D. Schwindt, and E. A. Cornell, Phys. Rev. Lett. **83**, 5194 (1999).

[17] N. H. Dekker, C. S. Lee, V. Lorent, J. H. Thywissen, S. P. Smith, M. Drndić, R. M. Westervelt, and M. Prentiss, Phys. Rev. Lett. **84**, 1124 (2000).

[18] P. Engels, W. Ertmer, and K. Sengstock, Opt. Comm. **204**, 185 (2002).

[19] M. Key, I. G. Hughes, W. Rooijakkers, B. E. Sauer, and E. A. Hinds, Phys. Rev. Lett. **84**, 1371 (2000).

[20] G. Birkl, F. B. J. Buchkremer, R. Dumke, and W. Ertmer, Opt. Comm. **191**, 67 (2001).

[21] H. P. Herzig (ed.), *Micro-Optics*, (Taylor & Francis, London, 1997).

[22] S. Sinzinger and J. Jahns, *Microoptics*, (Wiley-VCH Verlag, Weinheim, 1999).

[23] F. B. J. Buchkremer, R. Dumke, M. Volk, T. Müther, G. Birkl, and W. Ertmer, Laser Physics **12**, 143 (2002).

[24] R. Dumke, M. Volk, T. Münther, F. B. J. Buchkremer, G. Birkl, W. Ertmer, Phys. Rev. Lett. **89**, 097903 (2002).

[25] S. Chu, J. E. Bjorkholm, A. Ashkin, and A. Cable, Phys. Rev. Lett. **57**, 314 (1986).

[26] for an overview see R. Grimm, M. Weidemüller, and Y. B. Ovchinnikov, Adv. At. Mol. Opt. Phys. **42**, 95 (2000).

[27] N. Schlosser, G. Reymond, I. Protsenko, and P. Grangier, Nature **411**, 1024 (2001).

[28] S. J. M. Kuppens, K. L. Corwin, K. W. Miller, T. E. Chupp und C. E. Wieman, Phys. Rev. A **62**, 013406 (2000).

[29] S. Dürr, K. W. Miller, C. E. Wieman, Phys. Rev. A **63**, 011401(R)

[30] Z. Hu and H. J. Kimble, Opt. Lett. **19**, 1888 (1994); F. Ruschewitz, D. Bettermann, J. L. Peng, and W. Ertmer, Europhys. Lett. **34**, 651 (1996); D. Frese, B. Ueberholz, S. Kuhr, W. Alt, D. Schrader, V. Gomer, D. Meschede, Phys. Rev. Lett. **85**, 3777 (2000).

[31] S. Kuhr, W. Alt, D. Schrader, M. Müller, V. Gomer, Science **293**, 278 (2001).

[32] We measure this frequency by parametrically heating the atoms and observing the number of remaining atoms.

[33] We could observe an increase in trapping time to more than 2 seconds by switching to a Ti:Saphire laser as light source.

[34] M. D. Barrett, J. A. Sauer, and M. S. Chapman, Phys. Rev. Lett. **87**, 010404 (2001).

[35] R. Scheunemann, F. S. Cataliotti, T. W. Hänsch, and M. Weitz, Phys. Rev. A **62**, 051801 (2000).

[36] Th. Hessler, M. Rossi, J. Pedersen, M. T. Gale, M. Wegner, D. Steudle, and H. J. Tiziani, Pure Appl. Opt. **6** 673 (1997).

[37] F. B. J. Buchkremer, R. Dumke, H. Levsen, G. Birkl, and W. Ertmer, Phys. Rev. Lett. **85**, 5438 (2000).

[38] I. E. Protsenko, G. Reymond, N. Schlosser, and P. Grangier, Phys. Rev. A. **65**, 052301 (2002).

[39] M. Greiner, O. Mandel, T. Esslinger, T. W. Hänsch, and I. Bloch, Nature **415**, 39 (2002).

23 Quantum Information Processing with Neutral Atoms on Atom Chips

Peter Krüger, Albrecht Haase, Mauritz Andersson, and Jörg Schmiedmayer

Physikalisches Institut
Universität Heidelberg
Heidelberg, Germany

23.1 Introduction

Neutral atoms are promising candidates for the physical implementation of quantum information processing (QIP). The advantages of neutral atom implementations of quantum bits (qubits) are the weak interaction of neutral atoms with their environment, leading to long coherence times, and the mature experimental techniques from quantum optics and atomic physics to prepare and manipulate atomic quantum systems. Bose–Einstein condensation and precision measurements are some prominent examples. Our approach is to combine the quantum manipulation expertise of controlling atoms with the advantages of micro-fabrication technology from micro-electronics and micro-optics. This promises to build integrated devices (atom chips [1]) for complex and robust quantum manipulation of ultra-cold atomic samples and a collection of single atoms.

The requirements for an implementation of quantum information processing are discussed at length in [2]. Here we will focus on implementations in the framework of a simplified subset of these requirements: (i) storage of the quantum information in a set of two-level systems (qubits), (ii) processing of this information using quantum gates, and (iii) readout of the results.

The chapter is structured as follows. We first describe the atom chip and its implementation, and will then discuss qubits encoded in neutral atoms and their one- and two-qubit operations. Even though most of the neutral atom QIP proposals can in principle be implemented on the atom chip, we will concentrate mainly on the 'collisional' gate implementations [3–6]. The atom chip adds the advantage of robust manipulation of individual qubit sites at submicrometer precision. This technology has the potential of a massively parallel system.

23.2 The Atom Chip

Miniaturization of the structures used as atom optical components for atom manipulation is advantageous for a number of reasons: (i) Large electric and magnetic field gradients and field curvatures near microscopic conductors lead to tight confinement and large energy level spacings for the trapped atoms. (ii) The resolution of the potentials is given by the resolution

Quantum Information Processing, 2nd Edition. Edited by Thomas Beth and Gerd Leuchs

ISBN: 3-527-40541-0

of the structures used to create them (down to 1 μm used in experiments so far and close to 100 nm seems feasible). This is important for QIP proposals in which the distances between individual trapping sites is required to be in the submicrometer regime to achieve sizable qubit coupling and fast gate operations (see Sec. 23.4). (iii) Nanofabrication techniques on surfaces facilitate accurate and robust placement of structures. This enables the realization of nearly arbitrary potential configurations. The substrate serves as heat sink for the power dissipated in the current-carrying wires. (iv) Integration of new components, for example micro-optics and micro-cavities for preparation, manipulation, and detection of qubits are possible. (v) Last, but not least, nanofabrication schemes are particularly well suited for production of multiple structures (a prerequisite for potential scalability).

Starting from the early 1990s, the fundamental concepts of microscopic atom optics were developed (see [1] and references therein): The basic concepts of creating steep and spatially strong confining potentials for atoms by bringing them close to current- and/or charge-carrying microstructures were first validated experimentally as early as 1992, using free-standing wires and atom beams [7]. Later, the experiments were extended down to the submicrometer range. Guides and traps were then transferred to nanofabricated surfaces, and the concept of an integrated atom chip was developed [8–12]. Standard nanofabrication technology of semiconductor research laboratories is employed in the production of the atom chip (Fig. 23.1).

Even though the vast capabilities of atom chips [1] have by far not been explored exhaustively to date, a set of tools useful for QIP has been established (Fig. 23.2): The tightly confining chip traps are ideal to create Bose–Einstein condensates (BECs) [13–16] as a coherent source of atoms for QIP experiments. Atomic samples were loaded into a great variety of traps and guides that achieved strong confinement and proximity to the chip surface (for an overview see [1]). Trap frequencies in the range of hundreds of kHz were observed at trap surface distances down to a few micrometers. More complex static and time-dependent potentials were used to move, split, and recombine atomic samples [1, 17, 18].

In the context of QIP, qubit state selective manipulation is especially interesting. In the next two subsections we describe two different ways of implementing such state-selective potentials: first in a combination of electric and magnetic fields; second, by creating adiabatic RF- or microwave-induced potentials.

23.2.1 Combined Magneto–Electric Traps

Combining electric ($U_{\text{el}} = -\frac{1}{2}\alpha E^2$) and magnetic ($U_{\text{mag}} = -\vec{\mu} \cdot \vec{B}$) potentials allows state-dependent trapping and manipulation. The magnetic part of these potentials depends on the internal (magnetic) hyperfine state of the atoms while the electric part does not. The feasibility of a combination of electric and magnetic fields for the manipulation of matter waves on atom chips has recently been demonstrated [19], first in experiments with thermal ^{7}Li atoms, and later with BECs. The transverse confinement is provided by a magnetic (side guide) potential and the longitudinal confinement is introduced by an appropriately modulated electric field. The electric field was formed by a number of charged electrodes placed near the guiding wire.

In addition to simple trapping, the additional degrees of freedom allow atom clouds to be transported or split along the guide in a controlled way when time-dependent voltages are applied to the electrodes. An experiment with atoms in two different magnetic substates

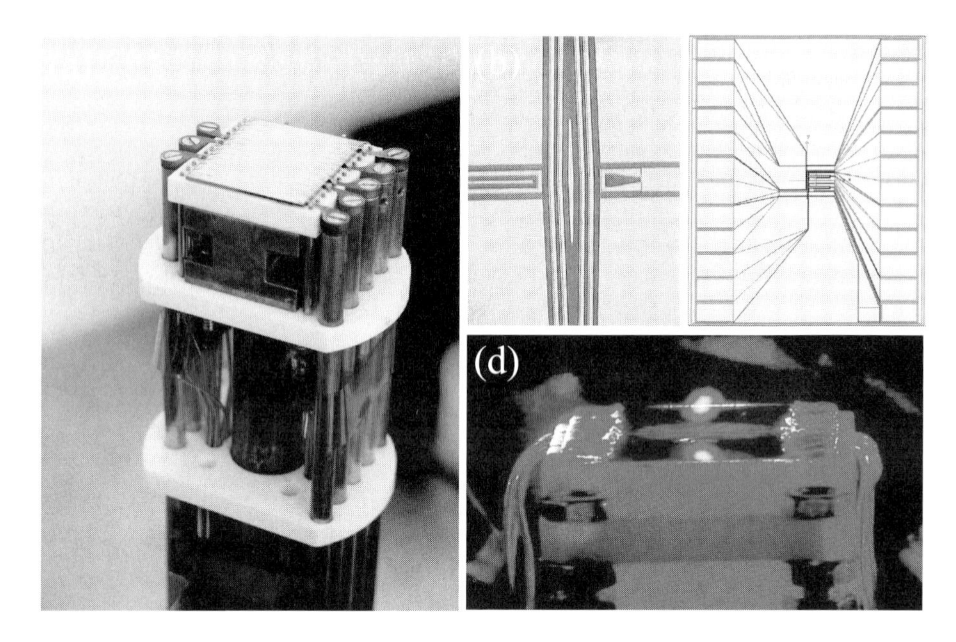

Figure 23.1: Atom chips: (a) Mounted chip just before it is introduced into the vacuum chamber. (b) Microscope picture of an atom chip surface: Wires are defined by electrically isolating grooves fabricated into a gold surface of a few micrometers thickness. In this case they form a four-wire interferometer. (c) A chip lithography mask containing various structures such as Z-traps and a guided interferometer. (d) The reflectivity of the gold surface of the chip is hardly obstructed by the wire defining grooves. This allows the chip itself to be used as a mirror for laser cooling, facilitating a simple loading of the chip traps and guides. Here, the fluorescence light of a cloud of lithium atoms in a magneto–optical trap is shown.

($F = 1, m_F = -1$ and $F = 2, m_F = 2$) showed that the voltage ranges for optimal trapping depend on m_F as expected for state-dependent operation [20].

23.2.2 RF-induced Adiabatic Potentials for Manipulating Atoms

An additional degree of freedom is obtained by combining static magnetic potentials with local radio-frequency (RF) or microwave (MW) fields. Tuning the RF or MW close to a magnetic or hyperfine transition creates modified potential surfaces, which can be understood as adiabatic potentials in a dressed state picture. For example, one can realize a double-well potential by splitting an elongated trap in a transverse direction using properly polarized RF magnetic fields (see Fig. 23.3).

RF fields couple different magnetic substates within a hyperfine manifold; microwaves couple different hyperfine states. Since only the near field of the RF (MW) and not the radiating field is of importance, one can realize such adiabatic potentials in a localized way by using chip wires for the RF (MW) currents. This leads directly to new state-dependent local potentials for qubit manipulation and gate operations.

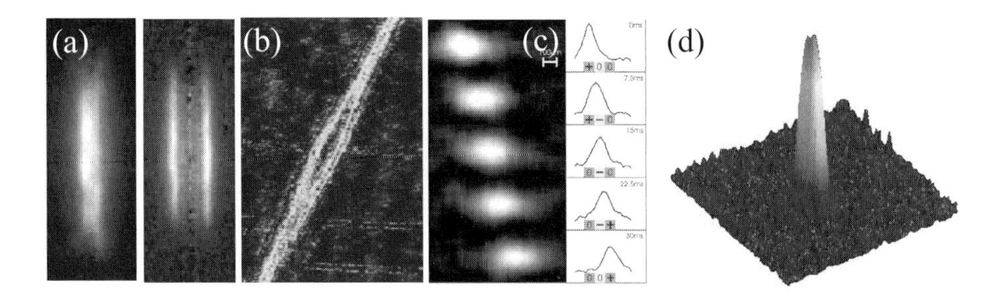

Figure 23.2: Examples of atom chip experiments: Several ways of splitting atomic clouds have been demonstrated. Such beam splitters are the building blocks of atom interferometers, which will be used to assess the coherence properties of the evolution of atomic wave functions. The splitting can be done by employing either (a) time-dependent potentials or (b) spatially varying potentials. (c) Combined magneto–electric potentials are used for moving an atom cloud. (d) Bose–Einstein condensation in a surface trap constitutes an important step toward the realization of QIP on atom chips, since a large sample of atoms in a defined quantum state is available.

23.2.3 Imperfections in the Atom Chip: Disorder Potentials

For efficient QIP on atom chips, it is important that the manipulation potentials can be structured with a spatial resolution of micrometers or better. This is only possible if the atom–surface distances can be reduced to the same order. However, in a number of experiments it was found that, even for comparably warm atom clouds ($T \sim 5$ μK), uncontrolled disorder potentials start to affect the trapping potentials at large distances ($d \lesssim 100$ μm) [21–24]. These disorder potentials have been considered to severely limit the scope of the atom chip concept.

Recently, we were able to show that the fabrication process of the chip is critical in this context. We use a purely lithographic lift-off procedure, which yields much smoother wire structures than the alternative electroplating techniques [25]. With our chips the disorder potentials are about 100 times weaker [20]. We find that atoms can be confined to the transverse zero-point oscillation in a potential whose longitudinal roughness is significantly smaller than the transverse level splitting.[1] In addition the roughness is dominated by large-scale fluctuations. A first analysis indicates a power spectrum quadratic in the length scale of the potential roughness. Within current limits, controlled quantum manipulation seems to be feasible. Further improvement by at least one order of magnitude was achieved by separating the trapping from the the high-resolution structuring of the potentials. This is possible when the actual trapping configuration (on larger scales) is obtained by using larger cross-section wires at safe distances from the atoms (typically tens of micrometers).

[1]The magnetic potential has a relative longitudinal variation of the magnetic field $\Delta B/B$ that is smaller than 10^{-5} on a length scale of 50 μm.

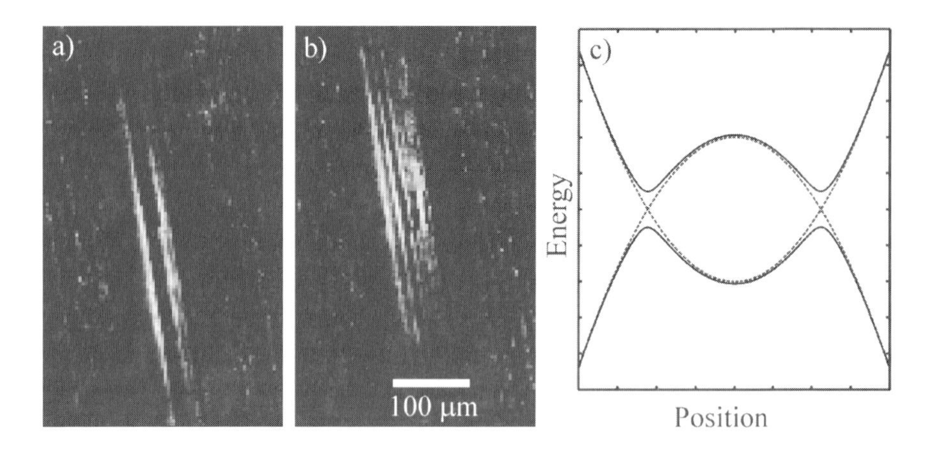

Figure 23.3: In-situ absorption picture of a BEC in a 100 μm wide wire trap 95 μm above the surface. The imaging axis has a small tilt with respect to the surface normal, hence two images are seen (image (b) is affected by artifacts due to structures on the mirror surface). The parameters for the trap are trap bottom 620 kHz, cooling RF 650 kHz. In (a), there is no RF modification of the trap. In (b), with an RF ramp from 500 kHz to 1.2 MHz, the trap is split into a wide double well and the BEC is split in two. (c) Example of RF-induced adiabatic potentials (solid lines) and uncoupled but RF frequency-shifted magnetic potentials (dashed). Atoms can be trapped in the upper adiabatic potential, which is split.

23.3 The Qubit

There are two distinct ways to encode a qubit into a neutral atom: by using (i) a long-lived internal state such as a hyperfine ground state; and (ii) external, motional states of the atom. These can be either the ground and an excited state in a trap, or the left and right states of a double well. In most proposed realizations, each qubit is written into a single atom, which requires selective cooling and deterministic filling of single atoms into the qubit sites. For this qubit loading, most proposals rely on the BEC as a coherent source of atoms from where one can deterministically extract single atoms in well defined states by, e.g., atom tweezers [26,27] or the use of the Mott-insulator phase transition [28–30].

We first discuss the qubit encoded in two internal, long-lived states (e.g. two different hyperfine electronic ground states) of an atom. An example would be the $|F = 1\rangle$ and $|F = 2\rangle$ ground states of the ^{87}Rb atom. To achieve long coherence times, it is advisable to use atomic states where the energy difference is to first order independent of external fields (clock states). In free space these would be the $|F = 2, m_F = 0\rangle$ and $|F = 1, m_F = 0\rangle$ states, which show only quadratic Zeeman and Stark shifts. On the atom chip we need magnetically trappable states. In ^{87}Rb one finds the $|F = 2, m_F = 1\rangle$ and $|F = 1, m_F = -1\rangle$ states to have the same magnetic moment at a magic magnetic field of $B = 3.23$ G. Operating a trap at this field, this pair of trappable states shows no first order dependence on external field fluctuations as the $m_F = 0$ clock states.

Single-qubit operations are induced as transitions between the hyperfine states of the atoms. These can be driven by external fields, using RF and/or MW pulses or optical Raman transitions. As an example we mention the atom chip experiment in Munich [31], which demonstrated very long coherence times in a magnetic chip trap for the special states $|F = 2, m_F = 1\rangle$ and $|F = 1, m_F = -1\rangle$.

Another possibility to encode qubits would be to use the motional states of the trapped atom [5, 32], e.g., the $|0\rangle$ and $|1\rangle$ vibrational levels of a potential well, or the $|left\rangle$ and $|right\rangle$ state in a double-well potential. Such realizations have the advantage that atoms are in the same internal state, and therefore better isolated from external fluctuations. Single-qubit operations are then Rabi rotations between the motional states, and can be seen like trapped atom interferometers. Atoms trapped in different motional states are easier to state-selectively manipulate, but much less is known on external state decoherence, and how to prevent it.

The above realizations of a qubit are based on single atoms. However, it was proposed that a single qubit can also be written into an ensemble of atoms using the dipole blockade [33]. This may be simpler, as it avoids the need for single atom loading into traps. On the other hand, it may lead to strong undesired couplings within the sample. The dipole blockade mechanism can also be used to manipulate the qubit.

23.4 Entangling Qubits

The fundamental two-qubit quantum gate requires state-selective interaction between two qubits, which is more delicate to implement. A two-qubit quantum gate is a state-dependent operation such as a phase gate where $|1\rangle|1\rangle \rightarrow e^{i\phi}|1\rangle|1\rangle$ and the other two-qubit states should remain unaffected.

To achieve different evolutions for different qubit states, either the interaction between the qubits has to be state-selective, or it has to be turned on conditioned on the qubit state. There are different ways to implement quantum gates in atom optics, depending on the type of interaction. Here we distinguish between (i) generic interactions between the atoms, like scattering interaction [3, 4], and (ii) interactions that can be switched on and off, e.g., dipole–dipole interactions [33–35] or locally tuned Feshbach resonance for state-dependent phase shift. In addition one can use a data bus to mediate the interactions, like in the cavity QED-based protocols for two-qubit gate operations, which involve photons inside a cavity directly as a data bus mediating the state-selective interaction between two atoms trapped inside the resonator [36]. Another method to mediate selective interactions is to use a marker atom and increase the interaction at that site [4], realizing a selective and state-dependent interaction. In the subsequent discussion we will focus on controlled collisions since they are the most straightforward to implement on atom chips.

23.4.1 Quantum Gate via Cold Controlled Collisions

One of the first proposed schemes is based on entangling atoms carrying internal state qubits by controlled collisions. State-dependent time-varying potentials will allow only one pair of states to collide and thereby obtain a phase shift [3, 4] (Fig. 23.4). Such collisional gates require selectively addressable traps and a strong transverse confinement to ensure a sizable phase shift for each collision and are therefore well suited for atom chips. The qubits are atoms in different (magnetic) substates held in state-dependent traps, which are formed by superposing magnetic, electric, and RF potentials.

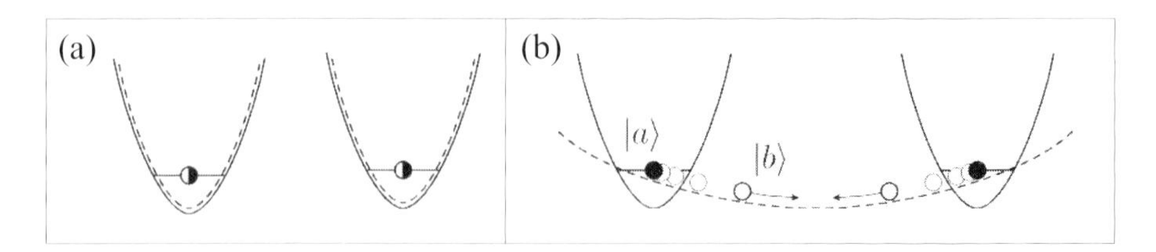

Figure 23.4: Switching potentials for QIP: configurations (a) before and after, and (b) during the gate operation. The solid (dashed) curves show the potentials for atoms in different internal states $|a\rangle$ and $|b\rangle$.

These collisional gates can reach high fidelity in idealized settings [4], but imperfections were found to be relevant. Timing errors as well as non-perfect harmonic traps can lead to states outside of the computational basis of the gate. The two effects cancel in first order if one adjusts the switching time according to the trap anharmonicity. In addition, the conditional phase induced in the two-qubit gate depends on additional parameters that have to be kept under precise control, i.e., trap frequency. Thus at least two parameters have to be precisely controlled in order to have a high fidelity. Finding ways to reduce this to a single or no parameter is important if a robust gate is to be implemented. More sophisticated methods from quantum control can be applied to keep high fidelity in realistic atom chip settings.

An important parameter to consider in the collisional gate are loss and decoherence mechanisms due to the two-body collision. Most significantly these are internal state-changing collisions, which can happen for atoms in different internal states. For ^{87}Rb these are negligible due to cancellation [37], leaving three-body collisions and surface effects as the dominant loss mechanisms. In the context of very tight confinement in two or in all three dimensions, as can be achieved on an atom chip (ground-state sizes on the order of the scattering length and interaction energies between two atoms larger than the level spacing), the collision properties of the atoms may change radically as discussed in one-dimensional physics [38]. This will be a critical issue to be studied in the future.

State-selective potentials can be realized with a combination of magnetic and electric interaction as discussed above, provided that the two internal states for the qubit have different magnetic moments. In these implementations, the decohering effect of potential differences for the two qubit states needs to be thoroughly reviewed.

By combining RF and microwaves, it will be possible to implement the above schemes also for states having the same magnetic moment, such as, e.g., the $|F = 2, m_F = 1\rangle$ and $|F = 1, m_F = -1\rangle$ states in ^{87}Rb. Using different local polarizations and detunings of the RF and MW fields allows one to create state-dependent adiabatic potentials and perform state-selective controlled collisions. The atom chip gives the possibility of a localized application of the RF and MW fields by integrating waveguides on the chip.

23.4.2 Motional Qubit Gates with Controlled Collisions

If the qubit is encoded in motional states (e.g. in the ground and first excited state of a trap), a conditional two-qubit operation can be achieved by lowering and raising the barrier between the two individual trapping sites of a double-well potential [5,32]. As the tunneling probabilities between the two wells depend exponentially on the relative difference between qubit state energy and barrier height, this allows state-conditional atom–atom interactions to be induced. A precise interferometric scheme ensures that the two atoms end up in their original individual wells. For this to work with high fidelity, the confinement has to be very strong and care has to be taken that states do not mix. Using the atom chip gives the advantage of selectively manipulating the potentials.

23.5 Input/Output

Input (setting a qubit to the initial condition) and output (reading a qubit) are elementary processes in every QIP scheme. The input of classical information, i.e. information that one can write down on a piece of paper, can be solved by setting each qubit to its ground state (ground-state cooling), with subsequent single qubit operations. In contrast, the input of quantum information is a much more difficult task due to the complicated nature of entangled states. Similarly, (destructive) readout of a classical bit from a quantum register is a much easier task than the direct conversion of a stationary qubit (atom) to a flying qubit (photon). In this section we will only be concerned with the requirements for reading a qubit.

23.5.1 Qubit Detection

It has been shown theoretically that cavities (even with a low finesse of a few hundred) can form very efficient, state-sensitive single atom detectors, if their mode waist can be kept small, typically a few wavelengths [39]. This small cross-section of the mode results in a sizable fraction of the light getting absorbed even by a single atom. This absorption signal is enhanced by the number of round trips in the cavity [39]. Atom chips and the technologies of micro-optics promise the fabrication of small cavities with intrinsically small beam waist by placing optical waveguides onto the atom chip surface. Introducing a small gap into the optical waveguides allows the atoms to be placed into the detection light. The cavities can then be formed either by coating the waveguide ends with high reflective coatings, or by fabricating mirrors in the waveguide itself.

Very recently we fabricated various micro-cavities and integrated them on an atom chip. Our cavities are formed by two optical fibers, mounted face to face, with coated front ends or mirrors inserted into the fiber (see Fig. 23.5). Various combinations of planar and concave mirrors have been tested.

The setups with front surface mirrors allow for a relatively high finesse. In a plano-convex geometry a finesse $\mathfrak{F} \approx 1200$ and a waist size $w_0 \approx 3$ μm were achieved for cavity length $L \approx 10$ μm. For larger gap sizes up to $L = 300$ μm, the finesse stayed well above 500. Such a cavity has a very wide spectral range of several THz and a large linewidth, but one can reach

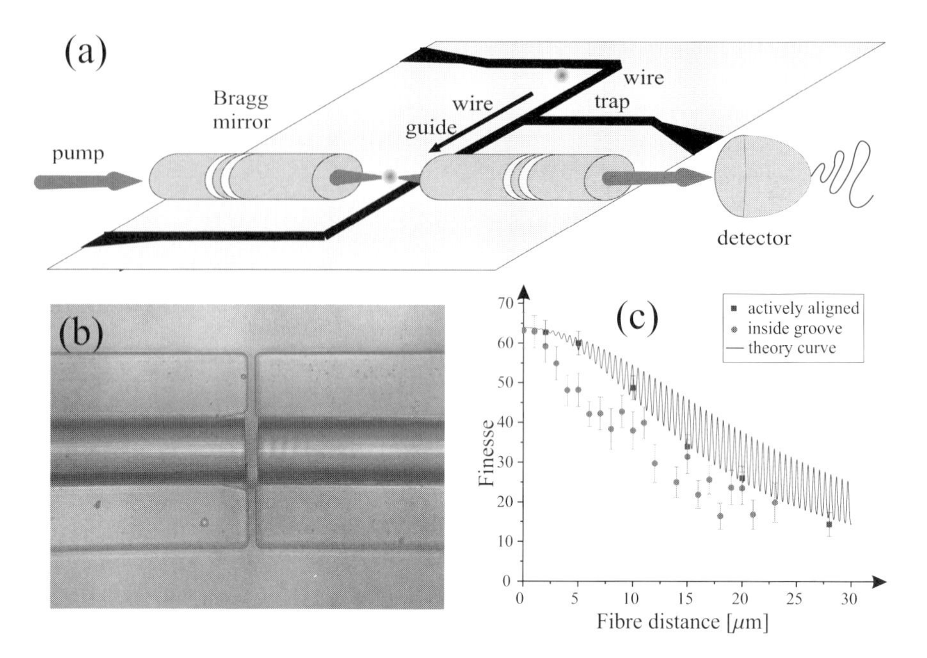

Figure 23.5: (a) Schematic setup of an on-chip micro-cavity for single atom detection. (b) Self-aligning cavity setup: coated single-mode fiber inside mounting grooves. (c) The finesse of such a cavity with mirrors fabricated into the fiber is strongly dependent on the gap size; the theory is taken from [39]

extremely high coupling parameters, which would allow even 'non-destructive' single atom detection and single atom–single photon coupling [39].

Setups with the mirrors fabricated into the optical waveguides have the advantage of easy tunability and of a smaller linewidth for state-selective detection. On the other hand, the gap cannot be too large, to avoid diffraction losses. Even in this geometry we have obtained cavity finesses $\mathfrak{F} \approx 100$ and waist sizes of $w_0 < 3$ μm, which, according to Ref. [39], should allow for single atom detection with $> 5\sigma$ in less than 10 μs.

To demonstrate the feasibility of detecting magnetically trapped neutral atoms with 'bad' cavities, we recently built a macroscopic nearly concentric cavity with a beam waist of $w_0 \approx 13$ μm and a finesse of $\mathfrak{F} \approx 1000$. Using this cavity, we detected atoms guided with a free-standing current-carrying wire.

Aside from high-sensitivity detection, an atom chip with atoms trapped in well controlled micro-traps and integrated with individual-site light elements can provide input/output processes by making use of techniques such as light scattering from trapped atomic ensembles [40], slow light [41, 42], stopped light [43–45], or macroscopic spin states [46, 47]. Furthermore, the accurate positioning of atoms at submicrometer precision above a surface should allow the use of near-field optics for detection using high-finesse micro-disk or micro-sphere resonators.

The atom chip also allows individual qubit sites to be addressed with large beams, by moving the atoms in each site individually in and out of resonance [48] by, e.g., localized electric fields giving rise to Stark shifts. Such fields have already been demonstrated on atom chips [19]. They can originate from electrodes as small as 50 nm. These schemes can also serve for single-qubit operations.

Last, single atoms can also be detected by selective ionization. This may be achieved using a multi-step process up to a Rydberg state. The electron and the location from where it came can then be detected with a simple electron microscope. Using a dipole blockade mechanism as discussed above, one should be able to implement an amplification mechanism, which will allow 100 % detection efficiency [6].

23.5.2 Quantum Input/Output

Another advanced application is the direct conversion of atomic to photonic qubits and vice versa. The photons, acting as flying qubits, offer the possibility of low-loss transfer of quantum information between distant atomic qubits for entangling atoms on different sites for a 'distributed' computation process [49, 50]. The conversion between atomic and photonic qubits requires either high-finesse cavities and CQED or low-finesse cavities and the qubit stored in an ensemble of atoms [51]. An atom chip setup could involve light from waveguides coupled in a mode-matched way into high-finesse cavities by integrated micro-lenses. These systems would allow single atom qubits and mesoscopic atomic ensembles to be connected by photonic links. In all of the above, the atom chip promises to enhance the feasibility of accurate atom–light systems.

23.6 Noise and Decoherence

One of the main questions regarding the implementation of QIP on atom chips is this: Can the coherence be preserved for atoms hovering micrometers above a 300 K hot surface (also a 4 K surface is "hot" for a nK atomic qubit). One must ensure that even in this environment the coherent evolution of atomic wave functions is undisturbed by uncontrolled external influences. The three main destructive processes that have to be dealt with are trap losses, heating, and decoherence. While trap losses are equivalent to losses of qubits, transfer of energy to the quantum system (heating) may result in excitations of motional degrees of freedom (i.e. trap vibrational levels), which would render the evolution of the system ill-defined. Decoherence does not require the transfer of energy but is still just as harmful because superpositions with a definite phase relation between different quantum states are destroyed.

In the atom chip environment it will be important to select the qubit states in such a way as to minimize decoherence. In an atomic clock these are usually the $m_F = 0$ clock states. For magnetic trapping in atom chips, one can use, for example, the $|F = 2, m_F = 1\rangle$ and the $|F = 1, m_F = -1\rangle$ states in Rb at a magic field of $B = 3.23$ G, as discussed above. Operating a trap at this field makes the qubit states insensitive to small magnetic field noise. This was nicely demonstrated in a recent chip experiment in Munich [31], where coherence times longer than 2 s were achieved in Ramsey-type interference with trapped Rb atoms.

Extensive work by C. Henkel et al. [52–54] has shown that the main coupling of the cold neutral atom to the chip surface is through its magnetic moment interacting with the fluctuating magnetic fields. Two important sources of magnetic noise that couple the surface to the atom are near-field noise induced by thermal random currents in metallic solids (Johnson noise), and noise in the currents that create the magnetic trapping fields (technical noise), where shot noise sets perhaps a fundamental limit. At present, theoretical predictions (number and scaling estimates can be found in Refs. [1, 54]) and experimental data show quantitative agreement [55, 56] for the spin-flip loss rates, which gives good confidence in the models and tells us how to reduce the unwanted surface noise: The fluctuating magnetic fields from the Johnson noise can be reduced by going to thinner wires, higher-resistivity materials, or keeping the atoms further away from the metal surface. Judging from these models, our own experimental data, and the characteristics of our fabrication processes [25], we can estimate that there will be a window of at least 100 ms coherence at micrometer distance from the wires for magnetic manipulation, and much longer for electrical manipulation even for 300 K atom chip surfaces.

Using superconducting wires or permanently magnetic structures as sources of the magnetic fields could solve some of the noise problems, but much more theoretical and experimental work is needed.

23.7 Summary and Conclusion

In the six years since their first realization, atom chips have come a long way. Numerous atom optics groups are now operating atom chip experiments, and BEC on an atom chip is now a standard tool. Many potential configurations and a variety of optical elements, such as traps, guides, mirrors, time-dependent and time-independent beam splitters, and conveyor belts, have been realized. The atom chip is now a powerful tool for neutral atom manipulation. To implement QIP on an atom chip, there are still many challenges ahead. Most notably, all the beautiful experiments were done with ensembles of atoms. One of the big challenges will be to deterministically load a single atom (qubit), manipulate it, and detect it with high fidelity. The road to loading a single atom will be through implementing 'quantum tweezers' [26, 27] using the Mott-insulator transition [28–30]. The single-qubit detection can be achieved with integrated micro-cavities on the chip [39]. Experiments to tackle both challenges are well under way, and we think that in a few years single atom micro-manipulation will be the same standard procedure as the creation of a BEC is today.

In addition, fabrication techniques will be improving continuously, with more and more functions being integrated onto a single chip. For example, several experiments concerning the integration of light are under way, involving micro-spheres, -disks, -cavities, and -lenses, in various laboratories. Altogether, this will allow us to create a monolithic device of interconnected webs of photons and atoms, able to prepare, manipulate, read and transmit quantum states, thus eventually leading to a compact quantum processor.

Acknowledgements

We are indebted to our many past and present collaborators on the atom chip experiments in Innsbruck and Heidelberg. The atom chips were fabricated by S. Groth in collaboration with I. Bar-Joseph at the Sub-Micron Center, Weizmann Institut of Science, Israel. We would like to personally thank our long-time theoretical collaborators P. Zoller, T. Calarco, P. Horak, H. Ritsch, and C. Henkel.

Our work was supported by many sources, most notably the Deutsche Forschungsgemeinschaft (Schwerpunktprogramm: Quanteninformationsverarbeitung), Landesstiftung Baden-Württemberg (Kompetenznetzwerk Quanteninformationsverarbeitung), and the European Union, contract numbers IST-1999-11055 (ACQUIRE), IST-2001-38863 (ACQP), HPRI-CT-1999-00069 (LSF), and HPMF-CT-2002-02022.

References

[1] Folman, R., Krüger, P., Schmiedmayer, J., Denschlag, J., and Henkel, C. *Adv. At. Mol. Opt. Phys.* **48**, 263 (2002).

[2] DiVincenzo, D. P. *Fortschr. Phys.* **48**, 771 (2000).

[3] Jaksch, D., Briegel, H.-J., Cirac, J. I., Gardiner, C. W., and Zoller, P. *J. Mod. Opt.* **47**, 415 (2000).

[4] Calarco, T., Hinds, E. A., Jaksch, D., Schmiedmayer, J., Cirac, J. I., and Zoller, P. *Phys. Rev. A* **61**, 022304 (2000).

[5] Charron, E., Tiesinga, E., Mies, F., and Williams, C. *Phys. Rev. Lett.* **88**, 077901 (2002).

[6] Schmiedmayer, J., Folman, R., and Calarco, T. *J. Mod. Opt.* **49**, 1375 (2002).

[7] Schmiemdayer, J. In *XVIII Int. Conf. on Quantum Electronics*, Technical Digest, 13 (1992).

[8] Schmiedmayer, J. *Eur. Phys. J. D* **4**, 57 (1998).

[9] Reichel, J., Hänsel, W., and Hänsch, T. W. *Phys. Rev. Lett.* **83**, 3398 (1999).

[10] Müller, D., Anderson, D. Z., Grow, R. J., Schwindt, P. D. D., and Cornell, E. A. *Phys. Rev. Lett.* **83**, 5194 (1999).

[11] Dekker, N. H., Lee, C. S., Lorent, V., Thywissen, J. H., Smith, S. P., Drndić, M., Westervelt, R. M., and Prentiss, M. *Phys. Rev. Lett.* **84**, 1124 (2000).

[12] Folman, R., Krüger, P., Cassettari, D., Hessmo, B., Maier, T., and Schmiedmayer, J. *Phys. Rev. Lett.* **84**, 4749 (2000).

[13] Ott, H., Fortagh, J., Schlotterbeck, G., Grossmann, A., and Zimmermann, C. *Phys. Rev. Lett.* **87**, 230401 (2001).

[14] Hänsel, W., Hommelhoff, P., Hänsch, T. W., and Reichel, J. *Nature* **413**, 498 (2001).

[15] Leanhardt, A. E., Chikkatur, A. P., Kielpinski, D., Shin, Y., Gustavson, T. L., Ketterle, W., and Pritchard, D. E. *Phys. Rev. Lett.* **89**, 040401 (2002).

[16] Schneider, S., Kasper, A., v. Hagen, C., Bartenstein, M., Engeser, B., Schumm, T., Bar-Joseph, I., Folman, R., Feenstra, L., and Schmiedmayer, J. *Phys. Rev. A* **67**, 023612.

[17] Cassettari, D., Hessmo, B., Folman, R., Maier, T., and Schmiedmayer, J. *Phys. Rev. Lett.* **85**, 5483 (2000).

[18] Hänsel, W., Reichel, J., Hommelhoff, P., and Hänsch, T. W. *Phys. Rev. Lett.* **86**, 608 (2001).

[19] Krüger, P., Luo, X., Klein, M. W., Brugger, K., Haase, A., Wildermuth, S., Groth, S., Bar-Joseph, I., Folman, R., and Schmiedmayer, J. *Phys. Rev. Lett.* **91**, 233201 (2003).

[20] Krüger, P. PhD thesis, Univ. Heidelberg (2004).

[21] Fortagh, J., Ott, H., Kraft, S., Günther, A., and Zimmermann, C. *Phys. Rev. A* **66**, 041604(R) (2002).

[22] Leanhardt, A. E., Shin, Y., Chikkatur, A. P., Kielpinski, D., Ketterle, W., and Pritchard, D. E. *Phys. Rev. Lett.* **90**, 100404 (2003).

[23] Jones, M. P. A., Vale, C. J., Sahagun, D., Hall, B. V., and Hinds, E. A. *Phys. Rev. Lett.* **91**, 080401 (2003).

[24] Estève, J., Aussibal, C., Schumm, T., Figl, C., Mailly, D., Bouchoule, I., Westbrook, C. I., and Aspect, A. *Phys. Rev. A* **70**, 043629 (2004).

[25] Groth, S., Krüger, P., Wildermuth, S., Folman, R., Fernholz, T., Mahalu, D., Bar-Joseph, I., and Schmiedmayer, J. *Appl. Phys. Lett.* **85**, 2980 (2004).

[26] Diener, R. B., Wu, B., Raizen, M. G., and Niu, Q. *Phys. Rev. Lett.* **89**, 070401 (2002).

[27] Dudarev, A. M., Diener, R. B., Wu, B., Raizen, M. G., and Niu, Q. *Phys. Rev. Lett.* **91**, 010402 (2003).

[28] Jaksch, D., Bruder, C., Cirac, J. I., Gardiner, C. W., and Zoller, P. *Phys. Rev. Lett.* **81**, 3108 (1998).

[29] Orzel, C., Tuchman, A. K., Fenselau, M. L., Yasuda, M., and Kasevich, M. A. *Science* **291**, 2386 (1998).

[30] Greiner, M., Mandel, O., Hänsch, T. W., and Bloch, I. *Nature* **419**, 51 (2002).

[31] Treutlein, P., Hommelhoff, P., Steinmetz, T., Hänsch, T. W., and Reichel, J. *Phys. Rev. Lett.* **92**, 203005 (2004).

[32] Mompart, J., Eckert, K., Ertmer, W., Birkl, G., and Lewenstein, M. *Phys. Rev. Lett.* **90**, 147901 (2003).

[33] Lukin, M. D., Fleischauer, M., Cote, R., Duan, L. M., Jaksch, D., Cirac, J. I., and Zoller, P. *Phys. Rev. Lett.* **87**, 037901 (2001).

[34] Brennen, G. K., Caves, C., Jessen, P. S., and Deutsch, I. *Phys. Rev. A* **61**, 062309 (2000).

[35] Jaksch, D., Cirac, J. I., Zoller, P., Rolston, S. L., Côté, R., and Lukin, M. D. *Phys. Rev. Lett.* **85**, 2208 (2000).

[36] Pellizzari, T., Gardiner, S. A., and Zoller, P. *Phys. Rev. Lett.* **75**, 3788 (1995).

[37] Myatt, C., Burt, E., Ghrist, R., Cornell, E., and Wieman, C. *Phys. Rev. Lett.* **78**, 586 (1997).

[38] Bergeman, T., Moore, M. G., and Olshanii, M. *Phys. Rev. Lett.* **91**, 163201 (2003).

[39] Horak, P., Klappauf, B. G., Haase, A., Folman, R., Schmiedmayer, J., Domokos, P., and Hinds, E. A. *Phys. Rev. A* **67**, 043806 (2003).

[40] Duan, L.-M., Lukin, M., Cirac, J. I., and Zoller, P. *Nature* **414**, 413 (2001).

[41] Hau, L. V., Harris, S. E., Dutton, Z., and Behroozi, C. H. *Nature* **397**, 594 (1999).

[42] Vitali, D., Fortunato, M., and Tombesi, P. *Phys. Rev. Lett.* **85**, 445 (2000).

[43] Phillips, D. F., Fleischhauer, M., Mair, A., Walsworth, R. L., and Lukin, M. D. *Phys. Rev. Lett.* **86**, 783 (2001).

[44] Liu, C., Dutton, Z., Behroozi, C. H., and Hau, L. V. *Nature* **409**, 490 (2001).

[45] Fleischhauer, M., and Lukin, M. D. *Phys. Rev. A* **65**, 022314 (2002).

[46] Duan, L.-M., Cirac, J. I., Zoller, P., and Polzik, E. S. *Phys. Rev. Lett.* **85**, 5643 (2000).

[47] Julsgaard, B., Kozhekin, A., and Polzik, E. S. *Nature* **413**, 400 (2001).

[48] Schrader, D., Dotsenko, I., Khudaverdyan, M., Miroshnychenko, Y., Rauschenbeutel, A., and Meschede, D. *Phys. Rev. Lett.* **93**, 150501 (2004).

[49] van Enk, S. J., Cirac, J. I., and Zoller, P. *Science* **279**, 205 (1998).

[50] van Enk, S. J., Kimble, H. J., Cirac, J. I., and Zoller, P. *Phys. Rev. A* **59**, 2659 (1999).

[51] Lukin, M. D., Yelin, S. F., and Fleischhauer, M. *Phys. Rev. Lett.* **84**, 4232 (2000).

[52] Henkel, C., and Wilkens, M. *Europhys. Lett.* **47**, 414 (1999).

[53] Henkel, C., Pötting, S., and Wilkens, M. *Appl. Phys. B* **69**, 379 (1999).

[54] Henkel, C., Krüger, P., Folman, R., and Schmiedmayer, J. *Appl. Phys. B* **76**, 173 (2003).

[55] Y. J. Lin, Teper, C. C., and Vuletic, V. *Phys. Rev. Lett.* **92**, 050404 (2004).

[56] Haller, E. Diplomarbeit, Univ. Heidelberg (2004).

24 Quantum Gates and Algorithms Operating on Molecular Vibrations

Ulrike Troppmann[1], *Carmen M. Tesch*[2], *and Regina de Vivie-Riedle*[1,2]

[1]Department Chemie, LMU München
Butenandtstr. 11
München, Germany

[2] Max-Planck-Institut für Quantenoptik
Garching, Germany

24.1 Introduction

Quantum information processing is a topic of growing interest in various research fields [1]. First principles for a realization of quantum gates and algorithms have been proposed in quantum optics, solid state physics, nuclear magnetic resonance (NMR), and molecular physics. Successful experimental realizations have been demonstrated for a limited number of qubits using cavity quantum electrodynamics [2], trapped ions [3,4], and NMR [5,6]. Recently, we proposed to use the vibrational modes of polyatomic molecules to encode qubit systems [7].

In an N-atom molecule, $3N-5$ or $3N-6$ different normal modes can be identified, and, depending on the spectroscopic method applied, a selection of them can be used to encode qubits. Excitations of two different quanta in each mode are referred to as $|0\rangle$ and $|1\rangle$. The qubit states are embedded in a vibrational spectrum with increasing density for higher quantum numbers. The interaction between single vibrational qubits is system inherent and mediated via the molecular bonds. Short laser light pulses induce special vibrational dynamics for each eigenstate addressed. These typical molecular properties have to be controlled when quantum logic operations are implemented and require shaped femtosecond laser pulses.

The proof of principle was given for acetylene, selected as a molecular model system. A set of global quantum gates that form a desired universal quantum gate [8] was calculated [7]. The unitary transformations could be handled correctly with a newly developed version of optimal control theory (OCT) [7,9]. Within the DFG-Schwerpunkt "Quantum Information Processing" two topics were addressed in particular. A complete quantum algorithm, the Deutsch–Jozsa algorithm, was simulated with the calculated quantum gates [9]. Also, the first steps toward more complex molecular qubit systems were taken by extending the acetylene model to include a third IR-inactive mode [10]. Thereby, a perturbation was introduced to the qubit basis due to anharmonic resonances, one possible source of decoherence in molecular systems.

Quantum Information Processing, 2nd Edition. Edited by Thomas Beth and Gerd Leuchs

ISBN: 3-527-40541-0

24.2 Qubit States Encoded in Molecular Vibrations

Our first investigations on molecular quantum computing were performed for acetylene, C_2H_2, a linear molecule with seven vibrational degrees of freedom and a rich vibrational spectrum that has partially been observed using direct photoabsorption spectroscopy or stimulated emission pumping [11–16].

The vibrational characteristics can be allocated in the normal mode notation $(n_1n_2n_3n_4n_5)$, where n_1 and n_3 represent the number of vibrational quanta in the symmetric and asymmetric CH stretching modes, n_2 corresponds to the symmetric CC stretching mode, and n_4 and n_5 are doubly degenerate and describe the degree of excitation in the *trans* and *cis* bending modes. The two IR-active modes in the electronic ground state, the asymmetric CH stretching mode n_3 and the *cis* bending mode n_5, are selected to encode qubits. In the general qubit definition, n quanta of excitation are labeled as logic 0, while $n+1$ quanta of excitation are labeled as logic 1. One possible qubit definition is to start with zero quanta of excitation in each vibrational mode. In this case the complete 2^2-dimensional qubit basis (lower basis) reads:

$$(00000) \equiv |00\rangle, \qquad (00001) \equiv |01\rangle, \qquad (00100) \equiv |10\rangle, \qquad (00101) \equiv |11\rangle.$$

Another possibility is to start with the higher-lying state (00101) labeled as $|00\rangle$. In this upper qubit basis, two quanta of excitation encode the qubit state $|1\rangle$. Both qubit basis sets are embedded in the vibrational spectrum of the 2D acetylene model as shown schematically in Fig. 24.1. The corresponding wavefunctions of the lower qubit system are shown on the right-hand side; the wavefunctions of the upper qubit system are on the left. The freedom of choice for the qubit basis helps to find an almost decoherence-free subspace within the vibrational manifold by avoiding basis states that couple to other normal modes. Even so, the issue of vibrational coupling and its consequences for the qubit basis is addressed in Sec. 24.6.

Motivated by stability arguments, we selected low-lying vibrational states to stay close to the molecular ground state in the less dense spectral region. Thereby the coupling to the molecular surroundings as well as to the other vibrational modes is kept controllable and the long lifetimes in the nanosecond regime are conserved. The quantum logic gates operate on the femtosecond timescale, providing a favorable relation between switching and decoherence time.

The actual proposal for molecular quantum computing relies on IR-active vibrational modes; however, there is no principal restriction. One could think of using Raman-active modes [17] as well as vibrational modes in electronically excited states [18]. The only requirement is optical addressability.

24.3 Optimal Control Theory for Molecular Dynamics

Optimal control theory (OCT) [19, 20] is a very powerful tool for calculating laser pulses that guide a quantum system to any selected objective. Possible objectives are the selective preparation of specific eigenstates of the quantum system, a well-defined transition between different states, or the control of molecular motion. They can be realized using the interaction between the electric field of the laser light and the quantum system. In the optimized laser field

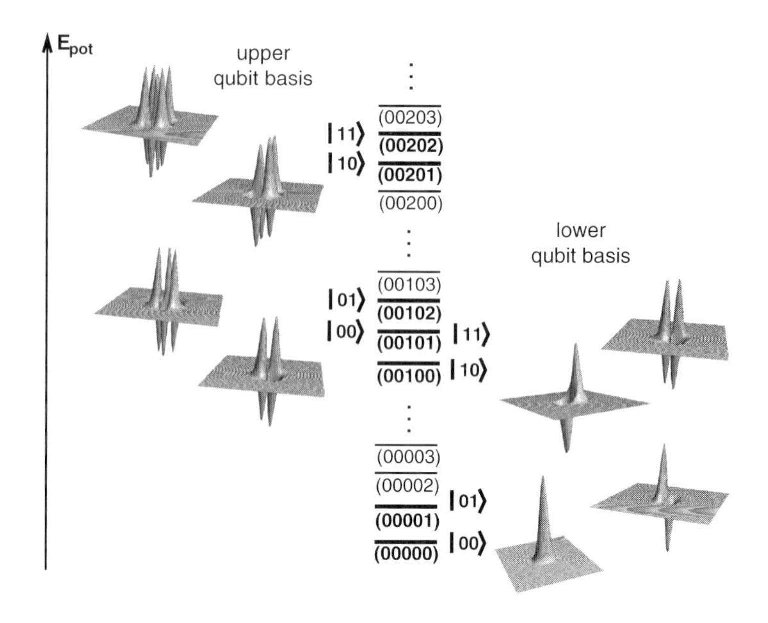

Figure 24.1: Schematic representation of the lower and upper qubit basis in the vibrational manifold of the 2D acetylene model. Also depicted are the corresponding vibrational eigenfunctions.

the constructive and destructive interference that occurs is manipulated to obtain the desired target state. OCT has already been applied successfully to several molecular systems [21–24].

The electric field of a femtosecond laser pulse and the variation of its numerous parameters is our optical tool to manipulate molecular processes. In general, optimal laser fields for different control objectives can be found by variation of the functional K:

$$
\begin{aligned}
K\left(\psi_i(t), \psi_f(t), \varepsilon(t)\right) &= \left|\langle\psi_i(T)|\phi_f\rangle\right|^2 - \alpha_0 \int_0^T \frac{|\varepsilon(t)|^2}{s(t)}\,\mathrm{d}t \\
&-2\,\mathrm{Re}\left\{\langle\psi_i(T)|\phi_f\rangle \int_0^T \langle\psi_f(t)| \left(\frac{\mathrm{i}}{\hbar}[H_0 - \mu\varepsilon(t)] + \frac{\partial}{\partial t}\right) |\psi_i(t)\rangle\,\mathrm{d}t\right\}. \qquad (24.1)
\end{aligned}
$$

The resulting laser field $\varepsilon(t)$ guides the system from an initial state $\psi_i(0) = \phi_i$ at $t = 0$ to the selected target state $\psi_i(T) = \phi_f$ at time $t = T$. Variation of the functional K leads to a set of three coupled differential equations:

$$
\begin{aligned}
\varepsilon(t) &= -\frac{s(t)}{\hbar\alpha_0}\,\mathrm{Im}\big\{\langle\psi_i(t)|\psi_f(t)\rangle\,\langle\psi_f(t)|\,\mu\,|\psi_i(t)\rangle\big\}, \\
\mathrm{i}\hbar\frac{\partial}{\partial t}\psi_i(t) &= [H_0 - \mu\varepsilon(t)]\psi_i(t), \qquad \psi_i(0) = \phi_i, \qquad (24.2)\\
\mathrm{i}\hbar\frac{\partial}{\partial t}\psi_f(t) &= [H_0 - \mu\varepsilon(t)]\psi_f(t), \qquad \psi_f(T) = \phi_f.
\end{aligned}
$$

The first equation in Eqs. (24.2) represents the optimal electric field in terms of the evolving wavepackets $\psi_i(t)$ and $\psi_f(t)$. The two subsequent equations guarantee the compliance of the Schrödinger equation for $\psi_i(t)$ and $\psi_f(t)$ under the influence of the laser field $\varepsilon(t)$. In the equations, H_0 is the time-independent Hamiltonian of the molecule, and μ is the dipole moment vector field. Both have been calculated ab initio. As a consequence, all molecular characteristics of the electronic and nuclear motion and their interactions are included in the calculations. The overall shape function $s(t)$ satisfies the demand of smooth switch-on and -off behavior of experimental laser pulses [21]. The penalty factor α_0 limits the time-averaged laser intensity, and T denotes the overall pulse duration of the shaped pulse.

OCT allows one to answer the question of controllability of the selected objective. Still, approximations in the system Hamiltonian and in the theoretical description of the experimental environment are necessary. Thus, the calculated laser pulses will not be completely identical to the optimal ones needed in an experiment, but the main characteristics, such as the underlying control mechanism, can be extracted. Furthermore, the calculated laser field can be used as a first guess for a control pulse in the experiment. The laboratory implementation of OCT is the optimal control experiment (OCE) realized in a closed loop setup that combines three main parts: a genetic algorithm for the generation of the optimal pulse forms; a spatial light modulator (SLM; e.g. liquid crystal devices) to realize the proposed pulse shapes; and the molecular probe itself. The SLM acts as a mask in the frequency domain of the laser pulse. The complex mask function

$$M(\omega) = \frac{|\varepsilon^{\mathrm{opt}}(\omega)|}{|\varepsilon^{\mathrm{FL}}(\omega)|} \tag{24.3}$$

is the direct interface between theory and experiment. It can be extracted from the calculated optimal laser fields with $\varepsilon^{\mathrm{opt}}(\omega)$ the frequency spectrum of the optimized laser field and $\varepsilon^{\mathrm{FL}}(\omega)$ a Gaussian spectrum of a Fourier limited pulse that encompasses the modulated spectrum completely [25]. The resulting mask pattern $M_n = T_n \mathrm{e}^{\mathrm{i}\phi_n}$, where n is the pixel index and each pixel sees a bandwidth $\mathrm{d}\omega$, can be put directly on the pulse modulator to produce an "intelligent" initial guess for an experimental closed loop setup. Here T_n is the amplitude modulation, and ϕ_n is the phase of the contributing frequencies.

24.3.1 Local Quantum Gates

In the following sections we will focus our discussion on the Hadamard gate, one of the fundamental gates in quantum information processing. To demonstrate the functioning of OCT, we search for a laser pulse that performs a Hadamard transformation in the upper qubit system (see Fig. 24.1). As an example, we show the Hadamard transformation on the second qubit, the *cis* bending mode. A laser pulse acting as a global gate on the second qubit performs all of the following eight transformations:

$$\begin{aligned} |00\rangle &\leftrightarrow \frac{1}{\sqrt{2}}(|00\rangle + |01\rangle), \qquad |10\rangle \leftrightarrow \frac{1}{\sqrt{2}}(|10\rangle + |11\rangle), \\ |01\rangle &\leftrightarrow \frac{1}{\sqrt{2}}(|00\rangle - |01\rangle), \qquad |11\rangle \leftrightarrow \frac{1}{\sqrt{2}}(|10\rangle - |11\rangle), \end{aligned} \tag{24.4}$$

which can be formally separated into two groups — transformations with the first, inactive qubit set to $|0\rangle$, and transformations with the first qubit set to $|1\rangle$.

In a first step local operations are optimized using Eqs. (24.2). The $|00\rangle$ state is chosen as initial state and the $\frac{1}{\sqrt{2}}(|00\rangle + |01\rangle)$ superposition as target state. The convergence of the iteration process is extremely fast: only nine iterations are needed to reach a fidelity of 99.2 %. A fast convergence with an efficiency of 98.1 % is also achieved for the second transition $|10\rangle \to \frac{1}{\sqrt{2}}(|10\rangle + |11\rangle)$. Similar fidelities are obtained starting separately from states $|01\rangle$ and $|11\rangle$. To get the two best pulses for both groups in Eqs. (24.4) (left- and right-hand sides), the local pulses are symmetrized, applying several simple mathematical techniques (averaging, mirroring, etc.). The crossed frequency-resolved optical gating (XFROG) representations of the two best pulses are shown in Fig. 24.2a. The pulse on the left produces back-and-forth transformations for the first qubit in $|0\rangle$, and that on the right for the first qubit in $|1\rangle$. Both pulses are symmetric in time, which means that they satisfy the Hermitian condition for the Hadamard transformation. The ratio of both best pulses is over 88 %.

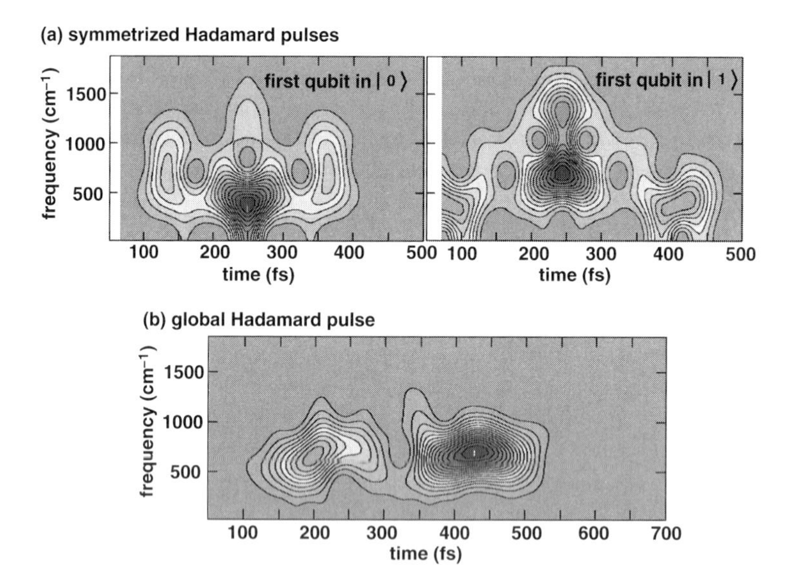

Figure 24.2: Local and global Hadamard gates on the second qubit in the upper basis. (a) XFROG representations of the local Hadamard gates with the first qubit in $|0\rangle$ and $|1\rangle$, respectively. (b) XFROG representation of the laser pulse that acts as a global Hadamard gate.

The two pulses are not yet global Hadamard transformations, however. They still depend on the influence of the inactive first qubit. This influence is understandable already from the differently shaped wavefunctions of the contributing qubit basis states (see Fig. 24.1). The hardest step in order to get a global Hadamard pulse is to suppress this influence. It can be done by hand. A much more efficient and general alternative to this tedious procedure, valid for all quantum gates, is to introduce additional constraints in the optimal control functional.

24.4 Multi-target OCT for Global Quantum Gates

For the application of OCT in molecular quantum computing, the goal is to find a laser field $\varepsilon(t)$ that drives a system from a set of initial states ψ_{ik} at time $t = 0$ to multiple final target states ψ_{fk} at a fixed time $t = T$. The initial (index i) and target states (index f) correspond to the eigenstates defining our qubits or to superpositions. The calculated laser pulse $\varepsilon(t)$ represents the global quantum gate. Accordingly the modified maximization functional can be formulated as [7, 9, 26]:

$$K\big(\psi_{ik}(t), \psi_{fk}(t), \varepsilon(t)\big) = \sum_{k=1}^{z} \Bigg\{ \left|\langle\psi_{ik}(T)|\phi_{fk}\rangle\right|^2 - \alpha_0 \int_0^T \frac{|\varepsilon(t)|^2}{s(t)}\,\mathrm{d}t$$
$$-2\,\mathrm{Re}\left[\langle\psi_{ik}(T)|\phi_{fk}\rangle \int_0^T \langle\psi_{fk}(t)| \left(\frac{\mathrm{i}}{\hbar}[H_0 - \mu\varepsilon(t)] + \frac{\partial}{\partial t}\right) |\psi_{ik}(t)\rangle\,\mathrm{d}t\right]\Bigg\}. \quad (24.5)$$

Varying the above functional with respect to its variables $\psi_{ik}(t)$, $\psi_{fk}(t)$, and $\varepsilon(t)$ leads to a set of $2z + 1$ coupled differential equations. This set of differential equations is solved iteratively using forward/backward propagation and results in one optimal quantum gate $\varepsilon(t)$ as a self-consistent solution to this system. Of highest importance for the application of OCT in the context of molecular quantum computing is the fact that the multi-target functional enables us to optimize several transitions within the molecule simultaneously with the same laser pulse. A global quantum gate has to operate correctly on each possible molecular state, and in general this quantum state is unknown. But if a quantum gate operates correctly on each basis state, it acts correctly on each possible quantum state, which can always be completely decomposed into basis states. The key effect of the multi-target algorithm is to extract the characteristic features of single transitions and to combine them into one global pulse.

24.4.1 Global Quantum Gates for Molecular Vibrational Qubits

We were able to calculate all elementary one- and two-qubit gates for the selected model system, acetylene, with an efficiency of 95–99 % using the multi-target OCT (Eq. 24.5). In continuation of the previously discussed local Hadamard gate in the upper qubit, we show the corresponding global Hadamard gate in Fig. 24.2b. Again the XFROG representation of the laser field is displayed. The carrier frequency lies around 700 cm^{-1}, red-detuned from the transition frequency of the *cis* bending mode on which the Hadamard gate operates.

The global Hadamard differs significantly from the previously "hand"-optimized pulses starting from local gates. Surprisingly, its substructure is much simpler, the frequency range is smaller, and the efficiency is enhanced, over 90 % for all eight transformations. The multi-target algorithm is able to extract the most relevant features for all transitions and combine them into the most simple and efficient solution. The control mechanism relies on intermediate overtone excitations [27].

The Hadamard gate changes its structure when optimized for the lower qubit system (Fig. 24.3a), while the intensity of the laser pulse (4.24×10^{12} W cm^{-2}) stays the same. Again, the spectrum is centered around one main frequency (see XFROG, Fig. 24.3a). The efficiency for all target states is enhanced ($\geq 95\,\%$). The corresponding mask functions for

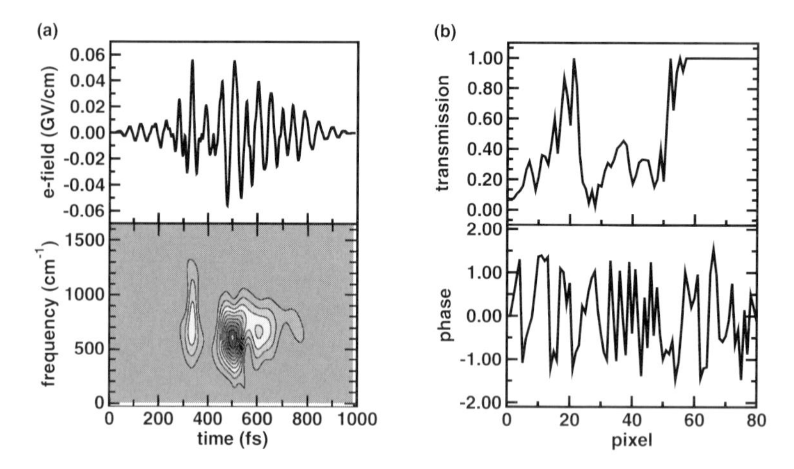

Figure 24.3: Global Hadamard gate on the second qubit in the lower basis: (a) e-field and XFROG representations; (b) transmission and phase mask functions.

the amplitude (top) and phase modulation (bottom) are shown in Fig. 24.3b. Only 81 pixels are needed to resolve the mask function. For comparison, 128 pixels can be addressed in the mask of a standard SLM; special designs allow 640 pixels [28].

24.5 Basis Set Independence and Quantum Algorithms

The multi-target OCT (Eq. 24.5) allows the calculation of global quantum gates operating on molecular vibrations. In this setup the optimization target defined as

$$\left|\langle\psi_{ik}(T)|\phi_{fk}\rangle\right|^2, \qquad k = 1, \ldots, z \tag{24.6}$$

is insensitive to the phase occurring between $|\psi_{ik}\rangle$ and $|\phi_{fk}\rangle$, because transitions of the following form will be optimized:

$$|n_{ik}\, m_{ik}\rangle \longrightarrow |n_{fk}\, m_{fk}\rangle\, \mathrm{e}^{\mathrm{i}\varphi_k}. \tag{24.7}$$

In this very general notation, n and m are arbitrary states of the first and second qubit. Again the indices i and f indicate the initial state and final state; and k is the index of a special transition out of z possible ones. The φ_k can differ as no pressure has so far been put on the development of the phase during the optimization.

For pure basis states exclusively, a sequence of qubit flips is realizable without any problems, because transitions of the form:

$$|n_{ik}\, m_{ik}\rangle\, \mathrm{e}^{\mathrm{i}\varphi_k} \longrightarrow |n_{fk}\, m_{fk}\rangle\, \mathrm{e}^{\mathrm{i}\theta_k} \tag{24.8}$$

can be operated with the same pulses as transitions in Eq. (24.7). Superposition states composed of two qubits can also be generated without trouble, because these originate from a

Hadamard transformation and therefore have the same phase:

$$|n_{ik}\, m_{ik}\rangle\, \mathrm{e}^{\mathrm{i}\varphi_k} \longrightarrow \frac{1}{\sqrt{2}}(|n_{fkl}\, m_{fkl}\rangle \pm |n_{fkm}\, m_{fkm}\rangle)\, \mathrm{e}^{\mathrm{i}\theta_k}. \tag{24.9}$$

The third indices l and m are introduced for distinguishability of the states on the right-hand side of the equation above. Critical is another transformation subsequent to

$$\frac{1}{\sqrt{2}}(|n_{ikl}\, m_{ikl}\rangle \pm |n_{ikm}\, m_{ikm}\rangle)\, \mathrm{e}^{\mathrm{i}\varphi_k} \longrightarrow \frac{1}{\sqrt{2}}\, |n_{fkl}\, m_{fkl}\rangle\, \mathrm{e}^{\mathrm{i}\alpha_k} \pm \frac{1}{\sqrt{2}}\, |n_{fkm}\, m_{fkm}\rangle\, \mathrm{e}^{\mathrm{i}\beta_k} \tag{24.10}$$

or

$$\frac{1}{\sqrt{2}}(|n_{ikl}\, m_{ikl}\rangle \pm |n_{ikm}\, m_{ikm}\rangle)\, \mathrm{e}^{\mathrm{i}\varphi_k} \longrightarrow \frac{1}{2}(|n_{fklx}\, m_{fklx}\rangle \pm |n_{fkly}\, m_{fkly}\rangle)\, \mathrm{e}^{\mathrm{i}\alpha_k} \pm \frac{1}{2}(|n_{fkmx}\, m_{fkmx}\rangle \pm |n_{fkmy}\, m_{fkmy}\rangle)\, \mathrm{e}^{\mathrm{i}\beta_k}. \tag{24.11}$$

Now the phases α_k and β_k can be different and the following application of quantum gates becomes phase-dependent. Thus one has to develop concepts for relative phase control.

One possible way is the simultaneous optimization of an additional superposition state [9]. In principle, each of the so-called Fourier basis states, formed as superpositions of the standard basis states $|nm\rangle$, is possible. This method guarantees that the obtained global phase-corrected quantum gates can be operated successively without any time delay. Such phase-corrected quantum gates can be very similar to the phase-free optimized ones as is the case for the Hadamard gate presented in Fig. 24.3a. Other phase-sensitive methods are under development where fixed time delays for the pulse sequences are used. For our method the relative phase control is directly correlated to the basis set independence, which is a prerequisite for the successful application of quantum algorithms. Basis set independence means that equivalent quantum gates can be obtained for all logic transformations, regardless of which qubit basis set is used.

As an example we regard the Fourier basis state $|q_1\rangle$ under a Hadamard operation on the second qubit:

$$|q_1\rangle \equiv \frac{1}{2}(|00\rangle + |01\rangle + |10\rangle + |11\rangle)\, \mathrm{e}^{\mathrm{i}\phi_k} \leftrightarrow \frac{1}{\sqrt{2}}(|00\rangle + |10\rangle)\, \mathrm{e}^{\mathrm{i}\theta_k}. \tag{24.12}$$

Here, all standard qubit basis states have the same global phase in the superposition states and their phases are locked together in the transition. Exactly this kind of transition is important for the optimization with phase control and underlines the strong correlation between phase control and basis set independence.

The construction of phase-corrected global quantum gates permits the realization of each quantum algorithm in the molecular quantum computing framework. We selected the well-known Deutsch–Jozsa algorithm as proof of principle [9]. Regarded are four different one-bit functions

$$f : B\{0,1\} \mapsto B\{0,1\}. \tag{24.13}$$

There exist two constant functions

$$\begin{aligned} &f_1(0) = 0 \text{ and } f_1(1) = 0, \\ &f_2(0) = 1 \text{ and } f_2(1) = 1, \end{aligned} \tag{24.14}$$

and two balanced functions in the sense that one gets the values 0 and 1 equally often:

$$\begin{aligned} &f_1(0) = 0 \text{ and } f_1(1) = 1, \\ &f_2(0) = 1 \text{ and } f_2(1) = 0. \end{aligned} \tag{24.15}$$

The goal of the Deutsch–Jozsa algorithm is to distinguish between constant and balanced functions using one single measurement in the two-bit problem. This becomes feasible in a two-qubit system due to the special state preparation and to quantum parallelism, which does not exist classically.

To demonstrate the theoretical realization of the Deutsch–Jozsa algorithm in our 2D acetylene model system, we have to translate the algorithm for the molecular model system. The preparation of the superposition state that is presented to the oracle is straightforward: first a NOT gate is applied to the second qubit, and then two Hadamard transformations are applied, one in each qubit. The quantum state we obtain is

$$\left(\frac{1}{\sqrt{2}}(|0\rangle + |1\rangle)\right) \otimes \left(\frac{1}{\sqrt{2}}(|0\rangle - |1\rangle)\right) = \frac{1}{2}(|00\rangle - |01\rangle + |10\rangle - |11\rangle). \tag{24.16}$$

Now a possible implementation for the unknown function f must be found. The evaluation of a corresponding unitary transformation U_f follows the rule

$$U_f : |x\rangle|y\rangle \mapsto |x\rangle|y \oplus f(x)\rangle \tag{24.17}$$

and has to be applied for all four possible definitions of f to all four qubit basis states. The oracle U_f operates on the second qubit, or in general on the last qubit. For the first case, if U_f corresponds to $f_1(0) = 0$ and $f_1(1) = 0$, in detail we get for the four qubit basis states:

$$\begin{aligned} &|0\rangle|0\rangle \mapsto |0\rangle|0 \oplus f(0)\rangle = |0\rangle|0 \oplus 0\rangle = |0\rangle|0\rangle, \\ &|0\rangle|1\rangle \mapsto |0\rangle|1 \oplus f(0)\rangle = |0\rangle|1 \oplus 0\rangle = |0\rangle|1\rangle, \\ &|1\rangle|0\rangle \mapsto |1\rangle|0 \oplus f(1)\rangle = |1\rangle|0 \oplus 0\rangle = |1\rangle|0\rangle, \\ &|1\rangle|1\rangle \mapsto |1\rangle|1 \oplus f(1)\rangle = |1\rangle|1 \oplus 0\rangle = |1\rangle|1\rangle. \end{aligned} \tag{24.18}$$

A comparison of the initial and final states shows that the function unitary operation U_{f_1} operates as the identity gate on the second qubit. The second function $f_2(0) = 1$ and $f_2(1) = 1$ can be described with a NOT gate on the second qubit. The third case is that of a balanced function $f_3(0) = 0$ and $f_3(1) = 1$ and translates into a CNOT gate with the first qubit as control and the second one to be flipped. The last function $f_4(0) = 1$ and $f_4(1) = 0$ is again balanced and can be described using an ACNOT (alternative CNOT) gate. Here, in contrast to the CNOT gate, the flipping condition is fulfilled when the first qubit is in $|0\rangle$.

All quantum gates needed to simulate the oracle U_f have been calculated globally and correct in phase and therefore qualify for quantum algorithms. The final Hadamard gate in the first qubit has also been optimized; it is responsible for preparing a pure state in the first qubit.

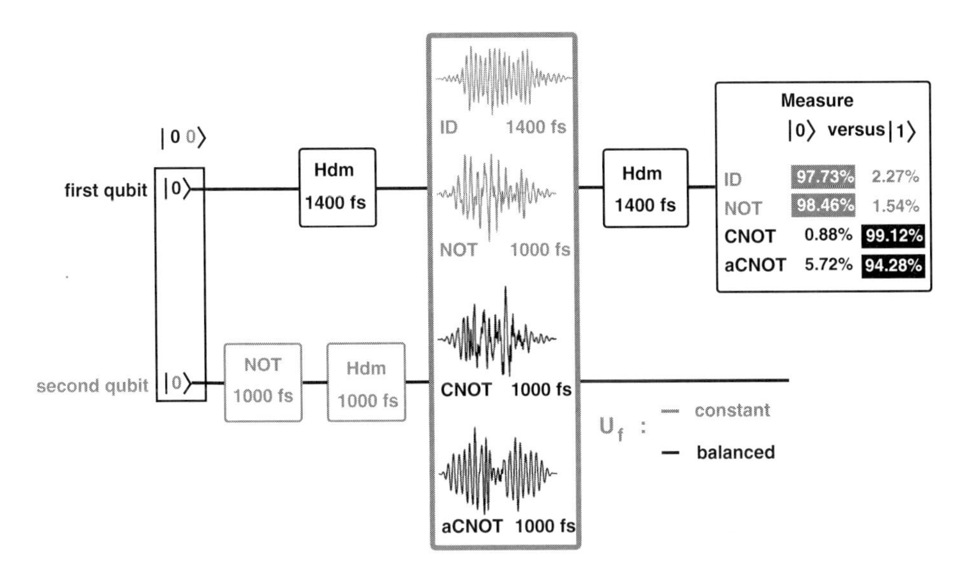

Figure 24.4: Gate sequence for the Deutsch–Jozsa algorithm in the two-qubit model system of acetylene. The gates that represent the unknown operation U_f are displayed as e-fields with their duration. The outcomes of the final measurement are displayed in the rightmost box.

The final measurement in the first qubit will be simulated by projection onto the qubit basis states. The complete gate sequence is shown in Fig. 24.4.

The quality of our realization of the Deutsch–Jozsa algorithm is also shown in Fig. 24.4. The probabilities for a correct measurement have been calculated using a projection onto each basis state followed by a standardization to one of the complete projections. In practice, the measurement is done with an ensemble of molecules, and relative signal strengths are compared to one another. In more than $\geq 94.28\,\%$ of all cases the distinction between a balanced and constant function f will be correct. For the background of an ensemble measurement, this means that we can distinguish between the two types of function with certainty.

24.6 Towards More Complex Molecular Systems

Two normal modes of acetylene are IR-active; the remaining normal modes are referred to as spectator modes. Although not part of the qubit basis, they can have a certain impact on the qubit states, e.g., via anharmonic resonances. To investigate the relevance of these molecular features we extended the 2D model to additionally include the IR-inactive *trans* bending mode [10]. Its vibrational levels lie energetically close to the *cis* bending mode.

The 3D system eigenvalues and eigenfunctions were calculated ab initio and assigned to the *cis* bending, *trans* bending, and asymmetric CH stretching modes and their combinations (schematic representation in Fig. 24.5). Several resonances could be observed, including one anharmonic resonance that is very important to our qubit system. The combination mode

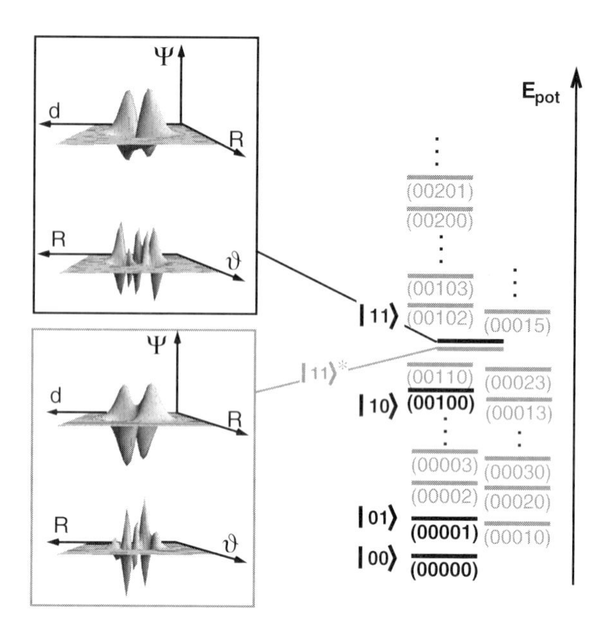

Figure 24.5: Lower qubit basis in the vibrational manifold of the 3D model, schematically depicted with its anharmonically resonant states $|11\rangle$ and $|11\rangle^*$ and 2D cuts of the corresponding 3D eigenfunctions.

(00101), the lower qubit basis state $|11\rangle$ in the 2D model, experiences a strong coupling with the (00014) combination state of *trans* and *cis* bending. This is reflected in the resulting two eigenstates obtained by a diagonalization of the 3D Hamiltonian. They display characteristic nodal features of both original states and are separated by an energy gap of only 37 cm^{-1}. We arbitrarily assigned the new $|11\rangle$ qubit state to the energetically higher-lying eigenstate, and the second one is referred to as $|11\rangle^*$. Both states are spectroscopically almost equally addressable. We examined the possibility of selectively switching the "resonant" qubit state $|11\rangle$. With the introduced optimal control tools (Eq. 24.5), we were able to calculate appropriately shaped femtosecond laser pulses to address the $|11\rangle$ state selectively as well as to realize global quantum gates [10, 27]. A net transition to the resonant $|11\rangle^*$ state is suppressed by the specially shaped laser fields; however, intermediately it is very important. Temporary population transfer to $|11\rangle^*$ during the pulse action replaces a great portion of the usual overtone excitation. Our example is again a global Hadamard pulse in the lower qubit system, this time on the first qubit, represented by the asymmetric CH stretching mode:

$$|00\rangle \leftrightarrow \frac{1}{\sqrt{2}}(|00\rangle + |10\rangle), \qquad |10\rangle \leftrightarrow \frac{1}{\sqrt{2}}(|00\rangle - |10\rangle),$$
$$|01\rangle \leftrightarrow \frac{1}{\sqrt{2}}(|01\rangle + |11\rangle), \qquad |11\rangle \leftrightarrow \frac{1}{\sqrt{2}}(|01\rangle - |11\rangle).$$

The e-field and the frequency spectrum of the optimized pulse are shown in Fig. 24.6 (upper panels, gray). The spectrum is composed of three frequency ranges. The main frequency around 3700 cm^{-1} excites the asymmetric CH stretching. Due to the low-frequency parts, the structure of the global Hadamard pulse is more complex compared to the previously shown gates in the 2D model, exhibiting only one frequency range (e.g. Fig. 24.3). Decomposition of the laser field [21] in the complex frequency domain allows one to test single contributions. The frequencies around 1900 cm^{-1} are dispensable. They can be eliminated from the Hadamard pulse without loss of performance. The fidelity of the resulting laser field (Fig. 24.6, lower panels, black) stays above 93 %. This can be realized by two phase-independent laser fields with two different wavelengths at 250 and 3700 cm^{-1}.

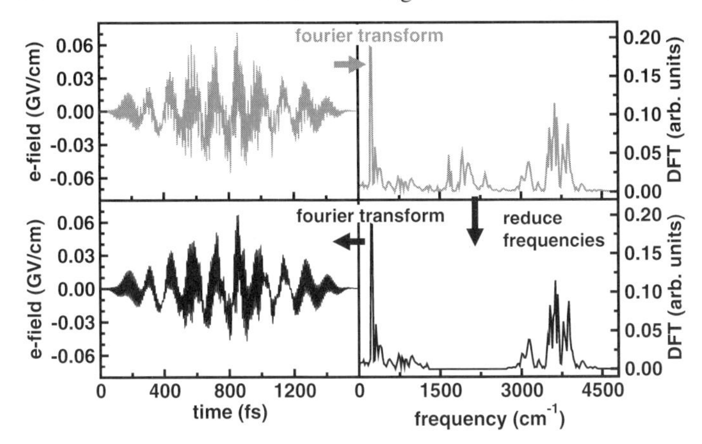

Figure 24.6: Laser pulse for the global Hadamard gate in the 3D model. Depicted are the e-field and frequency spectrum of the originally optimized laser field (upper panels, gray) and the reduced frequency spectrum as well as the reconstructed global Hadamard pulse (lower panels, black).

The quantum gate mechanism can be detected by monitoring the population development of the system eigenstates. As an example, in Fig. 24.7 the mechanisms of four transitions that are driven by the reduced laser field are plotted: the forward and backward transformations starting from $|00\rangle$ and $|01\rangle$. When the second, passive qubit is in $|0\rangle$, the predominant mechanism is an oscillating population exchange between $|00\rangle$ and $|10\rangle$, which finally leads to the superposition state of $\frac{1}{\sqrt{2}}(|00\rangle + |10\rangle)$ (Fig. 24.7, upper panels). The (00200) overtone of the asymmetric CH stretching mode is used only to a very small extent to intermediately "park" population outside of the computational space. The backward transformation works in an analogous, nearly time-reversed way. It uses the same intermediate states, a common feature of global quantum gates with respect to the forward/backward transformations.

When the passive qubit is in state $|1\rangle$ (Fig. 24.7, lower panels), the intermediate overtone excitation (here to (00201)) nearly vanishes and is replaced by transient population of the resonant $|11\rangle^*$ state as well as combinations of *trans* and *cis* bending modes $(000xy)$. The latter are the states controlled by the 250 cm^{-1} frequency component in the laser pulse. The backward transformation is again nearly the time reversal of the forward transformation.

From the investigations in the 3D model system, two fundamental conclusions can be drawn. An increased complexity of the molecular vibrational manifold, including anhar-

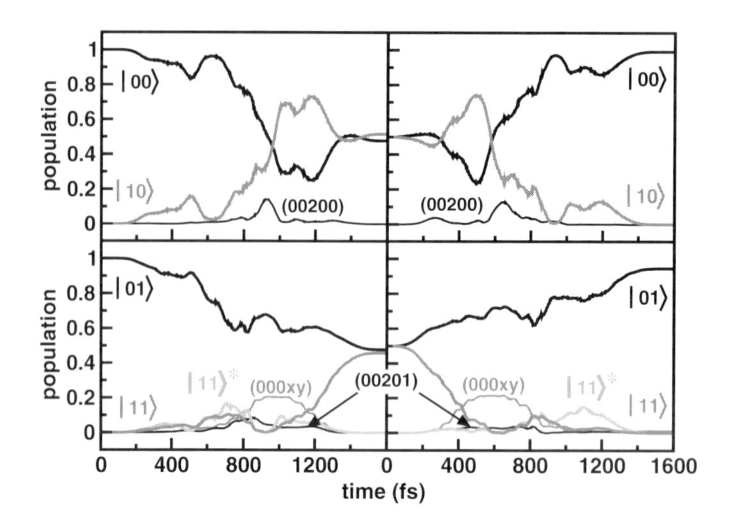

Figure 24.7: Mechanisms of four transitions of the global Hadamard pulse of Fig. 24.6. Shown as typical examples are the $|00\rangle \leftrightarrow \frac{1}{\sqrt{2}}(|00\rangle + |10\rangle)$ (upper panels) and $|01\rangle \leftrightarrow \frac{1}{\sqrt{2}}(|01\rangle + |11\rangle)$ (lower panels) transitions.

monic resonances, can be handled by optimally shaped laser pulses that act in the femtosecond regime. Furthermore, a resonant state like $|11\rangle^*$ will always be addressable by the laser used for the corresponding qubit state. It replaces the overtones in the intermediate coherent states during quantum operations. Additional frequencies that emerge during the optimization control extra pathways in the extended 3D vibrational manifold. However, it might be possible to suppress them and the associated pathways by adequate optimal control schemes that put pressure on the laser frequencies [26]. In combination with the decomposition of the laser pulse and the analysis of the single contributions, it should be possible to further simplify the laser fields.

24.7 Outlook

Our new proposal for quantum information processing was to use the vibrational eigenstates of a polyatomic molecule as qubits and specially shaped femtosecond laser pulses to implement quantum gates. Our modified OCT variant, the multi-target OCT, permits one to calculate the necessary global and phase-corrected quantum gates. The method was developed for molecules but is applicable also for other quantum systems interacting with laser fields. For a model system — the acetylene molecule — we proved the principles of functioning of quantum gates and algorithms. Molecules promise to be good candidates for quantum computation. They provide numerous vibrational states for encoding qubits. The number of potential qubits is proportional to the number of atoms N. The vibrational qubit states can be prepared with long lifetimes (ns) while the switching time is fast (fs). In principle, modulated laser pulses are realizable; also direct shaping in the IR is within reach. Based on these principal results, we recently started to investigate molecular systems with respect to favorable properties for quantum computing. Strong IR absorbers like carbonyl complexes seem to be promising. First experiments toward control of the CO stretching mode excitation have been successfully performed [29, 30]. Subunits containing a limited number of such appropriate vibrational qubits could be combined in a modular approach for general scalability. Work on information transfer between these subunits is in progress.

Acknowledgements

This work was supported by the Deutsche Forschungsgemeinschaft.

References

[1] D. Bouwmeester, A. Eckert, A. Zeilinger (Eds.), The Physics of Quantum Information (Springer, Berlin, 2000).

[2] M. Brune, F. Schmidt-Kaler, A. Maali, J. Dreyer, E. Hagley, J. M. Raimond, S. Haroche, Phys. Rev. Lett. **76**, 1800 (1996).

[3] J. I. Cirac, P. Zoller, Phys. Rev. Lett. **74**, 4091 (1995).

[4] F. Schmidt-Kaler, H. Häffner, M. Riebe, S. Gulde, G. P. T. Lancaster, T. Deuschle, C. Becher, C. F. Roos, J. Eschner, R. Blatt, Nature **422**, 408 (2003).

[5] R. Marx, A. F. Fahmy, J. M. Myers, W. Bermel, S. J. Glaser, Phys. Rev. A **62**, 012310-1-8 (2000).

[6] J. A. Jones, M. Mosca, J. Chem. Phys. **109**, 1648 (1998).

[7] C. M. Tesch, R. de Vivie-Riedle, Phys. Rev. Lett. **89**, 157901 (2002).

[8] T. Sleator, H. Weinfurter, Phys. Rev. Lett. **74**, 4087 (1995).

[9] C. M. Tesch, R. de Vivie-Riedle, J. Chem. Phys. 121 (2004) 12158–12168.

[10] U. Troppmann, C. M. Tesch, R. de Vivie-Riedle, Chem. Phys. Lett. **378**, 273 (2003).

[11] L. Halonen, S. Carter, M. Child, Mol. Phys. **47**, 1097 (1982).

[12] G. J. Scherer, K. K. Lehmann, W. Klemperer, J. Chem. Phys. **78**, 2817 (1983).

[13] B. C. Smith, J. S. Winn, J. Chem. Phys. **94**, 4120 (1991).

[14] M. J. Bramley, S. Carter, N. C. Handy, I. Mills, J. Mol. Spectrosc. **157**, 301 (1993).

[15] S.-F. Yang, L. Biennier, A. Campargue, M. A. Temsamani, M. Herman, Mol. Phys. **90**, 807 (1997).

[16] A. Campargue, L. Biennier, M. Herman, Mol. Phys. **93**, 457 (1998).

[17] Z. Bihary, D. R. Glenn, D. A. Lidar, V. A. Apkarian, Chem. Phys. Lett. **360**, 459 (2002).

[18] J. Vala, Z. Amitay, B. Zhang, S. R. Leone, R. Kosloff, Phys. Rev. A **66**, 062316 (2002).

[19] D. J. Tannor, R. Kosloff, S. A. Rice, J. Chem. Phys. **83**, 5850 (1986).

[20] W. Zhu, J. Botina, H. Rabitz, J. Chem. Phys. **108**, 1953 (1998).

[21] K. Sundermann, R. de Vivie-Riedle, J. Chem. Phys. **110**, 1896 (1999).

[22] H. Rabitz, R. de Vivie-Riedle, M. Motzkus, K. Kompa, Science **288**, 824 (2000).

[23] D. Geppert, A. Hofmann, R. de Vivie-Riedle, J. Chem. Phys. **119**, 5901 (2003).

[24] D. Geppert, L. Seyfarth, R. de Vivie-Riedle, Appl. Phys. B (2004), 987–992.

[25] T. Hornung, M. Motzkus, R. de Vivie-Riedle, J. Chem. Phys. **115**, 3105 (2001).

[26] T. Hornung, M. Motzkus, R. de Vivie-Riedle, Phys. Rev. A **65**, 021403 (2002).

[27] U. Troppmann, R. de Vivie-Riedle, J. Chem. Phys., submitted to JCP 2004.

[28] G. Stobrawa, M. Hacker, T. Feurer, D. Zeidler, M. Motzkus, F. Reichel, Appl. Phys. B **72**, 627 (2001).

[29] T. Witte, T. Hornung, L. Windhorn, R. de Vivie-Riedle, D. Proch, M. Motzkus, K. L. Kompa, J. Chem. Phys. **118**, 2021 (2003).

[30] T. Witte, K. L. Kompa, M. Motzkus, Appl. Phys. B **76**, 467 (2003).

25 Fabrication and Measurement of Aluminum and Niobium Based Single-Electron Transistors and Charge Qubits

Wolfram Krech[1], *Detlef Born, Marián Mihalik*[2], *and Miroslav Grajcar*[3]
Institute of Solid State Physics
Friedrich Schiller University
Jena, Germany

Thomas Wagner, Uwe Hübner
Quantum Electronics Division
Institute for Physical High Technology e.V.
Jena, Germany

25.1 Introduction

There is a growing interest in building of machines that actively exploit the fundamental properties of quantum physics (e.g. [1,2]). Quantum computers could perform certain calculations, which no classical computer can do in acceptable times, by exploiting the quantum coherent evolution of superpositions of states. The implementation of the basic elements of quantum computers, quantum bits (or "qubits"), remains an experimental challenge. Different physical realizations have been proposed including trapped ions which are manipulated by laser irradiation [3], nuclear magnetic resonance in molecules [4], quantum optical [5] and other systems.

Among the practical realizations of qubits, solid state devices appear most promising for large scale applications and integration in electronic circuits. However, unlike the electric dipoles of isolated atoms, the state variables of a circuit, like voltages and currents, usually undergo rapid quantum decoherence because they are strongly coupled to an environment with a large number of uncontrolled degrees of freedom [6]. Nevertheless, superconducting systems of small-capacitance Josephson tunnel junctions appear particular attractive because the coherence of the superconducting state can be exploited and the relevant technology is quite advanced [6–10]. Especially, in this context it should mentioned that superconducting tunnel junction devices have displayed quality factors of quantum oscillations in the range from a few dozen [11] up to an excess of 10^4 [10].

[1]Corresponding author. *E-mail address:* wolfram.krech@uni-jena.de.
[2]On leave from Institute of Experimental Physics, Slovak Academy of Sciences, Košice, Slovakia
[3]On leave from Department of Solid State Physics, Comenius University, Bratislava, Slovakia

Quantum Information Processing, 2nd Edition. Edited by Thomas Beth and Gerd Leuchs

ISBN: 3-527-40541-0

25.2 Motivation for this Work

Nearly all experiments in the field of single-charge tunneling phenomena in metallic structures have been done with Al/AlO$_x$/Al junctions, and the shadow evaporation technique [12, 13] is commonly used to fabricate ultrasmall tunnel junctions of size less than 0.01 μm^2. However, the preparation of miniaturized Nb based junctions by means of this well-established method is quite complicated. Otherwise, from the point of view of future applications niobium junctions are more desirable because of the large superconducting energy gap. For this purpose, Harada *et al.* [14] have developed a four-layer resist system to fabricate Nb/AlO$_x$/Al junctions which were used to make a superconducting single-electron transistor (SET) composed of two dc-SQUIDs in series. As an alternative to the shadow evaporation method, we have introduced the so-called self-aligned in-line technique that was applied, for the first time, to the fabrication of Nb/AlO$_x$/Nb based (or all-niobium) SETs [15].

In the recent past years, there was a strong motivation to develop further techniques for the reliable preparation of Josephson junctions of smaller sizes. Pavolotsky *et al.* [16] have introduced a multilayer technique based on spin-on glass planarization. Especially, here the Al-AlO$_x$ barrier of the Nb/Al-AlO$_x$/Nb trilayer is wet etched which prevents the formation of "fences" at the edges of the base electrode [17]. Having the analogous aim in mind, we established the so-called electron beam direct-writing concept [18]. Using e-beam lithography in conjunction with material deposition by sputtering, we found with respect to the line width of the base strip lines good reproducibility down to 60 nm. Hence, the areas of the tunnel junctions were in the order of a few 10^{-3} μm^2 and, consequently, the junction capacitances in the order of 10^{-16}F. Recently, deep sub-micron Al/AlO$_x$/Nb tunnel junctions and SETs with niobium islands were fabricated by electron beam gun shadow evaporation [19]. High-quality structures with a superconducting gap energy of up $2\Delta_{Nb} = 2.5$ meV were achieved.

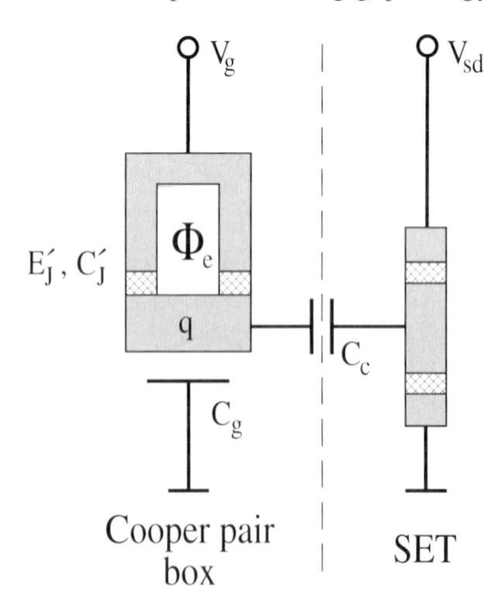

Figure 25.1: Superconducting charge-type qubit (on the left) capacitively coupled (via the capacitor C_c) to a single-electron transistor which is used as electrometer. The qubit's relevant quantum degree is the Cooper pair charge q on the island electrode (Cooper pair box) between gate capacitor C_g and Josephson junctions (with individual capacitances C'_J and coupling energies E'_J). The existing dc-SQUID can be tuned by the external flux Φ_e delivering the effective coupling term E_J (cf. Fig. 25.2).

The up to now by means of different preparation techniques fabricated Nb based small-tunnel structures are first efforts to apply the highly desirable Nb/AlO$_x$/Nb junctions for future mesoscopic devices. In the following, we describe as further step the introduction of the e-beam direct-writing concept [18], that is relevant to niobium, into quantum information technology. This will be illustrated by fabrication and measurement of SETs and qubit devices consisting of Cooper pair boxes with readout SETs (Fig. 25.1). For a rough explanation of the energy structure of such a box, its two-band model is depicted in Fig. 25.2. It is shown that in the vicinity of the degeneration points of charging energy the system effectively reduces to a two-state quantum system.

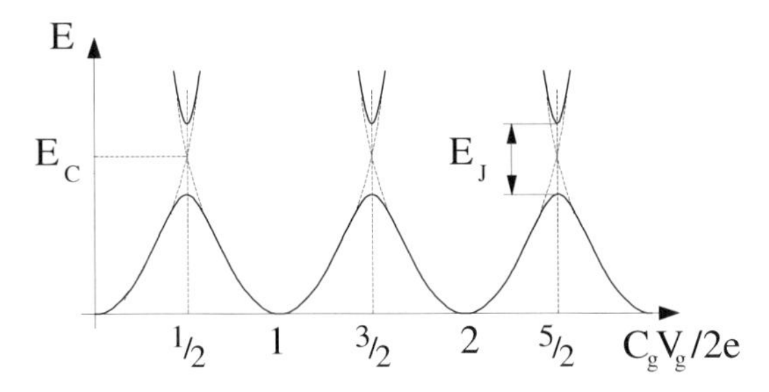

Figure 25.2: Two-band energy spectrum of a Cooper pair box. The total energy E (solid lines) is shown as a function of the applied gate charge, $n_g = C_g V_g/2e$. Near the degeneracy points the weak SQUID coupling energy, $E_J \ll E_C = e^2/(2C_\Sigma)$ (C_Σ: total qubit island capacitance, cf. Fig 25.1), mixes the charge states (dashed lines). In this regime the system reduces effectively to a two-state quantum system.

25.3 Sample Preparation

25.3.1 Scheme of the Junction Preparation Technique

In this work the e-beam direct-writing conception is applied to the fabrication of small-area tunnel junctions that may serve as relevant components of a quantum information technology based on charge-type qubits. The method is schematically depicted in Fig. 25.3 for a double junction with capacitive gate which represents the building block for both SETs and charge qubits.

The technique starts from narrow base striplines prepared by lateral patterning of a thin film (superconducting metal 1) in a lift-off process (a). Using e-beam lithography, one may reduce the line width of the strips down to typical dimensions of about 50 nm or less that are comparable with the film thickness. Next, a resist mask is made leaving parts of the base electrodes uncovered which have to be cleaned by ion beam etching. Now the generation of dielectric barriers on the surface of the striplines (b) and a subsequent deposition of a second film (metal 2) follow without breaking the vacuum (c). The above-mentioned resist mask serves in a final lift-off process as stencil for patterning the upper (island) electrode (d).

One is completely free in choosing the deposition process for all metal layers, both evaporation and sputtering are possible. In particular, the latter is more suitable for high-melting materials, especially the desirable niobium. To the best of our knowledge, also there are no details of this preparation technique as fragile as the suspended bridge mask required for shadow evaporation.

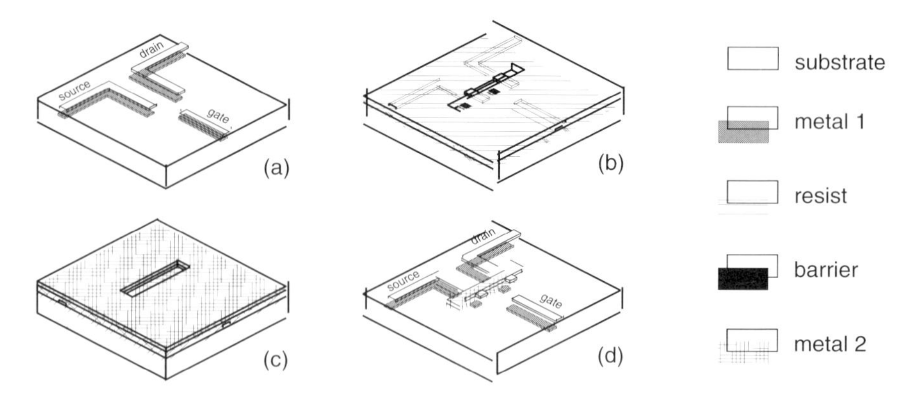

Figure 25.3: Scheme of the electron beam direct-writing preparation of a double-junction structure with gate (building block of both SETs and charge qubits, for explanation of the steps see text).

25.3.2 Fabrication of Tunnel Devices: SET and Charge Qubit Structures

Within the frame of current investigations, the metals Al (it serves as pilot material) and Nb for both base electrode striplines and islands between the junctions of the measured samples (see Sec. 25.4) were deposited by sputtering only. For this purpose, we used an Ar-ion beam with a current of 90 mA yielding, at a residual gas pressure of 10^{-5} Pa, and deposition rates of 15 nm/min and 5 nm/min for Al and Nb, respectively. The nonoxidized Si substrate, characterized by high resistivity at low temperature ($T < 20$ K), short-circuits the tunnel junctions at room temperature.

The respective pattern transfer was completed by subsequent lift-off processes. Because of its quality (good lift-off properties, high resolution) we applied to all lithography steps a special four-layer resist system [20] consisting of the following sequence (from bottom to top, cf. Fig. 25.4): PMMA/MMA ARP-610 (75 nm), PMMA ARP-671 (56 nm), ARP-610 (75 nm) again and, finally, ARP-671 (100 nm) as top layer (All Resist©). In each case, the ARP-610 and ARP-671 layers were tempered at 210°C and 180°C, respectively, for 10 min. The exposure was carried out with the Leica LION-LV1 e-beam lithographical system.

By e-beam writing and developing with a 1:3 solution of methylisobutylic ketone/ isopropyl alcohol a double undercut is generated (Fig. 25.4). This important detail of the patterning process permits the sputtering of flat metal strips with lens-shaped cross sections. Disturbing material redeposition lines (so-called fences that could be caused by back-sputtering) along the edges are thus avoided and, therefore, a perfect overlapping of subsequently deposited strips is guaranteed.

The film thickness of Al and Nb base striplines (metal 1 of Fig. 25.3) was about 50 nm. The second metal strip (island electrode) was characterized by similar dimensions (thickness 70 nm, width 60 nm). This way the nominal area of the tunnel junctions is in the order of a few 10^{-3} μm^2.

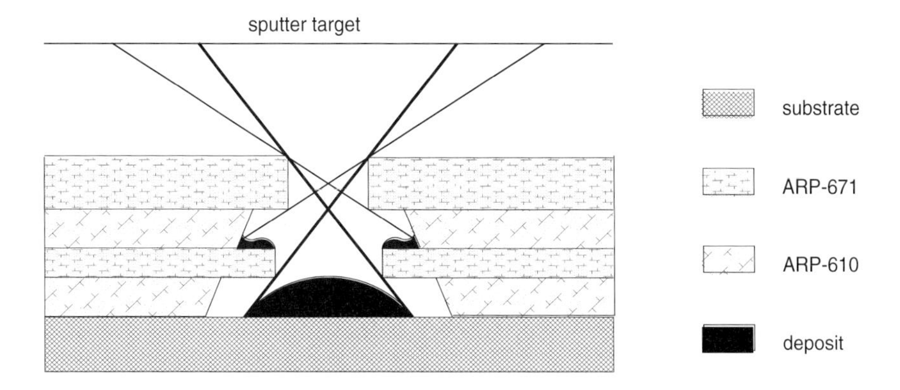

Figure 25.4: Scheme of the sputter deposition of metals by means of the four-layer resist technique.

The second resist mask covered the base electrodes only partially (Fig. 25.3 (b)). In the case of Al (for Al/AlO$_x$/Al and Al/AlO$_x$/Nb junctions) the unprotected area was cleaned by Ar-ion beam etching. Immediately after etching the dielectric barrier was generated by oxidation in an Ar/O_2 atmosphere in the loadlock of the high-vacuum facility. After reintroducing the wafer into the main chamber the device is completed (cf. Fig. 25.3 (d)) by sputtering the second Al film and lifting off.

In the case of Nb base striplines we did not use the natural oxide NbO$_x$ but AlO$_x$. After ion beam cleaning we applied the multi-oxide layer technique to obtain sufficiently thick barriers: We prepared stacked AlO$_x$ layers by sputtering several thin Al films (about 1 nm), each sputtering step followed by an oxidation step.

25.4 Experimental Results

We fabricated a multitude of chips with functioning SETs and single-charge boxes. In particular, more than 80% of the samples showed in fact the single-electron Coulomb blockade, but there was no evidence of Cooper pair charging effects. We found the remainder to be inoperative due to short-circuits and interrupted leads. The experiments were carried out in a ^{3}He evaporation cryostat at temperatures of about 300 mK as well as in a ^{3}He-^{4}He dilution refrigerator down to 10 mK.

First, we measured both $I_{sd} - V_{sd}$ (direct) and $I_{sd} - V_g$ (modulation) characteristics of Al and Nb based SETs in voltage-fed regime. (Here I_{sd}, V_{sd} and V_g denote the source-drain current, bias and gate voltage, respectively.) In order to demonstrate the usefulness of the described fabrication technique (Sec. 25.3.1), we present only a minor selection of the obtained data here. In Fig. 25.5 current-voltage curves and modulation characteristics of high-ohmic samples made of different (superconducting) material systems are shown. Charging effects are easy to recognize (cf. insets), even in the case of the Nb/AlO$_x$/Nb-type SET that is characterized by a rather small modulation amplitude (in the order of 10^{-13}A). Samples of this kind could admittedly be modulated at very low temperatures only (20 mK). There is no convincing interpretation of this phenomenon, probably, it is due to special features of the oxide/metal (AlO_x/Nb) interface, which have to be clarified by further investigations.

A supercurrent branch in the vicinity of zero voltage is visible in the low-voltage section of the $I_{sd} - V_{sd}$ curve of a mixed SET (Al/Nb sample, Fig. 25.6). It can be attributed to Cooper pair tunneling, its slope is determined by the input resistance of the preamplifier (10kΩ). The current peaks are related to the superconducting sum and difference energy gap of both metals. The analysis of the peak structure yields a gap value of about $2\Delta_{Nb} \simeq 0.79$ meV for the sputtered niobium that is considerable less than the bulk value, $2\Delta_{Nb}^{bulk} \simeq 3.0$ meV. However, it clearly exceeds the usual value, $2\Delta_{Al} \simeq 0.4$ meV, of aluminum based junctions.

Second, we have investigated the charging behaviour of superconducting qubit structures consisting of single-charge boxes with SETs (used as electrometers). We have tested two possible realizations with differently designed mutual couplings of the island electrodes (Fig. 25.7). The functioning of an Al/AlO_x/Al based structure is shown (Fig. 25.8) by means of the modulation of the electrometer current due to the step-by-step transfer of single charges onto the qubit island.

Here the gate voltage bias of the qubit was chosen to operate in a quasi-symmetric voltage-fed regime (Fig. 25.9).

This way the SET electrometer "feels" via the couple capacitor C_c potential variations on the qubit island, that are caused by the transfer of discrete charge carriers, in a noticeable manner. Applying increasing magnetic fields (up to $B \simeq 1$T) to the samples, we have observed no any change of the positions of the charge steps in the measured modulation curves (although their flanks became more and more pronounced). That is, the charge transfer of the qubit structures here constructed is not realized by Cooper pairs, but by single electrons.

In order to improve their superconducting properties, we have fabricated qubits with niobium islands. (By the way, the transistor central electrode was also made of Nb, because

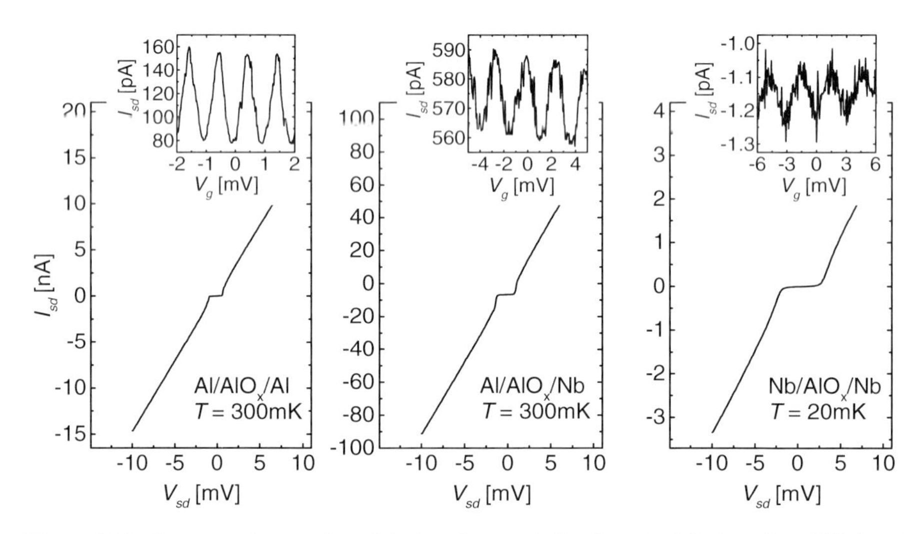

Figure 25.5: Current-voltage and modulation characteristics (insets) of distinct Al and Nb based single-electron transistors.

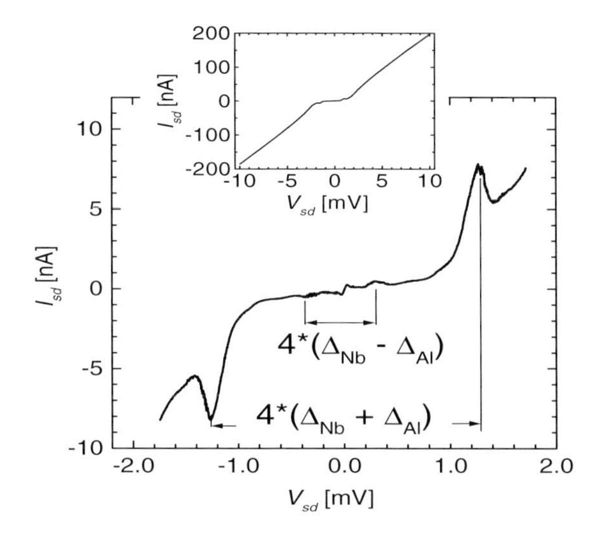

Figure 25.6: Current-voltage characteristic of a mixed Al/AlO$_x$/Nb SET ($T = 300$ mK, inset: large voltage and current scale).

along the lines of the preparation technique (cf. Sec. 25.3, Fig. 25.3) all islands were prepared in one step.) The Nb island was protected against possible destructive influences of the AlO$_x$ barrier by an additional thin Al film. The electrometer modulation characteristic of an Al/AlO$_x$/(Al+Nb)-type qubit structure is depicted in Fig. 25.8. The steps we looked for are visible in finite magnetic fields only (here $B \approx 300$ mT). They are rather weakly pronounced and caused, similarly to the above considered Al/AlO$_x$/Al qubits, only by single-e charge transfer.

25.5 Conclusions

In conclusion, we have presented a reliable process for fabrication of deep sub-micron (superconducting) metallic tunnel junctions with capacitances in the order of a few 10^{-16}F. It consists of the electron beam direct-writing conception, here supported by a special four-layer resist technique, in conjunction with material deposition by sputtering. We have applied the process to the preparation of aluminum and niobium based single-electron tunneling devices, i. e. single-electron transistors and single-electron boxes with readout transistors. The sputtered Nb of islands and wires, respectively, is characterized by a gap energy of up to $2\Delta_{Nb} \simeq 0.8$ meV that clearly exceeds the corresponding value of usual aluminum based junctions, $2\Delta_{Al} \simeq 0.4$ meV. Unexpectedly, the charge transfer onto the Nb islands of the designed Cooper pair boxes exhibits in the gate modulation characteristics of readout transistors a pseudo-periodicity corresponding to e (although the combination of Nb island and Al lead should operate as quasiparticle filter). Also other authors [16, 19] have reported on the "deficit" of stable Cooper pair states differing by 2e (on the islands of Nb based SETs).

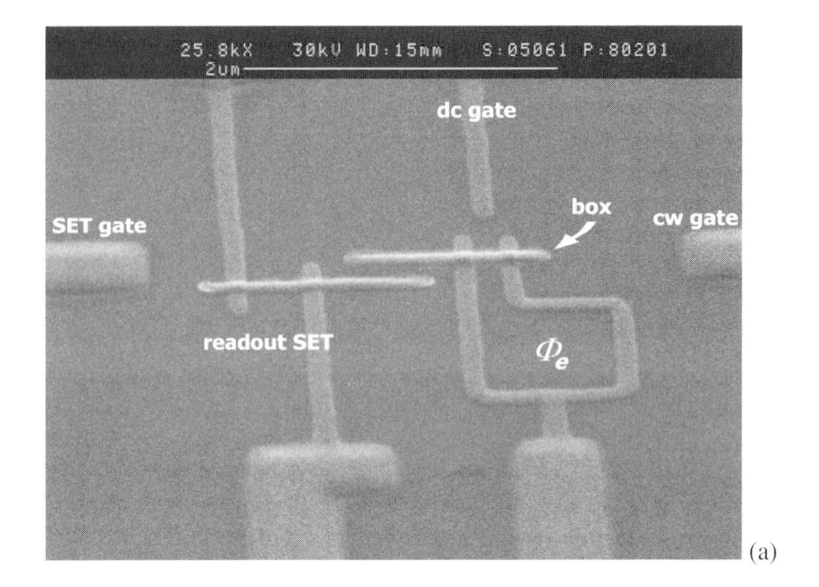

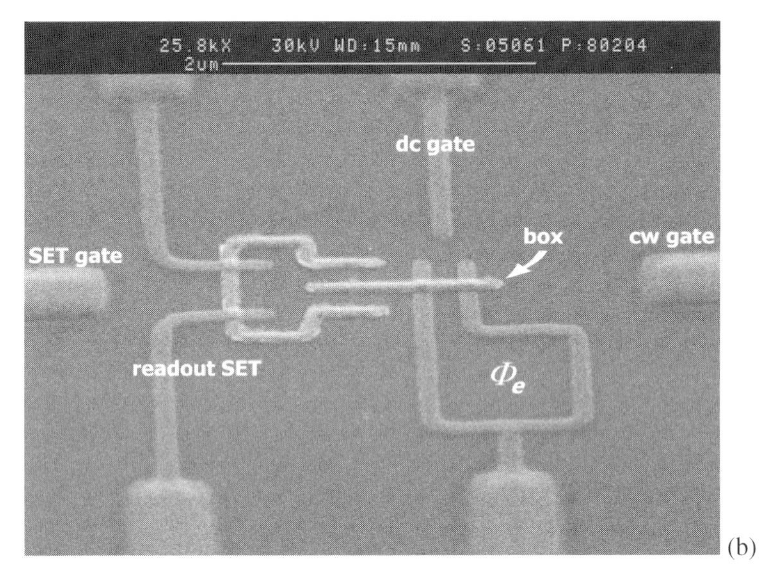

Figure 25.7: SEM micrographs of possible charge qubit realizations with readout SETs distinguished by different capacitive coupling layouts.

One could speculate about possibilities to overcome this situation (improved filter systems, use of qubit structures galvanically decoupled from outer electronics [21]), however, at the moment there is no convincing explanation of the described phenomenon, so that further investigations are called for. Nevertheless, the presented results are steps towards the application of highly desirable niobium based small-capacitance tunnel junctions in quantum information technology.

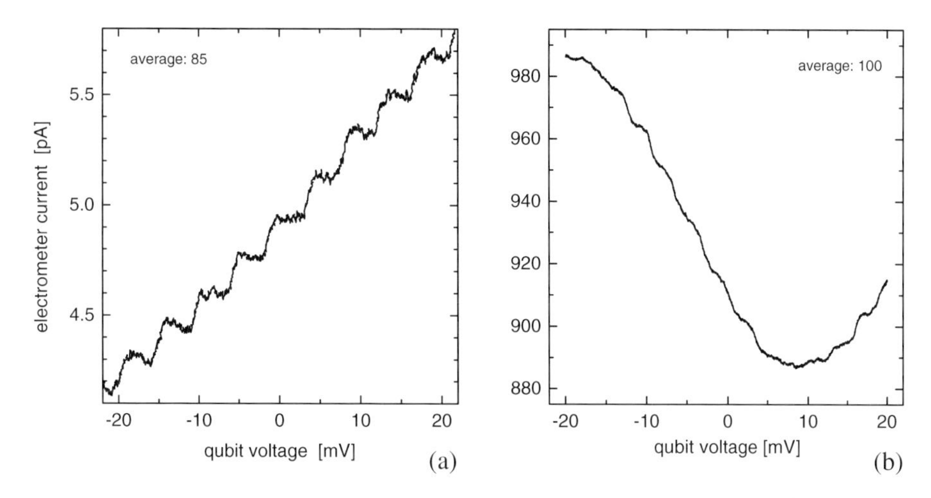

Figure 25.8: Modulation of the SET electrometer current due to single-electron transfer onto the qubit island: (a) Al/AlO_x/Al-type qubit. (b) Al/AlO_x/(Al+Nb)-type qubit (with Nb island); $T = 10$ mK.

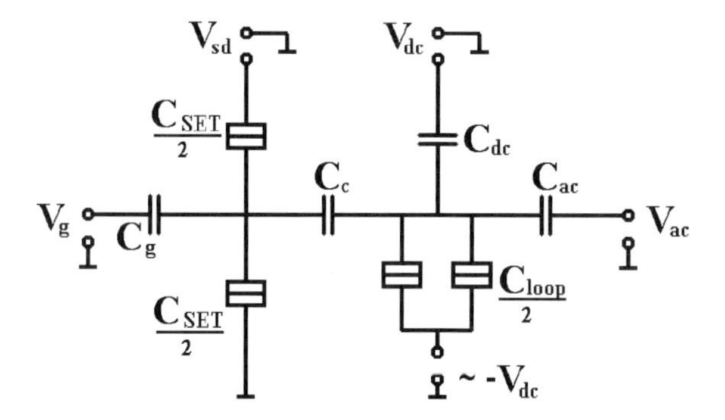

Figure 25.9: Scheme of the qubit device circuit with SET electrometer (on the left) and charge box. (Especially, consider the coupling capacitor C_c, cf. Figs. 25.1 and 25.7.)

Acknowledgements

The authors would like to acknowledge Mrs. W. Gräf, L. Fritzch, and H. Mühlig for actively supporting the preparation and measurement of samples.

References

[1] M. A. Nielsen and I. L. Chuang, *Quantum Computation and Quantum Information*, Cambridge University Press, 2000.

[2] D. Bouwmeester, A. Ekert, and A. Zeilinger (Eds.), *The Physics of Quantum Information: Quantum Cryptography, Quantum Teleportation, Quantum Computation*, Springer, 2000.

[3] J. A. Cirac and P. Zoller, *Quantum computation with cold trapped ions*, Phys. Rev. Lett. **74** (1974), 4091–4094.

[4] I. L. Chuang, N. A. Gershenfeld, and M. Kubinec, *Experimental implementation of fast quantum searching*, Phys. Rev. Lett. **80** (1998), 3408–3411.

[5] Q. A. Turchette, C. J. Hood, W. Lange, H. Mabuchi, and W. Kimble, *Measurement of conditional phase shifts for quantum logic*, Phys. Rev. Lett. **75** (1995), 4710–4713.

[6] Yu. Makhlin, G. Schön, and A. Shnirman, *Quantum-state engineering with Josephson-junction devices*, Rev. Mod. Phys. **73** (2001), 357–400.

[7] Y. Nakamura, Yu. A. Pashkin, and J. S. Tsai, *Coherent control of macroscopic quantum states in a single-Cooper-pair box*, Nature **398** (1999), 786-788.

[8] C. H. van der Wal, A. C. J. ter Haar, F. K. Wilhelm, R. N. Schouten, C. J. P. M. Harmans, T. P. Orlando, S. Lloyd, and J. E. Mooij, *Quantum Superposition of Macroscopic Persistent-Current States*, Science **290** (2000), 773–777.

[9] J. R. Friedman, V. Patel, W. Chen, S. K. Tolpygo, and J. E. Lukens, *Quantum superposition of distinct macroscopic states*, Nature **406** (2000), 43–46.

[10] D. Vion, A. Aassime, A. Cottet, P. Joyez, H. Pothier, C. Urbina, D. Esteve, and M. H. Devoret, *Manipulating the Quantum State of an Electrical Circuit*, Science **296** (2002), 886–889.

[11] Y. Nakamura, Yu. A. Pashkin, and J. S. Tsai, *Quantum coherence in a single-Cooper-pair box: experiments in the frequency and time domains*, Physica B **280** (2000), 405–609.

[12] J. Niemeyer, *Eine einfache Methode zur Herstellung kleinster Josephson-Elemente*, PTB-Mitteilgn. **84** (1974), 251–253.

[13] G. J. Dolan, *Offset masks for lift-off photoprocessing*, Appl. Phys. Lett. **31** (1977), 337–339.

[14] Y. Harada, D. B. Haviland, P. Delsing, C. D. Chen, and T. Claeson, *Fabrication and measurement of a Nb based superconducting single electron transistor*, Appl. Phys. Lett. **65** (1994), 636–638.

[15] K. Blüthner, M. Götz, A. Hädicke, W. Krech, Th. Wagner, H. Mühlig, H.-J. Fuchs, U. Hübner, D. Schelle, and E.-B. Kley, *Single-Electron Transistors Based on Al/AlO$_x$/Al and Nb/AlO$_x$/Nb Tunnel Junctions*, IEEE Trans. Appl. Supercond. **7** (1997), 3099–3102.

[16] A. B. Pavolotsky, Th. Weimann, H. Scherer, V. A. Krupenin, J. Niemeyer, and A. B. Zorin, *Multilayer technique for fabricating Nb junction circuits exhibiting charging effects*, J. Vac. Sci. Technol. B **17** (1999), 230–232.

[17] Th. Wagner, W. Krech, B. Frank, H. Mühlig, H.-J. Fuchs, and U. Hübner, *Fabrication and Measurement of Metallic Single Electron Transistors*, IEEE Trans. Appl. Supercond. **9** (1999), 4277–4280.

[18] D. Born, Th. Wagner, W. Krech, U. Hübner, and L. Fritzsch, *Fabrication of Ultrasmall Tunnel Junctions by Electron Beam Direct-Writing*, IEEE Trans. Appl. Supercond. **11** (2001), 373–376.

[19] R. Dolata, H. Scherer, A. B. Zorin, and J. Niemeyer, *Single electron transistors with high-quality superconducting niobium islands*, Appl. Phys. Lett. **80** (2002), 2776–2778.

[20] U. Hübner, L. Fritzsch, D. Born, Th. Wagner, H.-G. Meyer, and W. Krech, *Advanced lift-off-technique for the fabrication of ultrasmall tunnel junctions*, Extended Abstracts ISEC '01, Osaka (2001), 55–56.

[21] A. B. Zorin, *Cooper-pair qubit and Cooper-pair electrometer in one device*, Physica C **368** (2002), 284–288.

26 Quantum Dot Circuits for Quantum Computation

R. H. Blick, A. K. Hüttel, A. W. Holleitner, L. Pescini, and H. Lorenz

Center for Nanoscience & Sektion Physik
Ludwig-Maximilians-Universität
München, Germany

We discuss fundamental aspects how to realize circuits for quantum computational tasks based on quantum dots. A number of proposals have been suggested on how to integrate these artificial molecules for building quantum computing devices. Crucial for operating such circuits is the realization of wave function coherence as it can be established in coupled quantum dots. The dots discussed in this work are mostly formed in AlGaAs/GaAs-heterostructures and are generally treated as artificial atoms or molecules, being easily controlled by electrodes. We concentrate on how to create quantum bits in coupled dots and on spin effects in single dots.

26.1 Introduction

The ever increasing demand for computing power as well as theoretical considerations on the basic notions of information processing [1] have led to the development of new concepts of computing [2]. Making use of quantum mechanics different realizations of experimental systems have been suggested performing quantum computational tasks [3]. Among the most promising of these are quantum dots [4] which can by now be fabricated with great accuracy in a whole variety of circuits enabling not only probing molecular binding mechanisms in coupled dots [5], but also the definition of quantum bits (qubits).

A number of proposals have been introduced how to integrate these artificial molecules for building quantum computing devices [6, 7]. Such a quantum computer should be capable of solving specific tasks classical computing machines are not powerful enough for. While the main promise for quantum computing schemes based on dots is the scalability of semiconductor computing elements, yet the very basic circuits and operations with these specific qubits have to be devised and tested. The strength of applying quantum dots as quantum computational elements is the flexibility these systems offer as compared to real atoms.

Here we are pursuing the study of two coupled quantum dots forming a molecular mode which we treat as a qubit. This enables us to fully control and engineer the interaction of a qubit with the environment (dephasing) and with other qubits (communication). Dephasing is introduced via the electron-electron interaction, electron-spin and electron-phonon coupling. Naturally, the electron-electron and electron-photon interaction have to be considered as means for communicating, i.e. addressing, the indivivdual qubits. Another quite interesting aspect is the spin coupling of electrons and the nuclei of the semiconductor structure in which

Quantum Information Processing, 2nd Edition. Edited by Thomas Beth and Gerd Leuchs

ISBN: 3-527-40541-0

the quantum dots are embedded. In this context we will discuss preparation of electron spins in a single quantum dot.

The basis of all experimental approaches to quantum computation is the definition of a proper qubit: For quantum dot systems the qubit of choice is the coherent mode generated by discrete states in two quantum dots. The energy scales involved in this setup are the Coulomb interaction E_C of N electrons captured in a quantum dot, the electron spin interaction E_S, excited states ϵ^* of the N-electron system, and the tunnel coupled molecular or coherent state ϵ_t. In our experiments the foremost goal is to achieve large Coulomb energies $E_C \sim 1 \ldots 3$ meV in order to establish a sufficient signal/noise ratio. Connected to this energy scale are the excited states of the N-electron systems: Empirically it is found that these scale according to $\epsilon^* \sim E_C/5 - E_C/3$. This finally enables observation of tunnel coupled states ϵ_t of the order of $50 \ldots 200$ μeV. The advantage of using discrete states of two quantum dots instead of ground and excited states of the same artificial atom lies in the enhanced flexibility of gating two individual dots. At the same time variation of the coupling strength easily allows for activating or deactivating single qubits.

The material system of choice is AlGaAs/GaAs-heterostructures containing a high-mobility two-dimensional electron gas with a phase coherence length of typically 10–30 microns at low temperatures. Definition of quantum dots is usually achieved by electron beam lithography and adjacent deposition of Schottky field-effect gates. Quantum dots contain roughly 1–100 electrons and are attached to metallic leads by tunneling barriers. The main advantage of heterostructures is the high degree of perfection with which its electronic, photonic and phononic properties can be tailored. These materials already enabled ground breaking work which demonstrated that quantum dots in the few electron limit show not only charge quantization, but reveal a discrete energy spectrum similar to real atoms [4]. While it might not be possible with laterally integrated quantum dots to build a full scale quantum computer, these dot circuits are the ideal model systems for studying the principles of operation, which later on might be implemented in molecular devices.

In this work we focus on coupled dots forming the qubit as a core element for circuits such as the *capacitive qubit coupler* shown in Fig. 26.1. The single qubit in this approach is formed by a superposition of two discrete states in two coupled dots. Individual source and drain contacts allow simultaneous addressing of both qubits. Several gating electrodes enable to directly switch the entanglement in the qubit circuit. The qubits considered in this context are defined on the basis of electronic charge, i.e. on the spatial component of the wave functions. Although the electron's spin on the other hand can be manipulated in certain cases [8] as predicted earlier [9], spin resonance detection for single quantum dots is still in its infancy [10, 11].

26.2 Realizing Quantum Bits in Double Quantum Dots

Of prime interest here is entanglement of a qubit and back-action of the probing mechanism on the phase coherent quantum state. Entangling qubits will allow to introduce XOR-operations and hopefully error correction mechanisms. As introduced above dephasing of coherent electronic modes via coupling to the phonon bath has to be considered. After introducing measurements on tunnel coupled quantum dots, i.e. a single qubit, a phase sensitive detection

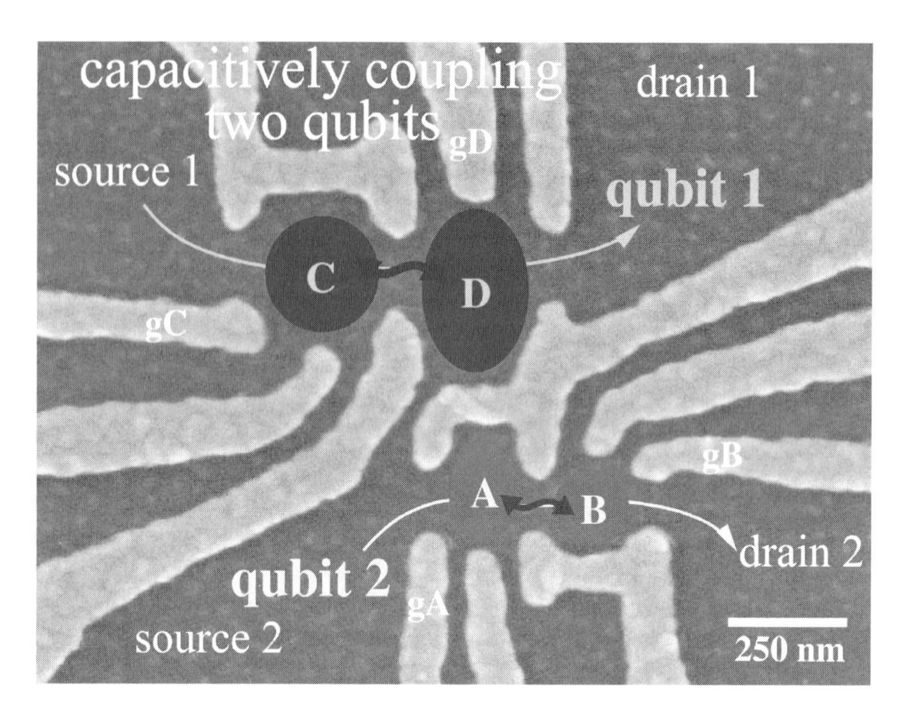

Figure 26.1: Quantum dot circuit with two qubits each formed by two dots – the qubits A-B and C-D are the key elements for a circuit such as this *capacitive qubit coupler*.

scheme for electronic wave functions is introduced [12]. This scheme is easily expanded for non-invasive probing truly molecular states in coupled dots [5]. We also want to point out a fabrication technique for high mobility low-dimensional electron systems in phonon cavities [13]. In the following we will not focus on the electron-photon interaction, however, it should be noted that we succeeded recently to probe electronic coherent modes in a qubit by microwave radiation in the frequency range of 1–40 GHz [14]. This gives sampling rates of 1–50 nsec with which qubits with life times of 1–100 nsec can be addressed.

The devices we use are realized in a two-dimensional electron gas (2DEG) being 90 nm below the surface of an AlGaAs/GaAs heterostructure. At a bath temperature of some 10 mK the electron mobility and the density are usually found to be $\mu = 8 - 12 \times 10^5$ m^2/Vs and $n_s = 1 - 3 \times 10^{15}$ m^{-2}, respectively. By electron beam writing and Au-evaporation Schottky-gates are defined. Under appropriate voltage bias these form the desired quantum dots. As seen in the scanning electron microscope micrograph of Fig. 26.1, 26.2, and 26.3, the gates also define the tunneling barriers between the two quantum dots. By pinching off the tunneling barriers of one of the dots, we first characterize each dot individually. From transport spectroscopy we find charging energies for the dots to be of the order of $E_C = e^2/2C_\Sigma = 1 - 2$ meV. Taking into account the electron density the number of electrons in the dots is estimated to be $10 - 50$ depending on the dot radii. For the actual measurements the mesoscopic system can be tuned

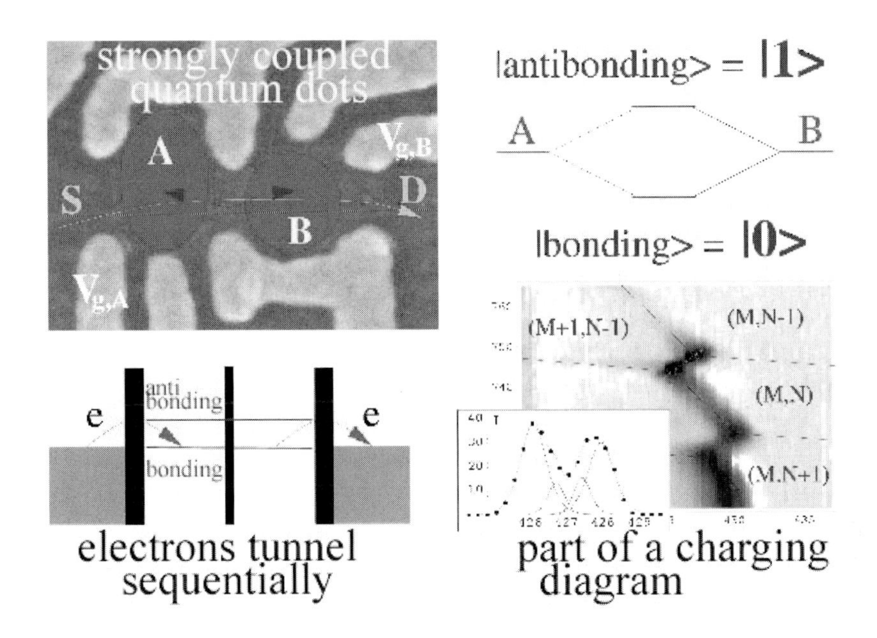

Figure 26.2: A single qubit A-B is tuned into resonance with the center gate regulating the overlap of the wave functions in A and B. For a strong overlap the tunnel splitting leads to the formation of a bonding and an anti-bonding state. In the lower right half of the figure a charging diagram of the two dot system is shown. The charge states are indicated by the number of (M, N) electrons in dot A and B. The dark lines represent a non-zero current through the dots, intersecting the different charging states. The inset is a single data trace taken from the main diagram, obviously not only two ground (binding) states of the $(M+1, N)$ and $(M, N-1)$ contribute to the electronic transport, but additionally the two anti-binding states are involved. The tunnel splitting in this case is of the order of 50 – 70 μeV.

into a regime $2E_C = U > \epsilon^*_{dots} \sim \Gamma > k_B T_e$ in which charge transport is dominated by tunneling through single particle levels with a width Γ (T_e is the electron temperature).

In Fig. 26.2 a single qubit is depicted with the two individual dots marked as A and B. When coupled properly through the tunneling barrier, a splitting of the two discrete states in A and B occurs into a bonding and an anti-bonding state, as shown in the upper right hand side. The bonding state was assumed to have a lower energy in this case what, however, depends on the spins states of the electrons involved. Such a qubit is then probed by determination of the charging diagram of the system: the gate voltages indicated alter the number of electrons in the two dots, e.g. from $(M = N_A, N = N_B)$ to $(M + 1, N)$, or $(M, N - 1)$, etc., see lower right side in Fig. 26.2. Each time the number of electrons is changed a conductance resonance appears, which is identified by the dark regions in the gray scale plot. Depending on the absolute number of electrons confined, the tunnel split resonance appears as an additional conductance channel for electrons.

As for single dots the dominant energy for coupled dots is the charging energy. This can be analyzed by writing the total electrostatic energy of the system, using a capacitance matrix C_{ij}.

The diagonal terms C_{AA}, C_{BB} give the single dot's capacitances and $C_{AB} = C_{BA}$ describe the electrostatic interaction [15]. The total energy is given by

$$E = \frac{1}{2} Q_i C_{ij}^{-1} Q_j, \tag{26.1}$$

with Q_A, Q_B being the charges $N_A e$ und $N_B e$ of dot A and B ($e = -|e|$). The capacitive coupling of the gate electrodes gives an additional term. As seen partly in Fig. 26.2 two gate voltages are used to span a charging diagram. In the following we use (Θ) and (Π) for the two gates ramped. This gives for the total electrostatic energy

$$E \equiv E(N_A, N_B) = \frac{1}{2} Q_i C_{ij}^{-1} Q_j + (V_\Pi \Pi_i + V_\Theta \Theta_i) Q_i \,, \tag{26.2}$$

with (Π_A, Π_B) and (Θ_A, Θ_B) being the capacitive coupling values of the single dots to the two gates with voltages V_Π und V_Θ applied. This yields for the charges on dots A and B:

$$Q_A = \sum_n C_{An} V_n = C_{AA} V_A + C_{AB} V_B + \sum_{n=\Pi,\Theta} C_{An} V_n, \tag{26.3}$$

$$Q_B = \sum_n C_{Bn} V_n = C_{BA} V_A + C_{BB} V_B + \sum_{n=\Pi,\Theta} C_{Bn} V_n. \tag{26.4}$$

From this the voltages V_A and V_B are obtained

$$V_A = \frac{1}{\det C_{ij}} \left(C_{BB} Q_A - C_{AB} Q_B + \sum_{n=\Pi,\Theta} \left(C_{AB} C_{Bn} - C_{BB} C_{An} \right) V_n \right), \tag{26.5}$$

$$V_B = \frac{1}{\det C_{ij}} \left(C_{AA} Q_B - C_{AB} Q_A + \sum_{n=\Pi,\Theta} \left(C_{AB} C_{An} - C_{AA} C_{Bn} \right) V_n \right). \tag{26.6}$$

After integration

$$\begin{aligned} E(Q_A, Q_B) &= \int_0^{Q_A} V_A(Q'_A, Q_B = 0) dQ'_A + \int_0^{Q_B} V_B(Q_A, Q'_B) dQ'_B \\ &= \frac{1}{\det C_{ij}} \left(\frac{1}{2} C_{BB} Q_A^2 + \sum_{n=\Pi,\Theta} \Big(C_{AB} C_{Bn} - C_{BB} C_{An} \Big) V_n Q_A \right. \\ &\quad \left. + \frac{1}{2} C_{AA} Q_B^2 - C_{AB} Q_A Q_B \right) + \frac{1}{\det C_{ij}} \sum_{n=\Pi,\Theta} \left(C_{AB} C_{An} - C_{AA} C_{Bn} \right) V_n Q_B. \end{aligned} \tag{26.7}$$

For the capacitively coupled gate potentials this results in

$$\Theta_i = \frac{1}{\det C_{ij}} \begin{pmatrix} C_{AB} C_{B\Theta} - C_{BB} C_{A\Theta} \\ C_{AB} C_{A\Theta} - C_{AA} C_{B\Theta} \end{pmatrix}, \tag{26.8}$$

$$\Pi_i = \frac{1}{\det C_{ij}} \begin{pmatrix} C_{AB} C_{B\Pi} - C_{BB} C_{A\Pi} \\ C_{AB} C_{A\Pi} - C_{AA} C_{B\Pi} \end{pmatrix}. \tag{26.9}$$

The conditions for charging an electron onto and off the two dots are for $\mu_S = \mu_D = \mu$:

$$(1)\quad E(N_A+1, N_B) - E(N_A, N_B) = \mu \tag{26.10}$$

$$(2)\quad E(N_A+1, N_B+1) - E(N_A+1, N_B) = \mu \tag{26.11}$$

$$(3)\quad E(N_A, N_B+1) - E(N_A, N_B) = \mu \tag{26.12}$$

$$(4)\quad E(N_A+1, N_B+1) - E(N_A, N_B+1) = \mu \tag{26.13}$$

The conditions (1)-(4) define the states in the charging diagram from which the phase boundaries can be derived. The only difference of the conditions (1) and (4) is the charging state of the whole system, the same holds for conditions (2) and (3). With the notation

$$\tilde{C}_{ij} = C_{ij}^{-1}, \tag{26.14}$$

condition (1) gives:

$$Q_i \tilde{C}_{ij} \begin{pmatrix} e \\ 0 \end{pmatrix} + \frac{1}{2} \begin{pmatrix} e \\ 0 \end{pmatrix} \tilde{C}_{ij} \begin{pmatrix} e \\ 0 \end{pmatrix} + (V_\Pi \Pi_i + V_\Theta \Theta_i) \begin{pmatrix} e \\ 0 \end{pmatrix} = \mu \tag{26.15}$$

which leads to

$$V_\Theta^{(1)} = -V_\Pi \frac{\Pi_A}{\Theta_A} + \frac{1}{\Theta_A} \left(\frac{\mu}{e} - \frac{e}{2} \tilde{C}_{AA} - Q_i \tilde{C}_{ij} \begin{pmatrix} 1 \\ 0 \end{pmatrix} \right) \tag{26.16}$$

and in the same way for conidition (4)

$$V_\Theta^{(4)} = -V_\Pi \frac{\Pi_A}{\Theta_A} + \frac{1}{\Theta_A} \left(\frac{\mu}{e} - \frac{e}{2} \tilde{C}_{AA} - Q_i' \tilde{C}_{ij} \begin{pmatrix} 1 \\ 0 \end{pmatrix} \right) \tag{26.17}$$

with $Q_i' = Q_i + \binom{0}{e}$. The distance of these linear traces is given by

$$\frac{1}{\Theta_A} \begin{pmatrix} 0 \\ e \end{pmatrix} \tilde{C}_{ij} \begin{pmatrix} 1 \\ 0 \end{pmatrix} = \frac{e}{\Theta_A} \tilde{C}_{AB}. \tag{26.18}$$

From condition (3) it is found:

$$V_\Theta^{(3)} = -V_\Pi \frac{\Pi_B}{\Theta_B} + \frac{1}{\Theta_B} \left(\frac{\mu}{e} - \frac{e}{2} \tilde{C}_{BB} - Q_i \tilde{C}_{ij} \begin{pmatrix} 0 \\ 1 \end{pmatrix} \right) \tag{26.19}$$

and finally for condition (3):

$$V_\Theta^{(2)} = -V_\Pi \frac{\Pi_B}{\Theta_B} + \frac{1}{\Theta_B} \left(\frac{\mu}{e} - \frac{e}{2} \tilde{C}_{BB} - Q_i'' \tilde{C}_{ij} \begin{pmatrix} 0 \\ 1 \end{pmatrix} \right) \tag{26.20}$$

with $Q_i'' = Q_i + \binom{e}{0}$ and the offset of the traces

$$\frac{e}{\Theta_B} \tilde{C}_{AB}. \tag{26.21}$$

These traces define charging diagrams with charges N_A, N_B on the dots. At the crossing points of the traces for (1) and (3), respectively (2) and (4) one finds:

$$E(N_A, N_B+1) = E(N_A+1, N_B), \tag{26.22}$$

and resonant tunneling occurs. The total charge is conserved, but only with the coupled dot system, i.e. the electrons are distributed over the whole system. From these conditions the following eqations are obtained:

$$V_\Theta = -V_\Pi \frac{\Pi_A - \Pi_B}{\Theta_A - \Theta_B} - \frac{1}{\Theta_A - \Theta_B}\left(Q_i \tilde{C}_{ij}\begin{pmatrix}-1\\+1\end{pmatrix} - \frac{e}{2}\tilde{C}_{AA} + \frac{e}{2}\tilde{C}_{BB}\right). \quad (26.23)$$

These define the charging diagram. Neglecting the interdot capacitance C_{AB} this results in the rectangular upper resonance pattern of Fig. 26.2. With a non-zero coupling C_{AB} conditions (1)–(4) can not be satisfied at the same time and the degeneracy of the resonances is partially lifted. The offset is determined by $e\tilde{C}_{AB}/\Theta_A$.

In addition to the pure capacitive interaction the two discrete resonances in two individual quantum dots can overlap depending on the coupling. This leads to the tunnel splitting indicated in Fig. 26.2 and can nicely be detected in the charging diagrams of coupled quantum dots. Following Pfannkuche [16] this can be derived from the Hamiltonian of the coupled dot system. In a simplified model the center of mass excitations with the quantum numbers k_A, k_B of N_A, N_B electrons in dot A and B are written

$$[|N_B k_B\rangle |N_A k_A\rangle]_{N_B, N_A, k_B, k_A}. \quad (26.24)$$

For decoupled dots the Hamiltonian is diagonal

$$\begin{aligned} &H_0^{DQP}|N_B k_B\rangle|N_A k_A\rangle = \quad (26.25)\\ &\quad \left[\frac{e^2}{2C_A}N_A^2 + \frac{e^2}{2C_B}N_B^2 + \frac{e^2 C_{AB}}{C^2}N_B N_A + eV_\Pi(\Pi_A N_A + \Pi_B N_B)\right.\\ &\quad \left. + eV_\Theta(\Theta_A N_A + \Theta_B N_B) + (k_B+1)\Omega_B + (k_A+1)\Omega_A\right]|N_B k_B\rangle|N_A k_A\rangle\\ &\quad = \left[E(N_A, N_B) + E^{cm}(k_B, k_A)\right]|N_B k_B\rangle|N_A k_A\rangle, \quad (26.26)\end{aligned}$$

where $\Omega_{B/A}$ characterizes the confinement potential. The coupling of states $N = N_B + N_A$ through the barrier is defined by

$$\begin{aligned} H_t^{DQP}|N_B k_B\rangle|N_A k_A\rangle = t\sum_{k_1,k_2}\Big[&|N_B+1, k_B+k_1\rangle|N_A-1, k_A+k_2\rangle\\ &+ |N_B-1, k_B+k_1\rangle|N_A+1, k_A+k_2\rangle\Big]. \end{aligned} \quad (26.27)$$

For simplicity it is assumed that the tunneling matrix element t is constant. If the states $|N_B+1,0\rangle|N_A,0\rangle$ and $|N_B,0\rangle|N_A+1,0\rangle$ are degenerated the two tunnel split states are a linear combination of both with a valence electron. When the electrostatic energy of the two dot system is termed $|N_B,0\rangle|N_A,0\rangle$ with $E(N_A,N_B)$, the energy of tunnel split state is

$$\begin{aligned}\epsilon_t = \frac{1}{2}&(E(N_A, N_B-1) + E(N_A-1, N_B))\\ &\pm \frac{1}{2}\sqrt{(E(N_A, N_B-1) - E(N_A-1, N_B))^2 + 4t^2}.\end{aligned}$$

A measurement of this tunnel split state is seen in Figs. 26.2 and 26.4: in the inset of the charging diagram in Fig. 26.2 a single line plot from the diagram is taken, crossing the charge

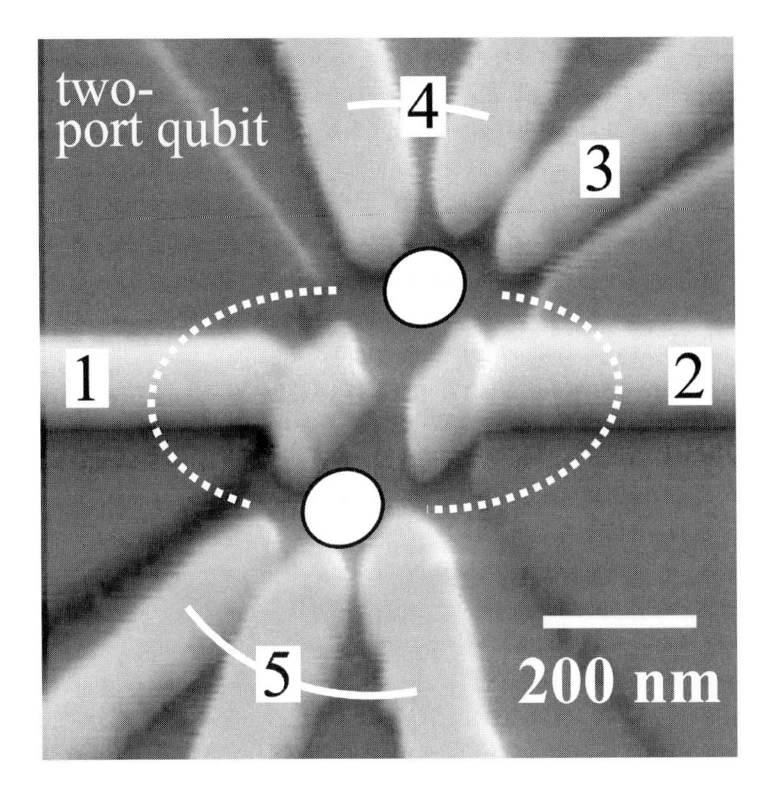

Figure 26.3: Two quantum dots arranged as a qubit each having individual source/drain contacts. Coupling to the leads and between the dots can be changed on a broad scale.

states $(M, N-1)$ and $(M+1, N)$. As seen we find a resonance trace which comprises not only the two ground states, but in addition two peaks hidden in the shoulder of the main resonance. These additional resonances are identified as the anti-bonding states of the $(M, N-1)$ and $(M+1, N)$. The energy splitting is of the order of $\epsilon_t \sim$ 50–70 μeV. Evidently, such a coherent mode can be monitored in a coupled dot system. Below we will see how to increase the coupling in order to achieve separate peaks for the bonding states.

In Fig. 26.3 another setup with two coupled dots is shown, which allows us to probe phase coherence of a single qubit. The two dots are located in the arms of an Aharonov–Bohm interferometer with tunable source and drain contacts [12]. Such a setup can be seen as a qubit formed by the two discrete dot states with individual drain/source contacts, allowing complete control of the both dots simultaneously. Another particular advantage of this setup is the ability to not only sample the electronic phase, but in addition to vary the overlap of the two discrete states in the dots: drain and source couple to dot1 and dot2 via Γ_{l_i,r_i}, while the interdot overlap is given by J. The magnetic field is oriented perpendicular to the plane of the sample. The measurements on this sample geometry of the tunnel coupling states (bonding & anti-bonding) are shown in Fig. 26.4: Since the dots in this case only contain 15 electrons

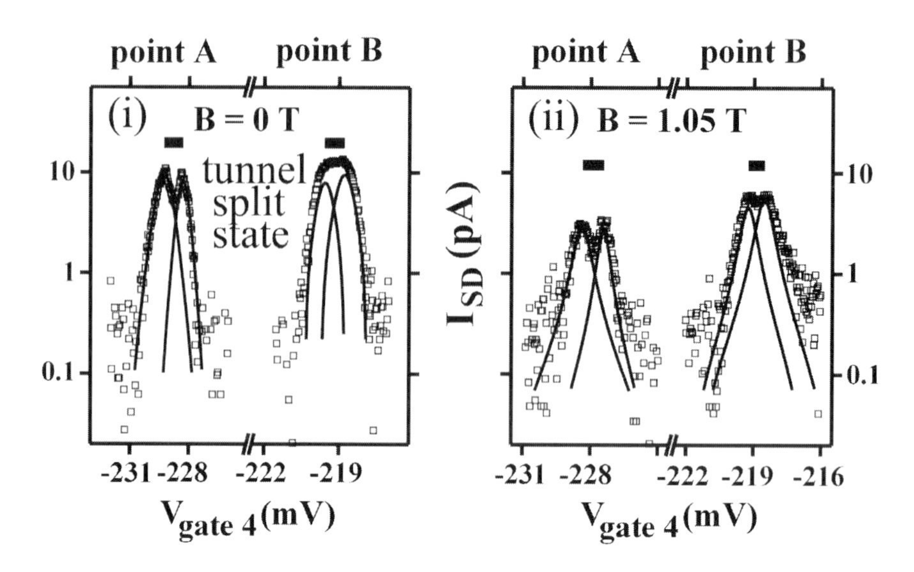

Figure 26.4: In the regime of strong interdot-coupling ($J > 0$) we find clear tunnel splitting, which can be tuned by switching on a magnetic field. Each dot contains about 15 electrons. each dot.

each instead of 50 as in measurements shown above, we find a clear splitting marked by the bold line. The coupling energy in this case is found to be of the order of 120 μeV – as before two resonances are shown. In terms of capacitances we obtain coupling strengths of about $C_{AB}/C_{\Sigma}^{dot1} \cong C_{AB}/C_{\Sigma}^{dot2} = 0.37 \pm 0.08$. This coupling can now be altered by a magnetic field as indicated in the right hand plot, where the tunnel splitting is measured with a magnetic field applied perpendicular to the plane of the dots. As mentioned earlier this phase coherent mode may be switched by microwave radiation in the range of $1 - 40$ GHz, depending on the life time. This in turn can be tuned by changing the barriers transmission coefficients $\Gamma_{i,j}$ and interdot exchange energy $\hbar J$.

26.3 Controlling the Electron Spin in Single Dots

The next major ingredient for controlling qubits is gathering detailed knowledge about the electron's spin properties confined in a quantum dot. Depending on barriers' opacity $\Gamma_{l,r}$ we distinguish two regimes: in the strong coupling limit when quantum dot wave functions overlap with electron wavefunctions in the leads we find the so-called Kondo effect, while in the opposite limit the electron spin quantum number determines transport through the dot in addition to Coulomb blockade. In both cases the electron spin is well controlled by external gate voltages and can directly be manipulated.

The Kondo effect is a well-known phenomenon in solid state physics: the hybridization of conduction electrons with the localized electron spin of a magnetic impurity atom in a metal leads to an enhancement of resistivity at low temperatures. The current interest stems from the fact that nanostructuring allows to study such complex solid state phenomena on highly controllable quantum dots. The Kondo effect in quantum dots was proposed in early theoretical work [17, 18] and then demonstrated in a series of detailed experiments [19]. The difference to the classical Kondo effect – causing a resistance increase – is that the one found in quantum dots leads to an enhancement of the conductance due to the dynamic spin-flip scattering between electrons in the contacts and on the dot, effectively leading to an additional conductance channel.

In other words the Kondo effect leads to an enhanced local density of states of the quantum dot at the Fermi levels of the electron reservoirs. The main features of the Kondo effect in quantum dots are a zero-bias conductance resonance, its specific temperature dependence and a resonance splitting in a magnetic field. In case of symmetric barriers this results in an enhanced conductance at zero bias which is rapidly decreasing for finite drain/source bias $V_{ds} > 0$. Following the idea of Kondo physics, a single quantum dot of the ones shown in Fig. 26.3 was defined by choosing appropriate gate voltages. As is shown in Fig. 26.5(a) first the conventional Coulomb blockade (CB) regime is determined with opaque barriers. Clear Coulomb diamonds can be identified in addition to the single electron tunneling regime with excited states (dark lines) and regions of negative differential conductance (NDC, bright white lines) indicating spin blockade of type-I [9], which will be discussed below. The determination of the total capacitance of the dot gives a dot radius of $r \approx 70$ nm with a number ~ 20 electrons and a charging energy of $U = 2E_C = 2.7$ meV. The spin-degenerate mean level spacing can be estimated to be $\Delta = \frac{2\hbar^2}{m^* r^2} \approx 500$ μeV [4]. From the temperature dependence of the lineshape the total intrinsic width of the resonances is found to be $h\Gamma = h\Gamma_L + h\Gamma_R =$ 100 μeV and the electron temperature $T_e \approx 100 - 120$ mK.

In Fig. 26.5(b) the dot is tuned into the Kondo regime: on the right the pair structure of the peaks and the Kondo resonance in the valley can clearly be seen. The diamond shape is strongly distorted in this case – the spin orientation is also given. In Fig. 26.6(a) a close-up of the grayscale plot of Fig. 26.5(b) at the position of the resonance around $V_g = -460$ mV is shown: In the CB region around zero bias and between the conductance peaks first-order tunneling processes are forbidden. However, due to the Kondo resonance at the Fermi level of the left barrier the conductance is enhanced on the negative bias side. The diamond shape is strongly distorted in the region where the quantum dot is occupied with an odd number of electrons whereas no effect is visible in the regions above and below where the dot contains an even number of electrons. In the latter regions spin flip processes are inhibited and therefore no Kondo resonance emerges at the Fermi level. The Kondo temperature $kT_K \approx (U\Gamma)^{1/2} \exp(-\pi|\mu_{L,R} - \mu_{dot}|/2\Gamma)$ varies from a minimum value of 20 mK in the valley to 6 K close to the conductance peaks. kT_K is equal to the width of the Kondo resonance. Comparing these values to T_e one finds $\Gamma/kT_e \approx 20$ and $T_K/T_e \approx 0.2 - 50$.

In Fig. 26.6(b) the evolution of the Kondo resonance with gate voltage is shown. From the slope $C_{res}/C_{gate} \approx 3.6$ of the linear shift of the resonance with gate voltage the capacitive coupling to the left reservoir can be estimated to be 60 aF. With an on peak resistance of ≈ 450 kΩ this yields an RC-time of 27 ps corresponding to a decay width of 150 μeV which is fully consistent with our estimation of the coupling to the left reservoir. In the lowest trace a parabola superposed by a Lorentzian has been used to fit the width of the resonance. It is found 0.42 meV, or $T_K = 4.9$ K, which is in agreement with an earlier estimation of T_K.

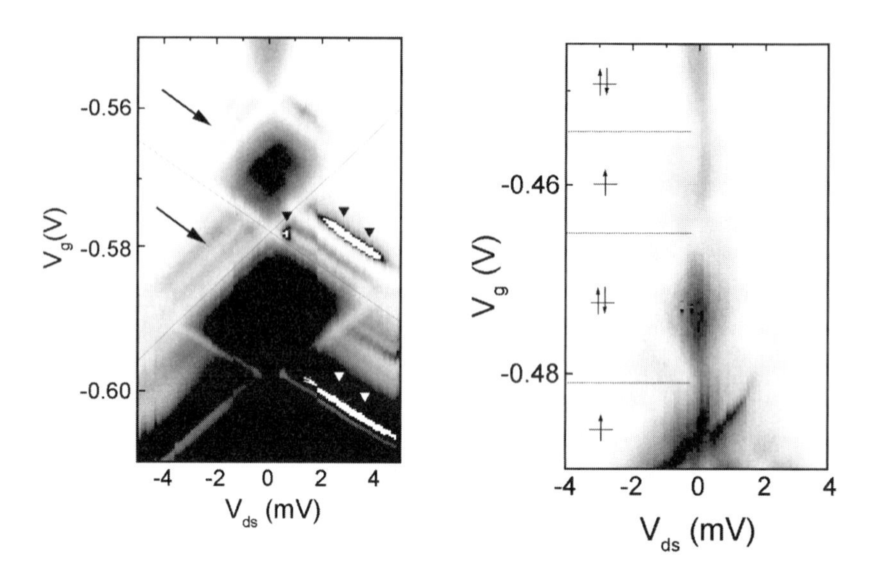

Figure 26.5: Grayscale plot of the bias and gate voltage dependence of the conductance (log-scale) (left) in the conventional CB and (right) in the Kondo regime (black: 0 μS $\leq dI/dV \leq 0.05$ μS, white: $dI/dV \geq 20$ μS or $dI/dV < 0$ μS). On the left the diamond structure and some excited states are well pronounced. The white arrows indicate regions of negative differential conductance related to spin blockade (type-I). The arrows on the left give the spin orientation, as derived from the meaurements of the Kondo resonance. On the right the pair structure of the peaks and the Kondo resonance in the valley can clearly be seen. The diamond shape is strongly distorted in this case – the spin orientation is also given.

Figure 26.7 (a) depicts the characteristic drain/source voltage dependence of a Kondo resonance. Towards larger voltages the resonance maximum is reduced. Increasing the temperature the Kondo effect vanishes and conventional conductance mechanisms take over. Figure 26.7 (b) shows the temperature dependence of the Kondo resonance in detail: as expected in the Kondo regime, with lower temperatures the conductance increases.

Finally, we want to address the opposite regime of opaque barriers: it was suggested by Weinmann *et al.* [9] that transport through single quantum dots can be blocked due to spin effects. Spin selection rules particularly prohibit single electron tunneling transitions between N and $N + 1$ electron ground states of a single dot which differ in total spin by $\Delta S > 1/2$. This phenomenon was termed spin blockade (SB) (type-II), occuring in addition to conventional CB and leading to a suppression of the corresponding conductance peak in linear transport.

In Fig. 26.8 we show such a spin blockade effect in a single quantum dot. The spin blockade energy found here is $E_S \cong 100$ μeV. Particularly shown is a detailed measurements on the bias and magnetic field dependence of the transport spectrum. In Fig. 26.8 (a) to

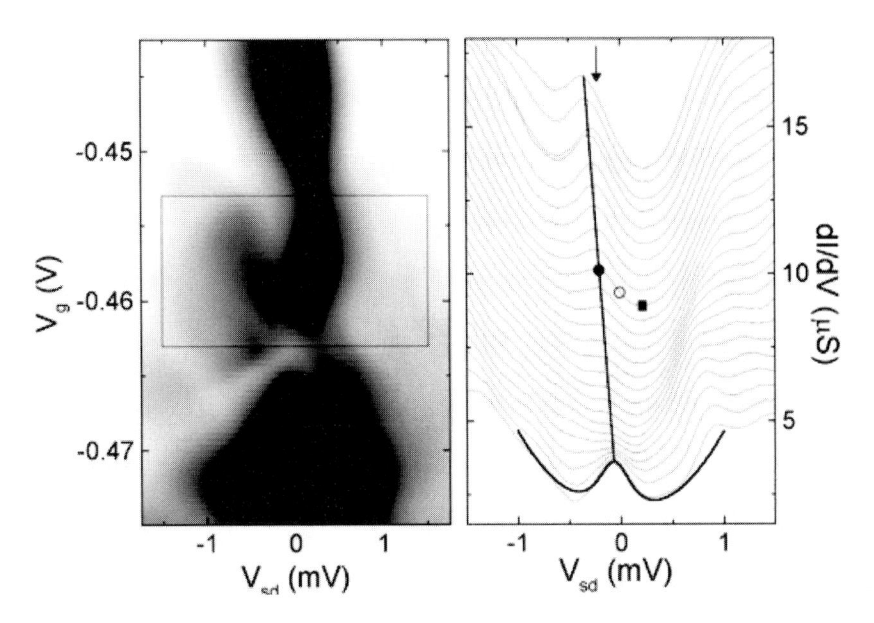

Figure 26.6: Left: Magnification of the Kondo resonance for the upper pair of peaks in Fig. 26.5 in a grayscale plot (linear scale, black: 0 μS, white: 8 μS). The resonance is offset to the negative bias region. Right: The boxed region of the left plot, displayed by line plots. The lowest trace is taken at $V_g = -0.4534$ V, the uppermost at $V_g = -0.4634$ V. The resonance close to zero bias shifts linearly with the gate voltage. A fit to the resonance in the lowest trace yields a Kondo temperature of 4.9 K.

(d) the transport-blocked conductance resonance is shown at 0 mT, 150 mT, 300 mT and 450 mT. As seen at a field strength of only 300 mT the quantum levels in the dot are already shifted sufficiently to reenable ground state transport. At $B = 300$ mT a strong increase in conductance is seen. Here, one quantum ground state participating in transport changes because of a level crossing; for higher B, $|\Delta S| = 1/2$ for the N and $N+1$ electron ground state and single electron tunneling transport takes place. For high B the overall conductance through the quantum dot is decreasing because of a gradual compression of the dot states by the magnetic field.

In our case of weak coupling to the reservoirs, where $\Gamma \ll k_B T \ll \epsilon < E_C$, the maximum conductance through the quantum dot is given by

$$\sigma_{\max} = \frac{e^2}{h} \frac{1}{4 k_B T_{\text{el}}} \frac{\Gamma}{2}.$$

The dwell time for an electron in the system is estimated with $\tau_b = h/\Gamma_b = 9.5\,\text{ns}$ in the case of transport blockade compared to $\tau_e = h/\Gamma_e = 0.8\,\text{ns}$ as extrapolated value without any spin effects. Therefore, coupling of the long-lived state to its environment corresponds to a time scale of $\Delta\tau = \tau_b - \tau_e \approx 8.7\,\text{ns}$, i.e. the high-spin state survives for several nanoseconds.

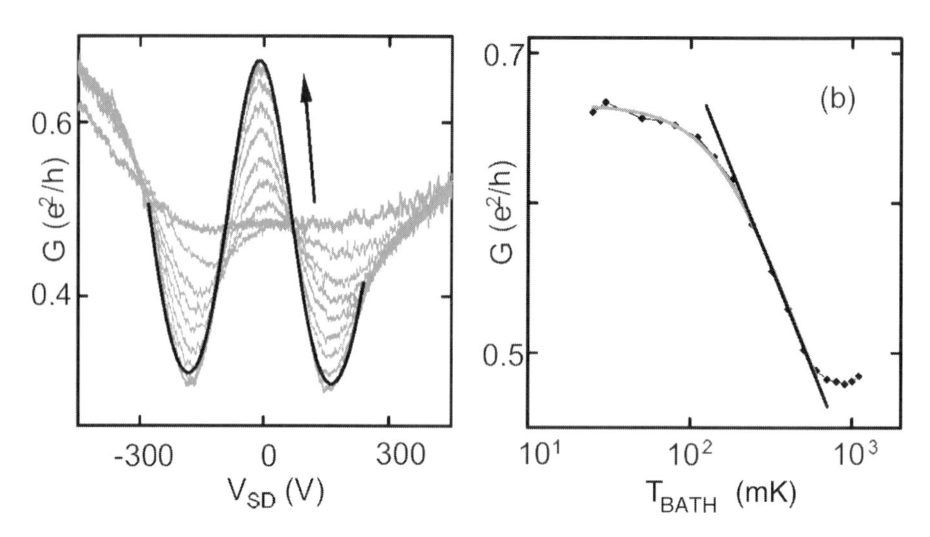

Figure 26.7: Drain/source (left) and temperature (right) dependence of the Kondo resonance.

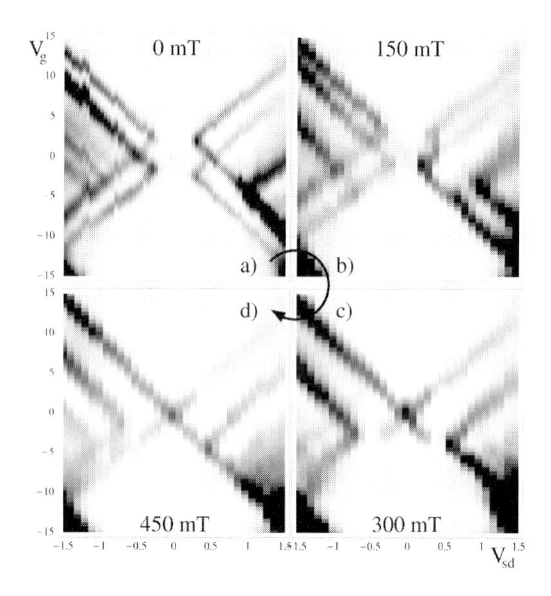

Figure 26.8: (a) to (d): Effect of an external perpendicular magnetic field on the spin blocked single electron tunneling resonance: (a) Detail of the spin blocked transition at $B = 0$ – white represents suppression of conductance. (b) At $B = 150$ mT the gap has partly closed. Conductance is still suppressed for $|V_{\mathrm{sd}}| < 200$ µV. (c–d) Blockade phenomena disappear completely at 300 mT. Simultaneously regions of NDC are evolving at 450 mT.

26.4 Summary

To summarize we have shown how to build and test qubits in coupled quantum dots. A single qubit is defined in two dots via tunnel splitting of the discrete states. This tunnel split state is traced in transport measurements. As we have seen the high degree of control which quantum dots are offering not only allows to manipulate single charges, but also gives control of the electron's spin. Although, the spin configurations in single quantum dots are not fully understood yet, the underlying principles are revealed. In general it can be stated that this approach gives experimental access to physics previously out of reach and enables testing concepts of spintronics. We found that laterally gated quantum dots are the ideal testbed for the physics of quantum computation, i.e. phenomena considering phase coherence, dissipation and the fundamentals of the measuring process.

References

[1] R. Landauer, Phys. Lett. A **217**, 188 (1996).

[2] M.A. Nielsen and I.L. Chuang, *Quantum Computation and Quantum Information*, Cambridge University Press (2000).

[3] D. Bouwmeester, A. Ekert, and A. Zeilinger (eds.), *The physics of quantum information: Quantum Cryptography, Quantum Teleportation, Quantum Computation*, Springer (2000).

[4] L.P. Kouwenhoven, C.M. Marcus, P.L. McEuen, S. Tarucha, R.M. Westervelt, and N.S. Wingreen, in Mesoscopic Electron Transport, edited by L.L. Sohn, L.P. Kouwenhoven and G. Schön, series E, vol. 345, p. 105, Kluwer Dordrecht, Netherlands (1997); R. Ashoori, Nature **379**, 413 (1996).

[5] A. W. Holleitner, R. H. Blick, A. K. Hüttel, K. Eberl, and J. P. Kotthaus, Science **297**, 70 (2002).

[6] D. Loss and D.P. DiVincenzo, Phys. Rev. A **57**, 120 (1998); P. Recher, E.V. Sukhorukov, and D. Loss, Phys. Rev. Lett. **85**, 1962 (2000); Xuedong Hu and S. Das Sarma, cond-mat/0101102 (2001).

[7] Robert H. Blick and Heribert Lorenz, Proc. IEEE Int. Symp. Circ. Sys. **II-245**, 28 (2000).

[8] A.K. Hüttel, H. Qin, A.W. Holleitner, R.H. Blick, K. Neumaier, D. Weinmann, K. Eberl, J.P. Kotthaus, submitted (2002); cond-mat/0109014.

[9] D. Weinmann, W. Häusler, and B. Kramer Phys. Rev. Lett. **74**, 984 (1995).

[10] R.H. Blick, V. Gudmundsson, R.J. Haug, K. von Klitzing, K. Eberl, Phys. Rev. Rap. Comm. **57**, R12685 (1998).

[11] A.K. Hüttel, J. Weber, A.W. Holleitner, D. Weinmann, K. Eberl, and R.H. Blick, submitted (2002).

[12] A.W. Holleitner, H. Qin, C.R. Decker, K. Eberl, and R.H. Blick, Phys. Rev. Lett. **87**, 256802 (2001).

[13] J. Kirschbaum, E.M. Höhberger, and R.H. Blick, W. Wegscheider, M. Bichler, Applied Physics Letters **81**, 280 (2002); E.M. Höhberger, F.W. Beil, R.H. Blick, W. Wegscheider, M. Bichler, J.P. Kotthaus, Physica E **12**, 487 (2002).

[14] H. Qin, A.W. Holleitner, K. Eberl, R.H. Blick, Phys. Rev. B **64**, R241302 (2001).

[15] R.H. Blick, PhD Thesis, *Künstliche Atome und Moleküle: Gekoppelte Quantenpunkte und Mikrowellenspektroskopie an Quantenpunkten*, Max-Planck-Institut für Festkörperphysik und Universität Stuttgart, December 1995; Publishing Company Harri-Deutsch, Physics **57**, Frankfurt am Main, Germany (1996), ISBN 3-8171-1507-5.

[16] D. Pfannkuche, Habilitation Thesis, *Aspects of Coulomb Interaction in Semiconductor Nanostructures*, Max-Planck-Institut für Festkörperforschung and Universität Karlsruhe, September 1999.

[17] J. Kondo, Prog. Theor. Phys. **32**, 37 (1964).

[18] T.K. Ng and P.A. Lee, Phys. Rev. Lett. **61**, 1768 (1988).

[19] D. Goldhaber-Gordon, H. Shtrikman, D. Mahalu, D. Abusch-Magder, U. Meirav, and M. A. Kastner, Nature **391**, 156 (1998); S.M. Cronenwett, T.H. Oosterkamp, and L.P. Kouwenhoven, Science **281**, 540 (1998); F. Simmel, R.H. Blick, J.P. Kotthaus, W. Wegscheider, and M. Bichler, Phys. Phys. Lett. **83**, 804 (1999).

[20] J. Schmid, J. Weis, K. Eberl, and K. von Klitzing, Phys. Rev. Lett. **84**, 5824 (2000).

27 Manipulation and Control of Individual Photons and Distant Atoms via Linear Optical Elements

XuBo Zou and Wolfgang Mathis

Theoretical Electrical Engineering Group
University of Hannover
Hannover, Germany

We present here an overview of our work concerning the manipulation and control of individual photons and distant atoms trapped in different optical cavities by using linear optical elements. These works include the linear optical implementation of the non-deterministic quantum logic operation and single-mode quantum filter, the generation of multiphoton polarization entangled states, and quantum entanglement between distant atoms trapped in different optical cavities.

27.1 Introduction

Quantum entanglement is one of the most striking features of quantum mechanics [1–3]. The recent surge of interest and progress in quantum information theory has allowed one to take a more positive view of entanglement and to regard it as an essential resource for many ingenious applications such as quantum cryptography [4], quantum dense coding [5], and quantum teleportation [6]. However, how to design and realize quantum entanglement is extremely challenging due to the coupling with an environmental degree of freedom. Recently, various quantum systems have been suggested as possible candidates for engineering quantum entanglement and implementing quantum information processing: linear optical systems [7], cavity QED systems [9], trapped ion systems [10], NMR systems [11], solid state systems with nuclear spins [12], quantum dot systems [13], and Josephson-junction device systems [14].

The realization of linear optical quantum information processing, quantum communication, and quantum computing is particularly appealing because of the robust nature of quantum states of light against decoherence. Based on the parametric down-conversion technique in a nonlinear optical crystal, experiments with polarization-entangled states has opened a whole field of research [8]. Currently, polarization-entangled states have only been produced randomly; this is a severe obstacle for further applications of this state to be used as the input state for subsequent experiments. Thus, there is a strong need for entangled photon sources that generate maximally polarization-entangled multiphoton states in a controllable way. Recently, significant progress has been achieved by a proposal to implement probabilistic quantum logic gates by the use of linear optical elements and photon detectors [15]. This seminal work demonstrates that a strong nonlinear effect can be implemented by exploiting post-selection strategies based on the single-photon technology. However, one disadvantage

Quantum Information Processing, 2nd Edition. Edited by Thomas Beth and Gerd Leuchs

ISBN: 3-527-40541-0

of linear optical quantum information processing and quantum computing is that photons are difficult to store.

On the other hand, cavity QED offers an almost ideal system for the generation of entangled states and the implementation of quantum information processing. This is due to the fact that cold and localized atoms in cavity QED are extremely suitable for storing quantum information in long-lived internal states, and the photons are a natural source for fast and reliable transport of quantum information over long distances. The entanglement generation and manipulation between many distant sites is an essential element of quantum network construction and quantum computing.

In this chapter, we present an overview of our work concerning the manipulation and control of individual photons and distant atoms trapped in various optical cavities by using linear optical elements [16–22]. Other works based on the quantum dot, Josephson junction and trapped ion systems [23] will be reviewed elsewhere. The present chapter is organized as follows. Section 27.2 considers how to use linear optical elements to manipulate and control individual photons. These works include linear optical implementation of the non-deterministic quantum logic operation of Knill, Laflamme, and Milburn, the single-mode quantum filter, and the generation of multiphoton polarization-entangled states. Section 27.3 considers how to combine cavity QED and linear optical elements to manipulate and control distant atoms trapped in various optical cavities. Finally, a conclusion is given in Section 27.4.

27.2 Manipulation and Control of Individual Photons via Linear Optical Elements

In Ref. [15], Knill, Laflamme, and Milburn demonstrate how to use linear optical elements and photon detectors to control individual photons. A non-deterministic gate, which is known as the nonlinear sign gate (NLS), was proposed. This gate transforms the quantum states, which are labeled by photon occupation number, according to

$$\alpha|0\rangle + \beta|1\rangle + \gamma|2\rangle \rightarrow \alpha|0\rangle + \beta|1\rangle - \gamma|2\rangle \tag{27.1}$$

The probability of success is rated as $1/4$. This seminal work demonstrates that a strong nonlinear effect can be implemented by exploiting post-selection strategies based on the single-photon technology. However, the optical network proposed is complex and would present major stability and mode-matching problems in their construction. In this section, we will propose several less complicated schemes with linear optical elements to control individual photons [16–21].

27.2.1 Teleportation Implementation of Non-deterministic NLS Gate and Single-mode Photon Filter

The experimental setup for teleportation implementation of the non-deterministic NLS gate is depicted in Fig. 27.1 [16]. The input quantum state is in the form

$$\Psi_{\text{in}} = \alpha|0\rangle_1 + \beta|1\rangle_1 + \gamma|2\rangle_1. \tag{27.2}$$

An ancilla quantum state in the form $\Psi_{\text{ancilla}} = |1\rangle_2|1\rangle_3$ is required. The asymmetric beam splitter BS_1 transforms this ancilla quantum state into

$$\Psi'_{\text{ancilla}} = \frac{\sin 2\theta}{\sqrt{2}}(|2\rangle_{2'}|0\rangle_{3'} - |0\rangle_{2'}|2\rangle_{3'}) + \cos 2\theta\, |1\rangle_{2'}|1\rangle_{3'}, \tag{27.3}$$

where $\sin\theta$ and $\cos\theta$ are the reflectance and the transmittance of the beam splitter BS_1. These coefficients will be determined later. The field state $\Psi_{\text{in}} \otimes \Psi'_{\text{ancilla}}$ will be used directly for the NLS gate implementation. The input modes of Alice's symmetric beam splitter BS_2 are the mode 1 and the mode $2'$. Thus, the quantum state $\Psi_{\text{in}} \otimes \Psi'_{\text{ancilla}}$ of the system transforms to

$$\begin{aligned}\Psi_{\text{o}} &= |2\rangle_{1''}|0\rangle_{2''}\left[(\sin 2\theta)\alpha|0\rangle_{3'} + 2(\cos 2\theta)\beta|1\rangle_{3'} - (\sin 2\theta)\gamma|2\rangle_{3'}\right]/2\sqrt{2} \\ &+|0\rangle_{1''}|2\rangle_{2''}\left[(\sin 2\theta)\alpha|0\rangle_{3'} - 2(\cos 2\theta)\beta|1\rangle_{3'} - (\sin 2\theta)\gamma|2\rangle_{3'}\right]/2\sqrt{2} + \Psi_{\text{other}}.\end{aligned} \tag{27.4}$$

The other terms Ψ_{other} of this quantum state do not contribute to the two events that we consider in the following. Alice makes a photon number measurement on the mode $1''$ and the mode $2''$ with the detectors D_1 and D_2. If D_1 detects two photons and D_2 does not detect any photon, the state is projected into

$$\Psi_{\text{out}} = \frac{1}{2\sqrt{2}}\left[(\sin 2\theta)\alpha|0\rangle + 2(\cos 2\theta)\beta|1\rangle - (\sin 2\theta)\gamma|2\rangle\right]. \tag{27.5}$$

If we choose the reflectance $\sin\theta$ of the beam splitter in order to fulfill $\sin 2\theta = 2/\sqrt{5}$, the output quantum state $\Psi_{\text{out}} = \alpha|0\rangle + \beta|1\rangle - \gamma|2\rangle$ is generated from the input quantum state Ψ_{in}. This is the transformation property (1) of the NLS gate. The probability of this outcome will be $10\,\%$. If D_2 detects two photons and D_1 does not detect any photon, the quantum state is projected into the form

$$\Psi_{\text{out}} = \alpha|0\rangle - \beta|1\rangle - \gamma|2\rangle. \tag{27.6}$$

The other terms Ψ_{other} of the quantum state (4) do not contribute to this event. In order to implement the NLS gate transformation property (1), Bob needs to apply a π phase shifter that changes the signs of the state $|1\rangle$. The probability of this outcome will be $10\,\%$. The total success probability of the gate is the probability of either of these two outcomes. Hence the desired gate will be implemented with $20\,\%$ probability of success.

We next demonstrate how to modify the experimental setup shown in Fig. 27.1 to implement the single-mode quantum filter, which eliminates the one-photon state in a particular input state $\alpha|0\rangle + \beta|1\rangle + \gamma|2\rangle$ [17]. This filter is applied to realize a two-qubit projective measurement [24] and generate multiphoton polarization-entangled states step by step. The experimental setup for implementing a single-mode quantum filter is similar to Fig. 27.1, except that now BS_1 is a symmetric beam splitter. In this case, by emitting two single photons $|\Psi_{\text{ancilla}}\rangle = |1\rangle_2|1\rangle_3$ into this symmetric beam splitter BS_1 an entangled photon channel is generated:

$$\Psi'_{\text{ancilla}} = \frac{1}{\sqrt{2}}(|2\rangle_{2'}|0\rangle_{3'} - |0\rangle_{2'}|2\rangle_{3'}). \tag{27.7}$$

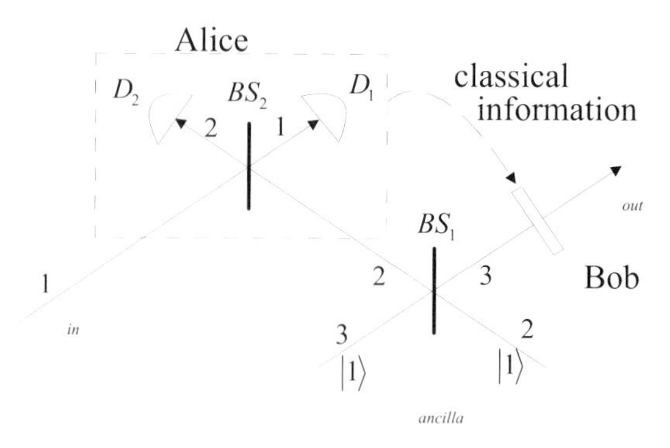

Figure 27.1: The schematic shows the implementation of the single-mode quantum filter. BS_i ($i = 1, 2$) denotes the symmetric beam splitters and D_i ($i = 1, 2, 3$) are photon number detectors. The $\pi/2$ phase shifter is denoted by P.

The output mode $2'$ of the beam splitter BS_1 is one of the input modes of the second symmetric beam splitter BS_2. The mode 1, which is chosen to be filtered out, is the second input mode of the beam splitter BS_2. Thus, the quantum state transforms to

$$\Psi_{\mathrm{o}} = |1\rangle_{1''}|1\rangle_{2''}(\alpha|0\rangle_{3'} + \gamma|2\rangle_{3'})/2 + |0\rangle_{1''}|2\rangle_{2''}(\alpha|0\rangle_{3'} - \gamma|2\rangle_{3'})/2\sqrt{2} + |2\rangle_{1''}|0\rangle_{2''}(\alpha|0\rangle_{3'} - \gamma|2\rangle_{3'})/2\sqrt{2} + \Psi_{\mathrm{other}}. \tag{27.8}$$

The other terms Ψ_{other} of the total quantum state Ψ_{o} do not contribute to the three events, which we consider by a photon number measurement on the modes $1''$ and $2''$ by the detectors D_1 and D_2. If each detector detects one photon, the state is projected into the (unnormalized) state

$$\Psi_{\mathrm{out}} = \alpha|0\rangle + \gamma|2\rangle. \tag{27.9}$$

If D_1 detects two photons and D_2 does not detect any photon or vice versa, the state is projected into the (unnormalized) state

$$\Psi'_{\mathrm{out}} = \alpha|0\rangle - \gamma|2\rangle. \tag{27.10}$$

In order to generate the quantum state (9), a $\pi/2$ phase shifter is needed to change the sign of the state $|2\rangle$. Then the desired transformation

$$\alpha|0\rangle + \beta|1\rangle + \gamma|2\rangle \longrightarrow \alpha|0\rangle + \gamma|2\rangle \tag{27.11}$$

will be obtained. This transformation demonstrates that the proposed setup, which is shown in Fig. 27.1, definitely implements the quantum state filter. The probability of this outcome is $(|\alpha|^2 + |\gamma|^2)/2$.

Now we use our quantum filter concept to implement the two-qubit projective measurement [24]. We will demonstrate that the input state

$$\Phi_{\mathrm{in}} = c_0|H\rangle_1|H\rangle_2 + c_1|H\rangle_1|V\rangle_2 + c_2|V\rangle_1|H\rangle_2 + c_3|V\rangle_1|V\rangle_2 \tag{27.12}$$

will be transformed into the (unnormalized) state

$$\Phi_{\text{out}} = c_0|H\rangle_1|H\rangle_2 + c_3|V\rangle_1|V\rangle_2. \tag{27.13}$$

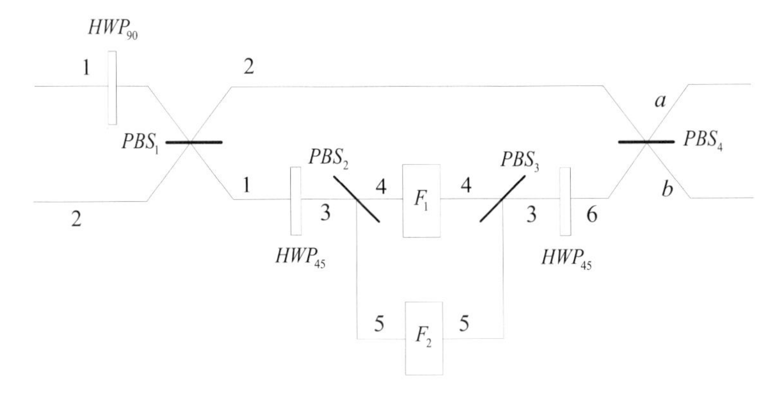

Figure 27.2: This scheme shows the implementation of the projective measurement. The half-wave plates HWP_{45} and HWP_{90} rotate the horizontal and vertical polarization by $\pi/4$ and $\pi/2$. The polarization beam splitters PBS_i ($i = 1, 2, 3, 4$) transmit H-photons and reflect V-photons. F_i ($i = 1, 2$) denotes the single-mode quantum filter, as shown in Fig. 27.1.

Figure 27.2 shows the required experimental setup to realize this projective measurement by two quantum filters, as shown in Fig. 27.1. The half-wave plate HWP_{90} rotates the polarization of the mode 1 by $\pi/2$. The state becomes

$$\Phi_1 = c_0|V\rangle_1|H\rangle_2 + c_1|V\rangle_1|V\rangle_2 + c_2|H\rangle_1|H\rangle_2 + c_3|H\rangle_1|V\rangle_2. \tag{27.14}$$

Then mode 1 and mode 2 are forwarded to a polarization beam splitter PBS_1 to transmit the H-polarized photon and to reflect the V-polarized photon. The system state evolves into

$$\Phi_2 = c_0|HV\rangle_{2'} + c_1|V\rangle_{1'}|V\rangle_{2'} + c_2|H\rangle_{1'}|H\rangle_{2'} + c_3|HV\rangle_{1'}. \tag{27.15}$$

The quantum state representation of two photons in the same mode i is defined as: $|HV\rangle_i = |VH\rangle_i = |H\rangle_i \otimes |V\rangle_i$. Another one half-wave plate (HWP_{45}) rotates the polarization of the mode $1'$ by $\pi/4$. This corresponds to the transformations $|H\rangle \longrightarrow \frac{1}{\sqrt{2}}(|H\rangle + |V\rangle)$ and $|V\rangle \longrightarrow \frac{1}{\sqrt{2}}(|H\rangle - |V\rangle)$. The quantum state of the system is transformed into

$$\begin{aligned}\Phi_3 = c_0|HV\rangle_{2'} &+ \frac{1}{\sqrt{2}}|H\rangle_3(c_1|V\rangle_{2'} + c_2|H\rangle_{2'}) \\ &+ \frac{1}{\sqrt{2}}|V\rangle_3(-c_1|V\rangle_{2'} + c_2|H\rangle_{2'}) + \frac{c_3}{\sqrt{2}}(|2H\rangle_3 - |2V\rangle_3).\end{aligned} \tag{27.16}$$

Mode 3 is injected into another polarization beam splitter PBS_2 so that the system state becomes

$$\begin{aligned}\Phi_4 &= c_0|HV\rangle_{2'} + \frac{1}{\sqrt{2}}|H\rangle_4(c_1|V\rangle_{2'} + c_2|H\rangle_{2'}) \\ &\quad + \frac{1}{\sqrt{2}}|V\rangle_5(-c_1|V\rangle_{2'} + c_2|H\rangle_{2'}) + \frac{c_3}{\sqrt{2}}(|2H\rangle_4 - |2V\rangle_5).\end{aligned} \tag{27.17}$$

Mode 4 and mode 5 pass through a single-mode quantum filter, as shown in Fig. 27.1, to eliminate the one-photon state of the mode 4 and mode 5. This filtering process transforms the state to the state

$$\Phi_5 = c_0|HV\rangle_{2'} + \frac{c_3}{\sqrt{2}}(|2H\rangle_{4'} - |2V\rangle_{5'}). \tag{27.18}$$

Mode $4'$ and mode $5'$ are two input modes of the polarization beam splitter PBS_3. That is why the system state transforms to

$$\Phi_6 = c_0|HV\rangle_{2'} + \frac{c_3}{\sqrt{2}}(|2H\rangle_{3'} - |2V\rangle_{3'}). \tag{27.19}$$

The polarization of mode $3'$ is rotated by $\pi/4$ to get

$$\Phi_7 = c_0|HV\rangle_{2'} + c_3|HV\rangle_6. \tag{27.20}$$

And finally mode 6 and mode $2'$ pass through a polarization beam splitter PBS_4 in order to obtain the output state of the implemented two-qubit projective measurement. The probability of this outcome is $(|c_0|^2 + |c_3|^2)/4$, which is four times greater than what the proposal of Ref. [24] makes possible.

Now we show how the multiphoton polarization-entangled states can be efficiently generated. First we consider the generation of the two-photon polarization-entangled state $(|H\rangle|H\rangle + |V\rangle|V\rangle)/\sqrt{2}$. We assume that the two polarized photons form the initial state $(|H\rangle_1 + |V\rangle_1)/\sqrt{2} \otimes (|H\rangle_2 + |V\rangle_2)/\sqrt{2}$. These two photons are injected into two input ports of the setup, which is shown in Fig. 27.2. Notice that this setup projects the input state into the two-dimensional subspace of the identical horizontal or vertical polarization. After passing through the setup, the initial state is projected into the two-photon polarization-entangled state $(|H\rangle|H\rangle + |V\rangle|V\rangle)/\sqrt{2}$. The probability of this outcome is $1/8$. In the following, we consider the generation of a multiphoton polarization Greenberger–Horne–Zeilinger (GHZ) state. We assume that an $(N-1)$-photon polarization GHZ state $(|H\rangle_1 \cdots |H\rangle_{N-1} + |V\rangle_1 \cdots |V\rangle_{N-1})/\sqrt{2}$ and the Nth independent polarization photon are prepared initially in the state $(|H\rangle_N + |V\rangle_N)/\sqrt{2}$. If the $(N-1)$th photon and the Nth polarization photon inject into two input ports of the setup, which is shown in Fig. 27.2, the N-photon polarization GHZ state $(|H\rangle_1 \cdots |H\rangle_N + |V\rangle_1 \cdots |V\rangle_N)/\sqrt{2}$ will be obtained. The probability of this outcome is $1/8$. This demonstrates that N-photon polarization-entangled GHZ states can be generated from single-photon sources step by step. The probability of the outcome of the total process is $1/8^{N-1}$.

27.2.2 Implementation of Non-deterministic NLS Gate via Parametric Amplifiers

Now we will describe the realization of the non-deterministic quantum logic operation (1), which uses the squeezed vacuum state as an auxiliary photon state [19]. The experimental setup is depicted in Fig. 27.3. It requires two non-degenerate squeezers and a two-photon coincidence detection. The input quantum state of the mode 1 is in state (2). An ancilla quantum state $\Psi_{\text{ancilla}} = |0\rangle_2|0\rangle_3$ in the vacuum state is required. The modes 2 and 3 pass through the first parametric amplifier, whose transformation is given by

$$S_1 = \exp[\theta_1(a_2^\dagger a_3^\dagger - a_2 a_3)]. \tag{27.21}$$

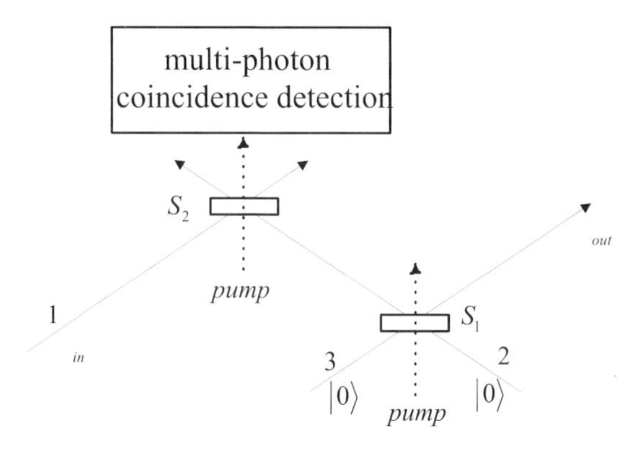

Figure 27.3: Schematic showing how to implement the non-deterministic quantum logic operation that is explained in the paper.

The Bose annihilation operators a_2 and a_3 are related to these modes. This expression can also be written in a disentangled form

$$U_3 = \exp(\sqrt{\gamma_1}\, a_2^\dagger a_3^\dagger)(1-\gamma_1)^{(a_2^\dagger a_2 + a_3^\dagger a_3 + 1)/2}\exp(-\sqrt{\gamma_1}\, a_2 a_3), \tag{27.22}$$

with $\gamma_1 = \tanh^2\theta_1$. The output state of the modes 2 and 3 is

$$\Psi'_{\text{ancilla}} = \sqrt{1-\gamma_1}\sum_{n=0}^{\infty}\gamma_1^{n/2}|n\rangle_2|n\rangle_3. \tag{27.23}$$

The coefficient γ_1 will be determined later. The field state $\Psi_{\text{in}} \otimes \Psi'_{\text{ancilla}}$ will be used directly for the NLS gate implementation. One of the two output modes of the first parametric amplifier is then used as one of the input modes of the second parametric amplifier, whose transformation is given by

$$S_2 = \exp[\theta_2(a_1^\dagger a_2^\dagger - a_1 a_2)]. \tag{27.24}$$

The quantum state $\Psi_{\rm in}$ is the input mode of the second parametric amplifier. The mode 1 and the mode 2 pass through the second parametric amplifier, which transforms the three-mode state to

$$\begin{aligned}\Psi_{\rm o} = \sqrt{(1-\gamma_1)(1-\gamma_2)}\,\big[\sqrt{\gamma_2}\,\alpha|0\rangle_3 + \sqrt{\gamma_1}\,(1-2\gamma_2)\beta|1\rangle_3 \\ + \gamma_1\sqrt{\gamma_2}\,(3\gamma_2-2)\gamma|2\rangle_3\big]|1\rangle_1|1\rangle_2 + \Psi_{\rm other}.\end{aligned} \tag{27.25}$$

The terms $\Psi_{\rm other}$ of this quantum state do not contribute to the events that correspond to a two-photon coincidence (one photon in each of the modes 1 and 2). Conditional on this two-fold coincidence detection, the output state of the system is projected into

$$\begin{aligned}\Psi_{\rm out} = \sqrt{(1-\gamma_1)(1-\gamma_2)}\,\big[\sqrt{\gamma_2}\,\alpha|0\rangle_3 + \sqrt{\gamma_1}\,(1-2\gamma_2)\beta|1\rangle_3 \\ + \gamma_1\sqrt{\gamma_2}\,(3\gamma_2-2)\gamma|2\rangle_3\big].\end{aligned} \tag{27.26}$$

The relation between the parameters γ_1 and γ_2,

$$\sqrt{\gamma_2} = \sqrt{\gamma_1}\,(1-2\gamma_2) = -\gamma_1\sqrt{\gamma_2}\,(3\gamma_2-2), \tag{27.27}$$

is satisfied with $\gamma_1 = (21-7\sqrt{2})/(9+4\sqrt{2}) \approx 0.757$ and $\gamma_2 = (3-\sqrt{2})/7 \approx 0.226$. With this choice the input quantum state $\Psi_{\rm in}$ is transformed to the output quantum state $\Psi_{\rm out} = \alpha|0\rangle + \beta|1\rangle - \gamma|2\rangle$. This is the transformation property (1) of the NLS gate. The probability of this outcome is approximately $4.25\,\%$. If the single-photon source is generated by using weak squeezing and photon detectors, an all-optical NLS gate can be implemented with the maximal probability of success $6.25\,\%$. But, four field modes and three photon detectors are needed. Our scheme is comparatively simple, since it requires only two parametric amplifiers, three field modes, and a two-photon coincidence detection. The parameter value $\gamma_1 \approx 0.757$ demonstrates that this scheme works in the strong coupling regime. Currently available parametric amplifiers can produce strongly squeezed light. Aytür and Kumar [25] reported a parametric gain $\gamma_1 \approx 0.9$.

27.2.3 Phase Measurement of Light and Generation of Superposition of Fock States

In this subsection, we will discuss the phase measurement of a single-mode light field and the generation of an arbitrary superposition of Fock states,

$$C_0|0\rangle + C_1|1\rangle + \cdots + C_N|N\rangle, \tag{27.28}$$

where C_i are arbitrary complex numbers [20].

Starting with Dirac's first paper on the quantum theory of radiation [26], various ingenious ways have been explored in order to understand the quantum phase of single-mode optical fields [27]. In Ref. [28], Pegg et al. introduce the truncated phase state

$$|\theta, N\rangle = \frac{1}{\sqrt{N+1}}\sum_{n=0}^{N}\exp(in\theta)|n\rangle, \tag{27.29}$$

and define the phase probability distribution for a single-mode field in state $|\Psi\rangle$ as a function of phase angle θ,

$$P_N(\theta) = \left|\langle\Psi|\theta, N\rangle\right|^2 = \frac{1}{2\pi}\left|\sum_{n=0}^{N}\langle\Psi|n\rangle \exp(in\theta)\right|^2 . \tag{27.30}$$

We propose several schemes to measure the phase probability distribution (27.30) of single-mode optical fields and demonstrate that the proposed scheme can be used to generate arbitrary single-mode states (27.28) from a squeezed vacuum state.

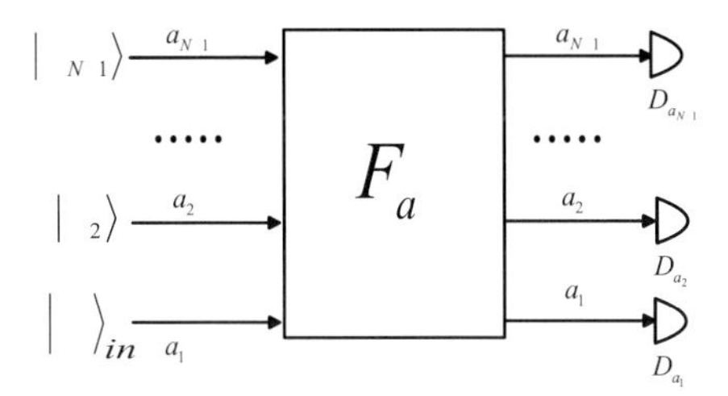

Figure 27.4: Schematic experimental setup for quantum phase measurement of a single-mode optical field $|\Psi\rangle_{\rm in}$, which includes two symmetric $(N+1)$-port devices F_a, and photon detectors D_{a_i}.

We first consider the experimental setup shown in Fig. 27.4, which consists of a symmetric $(N+1)$-port device F_a and $N+1$ photon detectors. An extended introduction to the symmetric multi-port device is given in Ref. [29]. The action of the symmetric $(N+1)$-port device can be described by the unitary operator U^F, which gives the transformation

$$U^{F\dagger} a_j^{\dagger} U^F = \sum_{i=1}^{N+1} U_{i,j}^F a_i^{\dagger}, \tag{27.31}$$

where

$$U_{i,j}^F = \frac{1}{\sqrt{N+1}} \gamma_N^{(i-1)(j-1)} \tag{27.32}$$

and $\gamma_N = \exp[i2\pi/(N+1)]$, and indices i, j denote the input and exit ports. Reck et al. [30] have shown that it is possible to construct a multi-port device from mirrors, beam splitters, and phase shifters that will transform the input modes into the output modes in accord with any unitary matrix.

Now we present a detailed analysis of the proposed scheme. As shown in Fig. 27.4, the mode a_1 of the measured single-mode state $|\Psi\rangle_{\rm in}$ is mixed with the N modes $a_2, \ldots, a_{N+1}$ at the symmetric device F_a. We assume that the mode a_i $(i = 2, \ldots, N+1)$ is prepared in the coherent state $|\alpha_i\rangle_{a_i}$, where parameters α_i will be determined later. After these field modes have passed through the symmetric device F_a, the state of the system becomes

$$U^F |\Psi\rangle_{\rm in} \prod_{i=2}^{N+1} |\alpha_i\rangle_{a_i}. \tag{27.33}$$

The photodetectors PD_{a_i} are used to measure the photon number in the modes a_i ($i = 1, \ldots, N+1$) respectively. We only consider the case that no photon is detected in detector PD_{a_1} and one photon is detected in the detectors PD_{a_i} ($i = 2, \ldots, N+1$). This means that all detectors can distinguish between zero photon, one photon, and more than one photon. The probability for the detection event is

$$\left| {}_{a_1}\langle 0| \prod_{j=2}^{N+1} {}_{a_j}\langle 1|U^F|\Psi\rangle_{\text{in}} \prod_{j=2}^{N+1} |\alpha_j\rangle_{a_j} \right|^2 = |{}_{\text{in}}\langle\Psi|M_s\rangle|^2 \tag{27.34}$$

with

$$|M_s\rangle = \left(\prod_{j=2}^{N+1} {}_{a_j}\langle\alpha_j|U^{F\dagger}|0\rangle_{a_1} \prod_{j=2}^{N+1} |1\rangle_{a_j} \right). \tag{27.35}$$

If we write the photon creation operators as $a_i^\dagger$, Eq. (27.31) can be rewritten as

$$|M_s\rangle = \prod_{j=2}^{N+1} {}_{a_j}\langle\alpha_j|\, U^{F\dagger} \prod_{j=2}^{N+1} a_j^\dagger U^F U^{F\dagger} \prod_{j=1}^{N+1} |0\rangle_{a_j}. \tag{27.36}$$

By using the Eq. (27.31), we obtain the (unnormalized) state

$$\begin{aligned} |M_s\rangle &= \prod_{j=2}^{N+1} {}_{a_j}\langle\alpha_j| \prod_{j=2}^{N+1} \left(\sum_{i=1}^{N+1} U_{i,j}^F a_i^\dagger \right) \prod_{j=1}^{N+1} |0\rangle_{a_j} \\ &= \prod_{j=2}^{N+1} \left(a_1^\dagger + \sum_{k=2}^{N+1} U_{k,j}^N \alpha_k \right) |0\rangle_{a_1}. \end{aligned} \tag{27.37}$$

In order to realize the quantum phase measurement, we require that the measurement probability $|{}_{\text{in}}\langle\Psi|M_s\rangle|^2$ is proportional to the probability distribution function $|{}_{\text{in}}\langle\Psi|\theta, N\rangle|^2$. This requirement can be satisfied by appropriately choosing parameters α_j. If the parameters $\gamma_2, \ldots, \gamma_{N+1}$ are the N complex roots of the characteristic polynomial

$$\sum_{n=0}^{N+1} \frac{\gamma^n}{\sqrt{n!}} (e^{i\varphi})^n = 0 \tag{27.38}$$

and α_i satisfy the relation

$$\begin{pmatrix} U_{2,2}^F & \cdots & U_{2,N+1}^F \\ \vdots & \ddots & \vdots \\ U_{N+1,2}^F & \cdots & U_{N+1,N+1}^F \end{pmatrix} \begin{pmatrix} \alpha_2 \\ \vdots \\ \alpha_{N+1} \end{pmatrix} = \begin{pmatrix} \gamma_2 \\ \vdots \\ \gamma_{N+1} \end{pmatrix}, \tag{27.39}$$

then expression (27.34) is proportional to the state $|\theta, N\rangle$. This demonstrates that the proposed setup, which is shown in Fig. 27.1, definitely implements the quantum phase measurement.

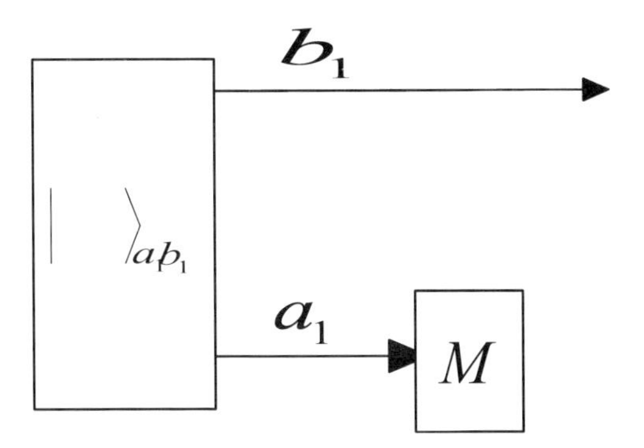

Figure 27.5: Schematic experimental setup for generation of arbitrary single-mode superposition states (1) from a squeezed state $|\Psi_s\rangle_{a_1,b_1}$. Here M denotes the setup proposed in Fig. 27.4.

In the following, we show that the experimental setup proposed in Fig. 27.4 can be used to generate arbitrary states (1) from a squeezed vacuum state. Figure 27.5 shows the required experimental setup, which generates arbitrary superposition states (1) from a squeezed state. Assume that we have generated a squeezed vacuum state $|\Psi_s\rangle_{a_1,b_1}$, which in the number basis has the form

$$|\Psi_s\rangle_{a_1,b_1} = \sqrt{1-\lambda^2}\sum_{n=0}^{\infty}\lambda^n|n\rangle_{a_1}|n\rangle_{b_1}, \tag{27.40}$$

where $\lambda < 1$ is the squeezing parameter. The mode a_1 is injected into the first input port of the experimental setup shown in Fig. 27.4. The other input modes are prepared in coherent states. We appropriately choose the parameters γ_i $(i = 2, \ldots, N+1)$ to satisfy

$$\sum_{n=0}^{N+1}\frac{C_n^*}{\sqrt{n!}\lambda^n}\gamma^n = 0, \tag{27.41}$$

the parameters of the coherent states satisfy Eq. (27.39), and the expression (27.37) is proportional to $\sum_{n=0}^{N}(C_n^*/\lambda^n)|n\rangle_{a_1}$. In this case, the state of the mode b_1 is projected into state

$$\langle M_s|\Psi_s\rangle_{a_1,b_1} = \sum_{n=0}^{N}C_n|n\rangle_{b_1}, \tag{27.42}$$

which is the desired state (27.28).

The above analysis and suggested procedure show that the coherent states and squeezed vacuum state are enough for quantum phase measurement and generation of arbitrary superposition state (1). The proposed scheme requires the implementation of the symmetric multi-port device. The number of beam splitters needed for such multi-port device increases quadratically with N, which limits the application of the scheme.

In the following we propose an alternative scheme for the same purpose. The number of beam splitters required only increases linearly with N. In order to present the principle of the

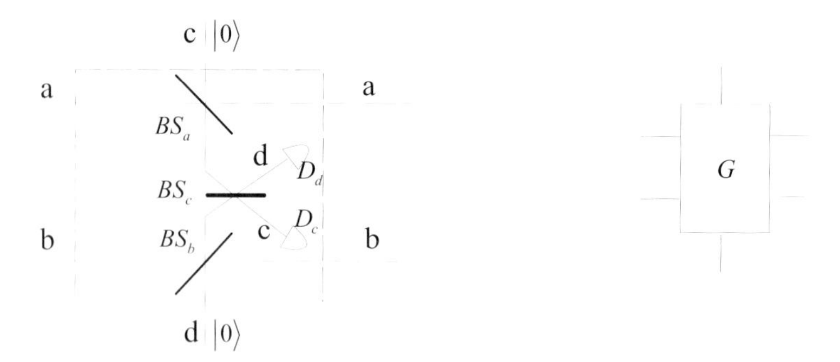

Figure 27.6: The basic building element G, which consists of three beam splitters BS_a, BS_b, and BS_c, and two photon detectors D_c and D_d.

scheme, we introduce simple building blocks G as optical four-mode devices (see Fig. 27.6), which consist of three beam splitters BS_a, BS_b, and BS_c. We label the four input modes by the letters a, b, c, d. The annihilation operators a, b, c, d and the creation operators $a^\dagger, b^\dagger, c^\dagger, d^\dagger$ are related to these modes. The beam splitters BS_a, BS_b, and BS_c can be formulated easily with the unitary operators

$$\begin{aligned} U_a &= \exp[\tfrac{1}{4}\pi(ac^\dagger - a^\dagger c)], \\ U_b &= \exp[\tfrac{1}{4}\pi(bd^\dagger - b^\dagger d)], \\ U_c &= \exp[\theta(cd^\dagger\, e^{i\varphi} - c^\dagger d\, e^{-i\varphi})]. \end{aligned} \tag{27.43}$$

In order to demonstrate the functionality of the basic block, we send any two-mode state of the form

$$|\Psi_{\text{two}}\rangle = \sum_{m,n} C_{m,n} |m\rangle_a |n\rangle_b \tag{27.44}$$

into the input ports a and b. The other input ports c and d are assumed to be in the vacuum state. After passing through these beam splitters, the state of the system becomes

$$\begin{aligned} |\Psi_{\text{mid}}\rangle &= U_c U_b U_a |\Psi_{\text{two}}\rangle |0\rangle_c |0\rangle_d \\ &= \sum_{m,n} C_{m,n} \frac{1}{\sqrt{m!n!2^{m+n}}} (a^\dagger + \cos\theta\, c^\dagger + \sin\theta\, e^{i\varphi} d^\dagger)^m \\ &\quad \times (b^\dagger + \cos\theta\, d^\dagger - \sin\theta\, e^{-i\varphi} c^\dagger)^n |\text{vac}\rangle, \end{aligned} \tag{27.45}$$

where $|\text{vac}\rangle = |0\rangle_a |0\rangle_b |0\rangle_c |0\rangle_d$. Then we make a photon number measurement on the modes c and d with the single-photon detectors D_c and D_d. If D_d detects one photon and D_c does not detect any photon, the state (27.44) is projected into the (unnormalized) state

$$\begin{aligned} |\Psi_{\text{out}}\rangle &= \sum_{m,n} C_{m,n} \frac{1}{\sqrt{m!n!2^{m+n}}} (\sqrt{n}\cos\theta\, |m\rangle_a |n-1\rangle_b + \sqrt{m}\sin\theta\, e^{i\varphi} |m-1\rangle_a |n\rangle_b) \\ &= K_{\theta,\varphi} |\Psi_{\text{two}}\rangle \end{aligned} \tag{27.46}$$

with

$$K_{\theta,\varphi} = (b\cos\theta + a\sin\theta\, e^{i\varphi})2^{-(a^\dagger a + b^\dagger b)/2}. \tag{27.47}$$

Now we apply the device G to demonstrate quantum phase measurement. The schematic is shown in Fig. 27.7. One input port of the setup is prepared in the measured state $|\Psi\rangle_{\rm in}$, and another input port is prepared in the coherent state $|\alpha\rangle_b$. After passing through N devices G, the state of the system becomes

$$K_{\theta_N,\varphi_N} K_{\theta_{N-1},\varphi_{N-1}} \cdots K_{\theta_2,\varphi_2} K_{\theta_1,\varphi_1} |\Psi\rangle_{\rm in} |\alpha\rangle_b. \tag{27.48}$$

We consider the event that the detectors D_a and D_b do not detect any photon. The probability for the detection event is

$$P = |\,{}_a\langle 0|\,{}_b\langle 0| K_{\theta_N,\varphi_N} \cdots K_{\theta_1,\varphi_1} |\Psi\rangle_{\rm in} |\alpha\rangle_b|^2 = |{}_{\rm in}\langle\Psi||M_a\rangle|^2 \tag{27.49}$$

with

$$|M_a\rangle = (a^\dagger + \tan\theta_N\, e^{-i\varphi_N}\alpha)\cdots(a^\dagger + \tan\theta_1\, e^{-i\varphi_1}\alpha)|0\rangle. \tag{27.50}$$

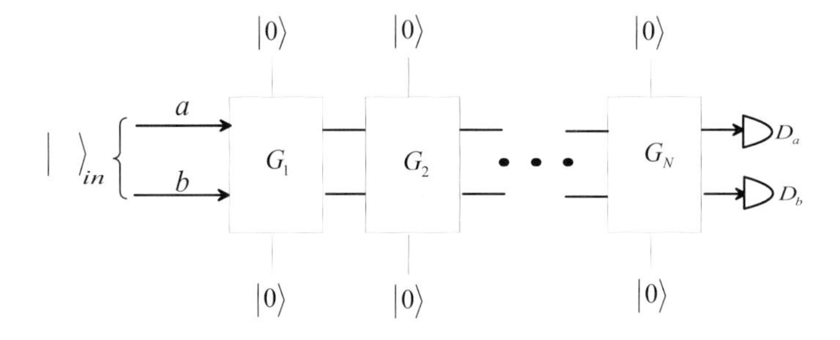

Figure 27.7: The second experimental setup for quantum phase measurement of single-mode optical field $|\Psi\rangle_{\rm in}$, which consists of N basic building blocks G_i proposed in Fig. 27.6.

If $\tan\theta_i\, e^{-i\varphi_i}\alpha$ are complex roots of the polynomial (27.38), the probability (27.49) is proportional to $|\langle\Psi_{\rm in}|\theta, N\rangle|^2$, which demonstrates that the experimental setup shown in Fig. 27.7 can be used to realize quantum phase measurement. If the parameters $\tan\theta_i\, e^{-i\varphi_i}\alpha$ are chosen to be complex roots of the polynomial (27.38), the experimental setup can be used to generate an arbitrary superposition state (27.28) from a squeezed vacuum state.

27.2.4 Joint Measurement of Photon Number Sum and Phase Difference Operators on a Two-mode Field

In this subsection, we will demonstrate how to make a joint measurement of photon number sum and phase difference operators on a two-mode field and generate two-mode N-photon entangled states from a pair of squeezed vacuum states [21].

In Ref. [31], Luis et al. introduce the Hermitian phase difference operator

$$\hat{\Phi}_{12} = \sum_{N=0}^{\infty} \sum_{r=0}^{N} \phi_r^N |\phi_r^N\rangle\langle\phi_r^N| \tag{27.51}$$

with

$$|\phi_r^N\rangle = \frac{1}{\sqrt{N+1}} \sum_{n=0}^{N} e^{in\phi_r^N} |n\rangle_1 |N-n\rangle_2 \tag{27.52}$$

and

$$\phi_r^N = \theta + \frac{2\pi r}{N+1}, \tag{27.53}$$

where $|n\rangle$ denotes the n-photon Fock state and θ is an arbitrary angle. It is obvious that the phase difference operator $\hat{\Phi}_{12}$ commutes with the total photon number operator $\hat{n} = \hat{n}_1 + \hat{n}_2$, where $\hat{n}_1$ and $\hat{n}_2$ are the photon number operators for each mode. Therefore, the joint measurement of the photon number sum and phase difference operators on a two-mode field $|\Psi\rangle$ projects the quantum state into the state $|\phi_r^N\rangle$. The success probability of the measurement is $|\langle\Psi|\phi_r^N\rangle|^2$, which is the joint probability distribution function for the total number and the phase difference [31]. In Ref. [32], it is shown that the joint measurement of the photon number sum and phase difference operators plays the role of Bell's measurement in quantum teleportation of photon number states.

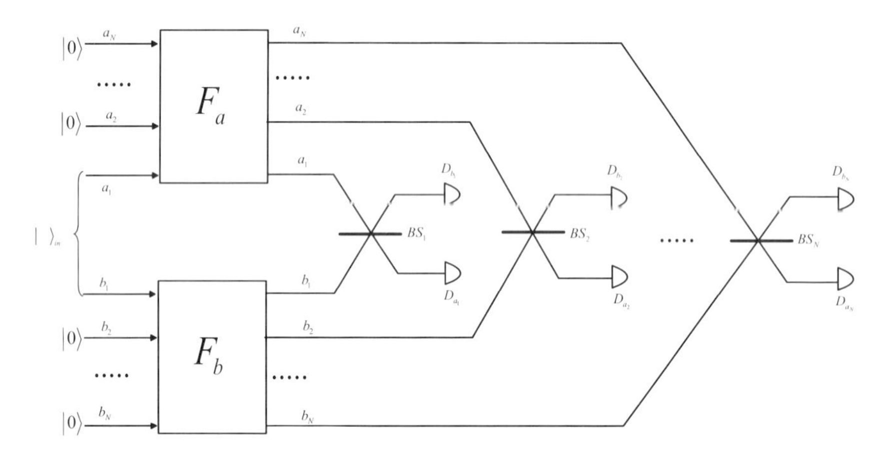

Figure 27.8: Schematic experimental setup for joint measurement of photon number sum and phase difference operators on two field modes $|\Psi_{\text{in}}\rangle$, which includes two symmetric N-port devices F_a and F_b, N beam splitters BS_i, and $2N$ photon detectors D_{a_i} and D_{b_i}.

The experimental setup is depicted in Fig. 27.8, which consists of two symmetric N-port devices F_a and F_b, N beam splitters, and $2N$ photon detectors. We assume that the mode

a_1 of the state $|\Psi_{\text{in}}\rangle = \sum_{n,m=0}^{\infty} C_{n,m}|n\rangle_{a_1}|m\rangle_{b_1}$ is mixed with the $N-1$ vacuum modes $a_2, \ldots, a_N$ at the symmetric N-port device F_a, while the mode b_1 of the state $|\Psi_{\text{in}}\rangle$ is mixed with the $N-1$ vacuum modes $b_2, \ldots, b_N$ at the symmetric N-port device F_b. After these modes have passed through two symmetric N-port devices F_a and F_b, the state of the system becomes

$$U_a^N U_b^N |\Psi_{\text{in}}\rangle \prod_{j=2}^{N} |0\rangle_{a_j}|0\rangle_{b_j} . \tag{27.54}$$

Then the modes a_i and b_i are mixed at beam splitters BS_i ($i = 1, \ldots, N$). The action of the beam splitters BS_i is described by the unitary operator

$$U_i^{bs} = \exp[\theta_i(a_i b_i^\dagger e^{\varphi_i} - a_i^\dagger b_i e^{-\varphi_i})], \tag{27.55}$$

where the parameters θ_i and φ_i characterize the beam splitter BS_i, which will be determined later. After these photon modes have passed through the beam splitters, we can obtain the state of the system

$$\left(\prod_{i=1}^{N} U_i^{bs}\right) U_a^N U_b^N |\Psi_{\text{in}}\rangle \prod_{j=2}^{N} |0\rangle_{a_j}|0\rangle_{b_j} . \tag{27.56}$$

Now let the photodetectors PD_{a_i} and PD_{b_i} measure the photon numbers in the modes a_i and b_i ($i = 1, \ldots, N+1$), respectively. Consider the case where one photon is detected in the detectors PD_{a_i} and the detectors PD_{b_i} do not detect any photon ($i = 1, \ldots, N$). This means that all detectors can distinguish between zero photon, one photon, and more than one photon. The probability for the detection event is

$$\begin{aligned} P_N &= \left| \left(\prod_{j=1}^{N} {}_{a_j}\langle 1|\, {}_{b_j}\langle 0|\right) \left(\prod_{i-1}^{N} U_i^{bs}\right) U_a^N U_b^N |\Psi_{\text{in}}\rangle \prod_{j=2}^{N+1} |0\rangle_{a_j}|0\rangle_{b_j} \right|^2 \\ &= \langle \Psi_{\text{in}}|\Psi_N\rangle\langle\Psi_N|\Psi_{\text{in}}\rangle , \end{aligned} \tag{27.57}$$

where

$$|\Psi_N\rangle = \left(\prod_{j=2}^{N} {}_{a_j}\langle 0|\, {}_{b_j}\langle 0|\right) U_b^{N\dagger} U_a^{N\dagger} \left(\prod_{i=1}^{N} U_i^{bs\dagger}\right) \left(\prod_{j=1}^{N} |1\rangle_{a_j}|0\rangle_{b_j}\right) . \tag{27.58}$$

If we write the photon creation operators acting on the modes a_i and b_i as $a_i^\dagger$ and $b_i^\dagger$, respectively, Eq. (27.58) can be rewritten as

$$\begin{aligned} |\Psi_N\rangle &= \left(\prod_{j=2}^{N} {}_{a_j}\langle 0|\, {}_{b_j}\langle 0|\right) \left[\prod_{j=1}^{N} \left(U_b^{N\dagger} U_a^{N\dagger} U_j^{bs\dagger} a_j^\dagger U_j U_a U_b\right)\right] \\ &\quad \times U_b^{N\dagger} U_a^{N\dagger} \prod_{j=1}^{N} \left(U_j^{bs\dagger}\right) \left(\prod_{j=1}^{N} |0\rangle_{a_j}|0\rangle_{b_j}\right) . \end{aligned} \tag{27.59}$$

By using the unitary transformation of the operators U_a^N, U_b^N and BS_j

$$\begin{aligned}
U_j^{bs\dagger} a_j^\dagger U_j^{bs} &= \cos\theta_j\, a_j^\dagger + \sin\theta_j\, e^{i\varphi_j} b_j^\dagger, \\
U_j^{bs\dagger} b_j^\dagger U_j^{bs} &= \cos\theta_j\, b_j^\dagger - \sin\theta_j\, e^{-i\varphi_j} a_j^\dagger, \\
U_j^{bs\dagger} b_k^\dagger U_j^{bs} &= b_k^\dagger, \qquad U_j^{bs\dagger} a_k^\dagger U_j^{bs} = a_k^\dagger, \qquad (j \neq k), \\
U_a^{N\dagger} a_j^\dagger U_a^N &= \sum_{i=1}^N U_{ij}^N a_i^\dagger, \\
U_b^{N\dagger} b_j^\dagger U_b^N &= \sum_{i=1}^N U_{ij}^N b_i^\dagger, \\
U_a^{N\dagger} b_j^\dagger U_a^N &= b_j^\dagger, \qquad U_b^{N\dagger} a_j^\dagger U_b^N = a_j^\dagger,
\end{aligned} \tag{27.60}$$

we obtain

$$\begin{aligned}
|\Psi_N\rangle &= \left(\prod_{j=2}^N {}_{a_j}\langle 0|\, {}_{b_j}\langle 0|\right) \left[\prod_{j=1}^N \left(\cos\theta_j \sum_{i=1}^N U_{ij}^N a_i^\dagger + \sin\theta_j\, e^{i\varphi_j} \sum_{i=1}^N U_{ij}^N b_i^\dagger\right)\right] \\
&\quad \times \left(\prod_{j=1}^N |0\rangle_{a_j} |0\rangle_{b_j}\right) \\
&= \prod_{j=1}^N \left(\cos\theta_j\, a_1^\dagger + \sin\theta_j\, e^{i\varphi_j} b_1^\dagger\right) |0\rangle_{a_1} |0\rangle_{b_1},
\end{aligned} \tag{27.61}$$

where we have used the relations

$$\begin{aligned}
&U_a^N U_b^N \prod_{j=1}^N (BS_j) \left(\prod_{j=1}^N |0\rangle_{a_j} |0\rangle_{b_j}\right) = \left(\prod_{j=1}^N |0\rangle_{a_j} |0\rangle_{b_j}\right), \\
&\left(\prod_{j=2}^N {}_{a_j}\langle 0|\right) a_k^\dagger = 0, \qquad \left(\prod_{j=2}^N {}_{b_j}\langle 0|\right) b_k^\dagger = 0, \qquad (k \geq 2).
\end{aligned} \tag{27.62}$$

In order to realize the joint measurement of photon number sum and phase difference operators on two field modes, we require that the measurement probability $|\langle\Psi_{\text{in}}|\Psi_N\rangle|^2$ is proportional to the joint probability distribution function $|\langle\Psi_{\text{in}}|\phi_r^N\rangle|^2$ for the total photon number N and the phase difference. This requirement can be satisfied by appropriately choosing parameters θ_j and φ_j. If the parameters $\tan\theta_1\, e^{i\varphi_1}, \ldots, \tan\theta_N\, e^{i\varphi_N}$ are the N complex roots of the characteristic polynomial

$$\sum_{n=0}^N \frac{e^{in\psi_r^N}}{\sqrt{n!(N-n)!}} (\tan\theta\, e^{i\varphi})^n = 0, \tag{27.63}$$

then expression (27.61) is proportional to the state $|\phi_r^N\rangle$. This demonstrates that the proposed setup, which is shown in Fig. 27.8, definitely implements the joint measurement of photon number sum and phase difference operators on the two-mode field $|\Psi_{\text{in}}\rangle$.

Further, if the parameters $\tan\theta_1\, e^{i\varphi_1},\ \cdots,\ \tan\theta_N\, e^{i\varphi_N}$ are the N complex roots of the characteristic polynomial

$$\sum_{n=0}^{N} \frac{C_n}{\sqrt{n!(N-n)!}} (\tan\theta\, e^{i\varphi})^n = 0, \tag{27.64}$$

in this case, the proposed setup can also be used to measure quantum state overlap of one two-mode field state and arbitrary two-mode N-photon entangled state.

We show that the experimental setup proposed can be used to generate two-mode N-photon entangled states

$$\sum_{n}^{N} C_n |n\rangle_{a_1} |N-n\rangle_{b_1} \tag{27.65}$$

from a pair of squeezed vacuum states. Figure 27.9 shows the required experimental setup, which can generate a two-mode N-photon entangled state from a pair of squeezed states. We assume that we have generated a pair of squeezed vacuum states $|\Psi_s\rangle_{a_1,b_1}$ and $|\Psi_s\rangle_{a_2,b_2}$. We rewrite the state of the total system as

$$\begin{aligned} |\Psi_s\rangle_{a_1,b_1} |\Psi_s\rangle_{a_2,b_2} &= (1-\lambda^2) \sum_{n,m=0}^{\infty} \lambda^{n+m} |n\rangle_{a_1} |n\rangle_{b_1} |m\rangle_{a_2} |m\rangle_{b_2} \\ &= (1-\lambda^2) \sum_{N=0}^{\infty} \lambda^N \sum_{j=0}^{N} |j\rangle_{a_1} |N-j\rangle_{a_2} |j\rangle_{b_1} |N-j\rangle_{b_2}. \end{aligned} \tag{27.66}$$

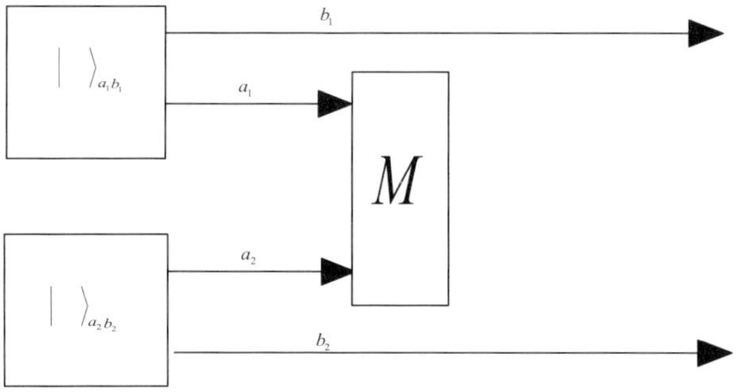

Figure 27.9: Schematic experimental setup for generation of N-photon entangled states from a pair of squeezed states $|\Psi_s\rangle_{a_1,b_1}$ and $|\Psi_s\rangle_{a_2,b_2}$. Here M denotes the setup proposed in Fig. 27.8.

The modes a_1 and a_2 of the state (27.66) act as input modes of the experimental setup shown in Fig. 27.8. If the parameters of the setup are chosen to satisfy

$$\sum_{n=0}^{N} \frac{C_n^*}{\sqrt{n!(N-n)!}} (\tan\theta\, e^{\varphi})^n = 0, \tag{27.67}$$

the state of the modes b_1 and b_2 is projected into state (27.67).

27.2.5 Remark

One of the difficulties of our scheme with respect to an experimental demonstration consists in the requirement on the sensitivity of the detectors. These detectors should be capable of distinguishing between one-photon events and two-photon events. Recently, the development of experimental techniques in this field has made considerable progress. A photon detector based on a visible light photon counter has been reported that can distinguish between the incidence of a single photon and two photons with a high quantum efficiency, a good time resolution and a low bit error rate [33]. Another difficulty is the availability of single-photon sources. Several triggered single-photon sources are available, which operate by means of fluorescence from a single molecule [34] or a single quantum dot [35, 36]. They exhibit a very good performance. More recently a deterministic single-photon source for distributed quantum networks was reported [37], which can emit a sequence of single photons on demand from a single three-level atom strongly coupled to a high-finesse optical cavity. However, our scheme needs the synchronized arrival of many single photons into input ports of many beam splitters. Of course, an experimental realization will be challenging.

27.3 Quantum Entanglement Between Distant Atoms Trapped in Different Optical Cavities

In Ref. [38], a scheme was proposed for implementation of teleportation of atomic states by detecting photons decaying from optical cavities. Further, several schemes have also been proposed for generating the EPR state between two distant atoms trapped in different optical cavities [38, 39]. In this section, we propose a unified scheme to generate W states, Greenberger–Horne–Zeilinger (GHZ) states, and cluster states of four distant atoms, which are trapped separately in leaky cavities. The proposed schemes require linear optical elements and photon detectors with single-photon sensitivity. Quantum noise influences the fidelity of the generated states. Further, based on the four-photon coincidence detection, we propose another scheme to generate W states and GHZ states of four distant atoms. The scheme is insensitive to quantum noise, but cannot be used to generate cluster states [22].

27.3.1 Generation of W States, GHZ States and Cluster States Based on Single-photon Detectors

The experimental setup is shown in Fig. 27.10a, which consists of four three-level atoms confined separately in four optical cavities, and a symmetric eight-port device and four single-photon detectors. Each of the three-level atoms (index j) has a ground state $|g_j\rangle$, a metastable state $|s_j\rangle$, and an excited state $|e_j\rangle$ (see Fig. 27.10b). The lifetimes of the atomic levels $|s_j\rangle$ and $|g_j\rangle$ are assumed to be comparatively long so that spontaneous decay of these states can be neglected. The $|s_j\rangle \longleftrightarrow |e_j\rangle$ transition is driven by the classical field and the $|g_j\rangle \longleftrightarrow |e_j\rangle$ transition is driven by the quantized cavity field. Both the classical laser field and the cavity field are assumed to be detuned from their respective transition by the same amount. In the case of large detuning, the excited state $|e_j\rangle$ can be eliminated adiabatically to obtain the

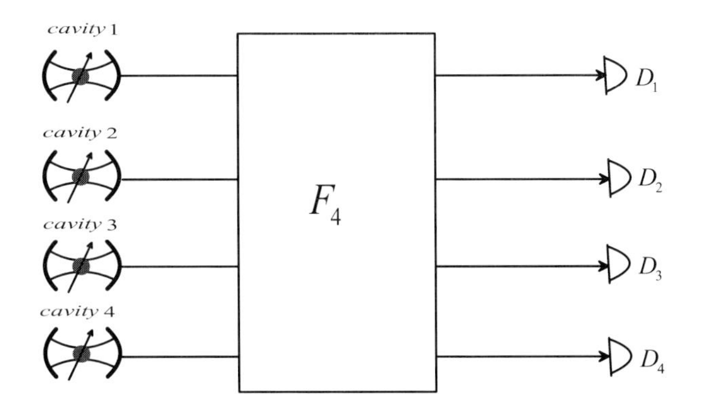

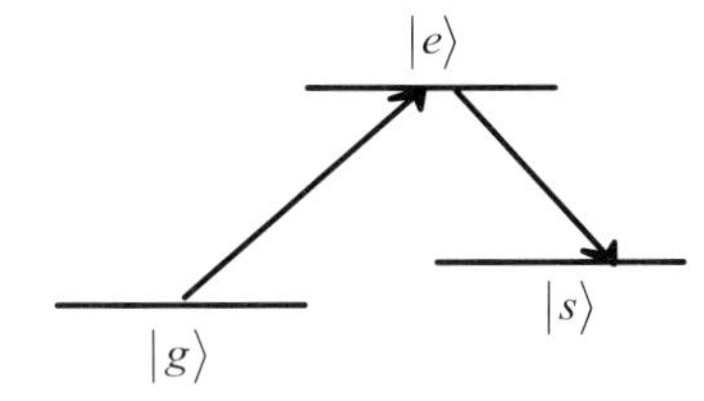

Figure 27.10: (a) The schematic setup for generating W states, GHZ states, and cluster states of four distant atoms based on single-photon detection. F_4 is a symmetric eight-port device [29] and D_i are photon detectors. (b) The relevant level structure with ground state $|g\rangle$, metastable state $|s\rangle$, and excited state $|e\rangle$.

effective interaction Hamiltonian (in the interaction picture)

$$H_j = i\Omega(a_j\sigma_{+j} - a_j^\dagger\sigma_{-j}), \tag{27.68}$$

with the operators $\sigma_{+j} = |s_j\rangle\langle g_j|$ and $\sigma_{-j} = |g_j\rangle\langle s_j|$. In Eq. (27.68), $a_j^\dagger$ and a_j are the creation and annihilation operators of the jth cavity field. Here we assume that the effective coupling constants of all the atoms coupled with their cavities is the same, denoted by Ω. In order to investigate the quantum dynamics of the system, it is convenient to follow a quantum trajectory description [40]. The evolution of the system's wave function is governed by a non-Hermitian Hamiltonian

$$H_j' = H_j - i\kappa_j a_j^\dagger a_j \tag{27.69}$$

as long as no photon decays from the cavity. Here $2\kappa_j$ is the decay rate of the jth cavity field, which is assumed to be the same for all four cavities: $\kappa_j = \kappa$. If a single-photon detector D_j ($j = 1, 2, 3, 4$) detects a photon, the coherent evolution according to H_j' is interrupted by a quantum jump. This corresponds to a quantum jump, which can be formulated with the operators b_j on the joint state vectors of four atom–cavity systems

$$\begin{aligned} b_1 &= \tfrac{1}{2}(a_1 + a_2 + a_3 + a_4), & b_2 &= \tfrac{1}{2}(a_1 + ia_2 - a_3 - ia_4), \\ b_3 &= \tfrac{1}{2}(a_1 - a_2 + a_3 - a_4), & b_4 &= \tfrac{1}{2}(a_1 - ia_2 - a_3 + ia_4). \end{aligned} \tag{27.70}$$

In the preparation stage the initial state of each atom–cavity system is $|s_j\rangle|0\rangle_j$ ($j = 1, 2, 3, 4$), i.e. each atom is initially in the metastable state and the cavity is in the vacuum state. Now we switch on the Hamiltonian (27.68) in each atom–cavity system for a time τ. If no photon is emitted from the cavity, the jth atom–cavity system is governed by the interaction H_j'. In this case the atom–cavity state evolves to the entangled state

$$|\Psi\rangle_j = \alpha|s_j\rangle|0\rangle_j + \beta|g_j\rangle|1\rangle_j \tag{27.71}$$

with

$$\begin{aligned} \alpha &= \frac{\cos(\Omega_\kappa\tau) - \sin(\Omega_\kappa\tau)\kappa/(2\Omega_\kappa)}{\sqrt{[\cos(\Omega_\kappa\tau) - \sin(\Omega_\kappa\tau)\kappa/(2\Omega_\kappa)]^2 + \sin^2(\Omega_\kappa\tau)\Omega^2/\Omega_\kappa^2}}, \\ \beta &= \frac{\sin(\Omega_\kappa\tau)\Omega}{\Omega_\kappa\sqrt{[\cos(\Omega_\kappa\tau) - \sin(\Omega_\kappa\tau)\kappa/(2\Omega_\kappa)]^2 + \sin^2(\Omega_\kappa\tau)\Omega^2/\Omega_\kappa^2}}, \\ \Omega_\kappa &= \sqrt{\Omega^2 - \kappa^2/4}. \end{aligned} \tag{27.72}$$

The probability that no photon is emitted during this evolution becomes

$$P_{\text{single}} = e^{-\kappa\tau}\left\{\left[\cos(\Omega_\kappa\tau) - \frac{\sin(\Omega_\kappa\tau)\kappa}{2\Omega_\kappa}\right]^2 + \frac{\sin^2(\Omega_\kappa\tau)\Omega^2}{\Omega_\kappa^2}\right\}. \tag{27.73}$$

We assume that the interaction Hamiltonian (27.68) is applied to each atom–cavity system simultaneously, so that the preparation of the atom–cavity states $|\Psi\rangle_j$ ends at the same time. This concludes the preparation stage of the protocol. The probability that this stage is equal to the probability that no photon decays during the preparation stage. This quantity is given by $P_{\text{suc}} = P_{\text{single}}^4$.

Now we consider the detection stage and demonstrate the generation of W states, GHZ states, and cluster states. In this stage, we assume that we have prepared the four atom–cavity systems in the form

$$|\Phi(0)\rangle = |\Psi\rangle_1|\Psi\rangle_2|\Psi\rangle_3|\Psi\rangle_4, \tag{27.74}$$

where the state $|\Psi\rangle_j$ is given by Eq. (27.71). When the state (27.74) has been prepared, we turn off the laser pulse and wait for the one or two of the four detectors D_1, D_2, D_3, and D_4 to click. We assume that photons are detected at the time t. This assumption is made in order to calculate the system's time evolution during this time interval in a way consistent with the "no-photon-emission Hamiltonian" (27.69). Notice that, in the detection stage, we turn off the laser pulses, and the atom–cavity interaction term H_j of Eq. (27.69) is set to zero. In this case, the state of the jth atom–cavity system at the time t evolves into

$$|\Psi(t)\rangle_j = \exp(-iH_j't)|\Psi\rangle_j = \frac{1}{\sqrt{\alpha^2 + \beta^2 e^{-2\kappa t}}}(\alpha|s_j\rangle|0\rangle_j + \beta e^{-\kappa t}|g_j\rangle|1\rangle_j). \tag{27.75}$$

The probability that no photon decay takes place during this evolution is given by $(\alpha^2 + \beta^2 e^{-2\kappa t})$. At the time t, the joint state of the total system becomes

$$|\Phi(t)\rangle = |\Psi(t)\rangle_1 |\Psi(t)\rangle_2 |\Psi(t)\rangle_3 |\Psi(t)\rangle_4 . \tag{27.76}$$

The detection of one photon with the detector D_l ($l = 1, 2, 3, 4$) corresponds to a quantum jump, which can be formulated with the operator b_l on the joint state $|\Phi(t)\rangle$. In order to generate the W states, we consider the events that one of the four detectors registers one photon and the other three detectors do not register any counts. For simplicity, we assume that the detector D_1 detects one photon, and the detectors D_2, D_3, and D_4 do not detect any photon. In this case, the combined state of the four atom–cavity systems is projected into

$$\begin{aligned} |\Psi_1\rangle &= b_1 |\Phi(t)\rangle \\ &= \tfrac{1}{2}\big[|g_1\rangle|0\rangle_1 |\Psi(t)\rangle_2 |\Psi(t)\rangle_3 |\Psi(t)\rangle_4 + |\Psi(t)\rangle_1 |g_2\rangle|0\rangle_2 |\Psi(t)\rangle_3 |\Psi(t)\rangle_4 \\ &\quad + |\Psi(t)\rangle_1 |\Psi(t)\rangle_2 |g_3\rangle|0\rangle_3 |\Psi(t)\rangle_4 + |\Psi(t)\rangle_1 |\Psi(t)\rangle_2 |\Psi(t)\rangle_3 |g_4\rangle|0\rangle_4\big] \end{aligned} \tag{27.77}$$

with the success probability $\beta^2(1 - e^{-2\kappa t})$. By tracing over the cavity field, we obtain the state of four atoms

$$\begin{aligned} \rho_W =& \frac{1}{(\alpha^2 + \beta^2 e^{-2\kappa t})^3} \Big\{ \alpha^6 |W_4\rangle\langle W_4| + \beta^6 e^{-6\kappa t} |g_1\rangle|g_2\rangle|g_3\rangle|g_4\rangle\langle g_1|\langle g_2|\langle g_3|\langle g_4| \\ &+ \tfrac{1}{4}\alpha^4\beta^2 e^{-2\kappa t} \Big[\big(|g_1\rangle|s_2\rangle|s_3\rangle|g_4\rangle + |s_1\rangle|g_2\rangle|s_3\rangle|g_4\rangle + |s_1\rangle|s_2\rangle|g_3\rangle|g_4\rangle\big) \\ &\qquad \times \big(\langle g_1|\langle s_2|\langle s_3|\langle g_4| + \langle s_1|\langle g_2|\langle s_3|\langle g_4| + \langle s_1|\langle s_2|\langle g_3|\langle g_4|\big) \\ &+ \big(|g_1\rangle|s_2\rangle|g_3\rangle|s_4\rangle + |s_1\rangle|g_2\rangle|g_3\rangle|s_4\rangle + |s_1\rangle|s_2\rangle|g_3\rangle|g_4\rangle\big) \\ &\qquad \times \big(\langle g_1|\langle s_2|\langle g_3|\langle s_4| + \langle s_1|\langle g_2|\langle g_3|\langle s_4| + \langle s_1|\langle s_2|\langle g_3|\langle g_4|\big) \\ &+ \big(|g_1\rangle|g_2\rangle|s_3\rangle|s_4\rangle + |s_1\rangle|g_2\rangle|g_3\rangle|s_4\rangle + |s_1\rangle|g_2\rangle|s_3\rangle|g_4\rangle\big) \\ &\qquad \times \big(\langle g_1|\langle g_2|\langle s_3|\langle s_4| + \langle s_1|\langle g_2|\langle g_3|\langle s_4| + \langle s_1|\langle g_2|\langle s_3|\langle g_4|\big) \\ &+ \big(|g_1\rangle|g_2\rangle|s_3\rangle|s_4\rangle + |g_1\rangle|s_2\rangle|g_3\rangle|s_4\rangle + |g_1\rangle|s_2\rangle|s_3\rangle|g_4\rangle\big) \\ &\qquad \times \big(\langle g_1|\langle g_2|\langle s_3|\langle s_4| + \langle g_1|\langle s_2|\langle g_3|\langle s_4| + \langle g_1|\langle s_2|\langle s_3|\langle g_4|\big)\Big] \\ &+ \tfrac{1}{4}\alpha^2\beta^4 e^{-4\kappa t} \Big[\big(|g_1\rangle|g_2\rangle|g_3\rangle|s_4\rangle + |s_1\rangle|g_2\rangle|g_3\rangle|g_4\rangle\big) \\ &\qquad \times \big(\langle g_1|\langle g_2|\langle g_3|\langle s_4| + \langle s_1|\langle g_2|\langle g_3|\langle g_4|\big) \\ &+ \big(|g_1\rangle|g_2\rangle|s_3\rangle|g_4\rangle + |s_1\rangle|g_2\rangle|g_3\rangle|g_4\rangle\big)\big(\langle g_1|\langle g_2|\langle s_3|\langle g_4| + \langle s_1|\langle g_2|\langle g_3|\langle g_4|\big) \\ &+ \big(|g_1\rangle|s_2\rangle|g_3\rangle|g_4\rangle + |s_1\rangle|g_2\rangle|g_3\rangle|g_4\rangle\big)\big(\langle g_1|\langle s_2|\langle g_3|\langle g_4| + \langle s_1|\langle g_2|\langle g_3|\langle g_4|\big) \\ &+ \big(|g_1\rangle|g_2\rangle|g_3\rangle|s_4\rangle + |g_1\rangle|s_2\rangle|g_3\rangle|g_4\rangle\big)\big(\langle g_1|\langle g_2|\langle g_3|\langle s_4| + \langle g_1|\langle s_2|\langle g_3|\langle g_4|\big) \\ &+ \big(|g_1\rangle|g_2\rangle|s_3\rangle|g_4\rangle + |g_1\rangle|s_2\rangle|g_3\rangle|g_4\rangle\big)\big(\langle g_1|\langle g_2|\langle s_3|\langle g_4| + \langle g_1|\langle s_2|\langle g_3|\langle g_4|\big) \\ &+ \big(|g_1\rangle|g_2\rangle|g_3\rangle|s_4\rangle + |g_1\rangle|g_2\rangle|s_3\rangle|g_4\rangle\big)\big(\langle g_1|\langle g_2|\langle g_3|\langle s_4| + \langle g_1|\langle g_2|\langle s_3|\langle g_4|\big)\Big]\Big\} \end{aligned} \tag{27.78}$$

where $|W_4\rangle$ is the W-type entangled states of the four atoms

$$|W_4\rangle = \frac{1}{2}\big(|g_1\rangle|s_2\rangle|s_3\rangle|s_4\rangle+|s_1\rangle|g_2\rangle|s_3\rangle|s_4\rangle+|s_1\rangle|s_2\rangle|g_3\rangle|s_4\rangle+|s_1\rangle|s_2\rangle|s_3\rangle|g_4\rangle\big). \quad (27.79)$$

If the time $t \gg \kappa^{-1}$, then $e^{-\kappa t} \approx 0$ and we need only retain the first term $|W_4\rangle\langle W_4|$ of Eq. (27.78) and neglect all the other terms. This shows the obvious physical result that the W state is created in the event that only one photon is registered during the time $t \gg \kappa^{-1}$. During this period, the event that two, three or photons are registered has to be dropped.

In order to quantify how close the state (27.78) comes to the W state, we calculate the fidelity

$$F = \langle W_4|\rho_W|W_4\rangle = \frac{\alpha^6}{(\alpha^2+\beta^2 e^{-2\kappa t})^3}. \quad (27.80)$$

The fidelity increases with increasing detection time and increasing ratio $|\alpha/\beta|$.

Now we show how to generate GHZ states and cluster states of four distant atoms. For this purpose, we consider the events that two of the four detectors detect one photon each, and the other two do not detect any photon. If detectors D_2 and D_4 detect one photon and D_1 and D_3 do not detect any photon, the state of the total system becomes projected into

$$\begin{aligned}|\Psi_{24}\rangle &= b_2 b_4|\Phi(t)\rangle \\ &= \frac{1}{\sqrt{2}}\big(|g_1\rangle|0\rangle_1|\Psi(t)\rangle_2|g_3\rangle|0\rangle_3|\Psi(t)\rangle_4 + |\Psi(t)\rangle_1|g_2\rangle|0\rangle_2|\Psi(t)\rangle_3|g_4\rangle|0\rangle_4\big),\end{aligned} \quad (27.81)$$

with the success probability $\beta^4(1-e^{-2\kappa t})^2$. If detectors D_1 and D_3 detect one photon and D_2 and D_4 do not detect any photon, the state of the total system becomes projected into

$$\begin{aligned}|\Psi_{13}\rangle &= b_1 b_3|\Phi(t)\rangle \\ &= \frac{1}{\sqrt{2}}\big(|g_1\rangle|0\rangle_1|\Psi(t)\rangle_2|g_3\rangle|0\rangle_3|\Psi(t)\rangle_4 - |\Psi(t)\rangle_1|g_2\rangle|0\rangle_2|\Psi(t)\rangle_3|g_4\rangle|0\rangle_4\big),\end{aligned} \quad (27.82)$$

which can be transformed into Eq. (27.81) by local operations. Thus we only consider the state (27.81). By tracing over the cavity field, we obtain the state of four atoms

$$\begin{aligned}\rho_{GHZ} =& \frac{1}{(\alpha^2+\beta^2 e^{-2\kappa t})^2}\Big[\alpha^4|GHZ_4\rangle\langle GHZ_4| \\ &+ \beta^4 e^{-4\kappa t}|g_1\rangle|g_2\rangle|g_3\rangle|g_4\rangle\langle g_1|\langle g_2|\langle g_3|\langle g_4| \\ &+ \tfrac{1}{2}\alpha^2\beta^2 e^{-2\kappa t}\Big(|s_1\rangle|g_2\rangle|g_3\rangle|g_4\rangle\langle s_1|\langle g_2|\langle g_3|\langle g_4| \\ &+ |g_1\rangle|s_2\rangle|g_3\rangle|g_4\rangle\langle g_1|\langle s_2|\langle g_3|\langle g_4| + |g_1\rangle|g_2\rangle|s_3\rangle|g_4\rangle\langle g_1|\langle g_2|\langle s_3|\langle g_4| \\ &+ |g_1\rangle|g_2\rangle|g_3\rangle|s_4\rangle\langle g_1|\langle g_2|\langle g_3|\langle s_4|\Big)\Big]\end{aligned} \quad (27.83)$$

with

$$|GHZ_4\rangle = \frac{1}{\sqrt{2}}\big(|g_1\rangle|s_2\rangle|g_3\rangle|s_4\rangle + |s_1\rangle|g_2\rangle|s_3\rangle|g_4\rangle\big). \tag{27.84}$$

We calculate the fidelity

$$F = \langle GHZ_4|\rho|GHZ_4\rangle = \frac{\alpha^4}{(\alpha^2+\beta^2 e^{-2\kappa t})^2}, \tag{27.85}$$

which again increases with increasing detection time and increasing ratio $|\alpha/\beta|$.

If detectors D_1 and D_2 detect one photon and D_3 and D_4 do not detect any photon, or vice versa, the state of the total system becomes projected into

$$\begin{aligned}|\Psi_{12}\rangle = \tfrac{1}{2}\Big(&e^{i\pi/4}|g_1\rangle|0\rangle_1|g_2\rangle|0\rangle_2|\Psi(t)\rangle_3|\Psi(t)\rangle_4\\ &- e^{i\pi/4}|\Psi(t)\rangle_1|\Psi(t)\rangle_2|g_3\rangle|0\rangle_3|g_4\rangle|0\rangle_4\\ &+ e^{-i\pi/4}|g_1\rangle|0\rangle_1|\Psi(t)\rangle_2|\Psi(t)\rangle_3|g_4\rangle|0\rangle_4\\ &- e^{-i\pi/4}|\Psi(t)\rangle_1|g_2\rangle|0\rangle_2|g_3\rangle|0\rangle_3|\Psi(t)\rangle_4\Big)\end{aligned} \tag{27.86}$$

with the probability $\beta^4(1-e^{-2\kappa t})^2$. If detectors D_1 and D_4 detect one photon and D_2 and D_3 do not detect any photon, or vice versa, the state of the total system becomes projected into

$$\begin{aligned}|\Psi_{14}\rangle = \tfrac{1}{2}\Big(&|g_1\rangle|0\rangle_1|g_2\rangle|0\rangle_2|\Psi(t)\rangle_3|\Psi(t)\rangle_4 - |\Psi(t)\rangle_1|\Psi(t)\rangle_2|g_3\rangle|0\rangle_3|g_4\rangle|0\rangle_4\\ &+ i|g_1\rangle|0\rangle_1|\Psi(t)\rangle_2|\Psi(t)\rangle_3|g_4\rangle|0\rangle_4 - i|\Psi(t)\rangle_1|g_2\rangle|0\rangle_2|g_3\rangle|0\rangle_3|\Psi(t)\rangle_4\Big),\end{aligned} \tag{27.87}$$

which can be transformed into Eq. (27.86) by local operations. The success probability of the outcome is $\beta^4(1-e^{-2\kappa t})^2$. By tracing over the cavity field of the state (27.20), we obtain the state of four atoms

$$\begin{aligned}\rho_C = \frac{1}{(\alpha^2+\beta^2 e^{-2\kappa t})^2}\Big[&\alpha^4|C_4\rangle\langle C_4| + \beta^4 e^{-4\kappa t}|g_1\rangle|g_2\rangle|g_3\rangle|g_4\rangle\langle g_1|\langle g_2|\langle g_3|\langle g_4|\\ &+ \frac{\alpha^2\beta^2 e^{-2\kappa t}}{2}\big(|s_1\rangle|g_2\rangle|g_3\rangle|g_4\rangle\langle s_1|\langle g_2|\langle g_3|\langle g_4|\\ &+ |g_1\rangle|s_2\rangle|g_3\rangle|g_4\rangle\langle g_1|\langle s_2|\langle g_3|\langle g_4| + |g_1\rangle|g_2\rangle|s_3\rangle|g_4\rangle\langle g_1|\langle g_2|\langle s_3|\langle g_4|\\ &+ |g_1\rangle|g_2\rangle|g_3\rangle|s_4\rangle\langle g_1|\langle g_2|\langle g_3|\langle s_4|\big)\Big]\end{aligned} \tag{27.88}$$

with

$$\begin{aligned}|C_4\rangle = \tfrac{1}{2}\big(&|g_1\rangle|g_2\rangle|s_3\rangle|s_4\rangle - |s_1\rangle|s_2\rangle|g_3\rangle|g_4\rangle\\ &+ i|g_1\rangle|s_2\rangle|s_3\rangle|g_4\rangle - i|s_1\rangle|g_2\rangle|g_3\rangle|s_4\rangle\big).\end{aligned} \tag{27.89}$$

Using the local operation, we can transform the state (27.89) into the normal cluster state

$$\tfrac{1}{2}\big(|g_1\rangle|g_2\rangle|s_3\rangle|s_4\rangle - |s_1\rangle|s_2\rangle|g_3\rangle|g_4\rangle + |g_1\rangle|s_2\rangle|s_3\rangle|g_4\rangle - |s_1\rangle|g_2\rangle|g_3\rangle|s_4\rangle\big). \quad (27.90)$$

We calculate the fidelity $F = \langle C_4|\rho|C_4\rangle$, which is equal to that in Eq. (27.85).

We now give a brief discussion on the influence of photon losses on the scheme. In the generation process, the dominant noise is photon loss, which includes the contribution from channel attenuation and inefficiency of the single-photon detectors. All these kinds of noise can be considered by an overall photon loss probability η [38]. For simplicity, we only consider the influence of noise on the preparation of the cluster state. In this case, the effective state of the four atoms is actually described by

$$\begin{aligned}\rho_{\text{loss}} = \frac{1}{N}\Big\{&(\alpha^2 + \beta^2 e^{-2\kappa t})^2 \rho_C \\ &+ \tfrac{1}{2}\eta\alpha^2\beta^2(1 - e^{-2\kappa t})\big(4\beta^2 e^{-2\kappa t}|g_1\rangle|g_2\rangle|g_3\rangle|g_4\rangle\langle g_1|\langle g_2|\langle g_3|\langle g_4| \\ &\quad + |s_1\rangle|g_2\rangle|g_3\rangle|g_4\rangle\langle s_1|\langle g_2|\langle g_3|\langle g_4| + |g_1\rangle|s_2\rangle|g_3\rangle|g_4\rangle\langle g_1|\langle s_2|\langle g_3|\langle g_4| \\ &\quad + |g_1\rangle|g_2\rangle|s_3\rangle|g_4\rangle\langle g_1|\langle g_2|\langle s_3|\langle g_4| + |g_1\rangle|g_2\rangle|g_3\rangle|s_4\rangle\langle g_1|\langle g_2|\langle g_3|\langle s_4|\big) \\ &+ \tfrac{3}{2}\eta^2\beta^4(1 - e^{-2\kappa t})^2|g_1\rangle|g_2\rangle|g_3\rangle|g_4\rangle\langle g_1|\langle g_2|\langle g_3|\langle g_4|\Big\}\end{aligned} \quad (27.91)$$

with

$$\begin{aligned}N = &(\alpha^2 + \beta^2 e^{-2\kappa t})^2 + 2(1 - \eta)\alpha^2\beta^2(1 - e^{-2\kappa t}) \\ &+ \eta\beta^4(1 - e^{-2\kappa t})[2e^{-2\kappa t} + \tfrac{3}{2}\eta(1 - e^{-2\kappa t})].\end{aligned} \quad (27.92)$$

The first term ρ_C of Eq. (27.91) is the expected state, which comes from the event that the two photons have been emitted from the cavities and two photons are detected at the detectors D_1 and D_2. The second (or third) term of Eq. (27.91) is the noise contribution caused by photon loss. It comes from the event that three (or four) photons have been emitted from the cavities, but only two photons have been detected at the detectors D_1 and D_2.

In order to quantify how close the state (27.91) is to the cluster state, we calculate the fidelity $F = \langle C_4|\rho_{\text{loss}}|C_4\rangle = \alpha^4/N$. The recalculated fidelity does not increases with increasing detection time. But this fidelity still increases with increasing ratio $|\alpha/\beta|$.

27.3.2 Generation of W States and GHZ States Based on Four-photon Coincidence Detection

In this section, we will present another unified scheme to generate the W states and GHZ states of four distant atoms. The scheme is based on four-photon coincidence detection, and is insensitive to quantum noise. But the scheme is not used to generate the cluster states of the four distant atoms.

The experimental setup is shown in Fig. 27.11a, and the atoms have the level structure shown in Fig. 27.11b. Each of the four-level atoms (index j) has the Zeeman sublevels $|c_j\rangle$, $|g_j\rangle$, and $|s_j\rangle$, and an excited state $|e_j\rangle$ ($j = 1, 2, 3, 4$). The lifetimes of the atomic levels $|c_j\rangle$, $|g_j\rangle$, and $|s_j\rangle$ are assumed to be comparatively long so that spontaneous decay of these states can be neglected. The transitions $|e_j\rangle \longleftrightarrow |g_j\rangle$ and $|e_j\rangle \longleftrightarrow |s_j\rangle$ are coupled to two

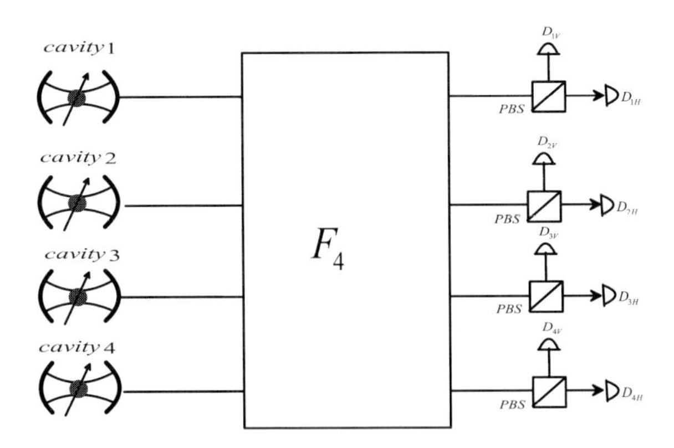

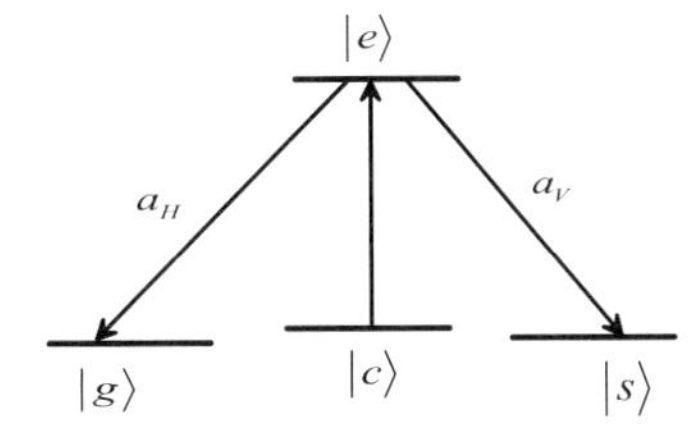

Figure 27.11: (a) The schematic setup for generating W states and GHZ states of four distant atoms based on four-photon coincidence detection. (b) The relevant level structure with ground state $|g\rangle$, $|s\rangle$, $|c\rangle$, and excited state $|e\rangle$.

degenerate cavity modes a_{jH} and a_{jV} with different polarizations H and V. The transition $|e_j\rangle \longleftrightarrow |c_j\rangle$ is driven by the classical field. We assume that the classical laser field and the cavity field are detuned from their respective transition by the same amount. In this case of large detuning the excited state $|e_j\rangle$ can be eliminated adiabatically to obtain the effective interaction Hamiltonian (in the interaction picture)

$$H_j = \Omega\big(a_{jH}|c_j\rangle\langle g_j| + a_{jH}^\dagger|g_j\rangle\langle c_j| + a_{jV}|c_j\rangle\langle s_j| + a_{jV}^\dagger|s_j\rangle\langle c_j|\big), \tag{27.93}$$

where a_{jH} and a_{jV} are the annihilation operators of the H and V polarization modes of the jth cavity. Here we assumed that the effective coupling constants of the atoms coupled with their cavity modes are the same, which are described by Ω. The evolution of the system's wave function is governed by a non-Hermitian Hamiltonian

$$H_j' = H_j - i\kappa\big(a_{jH}^\dagger a_{jH} + a_{jV}^\dagger a_{jV}\big) \tag{27.94}$$

as long as no photon decays from the cavity. Here we assume that the four cavities have the same loss rate κ for all the modes. If a single-photon detector D_{jH} or D_{jV} $(j = 1,2,3,4)$ detects a photon, the coherent evolution according to H_j' is interrupted by a quantum jump.

This corresponds to a quantum jump, which can be formulated with the operators b_{jH} or b_{jV} on the joint state vectors of four atom–cavity systems

$$\begin{aligned}
b_{1H} &= \tfrac{1}{2}(a_{1H} + a_{2H} + a_{3H} + a_{4H}), & b_{2H} &= \tfrac{1}{2}(a_{1H} + ia_{2H} - a_{3H} - ia_{4H}), \\
b_{3H} &= \tfrac{1}{2}(a_{1H} - a_{2H} + a_{3H} - a_{4H}), & b_{4H} &= \tfrac{1}{2}(a_{1H} - ia_{2H} - a_{3H} + ia_{4H}), \\
b_{1V} &= \tfrac{1}{2}(a_{1V} + a_{2V} + a_{3V} + a_{4V}), & b_{2V} &= \tfrac{1}{2}(a_{1V} + ia_{2V} - a_{3V} - ia_{4V}), \\
b_{3V} &= \tfrac{1}{2}(a_{1V} - a_{2V} + a_{3V} - a_{4V}), & b_{4V} &= \tfrac{1}{2}(a_{1V} - ia_{2V} - a_{3V} + ia_{4V}).
\end{aligned} \tag{27.95}$$

In the preparation stage the initial state of each atom–cavity system is $|c_j\rangle|0\rangle_{jH}|0\rangle_{jV}$ ($j = 1, 2, 3, 4$), i.e. each atom is initially in the Zeeman level $|c_j\rangle$ and the cavity modes are prepared in the vacuum states. Now we switch on the Hamiltonian (27.93) in each atom–cavity system for a time τ. If no photon is emitted from the cavity, the jth atom–cavity system is governed by the interaction H'_j. In this case the atom–cavity state evolves to the entangled state

$$|\Psi\rangle_j = \alpha|c_j\rangle|0\rangle_{jH}|0\rangle_{jV} - i\beta(|g_j\rangle|H\rangle_j + |s_j\rangle|V\rangle_j) \tag{27.96}$$

with

$$\begin{aligned}
\alpha &= \frac{\cos(\Omega_\kappa\tau) - \sin(\Omega_\kappa\tau)\kappa/(2\Omega_\kappa)}{\sqrt{[\cos(\Omega_\kappa\tau) - \sin(\Omega_\kappa\tau)\kappa/(2\Omega_\kappa)]^2 + 2\sin^2(\Omega_\kappa\tau)\Omega^2/\Omega_\kappa^2}}, \\
\beta &= \frac{\sin(\Omega_\kappa\tau)\Omega}{\Omega_\kappa\sqrt{[\cos(\Omega_\kappa\tau) - \sin(\Omega_\kappa\tau)\kappa/2\Omega_\kappa]^2 + 2\sin^2(\Omega_\kappa\tau)\Omega^2/\Omega_\kappa^2}}, \\
\Omega_\kappa &= \sqrt{2\Omega^2 - \kappa^2/4}.
\end{aligned} \tag{27.97}$$

The probability that no photon is emitted during this evolution becomes

$$P_{\text{single}} = e^{-\kappa\tau}\left\{\left[\cos(\Omega_\kappa\tau) - \frac{\sin(\Omega_\kappa\tau)\kappa}{2\Omega_\kappa}\right]^2 + \frac{2\sin^2(\Omega_\kappa\tau)\Omega^2}{\Omega_\kappa^2}\right\}. \tag{27.98}$$

We assume that the interaction Hamiltonian (27.93) is applied to each atom–cavity system simultaneously, so that the preparation of the atom–cavity states $|\Psi\rangle_j$ ends at the same time. This implements the preparation stage of the protocol.

Now we consider the detection stage. In this stage, the joint state of three atom–cavity systems becomes prepared in the form

$$|\Phi(0)\rangle = |\Psi\rangle_1|\Psi\rangle_2|\Psi\rangle_3|\Psi\rangle_4, \tag{27.99}$$

where the state $|\Psi\rangle_j$ is given by Eq. (27.96). We assume that photons are detected at the time t and the joint state of the total system evolves into

$$|\Phi(t)\rangle = |\Psi(t)\rangle_1|\Psi(t)\rangle_2|\Psi(t)\rangle_3|\Psi(t)\rangle_4 \tag{27.100}$$

with

$$|\Psi(t)\rangle_j = \frac{1}{\alpha^2 + 2\beta^2 e^{-2\kappa t}} \left[\alpha |c_j\rangle |0\rangle_j + \beta e^{-\kappa t} (|g_j\rangle |H\rangle_j + |s_j\rangle |V\rangle_j)\right]. \tag{27.101}$$

If the detectors D_{1H}, D_{2H}, D_{3H} and D_{4V} detect one photon respectively, the state of the total system becomes projected into the W state

$$\begin{aligned} |Dicke\rangle_1 = \tfrac{1}{2}\big(&|g_1\rangle|g_2\rangle|g_3\rangle|s_4\rangle + |g_1\rangle|g_2\rangle|s_3\rangle|g_4\rangle \\ &+ |g_1\rangle|s_2\rangle|g_3\rangle|g_4\rangle + |s_1\rangle|g_2\rangle|g_3\rangle|g_4\rangle\big). \end{aligned} \tag{27.102}$$

The success probability of the scheme is $P_{\text{coin}} = 2\beta^8(1 - e^{-2\kappa t})^4/9$.

If the detectors D_{1H}, D_{3H} and D_{2V} and D_{4V} detect one photon respectively, or D_{1V}, D_{3V} and D_{2H} and D_{4H} detect one photon respectively, the state of the total system becomes projected into the GHZ state

$$|Dicke\rangle_1 = \frac{1}{\sqrt{2}}(|g_1\rangle|s_2\rangle|g_3\rangle|s_4\rangle - |s_1\rangle|g_2\rangle|s_3\rangle|g_4\rangle). \tag{27.103}$$

The success probability of the scheme is $P_{\text{coin}} = \beta^8(1 - e^{-2\kappa t})^4/4$.

We now give a brief discussion on the influence of photon losses on the scheme. Since the scheme is based on four-photon coincidence detection, this requires that each of the four detectors has to detect a photon. If one photon is lost, a click from each of the detectors is never recorded and the scheme fails. Therefore photon loss has no influence on the fidelity of the generated states, but decreases the success probability P_{coin} by a factor $(1 - \eta)^4$.

27.4 Conclusion

In summary, we present here an overview of our work concerning the manipulation and control of individual photons and distant atoms trapped in various optical cavities by using linear optical elements. The entanglement generation and manipulation between many distant sites is an essential element of quantum network construction and quantum computing.

Finally it should be noted that spins in quantum dots are alternative good candidates for stationary qubits, which can be locally stored and manipulated. To connect, manipulate, and control distant quantum spin via linear optical networks is an important step toward realizing a quantum network and large-scale quantum computation, which will be further studied.

References

[1] A. Einstein, B. Podolsky, and N. Rosen, Phys. Rev. 47, 777 (1935).

[2] J. S. Bell, Physics (Long Island City, NY) 1, 195 (1965).

[3] D. M. Greenberger, M. Horne, A. Shimony, and A. Zeilinger, Am. J. Phys. 58, 1131 (1990).

[4] A. K. Ekert, Phys. Rev. Lett. 67, 661 (1991).

[5] C. H. Bennett, and S. J. Wiesner, Phys. Rev. Lett. 69, 2881 (1992).
[6] C. H. Bennett, G. Brassard, C. Crepeau, R. Jozsa, A. Peres, and W. Wootters, Phys. Rev. Lett. 70, 1895 (1993).
[7] D. Bouwmeester, A. Ekert, and A. Zeilinger, The Physics of Quantum Information (Springer-Verlag, Berlin, 2000).
[8] P. G. Kwiat, K. Mattle, H. Weinfurter, A. Zeilinger, A. V. Sergienko, and Y. H. Shih, Phys. Rev. Lett. 75, 4337 (1995).
[9] Q. A. Turchette, C. J. Hood, W. Lange, H. Mabuchi, and H. J. Kimble, Phys. Rev. Lett. 75, 4710 (1995).
[10] C. Monroe, D. M. Meekhof, B. E. King, W. M. Itano, and D. J. Wineland, Phys. Rev. Lett. 75, 4714 (1995).
[11] I. Chuang, N. Gershenfeld, and M. Kubinec, Phys. Rev. Lett. 80, 3408 (1998).
[12] B. E. Kane, Nature 393, 133 (1998).
[13] D. Loss, and D. P. Divincenzo, Phys. Rev. A. 57, 120 (1998).
[14] Y. Makhlin, and G. Schon, Nature 398, 305 (1999).
[15] E. Knill, R. Laflamme, and G. Milburn, Nature 409, 46 (2001).
[16] XuBo Zou, K. Pahlke, and W. Mathis, Phys. Rev. A 65, 064305 (2002).
[17] XuBo Zou, K. Pahlke, and W. Mathis, Phys. Rev. A 66, 064302 (2002).
[18] XuBo Zou, K. Pahlke, and W. Mathis, Phys. Lett. A 306, 10 (2002).
[19] XuBo Zou, K. Pahlke, and W. Mathis, Phys. Lett. A 311, 271 (2003).
[20] XuBo Zou, K. Pahlke, and W. Mathis, Phys. Lett. A 323, 329 (2004).
[21] XuBo Zou, K. Pahlke, and W. Mathis, Phys. Rev. A 68, 043819 (2003).
[22] XuBo Zou, K. Pahlke, and W. Mathis, Phys. Rev. A 69, 052314 (2004); Phys. Rev. A 69, 013811 (2004); Phys. Rev. A 68, 024302 (2003).
[23] XuBo Zou, K. Pahlke, and W. Mathis, Phys. Rev. A 70, 035802 (2004); Phys. Rev. A 69, 053608 (2004); Phys. Lett. A 324, 484 (2004); Phys. Rev. A 69, 015802 (2004); Phys. Rev. A 68, 034306 (2003); Phys. Rev. A 67, 044301 (2003); Phys. Rev. A 66, 044307 (2002).
[24] H. F. Hofmann, and S. Takeuchi, Phys. Rev. Lett. 88, 147901 (2002).
[25] O. Aytür, and P. Kumar, Phys. Rev. Lett. 65, 1551 (1990).
[26] P. A. M. Dirac, Proc. R. Soc. London A 114, 193 (1927).
[27] P. Carruthers, and M. M. Nieto, Rev. Mod. Phys. 40, 411 (1968);
D. T. Pegg, and S. M. Barnett, J. Mod. Opt. 44, 225 (1997).
[28] D. T. Pegg, and S. M. Barnett, Europhys. Lett. 6, 483 (1988); Phys. Rev. A 39, 1665 (1989);
S. M. Barnett, and D. T. Pegg, J. Mod. Opt. 36, 7 (1989).
[29] M. Zukowski, A. Zeilinger, and M. A. Horne, Phys. Rev. A 55, 2564 (1997).
[30] M. Reck, A. Zeilinger, H. J. Bernstein, and P. Bertani, Phys. Rev. Lett. 73, 58 (1994).
[31] A. Luis, and L. L. Sanchez Soto, Phys. Rev. A 48, 4702 (1993); Phys. Rev. A 53, 495 (1996).

[32] P. T. Cochrane, G. J. Milburn, and W. J. Munro, Phys. Rev. A 62, 062307 (2000);
P. T. Cochrane, and G. J. Milburn, Phys. Rev. A 64, 062312 (2001);
Ngoc-Khanh Tran et al., Phys. Rev. A 65, 052313 (2002).

[33] J. Kim, S. Takeuchi, Y. Yamamoto, and H. H. Hogue, Appl. Phys. Lett. 74, 902 (1999).

[34] C. Brunel, B. Lounis, P. Tamarat, and M. Orrit, Phys. Rev. Lett. 83, 2722 (1999).

[35] C. Santori et al., Phys. Rev. Lett. 86, 1502 (2001).

[36] P. Michler et al., Science 290, 2282 (2000).

[37] A. Kuhn, M. Hennrich, and G. Rempe, Phys. Rev. Lett. 90 (24), (2003).

[38] S. Bose, P. L. Knight, M. B. Plenio, and V. Vedral, Phys. Rev. Lett. 83, 5158 (1999).

[39] X. L. Feng et al., Phys. Rev. Lett. 90, 217902 (2003);
L. M. Duan, and H. J. Kimble, Phys. Rev. Lett. 90, 253601 (2003).

[40] M. B. Plenio, and P. L. Knight, Rev. Mod. Phys. 70, 101 (1998).

28 Conditional Linear Optical Networks

Stefan Scheel

Quantum Optics and Laser Science
Blackett Laboratory
Imperial College
London, UK

28.1 Introduction

Over recent years, considerable progress has been made in implementing quantum information processing in a variety of physical systems. All of them have in common that the physical systems under consideration exhibit some kind of (effective) nonlinearity, which is crucial for coupling subsystems (i.e. particles representing physical qubits) in a nontrivial way. What do we mean by nontrivial? We would like to be able to control the evolution of a subsystem by the quantum state of another subsystem.

Let us consider, for example, a quantum operation known as the controlled-Z gate in which the Pauli $\hat{\sigma}_z$ operator is applied to a target qubit only if the control qubit is in the logical state '1' (or $|1\rangle$ if the logical state is identical to the physical state). This operation can be characterized in several different ways. From the point of view of quantum logic, one can devise a truth table that is nothing but the transition matrix between a complete set of basis states. It is, however, more instructive to represent the action of the controlled-Z gate in terms of operators as

$$\hat{C}_\pi = |0_1\rangle\langle 0_1| \otimes \hat{I}_2 + |1_1\rangle\langle 1_1| \otimes \hat{Z}_2, \tag{28.1}$$

where the subscripts refer to the control (1) and target (2) qubits. In order to see the nonlinear nature of this quantum gate, it is even more instructive to use a more physical picture in terms of creation and annihilation operators of particles (for fermions and bosons alike). If the logical states $|0\rangle_i$ and $|1\rangle_i$ are labeled by the occupation number of the mode i, then the controlled-Z operator reads

$$\hat{C}_\pi = e^{i\pi\hat{n}_1\hat{n}_2} = (-1)^{\hat{n}_1\hat{n}_2}, \tag{28.2}$$

which is now an obviously nonlinear operator. Such a unitary operator has its origin in a quartic (effective) Hamiltonian.

Of particular interest to us and the focus of this chapter is the situation in which the basis states are labeled by the occupation numbers of bosonic modes, e.g. photonic Fock states. Under what circumstances can nonlinear operators such as Eq. (28.2) be obtained for single photons? For many kinds of interactions between particles, there exists a hierarchy of transformations of bilinear (or quadratic) type, followed by cubic or quartic type of lower strength

Quantum Information Processing, 2nd Edition. Edited by Thomas Beth and Gerd Leuchs

ISBN: 3-527-40541-0

(in a perturbation sense) and so on. For example, photon dynamics in vacuum quantum electrodynamics is usually written as a perturbation series in Sommerfeld's fine structure constant α whose leading term describes fermion pair creation and annihilation ($\propto \alpha^2$) followed by a fourth-order term ($\propto \alpha^4$) that describes photon–photon coupling via fermion exchange (see Fig. 28.1). In these perturbation series, the coupling strength in the nonlinear evolution is usually too weak to be useful for designing nontrivial quantum gates for single photons.

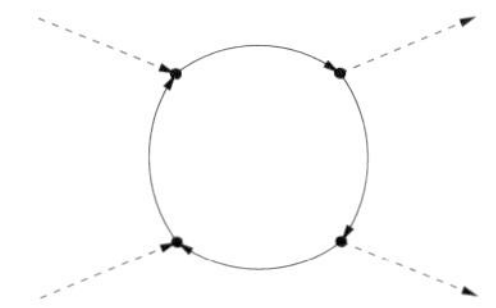

Figure 28.1: First photon–photon interaction term in vacuum QED perturbation series.

Another more efficient way of generating nonlinearities is by considering effective interactions in which a third (auxiliary) system is coupled to the qubits under study and later traced over. Typical examples for such processes are the Jaynes–Cummings model or (cross-)Kerr nonlinearities in certain materials. Here, the light interacts with the constituents of the material, which leads to an effective nonlinear interaction. But also in these cases, the achieved nonlinearities are usually too weak. A notable exception to that rule is provided by electromagnetically induced transparency (EIT) in atomic vapors, which can yield strong enough Kerr nonlinearities, at least for narrow-bandwidth light [1].

However, all of the above examples are either not very practicable for even medium-scale quantum information processing, or they imply the use of involved techniques from atomic physics (atoms trapped in high-Q cavities, atomic vapors, etc.). In view of all-optical quantum information processing, another method has to be used to generate effective nonlinearities. This is done by post-selection of a particular dynamics from the set of all possible evolutions of the system under consideration and some auxiliary system (commonly named ancilla). This scheme is very similar to the well-known technique of conditional quantum-state engineering in which a desired quantum state is being prepared by partial projective measurements after unitary evolution with known ancilla states [2].

In the following, we will describe the theory of measurement-induced nonlinearities or quantum-gate engineering (Sec. 28.2) and discuss the connection between the matrix elements of unitary operators and permanents (Sec. 28.3). We elaborate on the topic of maximizing success probabilities in Sec. 28.4. Finally, we mention a novel approach using weak nonlinearities in Sec. 28.5.

28.2 Measurement-induced Nonlinearities

Effective nonlinearities can be generated by performing partial projective measurements on selected subsystems. In the simplest case, two known states impinge onto a lossless beam splitter and a projective measurement, e.g. photon counting, is made on one output. The other output yields a transformed quantum state, which is conditioned on the measurement outcome. This idea is the basis of all quantum-state engineering schemes.

For the purpose of generating nonlinear quantum operations, this idea has been taken one step further by allowing one of the input states to be arbitrary. This amounts to looking at the set of (conditional nonlinear) operators that one can construct from a given set of resources. In view of experimentally accessible schemes, it seems reasonable to restrict our attention to single-photon sources and photon-number-resolving detectors that can distinguish between zero, one, and more than one photon. As will become clear later, more complicated, higher-order Fock states will not change our discussion substantially. However, notice that coherent states and homodyne detection have not been included in the list. If the purpose is to make contact with qubit encodings in single-photon states, then Gaussian operations as generated by homodyne measurement will introduce unwanted higher-order Fock states.[1]

28.2.1 Beam Splitters and Networks

Let us now turn to the lossless beam splitter and let us assume that a particular Fock state $|m\rangle$ impinges onto one of its ports. For the following discussion it is helpful to recall the exponential-operator disentangling theorem (or Euler decomposition) for SU(2)-generators [4] and to write the unitary beam splitter operator as

$$\hat{U} = T^{\hat{n}_1} e^{-R^* \hat{a}_2^\dagger \hat{a}_1} e^{R \hat{a}_1^\dagger \hat{a}_2} T^{-\hat{n}_2}, \tag{28.3}$$

where T and R are the complex transmittance and reflectance of the beam splitter, respectively, obeying $|T|^2 + |R|^2 = 1$. With the help of Eq. (28.3) one can calculate partial matrix elements. It has been shown in [5] that the following holds true:

$$\hat{Y} = \langle n_2 | \hat{U} | m_2 \rangle = \begin{cases} \sqrt{\dfrac{m!}{n!}} \dfrac{(-R^*)^{n-m}}{T^n} \hat{a}_1^{n-m} P_m^{(n-m,\hat{n}_1-n)}(2|T|^2-1) T^{\hat{n}_1}, & n \geq m, \\ \sqrt{\dfrac{n!}{m!}} \dfrac{R^{m-n}}{T^n} (\hat{a}_1^\dagger)^{m-n} P_n^{(m-n,\hat{n}_1-n)}(2|T|^2-1) T^{\hat{n}_1}, & m \geq n. \end{cases} \tag{28.4}$$

Here, the $P_m^{(a,b)}(x)$ denote Jacobi polynomials [6], which are orthogonal polynomials on the interval $(-1, 1)$ with weights $(1-x)^a(1+x)^b$. Clearly, from photon number conservation at the (unitary) beam splitter, the appropriate numbers of photons are either added to $(m \geq n)$ or subtracted from $(n \geq m)$ the quantum state in mode 1. In Table 28.1 we have collected some straightforward applications of Eq. (28.4) to the cases when only vacuum or a single-photon Fock state are present. Note that the "diagonal" elements $\langle k_2|\hat{U}|k_2\rangle$ lead to kth-order polynomials in the number operator $\hat{n}_1$. Note also that the argument of the Jacobi polynomials $P_m^{(a,b)}(x)$ is the permanent of the beam splitter ($x = 2|T|^2 - 1$). We will elaborate on this point in Sec. 28.3.

Obviously, the probability with which one of those conditional operators is obtained depends in general on the input state in mode 1 and is given by

$$p = \|\hat{Y}|\psi_1\rangle\|^2. \tag{28.5}$$

[1] Needless to say, the theory presented here could easily be extended to include Gaussian states and operations and has in fact been done in Ref. [3].

Table 28.1: Conditional operators obtained with either vacuum or single-photon Fock states in one input and one output arm of the beam splitter.

	$\|0_2\rangle$	$\|1_2\rangle$
$\langle 0_2\|$	$T^{\hat{n}_1}$	$T^{\hat{n}_1-1}R\hat{a}_1^\dagger$
$\langle 1_2\|$	$-T^{\hat{n}_1}R^*\hat{a}_1$	$T^{\hat{n}_1-1}(\|T\|^2-\hat{n}_1\|R\|^2)$

There are, however, special cases in which the conditional operator acts quasi-unitarily on certain classes of states, i.e. $\hat{Y}^\dagger\hat{Y}\propto\hat{I}$, such that the success probability becomes independent of $|\psi_1\rangle$. These operators will henceforth be called "quantum gates". Knowledge about the structure of the state is thereby of vital importance. By this statement we mean that the basis states describing the state should be known with their weights being arbitrary. In fact, for conditional operations on a single beam splitter, the state $|\psi_1\rangle$ can only be a superposition of at most two Fock states to ensure quasi-unitary transformations.

To give an example, let $|\psi_1\rangle = c_m|m\rangle + c_n|n\rangle$ and let the conditional operator be $\hat{Y} = \langle 0_2|\hat{U}|1_2\rangle$. Then it is easy to see that for the particular choice $|T|^{2(n-m)} = (m+1)/(n+1)$, with a probability $p = 1-|T|^2$, a photon is added, i.e. $\hat{Y}|\psi_1\rangle = c_m|m+1\rangle + c_n|n+1\rangle$. In general, when acting on an N-dimensional state $|\psi_1\rangle$, a network with at least $N-1$ independent beam splitters must be used.

When restricting ourselves to the resources mentioned above, i.e., single-photon sources and single-photon detectors, we see that a single beam splitter is of no great use. Instead, unitary networks that are made up of beam splitters, phase shifters and mirrors will be necessary. It is not difficult to generalize the results obtained so far to larger networks. However, the exponential-operator disentangling theorem (or Euler decomposition) as in Eq. (28.3) quickly becomes cumbersome to use. Instead, one uses the quantum-state transformation formula for the photonic amplitude operators $\hat{\boldsymbol{a}}' = \boldsymbol{\Lambda}^+\hat{\boldsymbol{a}}$, where $\boldsymbol{\Lambda}$ is the unitary matrix associated with the beam splitter network [7]. The most useful result derived with this restriction is that the conditional operator that is obtained when N single photons impinge on a network and N single photons are detected is an Nth-order polynomial in the number operator [8]. This becomes important when constructing quantum gates such as the nonlinear sign shift gate, which is one of the building blocks of linear optics quantum computing [9].

28.2.2 Post-processing of Single-Photon Sources and Number-Resolving Detectors

As we have seen in the examples in Sec. 28.2.1, even with the use of only linear elements such as beam splitters, effective nonlinearities can be generated by projective measurements. But this is equivalent to admitting that the nonlinearities are hidden in the process of generating single photons and the photon-number-resolving detectors. That is to say, the ability to produce single photons on demand as well as to build high-efficiency photodetectors is central to a successful linear optical scheme. One might be tempted to think that it would be possible to enhance the performance of either device by post-processing with linear optics. This, unfortunately, is not the case.

Let us start with single-photon sources and let us assume that these sources produce non-perfect photons in a statistical mixture $\hat{\varrho} = p|1\rangle\langle 1| + (1-p)|0\rangle\langle 0|$ with efficiency p. This is a rather realistic scenario as most single-photon sources such as atoms in cavities, NV color centers, or quantum dot structures may fail to deliver a photon without triggering a failure signal. Let us suppose we have several of those non-perfect sources to hand and we try to "purify" them by using a linear optical network and projective measurements. This turns out to be an impossible task. For example, suppose we had two independent sources of states of the form $\hat{\varrho}_i = (1-p_i)|0\rangle\langle 0| + p_i|1\rangle\langle 1|$ with respective single-photon efficiencies p_i. Let us assume that we mix $\hat{\varrho}_1$ and $\hat{\varrho}_2$ at a lossless beam splitter and perform a vacuum projection on mode 2. The resulting (un-normalized) output state reads

$$\hat{\varrho}_1^{(\mathrm{out})} \propto |0\rangle\langle 0| + \left(\frac{p_1}{1-p_1}|T|^2 + \frac{p_2}{1-p_2}|R|^2\right)|1\rangle\langle 1| + \text{two-photon term}. \tag{28.6}$$

Thus the ratio between the probabilities for obtaining one and zero photon is just the weighted average of $p_1/(1-p_1)$ and $p_2/(1-p_2)$ and cannot exceed either of these. A complete proof of the statement for all possible projections and network sizes can be found in Refs. [10, 11].

A similar situation is encountered when looking at the possibility of enhancing the performance of imperfect photon-number-resolving detectors. Let us suppose we were in possession of either of the two following photodetectors. On the one hand, let us take detectors that cannot discriminate between photon numbers at all, i.e., either they register any number of photons (equivalent to the POVM $\hat{I} - |0\rangle\langle 0|$) or they find no photon at all (thereby projecting as $|0\rangle\langle 0|$). On the other hand, let us take photon-number-resolving detectors that sometimes fail to distinguish between one and two photons but indicate failure. By employing similar types of arguments, it then turns out that neither of these detector types, together with linear optical elements and single-photon sources, can mimic a single-photon-resolving detector [12].

These two examples show the limitations on what linear optics can achieve in terms of simulating nonlinear devices such as single-photon sources and photon-number-resolving detectors. In particular, it should have become clear that both of these devices need to be available in order to make linear optical schemes work. The failure to produce either single photons on demand with high efficiency or to build single-photon-resolving detectors makes linear optical schemes not viable.

28.3 Probability of Success and Permanents

As we have seen in Sec. 28.2.1, the success probability p can be independent of the quantum state in mode 1 (in which case we can build a quantum gate), but it generally depends on the details of the linear optical network chosen to perform the task. In particular, it may depend on the size of the network and specifically on the number of auxiliary states fed into it. In any case, calculating the success probability requires the computation of matrix elements of the unitary operator $\hat{U}$ associated with the network in a multimode Fock basis, i.e., we are interested in objects of the type

$$\langle m_1, m_2, \ldots, m_N|\hat{U}|n_1, n_2, \ldots, n_N\rangle. \tag{28.7}$$

Let us first examine how the unitary operator acts on a multimode Fock state. Recall that the photonic amplitude operators transform as $\hat{\boldsymbol{a}}' = \boldsymbol{\Lambda}^{+}\hat{\boldsymbol{a}}$, which yields

$$\hat{U}|n_1, n_2, \ldots, n_N\rangle = \prod_{i=1}^{N} \frac{1}{\sqrt{n_i!}} \left(\sum_{k_i=1}^{N} \Lambda_{k_i,i} \hat{a}^{\dagger}_{k_i} \right)^{n_i} |0\rangle^{\otimes N}. \tag{28.8}$$

Using the multinomial expansion theorem and rearranging terms, we arrive at

$$\hat{U}|n_1, n_2, \ldots, n_N\rangle = \sum_{\{n_{ij}\}} \frac{\prod_{i=1}^{N} \sqrt{n_i!} \prod_{j=1}^{N} \sqrt{m_j!}}{\prod_{i,j=1}^{N} n_{ij}!} \left(\prod_{k,l=1}^{N} \Lambda_{lk}^{n_{kl}} \right) |m_1, m_2, \ldots, m_N\rangle, \tag{28.9}$$

where the n_{ij} have to be chosen such that $\sum_j n_{ij} = n_i$ and $\sum_i n_{ij} = m_j$. Now let us define multi-indices $\boldsymbol{\Omega}_n = (1^{n_1}, 2^{n_2}, \ldots, N^{n_N})$ and $\boldsymbol{\Omega}_m = (1^{m_1}, 2^{m_2}, \ldots, N^{m_N})$. Then, the prefactor in Eq. (28.9) is proportional to the permanent of the matrix $\boldsymbol{\Lambda}[\boldsymbol{\Omega}, \boldsymbol{\Omega}']$ [13]. More precisely,

$$\langle m_1, m_2, \ldots, m_N|\hat{U}|n_1, n_2, \ldots, n_N\rangle = \left(\prod_i n_i! \prod_j m_j! \right)^{-1/2} \operatorname{per} \boldsymbol{\Lambda}[\boldsymbol{\Omega}, \boldsymbol{\Omega}']. \tag{28.10}$$

This result should not come as a surprise. After all, photons are bosons and the Fock basis automatically takes care of symmetric ordering of multiparticle states. That is, an n-photon state is an element of the nth symmetric tensor power of the single-particle Hilbert space, not of the nth ordinary tensor power. Therefore, the transformation of a multiphoton state is described by an appropriate symmetric tensor power of the single-photon transformation matrix $\boldsymbol{\Lambda}$. But it is known from linear algebra [14] that the $(\boldsymbol{\Omega}, \boldsymbol{\Omega}')$ entry of the nth symmetric tensor power $\vee^n \boldsymbol{\Lambda}$ with $n = \sum_i n_i = \sum_j m_j$ is exactly Eq. (28.10).

Furthermore, it is known that the problem of calculating the permanent exactly is #P-complete[2] [15] (although effective approximation algorithms are also known [16]). What this means is that, generically, the calculation of matrix elements in a multimode Fock basis is computationally expensive. Since success probabilities, and hence the performance of linear optical networks, is given in terms of permanents, the optimization of such networks is in general a nontrivial task. Thus, before attempting to classify all possible linear optical networks with respect to their success probabilities, it is useful to reduce the complexity of the problem by studying more abstract networks. This will be the topic of Sec. 28.4.

We will remark briefly on a possible way of measuring permanents. From Eq. (28.10) it is clear that the permanent of any unitary matrix $\boldsymbol{\Lambda}$ of dimension N is given by the probability amplitude of finding one photon per output port if exactly one photon per input port was initially present, i.e.

$$\operatorname{per} \boldsymbol{\Lambda} = {}^{\otimes N}\langle 1|\hat{U}|1\rangle^{\otimes N}. \tag{28.11}$$

[2]#P-complete problems are at least as hard as the decision question for NP-complete problems.

Equation (28.11) suggests that the absolute value of the permanent could be approximated by sampling the photon statistics of a unitary network with a given input, in this case a tensor-product state with one photon per input mode. Unlike the situation in which the permanent of a Hermitian matrix is sampled [17], the absolute uncertainty of the sampled permanent does not grow with the size of the matrix. This is again due to Eq. (28.11), which states that the modulus of the permanent of a unitary matrix is bounded from above by one.[3]

Equation (28.10) also suggests that improvements to the sampling can be made when allowing for different measurement pattern and some classical post-processing. It is clear that, for a given input state, all possible outcomes yield some information about the permanent of the matrix $\mathbf{\Lambda}$. Using known inequalities for permanents [18], one can hope for a sampling algorithm that uses information from every possible detection outcome.

28.4 Upper Bounds on Success Probabilities

For a given quantum gate that one is trying to build, there exist many different networks that accomplish the task. They may differ in the number of single-photon sources used, the number of beam splitters involved, and the detection pattern that one regards as a successful event. The task is then to find the network with the highest probability of success. It seems hopeless to optimize all admissible schemes and pick out the one that works best. After all, calculating the success probability involves computing matrix elements in a multimode Fock basis, which, as we have seen in Sec. 28.3, rapidly becomes computationally expensive. It is therefore useful to find common features in linear optical networks and look at classes of schemes simultaneously.

Moreover, it will give some insight into the capabilities of linear optical networks if one can bound the probability of success for certain types of quantum gates from above. The first results were published in [19], where it was shown by very general arguments on photon number conservation of the conditional operator (28.2) that the success probability of any controlled-σ_z gate cannot exceed $3/4$ and that of the nonlinear sign shift gate cannot be more than $1/2$.

Here we will describe a more detailed route that eventually leads to tighter upper bounds. Our way to proceed is by noting that any $\mathrm{U}(N)$-network on photonic amplitude operators, i.e., a linear unitary transformation on N input modes, can be realized by at most $N(N-1)/2$ beam splitters [20] (including phase plates at one output port of each beam splitter) arranged in a triangular-shaped array.[4] It follows that, for single-mode quantum gates such as the nonlinear sign shift, which can be used as a primitive for building controlled-σ_z gates [9], there is only one beam splitter that mixes the signal with the ancilla modes.

Hence, we can divide every network into three blocks as shown in Fig. 28.2. The beam splitter that mixes the signal mode with one of the ancilla modes is denoted by "A". For multi-mode quantum gates acting on N qubits this single beam splitter would have to be replaced by

[3]This statement is trivial from the viewpoint of Eq. (28.11), but its general proof is rather involved [18]. This is an instance where quantum mechanics could be useful to develop new proof techniques.

[4]Note that this is equivalent to Householder transformations performed successively by each diagonal line of beam splitters. Each Householder transformation zeros a matrix column under the respective diagonal element. In general, that results in an upper triangular matrix. But for the latter to be unitary, it needs to be diagonal.

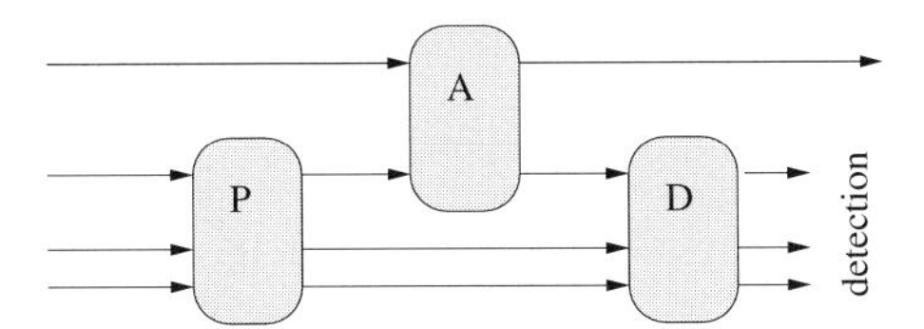

Figure 28.2: Abstract linear optical network consisting of a central beam splitter "A" that mixes signal and ancilla modes, a preparation stage "P" for generating the appropriate ancilla state from product-state inputs, and a detection stage "D" to disentangle the ancilla in preparation for photodetection.

a $U(2N)$-network. The auxiliary state itself can be thought of as being prepared from product-state inputs by block "P", and disentangled in preparation for photodetection by block "D". So far, we have not reduced the complexity of the network at all. This suddenly changes if we refrain from requiring that the preparation and detection stages "P" and "D" be realized by linear optics. It is clear that by allowing for any (admissible) auxiliary state the success probability obtained in this way could in fact be greater than with the restriction to linear optics [21].

Not all auxiliary states, however, are admissible. Let us concentrate on the simplest quantum gate, the single-mode nonlinear sign shift gate, which is defined by its action on superpositions with up to two photons as

$$c_0|0\rangle + c_1|1\rangle + c_2|2\rangle \mapsto c_0|0\rangle + c_1|1\rangle - c_2|2\rangle. \tag{28.12}$$

The requirement that the Hilbert spaces of input and output states be isomorphic, i.e., no photon is lost and no photon is added, restricts the ancilla state to a superposition of states with fixed total photon number. That is, an auxiliary state with N photons can be written as

$$|A\rangle = \sum_{k=0}^{N} \gamma_k |k\rangle |A_{N-k}\rangle, \tag{28.13}$$

where $|A_{N-k}\rangle$ denotes an arbitrary multimode state with the sole restriction that it must contain exactly $N - k$ photons. Then, it turns out that a two-dimensional ancilla state, i.e., a state with a superposition of only two terms, suffices to provide enough degrees of freedom to specify all relevant parameters [21]. In fact, it is also enough to work with an N-dimensional ancilla state to realize a generalized nonlinear sign shift on $(N + 1)$-dimensional signal states [22]. Moreover, a detailed analysis shows that, within the class of ancilla states of dimension two, the network with the lowest possible photon numbers, i.e. 0 and 1, leads to the highest success rates. Similarly, for the generalized nonlinear sign shift, it suffices to restrict the ancilla photon numbers to $0, \ldots, N - 1$.

Based on this abstract network, it has been proven in [23] using dual optimization theory that the upper bound on the success probability to achieve a nonlinear sign shift gate without the use of conditional dynamics is in fact $1/4$. That is, the computed upper bound is tight. Indeed, several schemes are known that saturate the $1/4$ bound [9, 24, 25]. Under the above restriction on the photon numbers, the success probability for generalized nonlinear sign shifts has been shown to scale as $1/N^2$ [22].

There are several interesting aspects to this result. First, it shows that the intuition that more ancilla modes yield higher success probabilities is a fallacy. Second, requiring only the least possible photon numbers means that the generated ancilla is least prone to decoherence. And last, the type of state required is experimentally accessible and indeed realizable with linear optics.

Between the bounds of $1/2$ and $1/4$ with and without conditional dynamics, respectively, there is a large gap, which it is hoped can be filled by concatenating several networks depending on their measurement results. Obviously, not all measurement patterns can be used for post-processing. Those detection patterns with more photons than in the original auxiliary state lead to loss of information about the signal state and have to be discarded. For example, an SU(3)-network with a $|10\rangle$ ancilla can result in detection patterns that contain more than one photon ($|11\rangle$, $|20\rangle$, $|02\rangle$, and those with three photons). In these cases, photons have been subtracted from the signal state resulting in an unrecoverable error. However, the measurement results $|01\rangle$ and $|00\rangle$, although not yielding the wanted transformation, have not destroyed the signal state. Such errors could in principle be recovered by another (conditional) network. However, preliminary results show that the total success probability after two rounds of networks does not exceed $1/3$, that is, convergence to a possible upper bound is slow.

28.5 Extension Using Weak Nonlinearities

Finally, we would like to mention an extension to the linear optical networks described so far that makes use of additional *weak* nonlinearities [27]. We have said earlier that naturally occurring nonlinearities are not strong enough to implement the controlled-Z gate trivially, that is, by using a cross-Kerr nonlinearity that realizes a π phase shift as required by Eq. (28.2). However, weaker nonlinearities can be used to effectively simulate photon-number-resolving detectors and to build parity and controlled-NOT gates.

The basic idea of a photon number measurement is to let the signal state of the form $c_0|0\rangle + c_1|1\rangle$ interact with a (sufficiently strong) coherent state $|\alpha\rangle$. Under the action of the cross-Kerr nonlinearity that produces a phase shift $\varphi = \chi^{(3)}t$ where $\chi^{(3)}$ is the strength of the nonlinearity and t the interaction time, the signal and probe beams evolve as

$$[c_0|0\rangle + c_1|1\rangle]\,|\alpha\rangle \stackrel{\chi^{(3)}}{\longmapsto} c_0|0\rangle|\alpha\rangle + c_1|1\rangle|\alpha\, e^{i\varphi}\rangle. \tag{28.14}$$

What we see is that the coherent probe beam picks up a phase that is directly proportional to the number of photons in the signal state. A subsequent phase measurement of the probe would then allow the projection of the signal mode onto a state with definite photon number. This, in essence, is a photon number quantum non-demolition measurement [26]. The required strength $\chi^{(3)}$ of the nonlinearity has to be determined by the necessity to distinguish between the states $|\alpha\rangle$ and $|\alpha\, e^{i\varphi}\rangle$, which can be done provided $|\alpha\varphi| \gg 1$. This condition can be satisfied with a strong probe beam, i.e., large $|\alpha|$, in which case the (scaled) nonlinearity φ can be taken to be small.

The idea of performing quantum non-demolition measurements can be taken further to design networks such as a two-qubit (polarization) parity detector or controlled-NOT gates [27].

The promised success probabilities are thereby much higher than for those conditional operations described in Sec. 28.2. It is hoped that the use of weak nonlinearities will have substantial impact on the experimental implementation of conditional quantum information processes.

Acknowledgements

I would like to thank K. Audenaert for sharing his insights into symmetric tensor product spaces, and J. Eisert, A. Gilchrist, P. L. Knight, P. Kok, N. Lütkenhaus, W. J. Munro, K. Nemoto, and M. B. Plenio for many fruitful discussions on the subject. This work was partly funded by the UK Engineering and Physical Sciences Research Council (EPSRC).

References

[1] H. Schmidt and A. Imamoglu, Opt. Lett. **21**, 1936 (1996).
[2] M. Ban, Phys. Rev. A **49**, 5078 (1994).
[3] J. Eisert, S. Scheel, and M. B. Plenio, Phys. Rev. Lett. **89**, 137903 (2002).
[4] K. Wodkiewicz and J. H. Eberly, J. Opt. Soc. Am. B **2**, 458 (1985).
[5] J. Clausen, M. Dakna, L. Knöll, and D.-G. Welsch, J. Opt. B: Quant. Semiclass. Opt. **1**, 332 (1999).
[6] I. S. Gradshteyn and I. M. Ryzhik, Table of Integrals, Series and Products (Academic Press, New York, 1965).
[7] W. Vogel, D.-G. Welsch, and S. Wallentowitz, Quantum Optics, An Introduction (Wiley-VCH, Berlin, 2001).
[8] S. Scheel, K. Nemoto, W. J. Munro, and P. L. Knight, Phys. Rev. A **68**, 032310 (2003).
[9] E. Knill, R. Laflamme, and G. J. Milburn, Nature (London) **409**, 46 (2001).
[10] D. W. Berry, S. Scheel, B. C. Sanders, and P. L. Knight, Phys. Rev. A **69**, 031806(R) (2004).
[11] D. W. Berry, S. Scheel, C. R. Myers, B. C. Sanders, P. L. Knight, and R. Laflamme, New J. Phys. **6**, 93 (2004).
[12] P. Kok, IEEE: Sel. Top. Quantum Electron. **9**, 1498 (2003).
[13] S. Scheel, quant-ph/0406127.
[14] R. Bhatia, Matrix Analysis (Springer, New York, 1997).
[15] L. G. Valiant, Theor. Comput. Sci. **8**, 189 (1979).
[16] M. Jerrum, A. Sinclair, and E. Vigoda, in: Proc. Thirty-Third Annual ACM Symp. on Theory of Computing, Hersonissos, Greece, pp. 712–721 (2001).
[17] L. Troyansky and N. Tishby, On the quantum evaluation of the determinant and permanent of a matrix, in: Proc. Physics of Computing, p. 96 (1996).
[18] H. Minc, Permanents, in: Encyclopedia of Mathematics and Its Applications, Vol. 6 (Addison-Welsey, Reading, MA, 1978).
[19] E. Knill, Phys. Rev. A **68**, 064303 (2003).
[20] M. Reck, A. Zeilinger, H. J. Bernstein, and P. Bertani, Phys. Rev. Lett. **73**, 58 (1994).
[21] S. Scheel and N. Lütkenhaus, New J. Phys. **6**, 51 (2004).

[22] S. Scheel, quant-ph/0410014.
[23] J. Eisert, quant-ph/0409156.
[24] T. C. Ralph, A. G. White, W. J. Munro, and G. J. Milburn, Phys. Rev. A **65**, 012314 (2001).
[25] G. G. Lapaire, P. Kok, J. P. Dowling, and J. E. Sipe, Phys. Rev. A **68**, 042314 (2003).
[26] G. J. Milburn and D. F. Walls, Phys. Rev. A **30**, 56 (1984).
[27] K. Nemoto and W. J. Munro, Phys. Rev. Lett. **93**, 250502 (2004).

29 Multiphoton Entanglement

Mohamed Bourennane[1,2], *Manfred Eibl*[1,2], *Sascha Gaertner*[1,2], *Nikolai Kiesel*[1,2], *Christian Kurtsiefer*[1,2], *Marek Żukowski*[3], *and Harald Weinfurter*[1,2]

[1] Max-Planck-Institut für Quantenoptik
Garching, Germany

[2] Department für Physik
Ludwig-Maximilians-Universität
München, Germany

[3] Instytut Fizyki Teoretycznej i Astrofizyki
Uniwersytet Gdański
Gdańsk, Poland

Multiphoton entanglement is the basis of novel quantum communication schemes, quantum cryptographic protocols, and fundamental tests of quantum theory. Spontaneous parametric down-conversion so far is the most effective source for polarization-entangled photon pairs. Here we show that an entangled four-photon state can be directly obtained from parametric down-conversion. This state exhibits perfect quantum correlations and a high robustness of entanglement against photon loss. We demonstrate these fundamental features using four-photon Bell operators and entanglement witnesses, and also analyze the persistent entanglement of reduced three- and two-photon states. Finally, we show its usefulness for new types of multi-party quantum communication and decoherence-free communication.

29.1 Introduction

Entangled states are key elements in the field of quantum information processing. The experimental preparation, manipulation, and detection of multiphoton entangled states is of great interest for implementations of quantum communication schemes, quantum cryptographic protocols, and for fundamental tests of quantum theory. Parametric down-conversion has been proven to be the best source of entangled photon pairs so far in an ever-increasing number of experiments on the foundations of quantum mechanics [1] and in the new field of quantum communication. Experimental realizations of concepts like entanglement-based quantum cryptography [2], quantum teleportation [3], and its variations [4] have demonstrated the usability of this source. More recent developments include improvement of the coupling efficiency [5], which, together with resonant enhancement of the pump beam [6], significantly increased the yield of photon pairs and made down-conversion pumped by frequency-doubled laser diodes possible [7].

Quantum Information Processing, 2nd Edition. Edited by Thomas Beth and Gerd Leuchs

ISBN: 3-527-40541-0

New proposals for quantum communication schemes [8–10] and, of course, for improved tests of local hidden variable theories initiated the quest for entangled multiphoton states. Interference of photons generated by independent down-conversion processes enabled the first demonstration of a three-photon Greenberger–Horne–Zeilinger (GHZ) argument [11] and, quite recently, even the observation of a four-photon GHZ state [12].

Instead of sophisticated but fragile interferometric setups, we utilize bosonic interference in a double-pair emission process. This effect causes strong correlations between measurement results of the four photons and renders type-II down-conversion a valuable tool for new multi-party quantum communication schemes. The analysis of the entanglement inherent in the four-photon emission leads us to a new form of inequality distinguishing local hidden variable theories from quantum mechanics, and demonstrates its potentiality for experiments on the foundations of quantum mechanics. This contribution summarizes a series of experiments: We start with the presentation of a new scheme to prepare four-photon entangled states (Sec. 29.2) and report the first experimental results (Sec. 29.3). Studies of quantum correlations and local realism versus quantum theory are presented in Secs. 29.7 and 29.5. We apply entanglement witnesses to prove the genuine four-photon entanglement (Sec. 29.6), and we show the persistence of the entanglement against photon loss in Sec. 29.7. In concluding (Sec. 29.8) we address possible applications for multi-party quantum communication and decoherence-free transmission of quantum information.

29.2 Entangled Multiphoton State Preparation

In type-II parametric down-conversion [13] multiple emission events during a single pump pulse lead to the following state:

$$C \exp\left[-i\alpha(a_V^* b_H^* - a_H^* b_V^*)\right] |0\rangle, \tag{29.1}$$

where C is a normalization constant, α is proportional to the pulse amplitude and the optical nonlinearity of the crystal, and a_V^* is the creation operator of a photon with vertical polarization in mode a, etc. The creation of one pair of entangled photons corresponds to the first-order term of the expansion on α of Eq. (29.1):

$$(a_H^* b_V^* - a_V^* b_H^*) |0\rangle. \tag{29.2}$$

This two-photon entangled state has become a routine in the laboratory and it has been used for tests of non-locality in quantum mechanics and in several experimental realizations of two-party communication protocols [2, 3].

The second-order term corresponds to the emission of four photons and it is proportional to

$$(a_H^* b_V^* - a_V^* b_H^*)^2 |0\rangle. \tag{29.3}$$

The particle interpretation of this term can be obtained by its expansion

$$(a_H^{*2} b_V^{*2} + a_V^{*2} b_H^{*2} - 2a_V^* a_H^* b_V^* b_H^*) |0\rangle, \tag{29.4}$$

and is given by the following superposition of photon number states:

$$|2H_a, 2V_b\rangle + |2V_a, 2H_b\rangle - |1H_a, 1V_a, 1H_b, 1V_b\rangle, \tag{29.5}$$

where, e.g., $2H_a$ means two horizontally polarized photons in the beam a and $2V_b$ means two vertically polarized photons in the beam b.

One should stress here that this type of description is valid only for down-conversion emissions, which are detected behind filters endowed with a frequency band that is narrower than that of the pumping fields [14]. If a wide band down-conversion is accepted then such a state is effective only if counts at the detectors are treated as coincidences when they occur within time windows narrower than the inverse of the bandwidth of the radiation [15]. If such conditions are not met, then the four-photon events are essentially emissions of two independent, entangled pairs, with the entanglement existing only within each pair.

Let us pass the four-photon state via two polarization-independent $50:50$ beam splitters. At the beam splitter a_i is transformed into $(a_i + a_i')/\sqrt{2}$ and b_i into $(b_i + b_i')/\sqrt{2}$ for $(i = H, V)$, with prime denoting the reflected output port. One can expand the expression (29.5), and then extract only those terms that lead to a four-photon coincidence behind the two beam splitters, i.e., only those terms for which there is one photon in each of the output ports. The resulting component of the full state is given by

$$\begin{aligned} \left|\Psi^{(4)}\right\rangle_{aa'bb'} = \sqrt{\tfrac{1}{3}}\Big[&|HHVV\rangle_{aa'bb'} + |HHVV\rangle_{aa'bb'} \\ &+ \tfrac{1}{2}\big(|HVHV\rangle_{aa'bb'} - |HVVH\rangle_{aa'bb'} \\ &- |VHHV\rangle_{aa'bb'} + |VHVH\rangle_{aa'bb'}\big)\Big], \end{aligned} \tag{29.6}$$

where we now use the notation of the first quantization with $|H\rangle_a$ describing a horizontally polarized photon in mode a, etc. The first term represents a four-photon GHZ state, whereas the second one is a product of two-photon entangled states. This shows that, from down-conversion, we are able to produce directly an entangled four-photon state without the need for fragile, interferometric setups [16]. One important property of this state is that the quantum state $\left|\Psi^{(4)}\right\rangle_{aa'bb'}$ is preserved if one applies the same unitary transformation to each of the four photons

$$\left|\Psi^{(4)}\right\rangle_{aa'bb'} = U_a \otimes U_{a'} \otimes U_b \otimes U_{b'} \left|\Psi^{(4)}\right\rangle_{aa'bb'}, \tag{29.7}$$

where $U_a = U_{a'} = U_b = U_{b'}$ are 2×2 unitary matrices.

29.3 Experiment

In the experiment [17], we used UV pulses from a frequency-doubled mode-locked Ti:sapphire laser with repetition rate of 76 MHz to pump a 2 mm thick BBO crystal. The degenerate down-conversion emission at the two characteristic type-II crossing points was coupled into single-mode fibers to exactly define the emission modes and then filtered with a narrow-band interference filter ($\Delta\lambda = 3$ nm). The output ports of the two $50:50$ beam splitters define four modes (see Fig. 29.1). We used quarter- and half-wave plates to set the orientation of the analyzers. The four photons were detected by single-photon Si avalanche photodiodes and analyzed with an eight-channel multi-coincidence logic. This specially designed system simultaneously registers any possible coincidence between the eight detectors and thus allows an efficient registration of the 16 relevant four-fold coincidences. Figure 29.2 shows the 16 possible four-fold coincidences when all four polarization analyzers are oriented along H/V, $+45°/-45°$, and left/right polarizations, respectively. We see clearly the superposition of a GHZ state and a product of two EPR states with weight ratio of $1/4$. The state shows the same distribution of the various terms in all three bases.

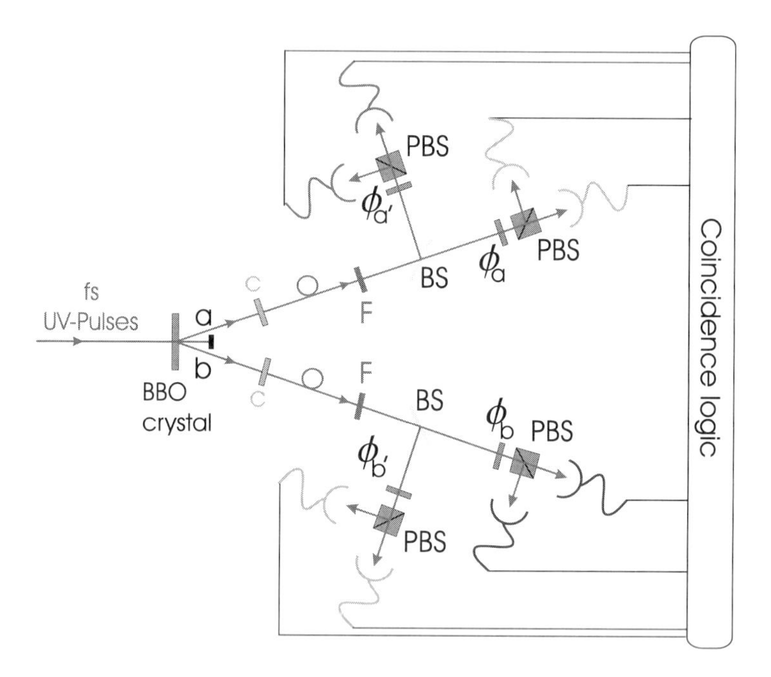

Figure 29.1: Experimental setup for the demonstration of four-photon entanglement where C, F, BS, and PBS stand for fiber coupler, filter, non-polarizing beam splitter, and polarizing beam splitter, respectively. The four photons emitted into modes a and b are split by BS. Four polarization analyzers with different settings (ϕ_i, $i = a, a', b, b'$) are used to investigate $|\Psi^{(4)}\rangle$.

29.4 Quantum Correlations

Quantum correlations are essential for experiments showing the violation of a Bell inequality and for quantum cryptographic protocols. Let us analyze polarization correlation measurements involving all four modes (a, a', b, b'), where the actual observables to be measured

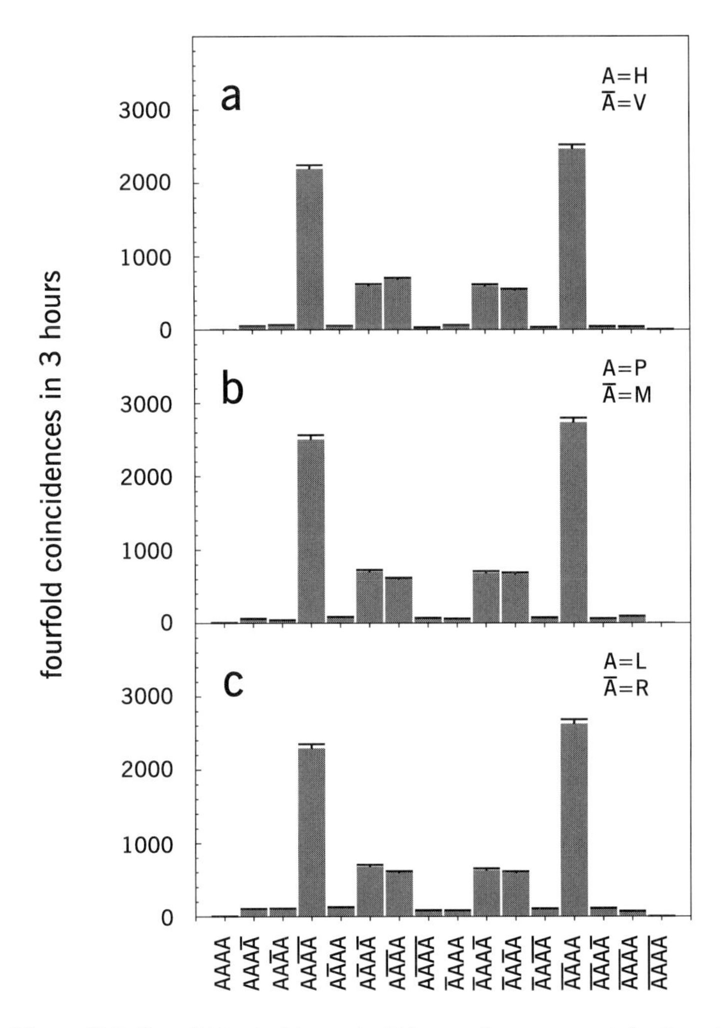

Figure 29.2: Four-fold coincidences in 16 hours of measurement, for the entangled four-photon state (Eq. (29.6)), with different polarization settings where A and $\bar{A}$ stand for $(H, +45^\circ, \text{left})$ and $(V, -45^\circ, \text{right})$ polarization, respectively.

are elliptic polarizations at 45°. Such observables are of dichotomic nature, i.e., endowed with a two-valued spectrum $k = +1, -1$, and are defined for each spatial propagation mode $x = a, a', b, b'$ by their eigenstates

$$\sqrt{\tfrac{1}{2}}\,|V\rangle_x + k\,e^{-i\phi_x}\sqrt{\tfrac{1}{2}}\,|H\rangle_x = |k, \phi_x\rangle. \tag{29.8}$$

The probability amplitudes for the results $k, l, m, n = \pm 1$ at the detector stations in the modes a, a', b, b', under local phase settings $\phi_a, \phi_{a'}, \phi_b, \phi_{b'}$, respectively, are given by

$$\frac{1}{4\sqrt{3}}\left[1 + klmn\, e^{i\sum\phi} + \tfrac{1}{2}(k\, e^{i\phi_a} + l\, e^{\phi_{a'}})(m\, e^{i\phi_b} + n\, e^{i\phi_{b'}})\right], \tag{29.9}$$

where $\sum\phi = \phi_a + \phi_a' - \phi_b - \phi_b'$. Therefore, the probability of getting a particular set of results (k, l, m, n) is given by

$$\begin{aligned} P(k, l, m, n \mid \phi_a, \phi_{a'}, \phi_b, \phi_{b'}) &= \tfrac{1}{16}\Big\{\tfrac{2}{3}\left(1 + klmn \cos \textstyle\sum\phi\right) \\ &+ \tfrac{1}{3}\left[1 + kl \cos(\phi_a - \phi_{a'})\right]\left[1 + mn \cos(\phi_b - \phi_{b'})\right] \\ &+ \tfrac{1}{3}\,\mathrm{Re}\left[(1 + klmn\, e^{i\sum\phi})(k\, e^{i\phi_a} + l\, e^{i\phi_{a'}})(m\, e^{i\phi_b} + n\, e^{\phi_{b'}})\right]\Big\}. \end{aligned} \tag{29.10}$$

The last term is written in the form of a real part of a complex function to shorten the expression. The correlation function is defined as the mean value of the product of the four local results

$$E(\phi_a, \phi_{a'}, \phi_b, \phi_{b'}) = \sum_{k,l,m,n=\pm 1} klmn\, P(k, l, m, n \mid \phi_a, \phi_{a'}, \phi_b, \phi_{b'}). \tag{29.11}$$

Its explicit form for the considered state (Eq. (29.6)) is thus

$$E(\phi_a, \phi_{a'}, \phi_b, \phi_{b'}) = \tfrac{2}{3}\cos(\phi_a + \phi_a' - \phi_b - \phi_b') + \tfrac{1}{3}\cos(\phi_a - \phi_{a'})\cos(\phi_b - \phi_{b'}). \tag{29.12}$$

The correlation function is a weighted sum of the GHZ correlation function (the first term) and a product of two EPR–Bell correlation functions. We note that our four-photon entangled state violates a generalized Bell inequality (for details see Ref. [16]). Figure 29.3 shows the dependence of the correlation function (Eq. (29.12)) on the angle ϕ_a when the other analyzers are fixed at angle $\phi_a' = \phi_b = \phi_b' = 0$. The data show a visibility of $V \simeq 92.3\,\% \pm 0.8\,\%$.

29.5 Bell Inequality

The strong correlations for numerous phase settings clearly indicate incompatibility with local realistic theories. However, here we present a reasoning, involving Bell inequalities of a new type, giving stronger inequalities for distinguishing the validity of the different theories in a four-photon experiment [16, 18, 19]. In a local hidden variable theory, a correlation function has to be modeled by a construction of the following form:

$$E_{LHV}(\phi_a, \phi_{a'}, \phi_b, \phi_{b'}) = \int d\lambda\, \rho(\lambda) I_a(\phi_a, \lambda) I_{a'}(\phi_{a'}, \lambda) I_b(\phi_b, \lambda) I_{b'}(\phi_{b'}, \lambda), \tag{29.13}$$

where λ represents an arbitrary set of values of local hidden variables, $\rho(\lambda)$ their probabilistic distribution, and $I_x(\phi_x, \lambda) = \pm 1$ $(x = a, a', b, b')$ represent the predetermined values of the

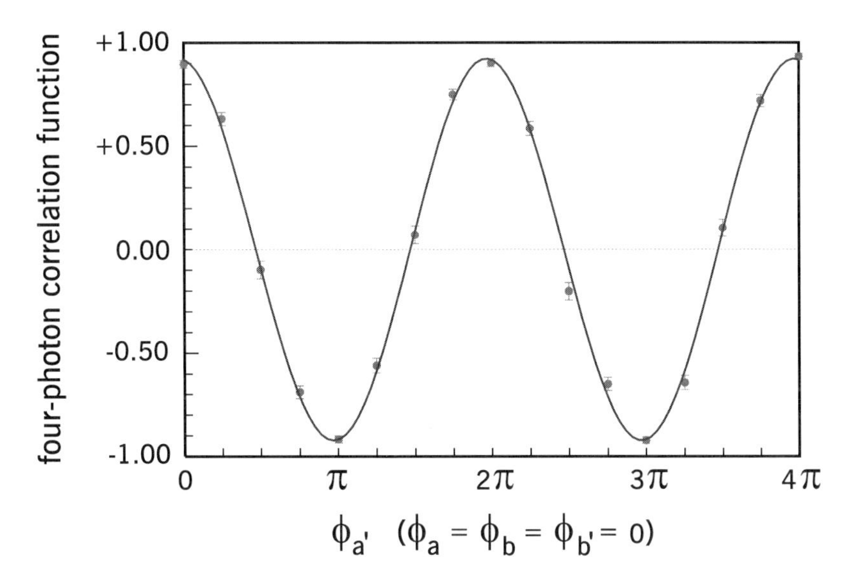

Figure 29.3: Quantum correlation function (Eq. (29.12)) on the angle ϕ'_a and the other analyzers are fixed at angle $\phi_a = \phi_b = \phi'_b = 0$. The data show a visibility of $V \simeq 92.3\,\% \pm 0.8\,\%$.

measurements. Their values depend on the set of hidden variables and on the value of the *local* phase settings. We allow each observer of mode x $(= a, a', b, b')$ to choose, just like in the standard cases of the Bell and GHZ theorems, between two values of the local phase settings [20]. A necessary and sufficient condition for the local realistic description is given by

$$\sum_{k,l,m,n=1,2} |c_{k,l,m,n}| \leq 1, \tag{29.14}$$

where the coefficients are defined by

$$\begin{aligned} c_{k,l,m,n} = p_{k,l,m,n} - p_{k+2,l,m,n} - p_{k,l+2,m,n} - \cdots + p_{k+2,l+2,m,n} + \cdots \\ - p_{k+2,l+2,m+2,n} + \cdots + p_{k+2,l+2,m+2,m+2}, \end{aligned} \tag{29.15}$$

with $k, l, m, n = +1, -1$. Here $p_{k,l,m,n}$ are the 256 probabilities that correspond to the 16 four-fold coincidences for each set of 16 directions. The maximum violation of the inequality (29.14) by a quantum prediction is obtained when the observer of mode a will be allowed to choose between $\phi_a^1 = 0$ and $\phi_a^2 = \pi/2$. The other observers $(y = a', b, b')$ can choose between $\phi_y^1 = -\pi/4$ and $\phi_y^2 = \pi/4$. Then the quantum prediction is $\frac{8}{3\sqrt{2}} \approx 1.89$. Our measured data lead to the following value:

$$\sum_{k,l,m,n=1,2} |q_{k,l,m,n}| = 1.664 \pm 0.028 > 1, \tag{29.16}$$

where $q_{k,l,m,n}$ are related to quantum correlation functions [16].

29.6 Genuine Four-photon Entanglement

The observed value clearly violates the boundary for local realistic theories and thus proves the entanglement of $|\Psi^{(4)}\rangle$. This value is also higher than the bound for bipartite entanglement ($S^{(4)}_{\text{bipartite}} \leq \sqrt{2}$) [21] and thus confirms that the observed state has at least tripartite entanglement. Yet, in order to test four-particle entanglement unambiguously, the Bell inequality is not suited, as there are tripartite entangled states giving values up to $S^{(4)}_{\text{tripartite}} \leq 2$. Although the possible tripartite entangled states do not exhibit the observed correlations and are thus ruled out by our measurements, the recently developed entanglement witnesses are the proper tool [22]. A genuine n-partite entanglement witness is an observable that has a negative expectation value on some n-partite entangled states, whereas for any state with only $(n-1)$-partite entanglement the expectation value is positive. A witness operator that detects genuine multipartite entanglement of $|\psi\rangle$ (and of states that are close to $|\psi\rangle$, e.g., in the presence of noise) is given by

$$\mathsf{W} = \alpha \mathbb{1} - |\psi\rangle\langle\psi|, \tag{29.17}$$

where

$$\alpha = \sup_{|\phi\rangle \in B} |\langle\phi|\psi\rangle|^2 \tag{29.18}$$

and B denotes the set of biseparable states. This construction guarantees that $\text{Tr}(\mathsf{W}\rho_B) \geq 0$ for all biseparable states ρ_B, and that $\text{Tr}(\mathsf{W}|\psi\rangle\langle\psi|) < 0$. Thus, a negative expectation value of the observable W clearly signifies that the state $|\psi\rangle$ carries multipartite entanglement. Determining α in Eq. (29.18) is a difficult task, when the maximization of the overlap with *any* biseparable state is performed explicitly. However, a simple method for the calculation of α is presented in Ref. [22]. For our state $|\Psi^{(4)}\rangle$ the witness W, given by

$$\mathsf{W} \; = \; \tfrac{3}{4}\,\mathbb{1} - |\Psi^{(4)}\rangle\langle\Psi^{(4)}|, \tag{29.19}$$

can prove the four-partite entanglement. Here, the theoretical expectation value is given by $\text{Tr}(\mathsf{W}_{|\Psi^{(4)}\rangle}\rho_{|\Psi^{(4)}\rangle}) = -1/4$, whereas it is positive on all biseparable, triseparable, and fully separable states.

The measurement of this observable requires 15 different analyzer settings, which is comparable with the 16 settings required for evaluating the four-photon Bell inequality, yet now with the possibility to clearly prove the type of entanglement. The observed four-fold coincidence probabilities result in an expectation value of $\text{Tr}(\mathsf{W}_{|\Psi^{(4)}\rangle}\rho_{|\Psi^{(4)}\rangle}) = -0.151 \pm 0.01$, which finally confirms the genuine multipartite entanglement beyond any doubt [23].

29.7 Entanglement Persistence

Entangled states are subject to decoherence and particle losses due to their interaction with the environment. For the four-photon entangled state we have analyzed two cases. The first case, when two photons are lost from the modes a' and b'; and the second, when they are lost from the modes b and b'. Mathematically this corresponds to tracing out two qubits from the four-photon density matrix $\rho^{(4)}_{aa'bb'} = |\Psi^{(4)}\rangle_{aa'bb'}\langle\Psi^{(4)}|$ and the density matrix of the remaining qubits becomes

$$\rho_{ab} = \mathrm{Tr}_{a'b'}[\rho^{(4)}_{aa'bb'}] = \tfrac{2}{3}\,|EPR\rangle_{ab}\langle EPR| + \tfrac{1}{3}(\tfrac{1}{4}I_{ab}) \tag{29.20}$$

or

$$\begin{aligned}\rho_{aa'} &= \mathrm{Tr}_{bb'}[\rho^{(4)}_{aa'bb'}] \\ &= \tfrac{1}{3}\,|EPR\rangle_{aa'}\langle EPR| + \tfrac{1}{3}[|HH\rangle_{aa'}\langle HH| + |VV\rangle_{aa'}\langle VV|],\end{aligned} \tag{29.21}$$

respectively, where I_{ab} is the unit density matrix

$$I_{ab} = |HH\rangle_{ab}\langle HH| + |VV\rangle_{ab}\langle VV| + |HV\rangle_{ab}\langle HV| + |VH\rangle_{ab}\langle VH|\,. \tag{29.22}$$

A strong criterion for the entanglement analysis is the Peres–Horodecki criterion, which states that, for separable states ρ, the partial transpose of the density matrix ρ must have non-negative eigenvalues [24]. We apply this criterion to the above reduced density matrices and obtain $\lambda_i = \{-1/4, 5/12, 5/12, 5/12\}$ and $\lambda_i = \{1/6, 1/6, 1/6, 1/2\}$, respectively. The first state (Eq. (29.20)) is partially entangled and there are purification procedures to transform this state to a pure entangled state [25] or one can obtain a state close to an EPR state by means of a filtering measurement [26]. The second state (Eq. (29.21)) is a separable state and possesses only classical correlations.

For the experimental analysis we have evaluated the two-photon density matrix ρ_{ab}, by making 16 polarization measurements in different bases of the linear and circular polarizations $(H/H,\ H/V,\ V/H,\ V/V,\ +45/H,\ +45/V,\ H/{+45},\ V/{+45},\ +45/{+45},\ R/H,\ R/V,\ +45/R,\ R/{+45},\ H/L,\ V/L, R/L)$. The results of the measurement allow us to tomographically reconstruct the density matrix of the reduced two-photon states [27]. Figure 29.4 shows the real parts of the elements of the density matrix ρ_{ab} in the H/V basis. The imaginary components are on the order of the noise in the real parts and therefore are neglected. The observed reduced two-photon states ρ^{exp} are compared with the theoretically expected ones ρ^{th} using the fidelity $F = \mathrm{Tr}[(\sqrt{\rho^{\mathrm{th}}}\rho^{\mathrm{exp}}\sqrt{\rho^{\mathrm{th}}})^{1/2}]^2$. It corresponds to the overlap between the theoretical and experimentally observed states, and we obtain $F_{ab} = 0.978 \pm 0.082$. For the partial transpose of the density matrix, we find negative eigenvalues $\lambda_{ab} = -0.252 \pm 0.024$, which clearly proves that the reduced state ρ_{ab} is entangled.

29.8 Conclusions

Quantum entangled states are used as resources for quantum information processing. Perfect correlations are present in our four-photon entangled state; for instance, when $\phi'_a = \phi_b = \phi'_b = 0$, the quantum correlation function becomes simply $\cos(\phi_a)$ and exhibits perfect correlation between the measurement outcomes. As these correlations are maintained when applying equal unitary transformations on each of the four photons, this state can be used to encode and protect quantum information in a decoherence-free subspace, which is spanned by the state $|\Psi^{(4)}\rangle$ and the product of two singlet states. Simulating the noisy transmission of the

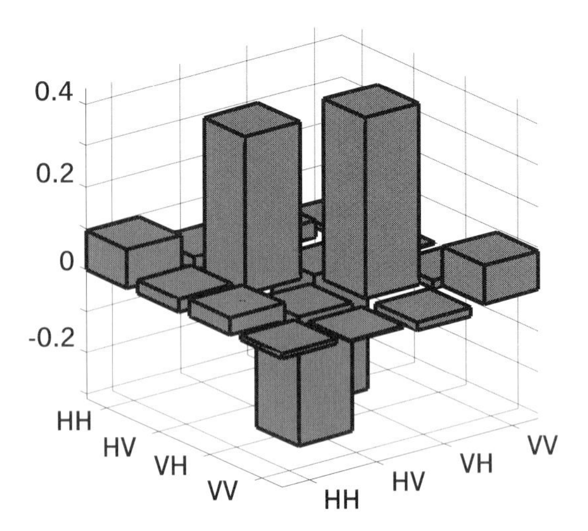

Figure 29.4: Real parts of the experimentally determined two-photon density matrix of the state ρ_{ab} in the $\{H, V\}$ polarization basis.

encoded qubit by applying equal unitary transformations, we demonstrated the stability of the encoded quantum information (Fig. 29.5) [28].

Perfect correlations together with the violation of a generalized Bell inequality are the ingredients for secure multi-party quantum communication. The two-photon EPR state was successfully employed for quantum teleportation — the transfer of an unknown quantum state to a remote location with perfect fidelity [29]. Multiphoton entangled states can be used for the generalization of the teleportation protocol. It is well known that the no-cloning theorem prohibits perfect copying of unknown quantum states, but it has been shown that this is a universal quantum cloning machine (UQCM), which can reproduce two copies with an optimal fidelity of $5/6$ [9]. Due to its persistent entanglement, the quantum state (Eq. (29.6)) is optimally suited to perform the so-called telecloning protocol, which enables optimal cloning of an unknown quantum state at different locations [10].

In conclusion, we have shown that the parametric type-II down-conversion produces not only entangled photon pairs, but also highly entangled four-photon states well suited for a new test of local realism and novel quantum communication schemes, like telecloning, four-party secret sharing or three-party secret key distribution.

Acknowledgements

This work was supported by the German DFG, BMBF, and DAAD, by the Polish KBN, and by the EU-FET-Projects QuComm and RAMBOQ.

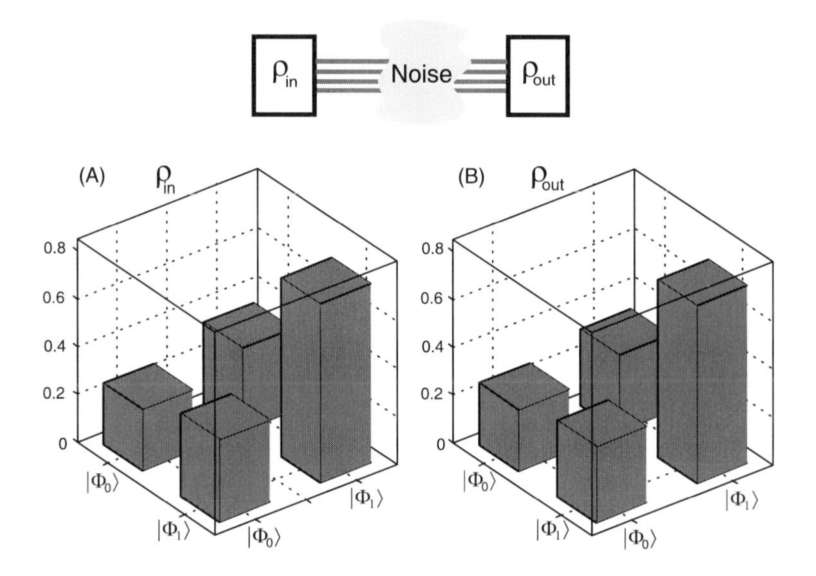

Figure 29.5: Propagation of the logical qubit $|\Phi_L\rangle = (\sqrt{3}\,|\Phi_0\rangle - |\Phi_1\rangle)/2$: (A) and (B) show the experimentally obtained density matrices before ($\rho_{\rm in}$) and after ($\rho_{\rm out}$) passage through a noisy quantum channel. The encoding in a DF subspace protected the transmission, leading to a fidelity of $F_{\rho_{\rm in},\rho_{\rm out}} = 0.9958 \pm 0.0759$ in the presence of noise (overall measurement time 12 hours).

References

[1] For reviews see:
D.M. Greenberger, M.A. Horne, A. Zeilinger, Phys. Today, August 1993, p. 22;
A.E. Steinberg, P.G. Kwiat, R.Y. Chiao, Atomic, Molecular and Optical Physics Handbook, ed. G. Drake (AIP Press, New York, 1996), ch. 77, p. 901;
L. Mandel, Rev. Mod. Phys. **71**, S274 (1999);
A. Zeilinger, Rev. Mod. Phys. **71**, S288 (1999).

[2] T. Jennewein, Ch. Simon, G. Weihs, H. Weinfurter, A. Zeilinger, Phys. Rev. Lett. **84**, 4729 (2000);
D.S. Naik, C.G. Peterson, A.G. White, A.J. Berglund, P.G. Kwiat, Phys. Rev. Lett. **84**, 4733 (2000);
W. Tittel, J. Brendel, H. Zbinden, N. Gisin, Phys. Rev. Lett. **84**, 4737 (2000);
G. Ribordy, J. Brendel, J.-D. Gautier, N. Gisin, H. Zbinden, Phys. Rev. A **63**, (2001).

[3] D. Bouwmeester, J.-W. Pan, K. Mattle, M. Eibl, H. Weinfurter, A. Zeilinger, Nature **390**, 575 (1997).

[4] D. Boschi, S. Branca, F. DeMartini, L. Hardy, S. Popescu, Phys. Rev. Lett. **80**, 1121 (1998);
J.-W. Pan, D. Bouwmeester, H. Weinfurter, A. Zeilinger, Phys. Rev. Lett. **80**, 3891 (1998);

A. Furusawa, J.L. Srenson, S.L. Braunstein, C.A. Fuchs, H.J. Kimble, E.S. Polzik, Science **282**, 706 (1998).
[5] C. Kurtsiefer, M. Oberparleiter, H. Weinfurter, Phys. Rev. A **64**, 023802 (2001).
[6] M. Oberparleiter, H. Weinfurter, Opt. Commun. **183**, 133 (2000).
[7] J. Volz, C. Kurtsiefer, H. Weinfurter, Appl. Phys. Lett. **79**, 869 (2001).
[8] R. Cleve, H. Buhrman, Phys. Rev. A **56**, 1201 (1997);
W. van Dam, P. Hoyer, A. Tapp, Los Alamos e-print quant-ph/9710054;
R. Cleve, D. Gottesman, H.-K. Lo, Phys. Rev. Lett. **83**, 648 (1999);
H.F. Chau, Phys. Rev. A **61**, 032308 (2000).
[9] V. Buzek, M. Hillery, Phys. Rev. Lett. **54** 1844 (1996).
[10] M. Murao, D. Jonathan, M.B. Plenio, V. Vedral, Phys. Rev. A **59** 156 (1999).
[11] D. Bouwmeester, J.-W. Pan, M. Daniell, H. Weinfurter, A. Zeilinger, Phys. Rev. Lett. **82**, 1345 (1999);
J.-W. Pan, D. Bouwmeester, M. Daniell, H. Weinfurter, A. Zeilinger, Nature **403**, 515 (2000).
[12] J.-W. Pan, M. Daniell, G. Weihs, A. Zeilinger, Phys. Rev. Lett. **86**, 4435 (2001).
[13] P.G. Kwiat, K. Mattle, H. Weinfurter, A. Zeilinger, A.V. Sergienko, Y.H. Shih, Phys. Rev. Lett. **75**, 4337 (1995).
[14] M. Żukowski, A. Zeilinger, H. Weinfurter, Ann. N.Y. Acad. Sci. **755**, 91 (1995).
[15] M. Żukowski, A. Zeilinger, M.A. Horne, A. Ekert, Phys. Rev. Lett. **71**, 4287 (1993).
[16] H. Weinfurter, M. Żukowski, Phys. Rev. A. **64**, 0101102 (2001).
[17] M. Eibl *et al.*, Phys. Rev. Lett. **90**, 200403 (2003);
S. Gaertner *et al.*, Appl. Phys. B **77**, 803 (2003).
[18] M. Żukowski, C. Brukner, Phys. Rev. Lett. **88**, 210401 (2002).
[19] R.F. Werner, M.M. Wolf, Phys. Rev. A **64**, 032112 (2001).
[20] J. Clauser, M. Horne, A. Shimony, R. Holt, Phys. Rev. Lett. **58**, 880 (1969).
[21] D. Collins *et al.*, Phys. Rev. Lett. **88**, 170405 (2002).
[22] B.M. Terhal, Phys. Lett. A **271**, 319 (2000);
M. Lewenstein *et al.*, Phys. Rev. A **62**, 052310 (2000);
D. Bruß *et al.*, J. Mod. Opt. **49**, 1399 (2002).
[23] M. Bourennane *et al.*, Phys. Rev. Lett. **92**, 087902 (2004).
[24] A. Peres, Phys. Rev. Lett **77**, 1413 (1996);
M. Horodecki, P. Horodecki, R. Horodecki, Phys. Lett. A. **233**, 1 (1996).
[25] C.H. Bennett *et al.*, Phys. Rev. A. **76**, 722 (1996).
[26] P.G. Kwiat, S. Barraza-Lopez, A. Stefanov, N. Gisin, Nature **409** 1014 (2001).
[27] W.J. Munro *et al.*, Phys. Rev. A **64**, 030302 (2001).
[28] M. Bourennane *et al.*, Phys. Rev. Lett. **92**, 107901 (2004).
[29] C.H. Bennett, G. Brassard, C. Crepeau, R. Jozsa, A. Peres, W. K. Wootters, Phys. Rev. Lett. **70**, 1895 (1993).

30 Quantum Polarization for Continuous Variable Information Processing

N. Korolkova

School of Physics and Astronomy
University of St. Andrews
St. Andrews
Scotland, UK

30.1 Introduction

Within the framework of quantum continuous variables (CV), nonclassical polarization states have recently attracted particular interest due to their compatibility with the spin variables of atomic systems and their simple detection schemes [1–4]. The relevant continuous polarization variables are hermitian Stokes operators (see Ref. [2] and references therein):

$$\begin{aligned} \hat{S}_0 &= \hat{a}_x^\dagger \hat{a}_x + \hat{a}_y^\dagger \hat{a}_y, \qquad \hat{S}_1 = \hat{a}_x^\dagger \hat{a}_x - \hat{a}_y^\dagger \hat{a}_y, \\ \hat{S}_2 &= \hat{a}_x^\dagger \hat{a}_y + \hat{a}_y^\dagger \hat{a}_x, \qquad \hat{S}_3 = i(\hat{a}_y^\dagger \hat{a}_x - \hat{a}_x^\dagger \hat{a}_y), \end{aligned} \tag{30.1}$$

where $\hat{a}_x$ and $\hat{a}_y$ denote the bosonic photon destruction operators associated with the x and y orthogonal polarization modes. The Stokes operator $\hat{S}_0$ commutes with all the others. The operators $\hat{S}_j$ ($j \neq 0$) obey the commutation relations of the SU(2) Lie algebra:

$$[\hat{S}_k, \hat{S}_l] = \epsilon_{klm}\, 2i\hat{S}_m, \qquad k, l, m = 1, 2, 3. \tag{30.2}$$

Simultaneous exact measurements of these Stokes operators are thus impossible in general and their variances are restricted by the uncertainty relations:

$$V_2 V_3 \geq |\langle \hat{S}_1 \rangle|^2, \qquad V_3 V_1 \geq |\langle \hat{S}_2 \rangle|^2, \qquad V_1 V_2 \geq |\langle \hat{S}_3 \rangle|^2, \tag{30.3}$$

where $V_j = \langle \hat{S}_j^2 \rangle - \langle \hat{S}_j \rangle^2$ is a shorthand notation for the variance of the quantum Stokes parameter $\hat{S}_j$. The angle brackets denote expectation values with respect to the state of interest.

Within the past few years, successful generation of polarization-squeezed [1, 3, 5, 6] and polarization-entangled [4,7] states has been reported (see also Chapter 32 in this book on quantum communication with fiber solitons, Sec. 32.7). The respective definitions of polarization squeezing [1, 8, 9] and entanglement [2, 4, 9] were formulated. While dealing with continuous variable polarization states, one should pay particular attention to the subtleties arising due to the q-number, i.e. operator-valued, commutator (cf. Eq. (30.2)). Along with nonclassical polarization states, coherent polarization states have proved to be very well suited for experimental quantum key distribution, in particular over a free-space link [10]. This chapter reviews the definitions of polarization squeezing and polarization entanglement, with the

Quantum Information Processing, 2nd Edition. Edited by Thomas Beth and Gerd Leuchs

ISBN: 3-527-40541-0

emphasis on the role of the q-number commutator and their applications in quantum communication. Furthermore, the open issues in a quantum description of polarization are briefly discussed, primarily with an example of defining the degree of polarization.

30.2 Nonseparability and Squeezing

30.2.1 Polarization Squeezing

The analogy between quadrature squeezing and polarization squeezing is both elucidating and misleading. Squeezing in general refers to the suppression — squeezing — of quantum uncertainty in a particular variable below the respective reference level at the cost of increasing uncertainty in the conjugate variable. For quadratures, the level of quantum fluctuations of the coherent state conveniently serves as such a reference, which corresponds to the minimal possible quantum uncertainty equally distributed between the relevant conjugate variables. The important difference between quadrature squeezing and polarization squeezing is the discrepancy between coherent and minimum uncertainty states for the latter due to the specific form of the commutation relation.

A coherent polarization state is defined as a quantum state with both polarization modes having a coherent excitation $\alpha_x, \alpha_y : \psi_{\rm coh} = |\alpha_x\rangle_x |\alpha_y\rangle_y$. The quantum uncertainty of such a state is equally distributed between the Stokes operators, and their variances are all equal to $V_j = V^{\rm coh} = |\alpha_x|^2 + |\alpha_y|^2 = \langle \hat{n} \rangle$. In analogy to quadrature squeezing, $V_j < V^{\rm coh}$ seems at first glance to be a natural definition for polarization squeezing. However, due to the SU(2) commutation algebra, a coherent polarization state is not a minimum uncertainty state for all three Stokes operators simultaneously. This was known for atomic states, i.e., for spin coherent states [11] and angular momentum coherent states [12]. The construction of the minimum uncertainty product for the SU(2) algebra and the properties of atomic coherent states were broadly studied around the early 1970s [11–14]. Although a polarization state with a sub-shot-noise variance $V_j < V^{\rm coh}$ is always a nonclassical state, it implies nothing more than conventional quadrature or single-mode squeezing observed through the measurement of the Stokes parameters. Hence a state is called *polarization squeezed* if.

$$V_k < |\langle \hat{S}_l \rangle| < V_m, \qquad k \neq l \neq m = 1, 2, 3. \tag{30.4}$$

Thus the reference state to quantify squeezing is chosen to be the corresponding minimum uncertainty state, not a coherent state.

Interestingly, polarization squeezing is strongly related to two-mode squeezing, i.e., quadrature entanglement. For two-mode squeezing, the nonclassical correlations are created between two spatially separated modes. For polarization squeezing, quantum correlations are created between two orthogonal polarization modes. However, by the appropriate choice of variables and basis, the correlations within a two-mode system can be redistributed so that polarization squeezing is transformed into two-mode squeezing and vice versa. This effect has already been observed in experiments: Polarization squeezing and quadrature entanglement were observed in the same nonlinear system of cold four-level atoms, depending of the choice of the mode basis [6]. Furthermore, the two schemes to generate continuous variable (CV) polarization entanglement have proven to be equivalent: superimposing two polarization-squeezed beams on a beam splitter [2, 7] or overlapping two quadrature-entangled beams with an orthogonally polarized coherent beam each [4]. As an example of a basis transformation that translates the two two-mode effects into each other, let us view quadrature entanglement in terms of new variables having the mathematical form of Stokes operators:

$$\begin{aligned} \hat{s}_0 &= \hat{a}_A^\dagger \hat{a}_A + \hat{a}_B^\dagger \hat{a}_B, \qquad \hat{s}_1 = \hat{a}_A^\dagger \hat{a}_A - \hat{a}_B^\dagger \hat{a}_B, \\ \hat{s}_2 &= \hat{a}_A^\dagger \hat{a}_B + \hat{a}_B^\dagger \hat{a}_A, \qquad \hat{s}_3 = i(\hat{a}_B^\dagger \hat{a}_A - \hat{a}_A^\dagger \hat{a}_B), \end{aligned} \tag{30.5}$$

where A and B are the two spatially separated output beams, the quadrature-entangled beams. To be specific, suppose that the quadrature entanglement emerges in the interference of two amplitude-squeezed beams with equal squeezing V^+ and coherent amplitude α on a beam splitter [15], where the input amplitude squeezing is quantified by the variances $V^+ < 1 < V^-$ of the quadrature operators $\hat{X}^+_{A,B} = \hat{a}^\dagger_{A,B} + \hat{a}_{A,B}$ and $\hat{X}^-_{A,B} = i(\hat{a}^\dagger_{A,B} - \hat{a}_{A,B})$. The variances of the new "Stokes" operators $\hat{s}_j$ of Eq. (30.5) for the noncommuting pair $\hat{s}_1, \hat{s}_3$ are equal to

$$v_1 = 2\alpha^2 V^- > |\langle \hat{s}_2 \rangle|, \qquad v_3 = 2\alpha^2 V^+ < |\langle \hat{s}_2 \rangle|. \tag{30.6}$$

Thus quadrature entanglement with anticorrelated amplitudes and correlated phases exhibits squeezing in the "Stokes parameter" $\hat{s}_3$.

30.2.2 Continuous Variable Polarization Entanglement

Along with polarization squeezing, CV polarization entanglement [2] has proven to be a useful tool in quantum communication. The nonseparability is a property of the state irrespective of the observables under consideration. All continuous variable entangled states have the same nature in this sense. In practice, some particular variables might be more advantageous to use. *Polarization entanglement* is a nonseparable state which implies correlations of the quantum uncertainties between one or more pairs of Stokes operators of two spatially separated optical beams. To quantify the degree of these quantum correlations and to verify the nonseparability of the state, different criteria can be used: There is no unique criterion to quantify CV entanglement in general, in particular for mixed states. A useful reference for the generalization and comparison of different sum and product entanglement criteria for CVs is the paper by Giovannetti *et al.* [16].

EPR criterion for the Stokes operators. The demonstration of the Einstein–Podolsky–Rosen (EPR) paradox for continuous variables takes place when measurements carried out on one subsystem can be used to infer the values of noncommuting observables of another, spatially separated subsystem to sufficient precision that an "apparent" violation of the uncertainty principle occurs [17]. The precision with which we can infer the value of an observable $\hat{Z}_D$ of subsystem D from the measurement of $\hat{Z}_C$ on subsystem C is given by the conditional variance

$$V_{\text{cond}}(\delta\hat{Z}_D \mid \delta\hat{Z}_C) = V(\delta\hat{Z}_D)\left(1 - \frac{|\langle \delta\hat{Z}_D\, \delta\hat{Z}_C \rangle|^2}{V(\delta\hat{Z}_D)\, V(\delta\hat{Z}_C)}\right). \tag{30.7}$$

Here the linearized approach is used, $\hat{Z}_j = \langle Z_j \rangle + \delta\hat{Z}_j$ $(j = C, D)$, and the variances $V(\delta\hat{Z}_j)$ are assumed to be inherently normalized to the shot-noise level. For the quadrature components this normalization is, for instance, just equal to unity, but note that in an experiment the shot-noise level should always be explicitly determined using respective measurement techniques. The EPR entanglement of the Stokes parameters in the sense of the EPR-like correlations of their uncertainties is realized if, for any of the conjugate pairs of Stokes operators, e.g. S_1, S_3, the following inequality holds [2]:

$$V_{\text{cond}}(\delta\hat{S}_{3D} \mid \delta\hat{S}_{3C})\, V_{\text{cond}}(\delta\hat{S}_{1D} \mid \delta\hat{S}_{1C}) < |\langle\hat{S}_{2C}\rangle|^2. \tag{30.8}$$

The two other inequalities for S_1, S_2 and S_2, S_3 are obtained by cyclic permutation of the indices.

Nonseparability criterion. The nonseparability criterion of Duan *et al.* [18] is derived for the canonical conjugate variables having a c-number commutator, like position and momentum or quadrature components of the light field. For a certain type of Gaussian state the criterion is necessary and sufficient. Unlike position and momentum, the Stokes operators are the variables with an operator-valued commutator. This provides a major obstacle in the derivation of the continuous variable nonseparability criterion in terms of polarization or spin. So far, it has only been possible to derive the sufficient criterion of nonseparability for such variables with a q-number operator [2,4,9,19].

The formulation of the sufficient nonseparability criterion [18] for the Stokes operators was first presented in Ref. [2] in terms of the quantities measured in an experiment and further elaborated in Ref. [4] on the basis of the generalized Heisenberg relation. The generalized Heisenberg uncertainty relation has the following form [20]:

$$V_A V_B \geq \tfrac{1}{4}|\langle\{\Delta\hat{A}, \Delta\hat{B}\}\rangle|^2 + \tfrac{1}{4}|\langle[\Delta\hat{A}, \Delta\hat{B}]\rangle|^2, \tag{30.9}$$

where the anticommutator and commutator of the two operators are defined by

$$\begin{aligned} \{\Delta\hat{A}, \Delta\hat{B}\} &= \hat{A}\hat{B} + \hat{B}\hat{A} - 2\langle\hat{A}\rangle\langle\hat{B}\rangle, \\ [\Delta\hat{A}, \Delta\hat{B}] &= \hat{A}\hat{B} - \hat{B}\hat{A} = [\hat{A}, \hat{B}]. \end{aligned} \tag{30.10}$$

This differs from the frequently used form

$$V_A V_B \geq \tfrac{1}{4}|\langle[\Delta\hat{A}, \Delta\hat{B}]\rangle|^2 \tag{30.11}$$

by the presence of the anticommutator term. For canonical observables like position and momentum, the commutator is a c-number, whereas the anticommutator $\{\Delta\hat{A}, \Delta\hat{B}\}$ is usually a q-number. Thus the *universal* general form of the Heisenberg inequality (30.11) can be obtained by retaining the state-independent part, the commutator term. If the commutator is an operator quantity, as in the case of the Stokes operators, both contributions on the right of (30.9) then depend on the state of the system and there is no reason to remove any of them. This justifies the use of the generalized Heisenberg uncertainty relation (30.10) as the starting point in the derivation of the nonseparability criterion [9] along the lines suggested by Duan *et al.* The inequality in (30.11) remains valid but the full form in (30.9) provides a stronger inequality with a higher minimum value of the variance product. There is no universal uncertainty relation in such cases, as in the examples of the angular momentum operators and of the Stokes parameter operators considered here.

The derivation of the nonseparability criterion for CV position x and momentum p having a c-number commutator [18] considers an overall system composed of two subsystems,

c and d, described by operators

$$\hat{A} = |a|\,\hat{x}_c + \frac{1}{a}\,\hat{x}_d, \qquad \hat{B} = |a|\,\hat{p}_c - \frac{1}{a}\,\hat{p}_d, \tag{30.12}$$

$$[\hat{x}_i, \hat{p}_j] = i\delta_{ij} \quad (i, j = c, d), \qquad [\hat{A}, \hat{B}] = i\left(a^2 - \frac{1}{a^2}\right). \tag{30.13}$$

The restrictions on the sum of the two variances are direct consequences of the uncertainty relation. With the use of the Schwarz inequality

$$V_A V_B \geq |\langle \Delta\hat{A}\,\Delta\hat{B}\rangle|^2, \qquad \Delta\hat{A} = \hat{A} - \langle\hat{A}\rangle, \qquad \Delta\hat{B} = \hat{B} - \langle\hat{B}\rangle \tag{30.14}$$

and the Cauchy inequality $V_A^2 + V_B^2 \geq 2V_A V_B$, it follows that

$$V_A + V_B \geq 2|\langle \Delta\hat{A}\,\Delta\hat{B}\rangle|. \tag{30.15}$$

Thus, with the Heisenberg uncertainty relation taken in the form (30.11), *all* states must satisfy

$$V_A V_B \geq \frac{1}{4}\left(a^2 - \frac{1}{a^2}\right)^2 \quad \text{and} \quad V_A + V_B \geq \left|a^2 - \frac{1}{a^2}\right|. \tag{30.16}$$

It is shown in Ref. [18] that *separable* states of the two subsystems must satisfy the stronger inequality

$$V_A + V_B \geq a^2 + \frac{1}{a^2}. \tag{30.17}$$

Nonseparable or *entangled* states thus exist in the region defined by

$$\left|a^2 - \frac{1}{a^2}\right| \leq V_A + V_B < a^2 + \frac{1}{a^2}, \tag{30.18}$$

where the lower limit on the left comes from the development of the Heisenberg uncertainty relation in (30.16) and the upper limit on the right comes from the nonseparability criterion in [18] in its sufficient form.

The derivation of [18] can be reworked for basic operator commutation relations more general than those given in (30.12) and (30.13) [9]:

$$\hat{A} = \hat{A}_c + \hat{A}_d, \qquad \hat{B} = \hat{B}_c - \hat{B}_d, \tag{30.19}$$

$$[\hat{A}_c, \hat{B}_d] = [\hat{B}_c, \hat{A}_d] = 0, \qquad [\hat{A}, \hat{B}] = [\hat{A}_c, \hat{B}_c] - [\hat{A}_d, \hat{B}_d]. \tag{30.20}$$

Here the nonzero commutators may themselves be operators. The uncertainty relations (30.16) are generalized to

$$V_A V_B \geq |\langle \Delta\hat{A}_c\,\Delta\hat{B}_c\rangle - \langle \Delta\hat{A}_d\,\Delta\hat{B}_d\rangle|^2, \tag{30.21}$$

$$V_A + V_B \geq 2|\langle \Delta\hat{A}_c\,\Delta\hat{B}_c\rangle - \langle \Delta\hat{A}_d\,\Delta\hat{B}_d\rangle|. \tag{30.22}$$

Note that these relations reduce to those in (30.14) and (30.15) when there is only a single system, c or d. By substitution of (30.19) into (30.12) and (30.15), Eqs. (3) and (4) in [18] can be reworked for the pair of variables with the q-number commutator giving the *sufficient* nonseparability criterion. The main difference from the derivation of [18] is the replacement of the universal limit in (30.18) by the state-dependent contribution containing the mean value of the operator-valued commutator (30.20) and the retention of the state-dependent anticommutator contribution. *Nonseparable* or *entangled* states must then satisfy the condition

$$2|\langle \Delta\hat{A}_c\, \Delta\hat{B}_c\rangle - \langle \Delta\hat{A}_d\, \Delta\hat{B}_d\rangle| \leq V_A + V_B \\ < 2|\langle \Delta\hat{A}_c\, \Delta\hat{B}_c\rangle| + 2|\langle \Delta\hat{A}_d\, \Delta\hat{B}_d\rangle|, \tag{30.23}$$

where the lower limit on the left comes from the development of the Heisenberg uncertainty relation in (30.22) and the upper limit on the right comes from the generalization of the *sufficient* nonseparability criterion. A derivation of the nonseparability criterion in its *necessary and sufficient form* in the case of the q-number commutator still remains a challenge. The sufficient general *product* criterion was obtained in Ref. [16], where the standard form of the Heisenberg uncertainty relation was used to derive an upper limit for the product of two variances.

30.3 Applications

Both the classical and quantum Stokes parameters represent useful tools for the description of the polarization of a light beam and also, more generally, of the phase properties of two-mode fields. They include explicitly the phase difference between the modes and they can be reliably measured in experiments. These features have triggered the use of Stokes operators for the construction of a formalism for the quantum description of relative phase [21]. The current development of methods for quantum information processing based on quantum continuous variables has stimulated interest in the nonclassical polarization states. The formalism of the quantum Stokes operators was used to describe the mapping of the polarization state of a light beam onto the spin variables of atoms in excited states [22, 23]. The correspondence between the algebras of the Stokes operators and the spin operators enables an efficient transfer of quantum information from a freely propagating optical carrier to a matter system. These developments pave the way towards the quantum teleportation of atomic states and towards the storage and read-out of quantum information [24].

One of the recent new developments in quantum communication are quantum key distribution systems based on continuous variables. Their potential advantages lie in the availability of high-bit-rate deterministic laser sources and of the fast and efficient homodyne detection instead of single-photon counters. Along with the originally proposed encoding by amplitude and phase modulations of light, polarization encoding is beneficial for continuous-variable quantum cryptography. The use of polarization variables allows one to dispense with the experimentally costly local oscillator techniques (Fig. 30.1). Thus implementation of the EPR-based protocols [25–28] using continuous-variable polarization entanglement combines the advantages of intense easy-to-handle sources of the EPR-entangled light and a efficient direct detection scheme (Fig. 30.1). Remarkably, secure quantum key distribution is possible not only with nonclassical continuous variable states, but also with quantum coherent states [10, 29–31]. Here again the Stokes polarization variables can offer a considerable advantage.

Experimental quantum key distribution (QKD) using coherent polarization states [10] has recently been implemented, which combines different features of the traditional discrete-variable BB84 [32, 33] and continuous variable coherent state cryptography [29–31, 34]. The

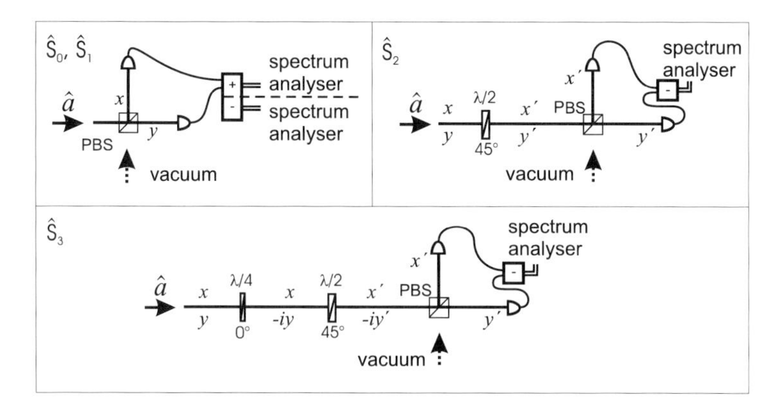

Figure 30.1: Measurement schemes for the Stokes operators. PBS: polarizing beam splitter; $\lambda/2$: half-wave plate; $\lambda/4$: quarter-wave plate. The two orthogonal polarizations are labeled x and y. No local oscillator is needed.

distinct difference from the discrete scheme is the use of homodyne detection. The particular properties of our system that mark it out from the related continuous-variable schemes are polarization encoding and, in contrast to [29, 30], the four-state protocol based on post-selection to ensure security and high loss tolerance. The scheme is illustrated in Fig. 30.2 (for details see Ref. [10]).

The sender Alice prepares randomly one of the four non-orthogonal coherent states by subjecting the linearly polarized beam with Stokes vector in the S_1 direction to sufficiently low, sub-shot-noise polarization modulations (Fig. 30.2, left). The non-orthogonality of the states is the key ingredient of the scheme ensuring its security: unknown non-orthogonal quantum states cannot be discriminated deterministically. For the scheme to be secure, the four signal quantum states should be generated with a substantial overlap, hence the modulation depth at the quantum noise level. An additional security mechanism is provided by the use of two conjugate bases at the detection side similar to the use of two non-orthogonal bases in the original BB84 protocol. The receiver Bob performs randomly measurement in either the S_2 or S_3 basis, assigning “1” (“0”) to his negative (positive) results (Fig. 30.2, right). Bob communicates to Alice his choice of basis at a given time slot but not his result. In this way two correlated bit strings are generated at sender and receiver, a raw key.

There is always potential information leakage to an eavesdropper and some errors. They occur due to (1) an imperfect channel, (2) interference from an eavesdropper, and (3) the impossibility also for the legitimate receiver to discriminate between the signal states deterministically. In the subsequent classical processing of the raw key, the errors are corrected and the privacy of the key is recovered. The crucial part of such a reconciliation procedure is the post-selection of data [10, 31]. It serves to increase the mutual information between Alice and Bob and decrease the potential mutual information with an eavesdropper and, not the least, it renders the QKD secure beyond the limit of 50 % loss in the transmission channel [31]. In the post-selection procedure, the measurement results falling within the central area around the origin in the graphs at the right-hand side of Fig. 30.2 are sorted out or post-selected after

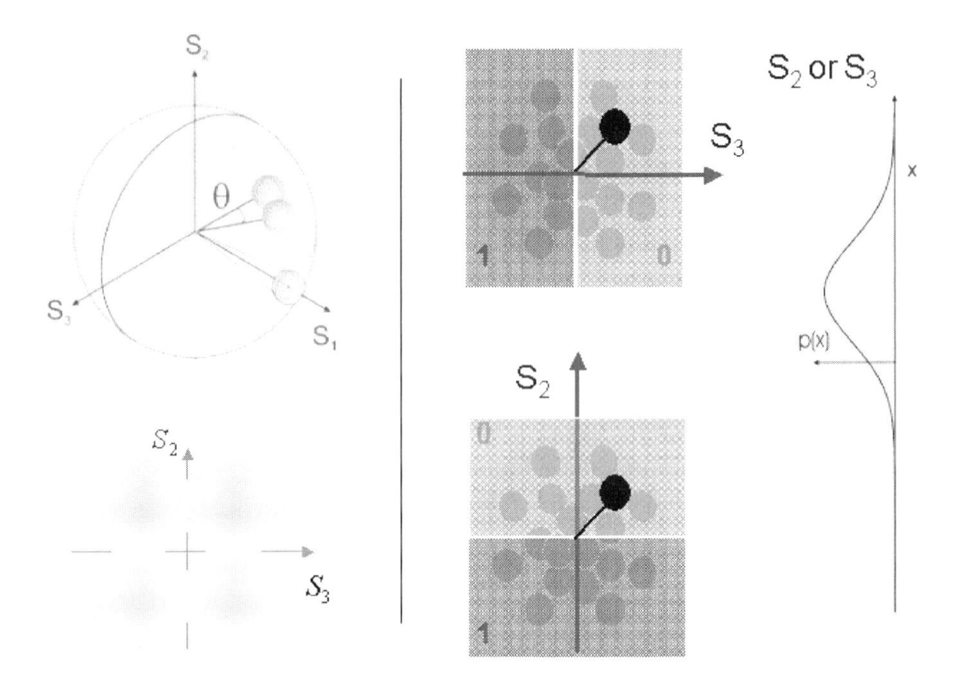

Figure 30.2: Quantum key distribution using coherent polarization states. Left: quantum state preparation. One of the four non-orthogonal coherent states is generated randomly in the polarization plane (S_2; S_3). Right: bit assignment. The signal is detected in one of the conjugate bases of the receiver's choice and evaluated as illustrated. The four signal states in the left bottom corner are shown with vanishing overlap for the sake of better visualization. (Part of this figure is courtesy of Ch. Silberhorn.)

the transmission is complete. This region corresponds to the maximal error probability in a decision about signal identity. As the measurement results of Bob and eavesdropper Eve are uncorrelated, this allows the mutual information between Alice and Bob to be increased. The particular value of the respective threshold should be evaluated on the basis of the estimated quantum channel properties.

The QKD scheme with coherent polarization states was recently implemented in the laboratory with transmission over a free-space link [10]. The use of polarization variables has provided substantial advantages on both the sender and receiver sides. These advantages are rooted in the two-mode structure of the Stokes operators and are surprisingly not only of merely technical type:

1. Signal preparation can be reduced to just four non-orthogonal states as opposed to the need to generate many states within a Gaussian distribution in the case of the amplitude and phase modulations [29,30]. This has a potential to increase the achievable bit rate and is more practicable. The possibility to contract the signal preparation to only four states is due to the following. The four polarization states in terms of the Stokes parameters can always be rewritten, by the choice of a basis, as a weak signal pulse containing the

information about the polarization encoding and a strong reference pulse on orthogonal polarization. The presence of the inherent strong reference prevents Eve from using the intercept–resend attack unnoticed and renders the protocol with just a few non-orthogonal states secure, exactly like the security of the original two-state B92 protocol with discrete variables was recovered by including an intensive reference pulse in the transmission along with the weak non-orthogonal signal states (see, e.g., Ref. [33]).

2. There is no need to send the phase reference, the local oscillator, along with the signal over the transmission link. With the Stokes parameters, the strong reference field for the homodyne detection is present within the pulse itself, hence the direct detection schemes of Fig. 30.1. These are basically homodyne detection schemes with the phase reference delivered by one of the orthogonal polarization modes after the appropriate basis change with the help of the wave plates, i.e., homodyne detection with the built-in local oscillator. Note that the transmission of a separate local oscillator is disadvantageous both because of the related technical difficulties and because the security of the signal transmission is impaired. For the polarization variables, these problems are avoided.

30.4 Stokes Operators Questioned: Degree of Polarization in Quantum Optics

We have used Stokes operators as the appropriate quantum polarization variables throughout this chapter. They are appealing variables for use in experimental quantum communication due to the particularly simple detection scheme and promise for creating an interface between photonic and atomic systems [1, 24]. However, whether the Stokes operators deliver an adequate description of quantum polarization properties is still open to question. The reasons for this are some odd quantum polarization properties [35, 36] that do not have analogues in our familiar classical picture of the polarization of light. The striking differences between the classical and quantum descriptions of polarization that can occur for discrete photon states have been explored in measurements on the pair states generated in spontaneous parametric down-conversion [36] and theoretically analyzed in the context of two-mode quantum number states with "hidden polarization" [37]. The primary concern is the degree of polarization, the definition of which is based on Stokes parameters and in its classical version seems to lead to discrepancies when the quantum properties are analyzed.

Classically, the state is "fully polarized" if the tip of the Stokes vector moves on the surface of the Poincaré (Bloch) sphere. Traditionally a beam is defined as "unpolarized" if it is invariant by geometrical rotations or if any two of its orthogonal plane polarized components have statistically independent phases. It is readily shown (Eq. (30.1) and Ref. [2]) that in quantum theory

$$\hat{S}_1^2 + \hat{S}_2^2 + \hat{S}_3^2 = \hat{S}_0^2 + 2\hat{S}_0, \tag{30.24}$$

and this is taken to define the quantum Poincaré sphere. The mean value of the sphere radius is given by the square root of the expectation value of either side of (30.24) and it generally has a nonzero variance. The particular form of the quantum Poincaré sphere thus depends on the particular polarization state under consideration (see examples below). If the classical degree of polarization is directly transferred to the quantum domain,

$$P = \frac{\sqrt{\langle \hat{S}_1^2 + \hat{S}_2^2 + \hat{S}_3^2 \rangle}}{\langle \hat{S}_0 \rangle}, \tag{30.25}$$

a number of problems arise and the description of quantum properties proves to be inadequate. Equation (30.25) then does not take full account of the quantum uncertainties of the Stokes operators described by the second moment and higher. Consider a linearly polarized light beam. If the light beam is linearly polarized, e.g., along x-direction, $|\psi\rangle = |\alpha\rangle_x |0\rangle_y$, classically the Stokes vector is aligned along S_1 axes and $S_1 = n$, $S_2 = S_3 = 0$. However, quantum noise measurements show nonzero values for the second-order moments, i.e., quantum uncertainties, of all three parameters S_1, S_2, S_3 whereas the mean values for S_2 and S_3 vanish exactly as in the classical picture. As a consequence, if (30.25) is applied, the light beam, which is linearly polarized in the classical theory, is linearly polarized in the quantum theory only in the limit of very large photon numbers (see examples below). On the other hand, the light may appear unpolarized in the classical theory but it is not so in the quantum theory — the quantum effect known as light with hidden polarization [35]. If the quantum optical properties are to be considered, Eq. (30.25) does not provide an appropriate definition and the classical condition $P < 1$ does not hold any more. We illustrate the problems arising with some examples of the "textbook" quantum states.

Number state. Consider a quantum state $|\psi\rangle = |n\rangle_x|0\rangle_y$, for which the mode x is excited in the number state and the mode y is in the vacuum state. For this state, all the uncertainty relations (30.3) are satisfied as equalities. The radius of the Poincaré sphere is then well-defined. For the particular state $|\psi\rangle = |n\rangle_x|0\rangle_y$, the $\hat{S}_1$ parameter is well-defined, while $V_2 = V_3 = n$. The state is linearly polarized in the classical picture. In the quantum picture the state stays fully polarized. The tip of the Stokes vector moves on the Poincaré sphere at $S_1 = n$, but not along the equator, as it would do for classical linearly polarized light. It moves in the (S_2, S_3) plane along a circle with radius $\sqrt{2n}$. So the light beam in a number state that is classically linearly polarized is fully polarized in the quantum picture, not purely linearly but with some ellipticity defined by the photon number in the beam (in other words by the uncertainties in S_2, S_3). This difference is more pronounced for a small number of photons. In the limit of large photon numbers $n \gg \sqrt{2n}$, the light beam is approximately linearly polarized (see also Ref. [2]).

Coherent state and squeezed coherent state (Gaussian states). Consider a quantum state $|\psi\rangle = |\alpha_x\rangle_x|\alpha_y\rangle_y$: modes x, y are excited in the coherent states with the respective amplitudes, $\langle n_x\rangle + \langle n_y\rangle = \langle n\rangle$. The Poincaré sphere takes the form:

$$\langle \hat{S}_1^2 + \hat{S}_2^2 + \hat{S}_3^2 \rangle = \langle \hat{n}^2 + 2\hat{n} \rangle = \langle \hat{n}\rangle^2 + 3\langle \hat{n}\rangle, \tag{30.26}$$

and the variance in the squared radius of the sphere is nonzero. The quantum Poincaré sphere for the coherent polarization state is therefore fuzzy, in contrast to that for the number state. The Poincaré sphere has the well-defined surface of its classical counterpart only in the limit of very large mean photon numbers, corresponding to bright coherent light, where the uncertainties in the Stokes parameters are negligible in comparison to the mean amplitude of the Stokes vector. The radius of the uncertainty sphere then shrinks relative to that of the Poincaré sphere and the tip of the Stokes vector approaches the surface of the Poincaré sphere. So for the coherent (squeezed) state with low photon number, light that is classically linearly polarized appears unpolarized in the quantum picture. In the limit of large photon numbers, as in the previous example, the light beam is approximately linearly polarized also in the quantum picture. According to the classical definition (30.25):

$$P^{\mathrm{coh}} = \frac{\sqrt{\langle \hat{S}_1^2 + \hat{S}_2^2 + \hat{S}_3^2 \rangle}}{\langle \hat{S}_0 \rangle} = \frac{\sqrt{\langle \hat{n} \rangle^2 + 3\langle \hat{n} \rangle}}{\langle n \rangle} > 1. \qquad (30.27)$$

It goes to ≈ 1 for $n \gg 1$ (see also Ref. [2]).

A single photon. This is an interesting limit of the number states. Classically we speak about, e.g., a horizontally (H-)polarized photon. It seems that all the Stokes parameters should be well-defined. However, this is not the case. The number state $|\psi\rangle = |n\rangle_x |0\rangle_y$ is an eigenstate of the operator $\hat{S}_1$, yet $\hat{S}_2$ and $\hat{S}_3$ are not well-defined but possess quantum uncertainties. This means that, if one "looks at" an H-polarized single photon $|\psi\rangle = |1\rangle_x |0\rangle_y$ in the right measurement basis (in the case of $|1\rangle_x |0\rangle_y$ the S_1 one), one can determine the polarization of the photon fully and deterministically. If one "looks" in any other basis, e.g., the S_2 basis, one gains only partial information on the photon polarization. This corresponds to rotating the polarizer by $45°$ and one would obtain H or V with $50\,\%$ probability. This is exactly the property used in the celebrated BB84 quantum cryptography protocol. The $0°-$ and $45°-$ bases are non-orthogonal (or S_1 and S_2 are conjugate bases). Thus choosing the inappropriate basis S_1 (S_2) delivers a probabilistic result, if the measured state is an eigenstate of S_2 (S_1). One has an uncertainty in determining polarization: the way you see the quantum uncertainties of the Stokes operators.

A quest for the "quantum" degree of polarization. Thus quantum optics entails polarization properties that cannot be fully described by the classical Stokes parameters. This statement emerged after light with hidden polarization was observed in an experiment [36] and the states with unexpected polarization behavior under geometrical rotations were demonstrated [38]. However, the Stokes operators obviously describe the real properties of the light beam, which was verified experimentally (see, e.g., Refs. [1, 3–5, 7]). All the discrepancies arise around determining the degree of polarization, which suggests rather that the classical definition of the degree of polarization is inappropriate in the quantum domain. So far, there have been only a few alternative suggestions on this subject. Luis [39] has considered a probability distribution on the Poincaré sphere, the Q-function based on the SU(2) coherent states. In his formalism the polarization degree is defined as a distance between a particular probability distribution and a uniform distribution of an unpolarized light. Very recently, quantum polarization properties have been discussed in the context of the two-mode number states with hidden polarization [37]. A state that is not invariant under all possible linear polarization transformations is said to have some degree of quantum polarization. An operator describing any SU(2) transformations was introduced and used to quantify the quantum degree of polarization.

Remarkably, the classical degree of polarization of a state does not give any indication of its degree of quantum polarization.

Acknowledgements

Over the years, the author has enjoyed fruitful collaboration on the subject with Gerd Leuchs, Rodney Loudon, Timothy C. Ralph, Ch. Silberhorn, N. Lütkenhaus, and S. Lorenz, who have contributed essentially to the different parts of this work. The corresponding experiments on the generation of nonclassical polarization states and coherent state cryptography were performed by the experimental Quantum Information Group, part of the Max-Planck Research Group, at the Institute for Optics, Information and Photonics, University of Erlangen–Nuremberg, Germany. Discussions with Ulf Leonhardt and Luciana C. Davilá Romero on the degree of polarization in quantum optics are gratefully acknowledged.

References

[1] J. Hald, J. L. Sorensen, C. Schori, E. S. Polzik, J. Mod. Opt. **47**, 2599 (2001).

[2] N. Korolkova, G. Leuchs, R. Loudon, T. C. Ralph, Ch. Silberhorn, Phys. Rev. A **65**, 052306 (2002).

[3] W. P. Bowen, R. Schnabel, H. A. Bachor, P. K. Lam, Phys. Rev. Lett. **88**, 093601 (2002).

[4] W. P. Bowen, N. Treps, R. Schnabel, P. K. Lam, Phys. Rev. Lett. **89**, 253601 (2002).

[5] J. Heersink, T. Gaber, S. Lorenz, O. Glöckl, N. Korolkova, G. Leuchs, Phys. Rev. A **68**, 013815 (2003).

[6] V. Josse, A. Dantan, A. Bramati, M. Pinard, E. Giacobino, Phys. Rev. Lett. **91**, 10360 (2003); Quantum Semiclass. Opt. **6**, S532 (2004).

[7] O. Glöckl, J. Heersink, N. Korolkova, G. Leuchs, S. Lorenz, Quantum Semiclass. Opt. **5**, 492 (2003).

[8] A. S. Chirkin, A. A. Orlov, D. Yu. Paraschuk, Quantum Electron. **23**, 870 (1993).

[9] N. Korolkova, R. Loudon, Nonseparability and squeezing of continuous polarization variables, quant-ph/0303135 (2003).

[10] S. Lorenz, N. Korolkova, G. Leuchs, Appl. Phys. B **79**, 273 (2004).

[11] J. M. Radcliffe, J. Phys. A **4**, 313 (1971).

[12] P. W. Atkins, J. C. Dobson, Proc. Roy. Soc. London A **321**, 321 (1971).

[13] R. Jackiw, J. Math. Phys. **9**, 339 (1968).

[14] F. T. Arecchi, E. Courtens, R. Gilmore, H. Thomas, Phys. Rev. A **6**, 2211 (1972).

[15] Ch. Silberhorn, P. K. Lam, O. Weiß, F. König, N. Korolkova, G. Leuchs, Phys. Rev. Lett. **86**, 4267 (2001).

[16] V. Giovannetti, S. Mancini, D. Vitali, P. Tombesi, Phys. Rev. A **67**, 022320 (2003).

[17] M. D. Reid, P. D. Drummond, Phys. Rev. Lett. **60**, 2731 (1988);
M. D. Reid, Phys. Rev. A **40**, 913 (1989).

[18] L.-M. Duan, G. Giedke, J. I. Cirac, P. Zoller, Phys. Rev. Lett. **84**, 2722 (2000).
[19] M. G. Raymer, A. C. Funk, B. C. Sanders, H. de Guise, Phys. Rev. A **67**, 052104 (2003).
[20] E. Merzbacher, Quantum Mechanics, 3rd edn (Wiley, New York, 1998), pp. 217–220.
[21] A. Luis, L. L. Sanchez-Soto, Prog. Opt. **41**, 421 (2000).
[22] L.-M. Duan, J. I. Cirac, P. Zoller, E. S. Polzik, Phys. Rev. Lett. **85**, 5643 (2000).
[23] A. Kurmich, E. S. Polzik, Phys. Rev. Lett. **85**, 5639 (2000).
[24] B. Julsgaard, J. Sherson, J. I. Cirac, J. Fiurasek, E. S. Polzik, Experimental demonstration of quantum memory for light, Nature **432**, 482–486 (2004).
[25] T. C. Ralph, Phys. Rev. A **61**, 010303 (1999).
[26] T. C. Ralph, Phys. Rev. A **62**, 062306 (2000).
[27] Ch. Silberhorn, N. Korolkova, G. Leuchs, Phys. Rev. Lett. **88**, 167902 (2002).
[28] S. Lorenz, C. Silberhorn, N. Korolkova, R. S. Windeler, G. Leuchs, Appl. Phys. B **73**, 855 (2001).
[29] F. Grosshans, P. Grangier, Phys. Rev. Lett. **88**, 057902 (2002).
[30] F. Grosshans, G. Van Assche, J. Wenger, R. Brouri, N. J. Cerf, P. Grangier, Nature **421**, 238 (2003).
[31] C. Silberhorn, T. C. Ralph, N. Lütkenhaus, G. Leuchs, Phys. Rev. Lett. **89**, 167901 (2002).
[32] C. H. Bennett, F. Bessette, G. Brassard, L. Salvail, J. Smolin, J. Cryptology **5**, 3 (1992); C. H. Bennett, G. Brassard, A. Ekert, Sci. Am. **267**, 50 (1992).
[33] B. Huttner, N. Imoto, N. Gisin, T. Mor, Phys. Rev. A **51**, 1863 (1995).
[34] R. Namiki, T. Hirano, Phys. Rev. A **67**, 022308 (2003).
[35] V. P. Karasev, A. V. Masalov, Opt. Spectrosc. **74**, 551 (1993); J. Opt. B **4**, 366 (2002).
[36] P. Usachev, J. Söderholm, G. Björk, A. Trifonov, Opt. Commun. **193**, 161 (2001).
[37] A. Sehat, J. Söderholm, G. Björk, L. L. Sanchez-Soto, Quantum polarization properties of two-mode number states, quant-ph/0407184 (2004).
[38] T. Tsegaye, J. Söderholm, M. Atatüre, A. Trifonov, G. Björk, A. V. Sergienko, B. E. A. Saleh, M. C. Teich, Phys. Rev. Lett. **85**, 5013 (2000).
[39] A. Luis, Phys. Rev. A **66**, 013806 (2002).

31 A Quantum Optical XOR Gate
How to Interfere Two Quantum Camels

H. Becker, K. Schmid, W. Dultz, W. Martienssen, and H. Roskos

Physikalisches Institut
Universität Franfurt (Main)
Germany

31.1 Introduction

When two indistinguishable photons are superimposed at a beamsplitter, an interference phenomenon called *photon pair interference*, *fourth order interference* or *HOM-interference* can be detected in the coincidence signal of two detectors at the outputs of the beamsplitter: the coincidence rate drops to zero [1] [2]. This is a consequence of the indistinguishability between the two processes that could lead to a coincidence event: both photons could be transmitted at the beamsplitter ('tt'), or both could be reflected ('rr'). Since these processes are indistinguishable in principle, they interfere, and due to the phase of $\frac{\pi}{2}$ per reflection there is a relative phase of π between these processes; thus they cancel each other out and the coincidence probability gets zero.

31.2 Double Bump Photons

We now modify the wave form of one of the pair photons by a quartz plate and a polarizer (Fig. 31.1): the quartz plate's birefringence splits the wave packet into two separated packets (bumps) wich have perpendicular polarization. In order that the two bumps do not overlap the quartz plate has to be thick enough to make the separation bigger than the photon's coherence length. A polarizer projects the polarization state of the two bumps back into the original orientation to make them indistinguishable [3] [4].

The resulting *double bump photon* (sometimes called a *quantum camel*) is in a temporal superposition state; it has a more complex wave form than before, and due to that it gains some additional degrees of freedom, for example the distance between the two bumps and thus the phase between them. This phase is an inner relative phase of that single photon, and it can be adjusted by varying the thickness of the quartz plate.

The photon pair interference of such a photon together with its unmodified pair photon is recorded using a setup schematically shown in Fig. 31.2. The pump laser (662 nm diode laser, 40 mW) is focused into a $LiNbO_3$ crystal of 5 mm thickness. The down-converted light is emitted noncollinearly and has a coherence length of about 10 µm. The two beams used in the experiment are usually named signal and idler; the signal beam passes through a 6 mm quartz plate and a polarizer. Signal and idler beam are then focussed by collimators into the

Quantum Information Processing, 2nd Edition. Edited by Thomas Beth and Gerd Leuchs

ISBN: 3-527-40541-0

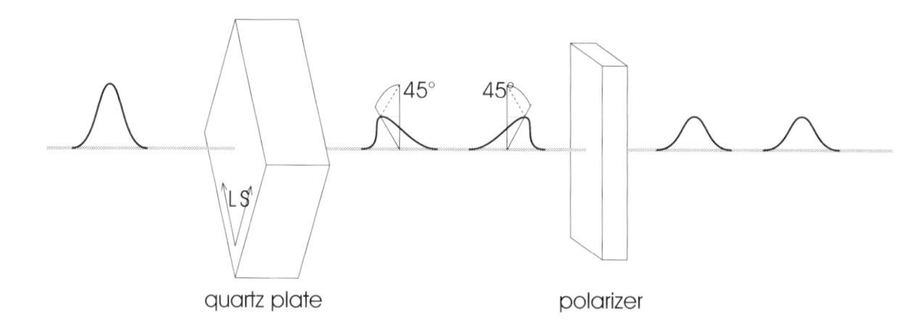

Figure 31.1: Generation of a double bump photon using a quartz plate and a polarizer (other interferometric setups are also possible).

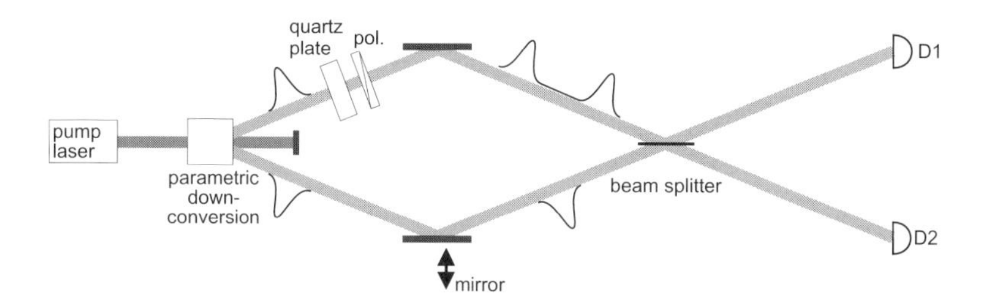

Figure 31.2: Schematic setup of photon pair interference using parametric fluorescence as source for entangled photon pairs, involving one double bump photon.

two input fibers of a 3 dB fiber coupler (the fiber optical equivalent of a 50/50 beamsplitter), and the coupler's output fibers are attached to passively quenched Ge-Avalanche-Photodiodes cooled by liquid nitrogen. The path length difference between signal and idler beam is varied by moving the collimator of the idler arm along the beam (denoted as the movable mirror in the schematic picture). A coincidence is counted when both detectors signal at the same time (within a certain time window), and the interference pattern is gathered by recording the coincidence rate in respect to the path length difference.

The resulting interference shows a more complex pattern (Fig. 31.3). Of special interest is the middle band of that pattern; this particular interference again results from the indistinguishability of the two processes tt and rr, but this time one of these processes involves the first bump of the double bump photon, whereas the other process involves the latter bump. Thus the phase between the bumps contributes to the phase between tt and rr, which allows to make this interference destructive (Fig. 31.3 top) or constructive (Fig. 31.3 bottom) or anything in between. This shows that the inner relative phase can be used as a carrier of information: one bit can be encoded into the photon by varying the birefringence (for instance electro-optically), and that bit can be regained by means of the photon pair interference.

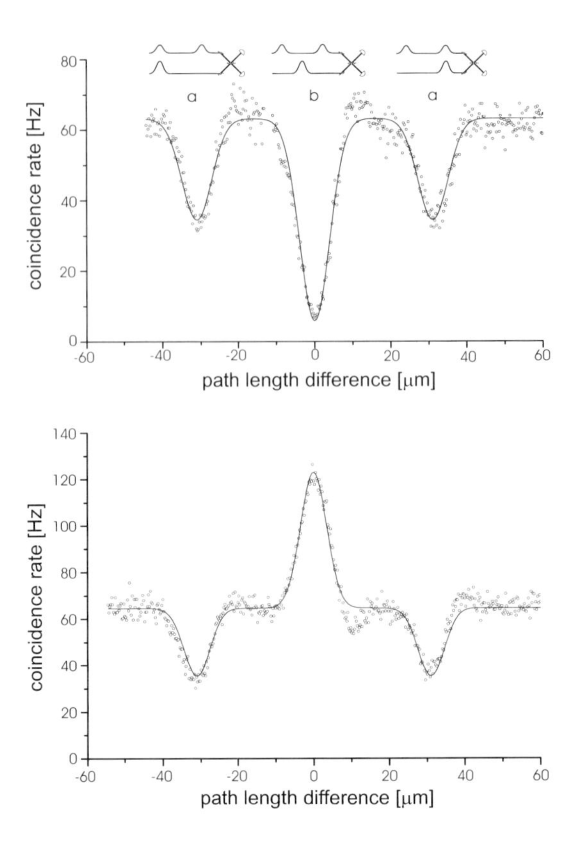

Figure 31.3: Photon pair interference involving one double bump photon; dots: measured data, solid line: calculated coincidence probability; top: destructive interference, (a) idler photon coincides with one of the signal photon's bumps, (b) idler photon coincides with the center of the signal photon; bottom: by changing the inner relative phase of the camel photon the central interference is changed from destructive to constructive.

31.3 The XOR Gate

We applied the same procedure of wave form modulation to the other photon (idler) of the photon pair, too, as shematically shown in Fig. 31.4. The signal photon's quartz plate is now about 10 mm, the idler photon's is 3 mm thick.

The resulting interference pattern is even more complex; depending on the parameters (i. e. the inner phases of both photons) nine interference features show up as in Fig. 31.5, the most prominent one of them again in the middle of the pattern and with (theoretically) full visibility. The phase of this interference (constructive or destructive) however depends on the inner phases of both pair photons in a way that resembles a logical XOR operation on the two logical bits encoded into the photons. If we identify the input bit "0" with an inner relative phase of zero (modulo 2π) and an input bit "1" as a phase of π, as well as a destructive interference (both photons in the same exit of the beamsplitter, no coincidence) as an output

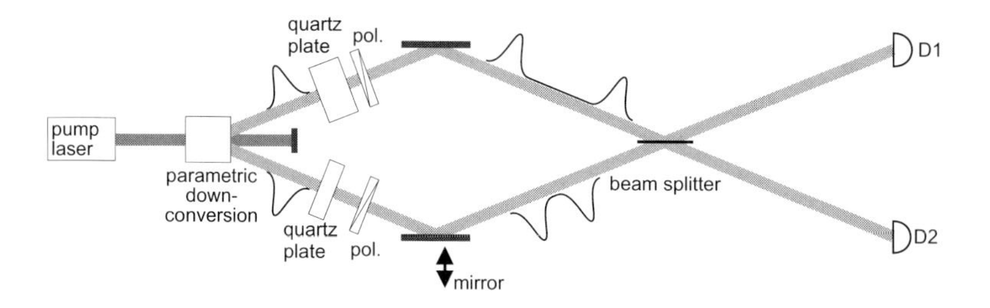

Figure 31.4: Schematic setup for photon pair interference involving two double bump photons.

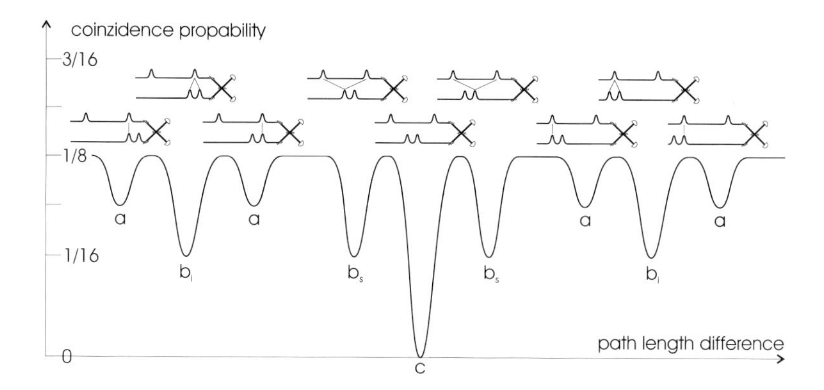

Figure 31.5: Schematic plot of the coincidence probability in a two photon interference pattern involving two double bump photons; (a) one bump of one photon coincides with one bump of the other, (b) one bump coincides with the center (between the bumps) of the other photon, (c) the centers of both photons coincide.

Table 31.1: How the interference shows the behaviour of the logical XOR operation

photon 1		photon 2		result	
bit	phase	bit	phase	interference	bit
0	0	0	0	destructive	0
0	0	1	π	constructive	1
1	π	0	0	constructive	1
1	π	1	π	destructive	0

of a logical "0" and constructive interference (coincidence) as an output of a "1", the logical XOR operation works as indicated in table 31.1.

The calculation of the coincidence probability follows a scheme by Steinberg, Kwiat, Chiao for chromatic dispersion [5]. The coincidence probability can be calculated as

$$P_c = \int d\omega_1 \int d\omega_2 \left| \langle 0 | \hat{a}_1(\omega_1)\, \hat{a}_2(\omega_2) | \Psi \rangle \right|^2 ,$$

where Ψ is the state of two photon light emerging from the parametric down-conversion with a gaussian spectral distribution $f(\omega)$ centered at ω_0,

$$\Psi = \int d\omega' \, f(\omega') \, |\omega_0 - \omega'\rangle_s |\omega_0 + \omega'\rangle_i \, .$$

The double bump state of the pair photons is represented by their annihilation operators

$$\begin{aligned} \hat{a}'_i(\omega) &= \frac{1}{2}\hat{a}_i(\omega)\, e^{i\omega\delta t_{iF}} + \frac{1}{2}\hat{a}_i(\omega)\, e^{i\omega\delta t_{iS}} \\ \hat{a}'_s(\omega) &= \frac{1}{2}\hat{a}_s(\omega)\, e^{i\omega\delta t_{sF}} + \frac{1}{2}\hat{a}_s(\omega)\, e^{i\omega\delta t_{sS}} \, , \end{aligned}$$

where the indices s and i indicate the signal or idler photon and the indices F and S indicate the faster and the slower bump of these photons, respectively. The signal and idler photons then are superimposed at the beam splitter

$$\begin{aligned} \hat{a}_1(\omega) &= \frac{i}{\sqrt{2}}\hat{a}'_s(\omega) + \frac{1}{\sqrt{2}}\hat{a}'_i(\omega) \\ \hat{a}_2(\omega) &= \frac{1}{\sqrt{2}}\hat{a}'_s(\omega) + \frac{i}{\sqrt{2}}\hat{a}'_i(\omega) \, . \end{aligned}$$

Calculation leads to

$$\begin{aligned} P_c = \; & \frac{1}{8} \quad + \frac{1}{8}e^{-\frac{\delta s^2}{a^2}}\cos\omega_0\delta s + \frac{1}{8}e^{-\frac{\delta i^2}{a^2}}\cos\omega_0\delta i \\ & + \frac{1}{16}e^{-\frac{(-\delta s-\delta i)^2}{a^2}}\cos\omega_0(\delta s - \delta i) + \frac{1}{16}e^{-\frac{(+\delta s-\delta i)^2}{a^2}}\cos\omega_0(\delta s + \delta i) \\ & - \frac{1}{32}e^{-\frac{4(\delta t_{sS}-\delta t_{iL})^2}{a^2}} - \frac{1}{32}e^{-\frac{4(\delta t_{sS}-\delta t_{iS})^2}{a^2}} \\ & - \frac{1}{32}e^{-\frac{4(\delta t_{sL}-\delta t_{iL})^2}{a^2}} - \frac{1}{32}e^{-\frac{4(\delta t_{sL}-\delta t_{iS})^2}{a^2}} \\ & - \frac{1}{16}\cos\omega_0\delta s\left[e^{-\frac{4(\delta t_s-\delta t_{iS})^2}{a^2}} + e^{-\frac{4(\delta t_s-\delta t_{iL})^2}{a^2}}\right] \\ & - \frac{1}{16}\cos\omega_0\delta i\left[e^{-\frac{4(\delta t_i-\delta t_{sS})^2}{a^2}} + e^{-\frac{4(\delta t_i-\delta t_{sL})^2}{a^2}}\right] \\ & - \frac{1}{8}e^{-\frac{4(\delta t_s-\delta t_i)^2}{a^2}}\cos\omega_0\delta s\cos\omega_0\delta i \end{aligned}$$

with the abbreviations $\delta t_s = \frac{\delta t_{sS}+\delta t_{sL}}{2}$ and $\delta t_i = \frac{\delta t_{iS}+\delta t_{iL}}{2}$ for the centers (between the bumps) of each photon; the path length difference used in the interference plots is $\delta t_i - \delta t_s$. The last term corresponds to the central interference feature, $\omega_0\delta s$ and $\omega_0\delta i$ are the inner phases of the signal and idler photon, respectively, and the term $\cos\omega_0\delta s\cos\omega_0\delta i$ is responsible for the phase dependence of this interference. If the idler photon is not a double bump photon (as in the previous section), δi is zero.

Experimental results for two exemplary cases are shown in Fig. 31.6.

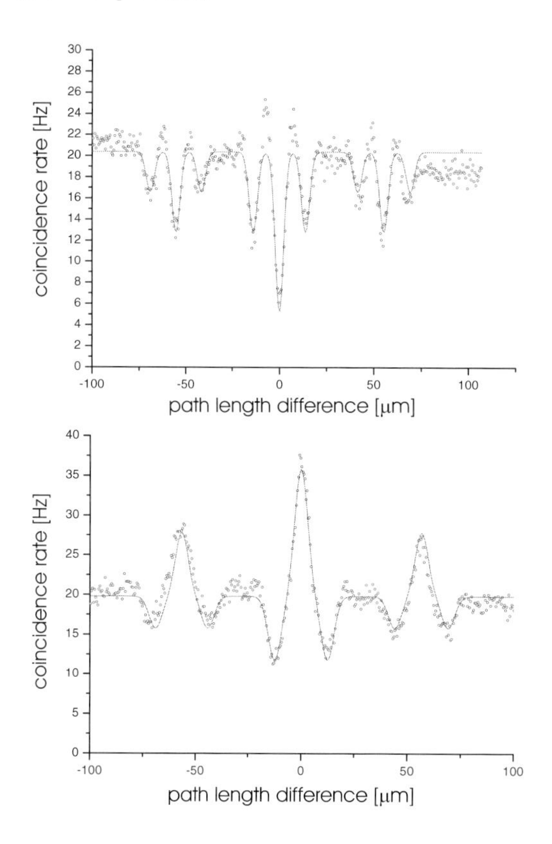

Figure 31.6: Two exemplary two photon interference patterns involving two double bump photons; top: both phases are zero (0 XOR 0 = 0); bottom: signal phase zero, idler phase π (0 XOR 1 = 1).

31.4 Quad Bump Photons

It is even possible to insert another quartz plate and polarizer into a photon's beam; therefor we placed both quartz plates (with the polarizers) into the signal beam. If the quartz plate's thicknesses differ from another, the result is a photon with four bumps (since each of the two bumps from the first quartz plate is doubled again). Such a quad bump photon possesses two inner relative phases, and thus two bits of information can be stored that way into one photon.

Interestingly, the interference of a quad bump photon (as signal photon) with the unmodified idler photon shows (nearly) the same patterns as the interference of two double bump photons, and the calculated coincidence probabilities differ only in two signs. This means that the XOR feature is contained in this interference, too, the difference lies only in the fact that both input bits now are encoded into the same photon.

31.5 Outlook

Further calculations show that the photon pair interference of a quad bump and a double bump photon shows an even more complex pattern, but still has a central band with full visibility. This interference depends on all three involved inner phases (two from the quad bump photon, one from the double bump photon) and shows a behaviour that mimics a logical XOR operation on all three logical bits encoded into these phases. Thus, a XOR gate with three input bits can be constructed, and further considerations show that this scheme can be expanded to even higher order XOR gates.

Quantum optical XOR gates may find applications in quantum cryptology, for instance in an identification scheme where two parties A and B have to identify themselves together at a third party C, which is not allowed to gain knowledge of the keys of A and B.

References

[1] H. Fern, R. Loudon: "Quantum Theory of the Lossless Beam Splitter", Optics Communications, Vol. 64, p. 485 (1987)

[2] C. K. Hong, Z. Y. Ou, L. Mandel: "Measurement of Subpicosecond Time Intervals between Two Photons by Interference", Phys. Rev. Lett., Vol. 59, p. 2044 (1987)

[3] D. V. Strekalov, T. B. Pittman, Y. H. Shih: "What can we learn about single photons in a two-photon interference experiment", Phys. Rev. A, Vol. 57, p. 567 (1998)

[4] H. Becker, "Wellenformgesteuerte Photonenpaarkorreleation", diploma thesis, Fachbereich Physik, Universität Frankfurt am Main (1998)

[5] A. M. Steinberg, P. G. Kwiat, R. Y. Chiao: "Dispersion cancellation and high-resolution time measurements in a fourth-order optical interferometer", Phys Rev. A, Vol. 45, p. 6659 (1992)

32 Quantum Fiber Solitons — Generation, Entanglement, and Detection

Gerd Leuchs, Natalia Korolkova, Oliver Glöckl, Stefan Lorenz, Joel Heersink, Christine Silberhorn, Christoph Marquardt, and Ulrik L. Andersen

Max-Planck Forschungsgruppe
Institut für Optik, Information und Photonik
Universität Erlangen–Nürnberg
Germany

32.1 Introduction

Solitons are a special wave phenomenon observed in systems exhibiting a pronounced non-linear dynamical behavior. One such system is the optical fiber. Optical fiber solitons show a remarkable stability. They can be viewed as quasi-particles composed of many photons, held together by a material-mediated nonlinear interaction between the photons, which is the optical Kerr effect in the case of fibers. One may wonder whether the particle-like nature of solitons can be exploited for quantum communication in analogy to single photons. The answer to this question is complex.

- In principle quantum solitons can be used for quantum communication, but in practice the losses in present-day fibers pose a severe problem. The soliton loses one photon every few millimeters of fiber length or less. In addition, appropriate optical elements are missing, e.g., no beam splitter that works for solitons in the same way that a regular beam splitter works for photons, transmitting or reflecting the photon but not splitting it.

- Another interesting aspect — still out of reach — is the production of a quantum superposition state, a so-called "cat" state. Here, too, losses are a prohibitive factor, the critical figure being the attenuation the soliton experiences for a propagation length that produces a π nonlinear phase shift before the soliton loses the first photon. Losses in today's best fibers are still many orders of magnitude too high for the observation of this effect.

- Solitons and their nonlinear interaction in a fiber may, however, be used to create quantum entanglement between different pulses. Interestingly, only quantum solitons with a well-defined photon number are stable. Coherent solitons, generally considered to be the analog of a classical soliton, are not stable but show a pronounced nonlinear evolution. This aspect is the basis for studies on continuous variable quantum communication reported in this chapter. There are three different avenues to achieve soliton entanglement. First, the soliton as a composite particle develops a particular internal quantum noise structure, which manifests itself in quantum correlations between different spectral

Quantum Information Processing, 2nd Edition. Edited by Thomas Beth and Gerd Leuchs

ISBN: 3-527-40541-0

components. Second, two solitons of different speed cross each other while propagating in a fiber. The amplitude of one modifies the phase of the other and vice versa. This effect produces a different type of entanglement. Third, the nonlinear interaction is used to produce two amplitude noise reduced, squeezed solitons, which are made to interfere at a beam splitter. A pair of quantum entangled pulses emerges, provided the interference phase is set to the right value.

This chapter focuses on the last point, the production of quantum entanglement and potential applications in quantum communication. In particular, the first and third schemes for entanglement generation will be discussed. Both schemes are closely related concerning the multimode quantum correlations involved. In view of applications, it is desirable to make use of another resource, the polarization. This opens the path towards continuous variable quantum communication requiring direct detection only — an advantage that should not be underestimated.

32.2 Quantum Correlations and Entanglement

Quantum correlation properties of a field made up of two local subsystems or modes are basic requisites for many quantum information protocols. During recent years much effort has therefore been devoted to the understanding of this phenomenon, in particular to its categorization and quantification. A major part of these studies in the continuous variable regime has been based on the assumption that the Wigner function associated with the state is Gaussian. These states are of particular importance since almost all proposed applications in quantum communication (based on continuous variables) use such states [1]. Furthermore, Gaussian states are also routinely prepared in the lab, whereas non-Gaussian states are much more experimentally challenging to produce [2, 3].

An entangled state is a non-separable quantum state of a system, which cannot be represented as a product of two wave functions describing the two subsystems (see e.g. Ref. [4]). For discrete variables, i.e., for the Hilbert space of 2×2 and 2×3 dimensions, a necessary and sufficient criterion for separability is the requirement that the partial transpose $\hat{\rho}^T$ of the density matrix is positive, which is known as the Peres–Horodecki criterion [5]. This basic tool to study the separability problem in the discrete variable regime was extended to the continuous variable regime [6–8]. In this regime the partial transposition criterion is always sufficient to prove entanglement but only necessary for bipartite systems having a single Gaussian mode in one of the systems and N Gaussian modes in the other one [7, 9]. Considering more general cases with an arbitrary number of modes in each system, the positivity of the partial transposition does not ensure separability. In such cases, a further categorization of the composite system into distillable and bound entangled states is needed, and the above transposition criterion is not complete [8]. However, in Sec. 32.5, where the partial transposition criterion will be applied to experimental results, only two Gaussian modes are considered (one in each subsystem), rendering the criterion sufficient and necessary [6, 7].

We consider a two-mode Gaussian state with the canonical conjugate coordinates, $\hat{o} = (\hat{X}_1, \hat{Y}_1, \hat{X}_2, \hat{Y}_2)$, where $\hat{X} = \hat{a} + \hat{a}^\dagger$ and $\hat{Y} = -i(\hat{a} - \hat{a}^\dagger)$ are the amplitude quadrature and phase quadrature, respectively, which satisfy the canonical commutation relations $[\hat{X}_i, \hat{Y}_j] =$

$2i\delta_{ij}$, and where $\hat{a}$ and $\hat{a}^\dagger$ are the annihilation and creation operators, respectively. Such a state is uniquely determined by the displacement operator and the covariance matrix defined as

$$\gamma = \begin{pmatrix} C_{11} & C_{12} & C_{13} & C_{14} \\ C_{21} & C_{22} & C_{23} & C_{24} \\ C_{31} & C_{32} & C_{33} & C_{34} \\ C_{41} & C_{42} & C_{43} & C_{44} \end{pmatrix}, \tag{32.1}$$

with $C_{ij} = \frac{1}{2}\langle\{\hat{o}_i, \hat{o}_j\}\rangle - \langle\hat{o}_i\rangle\langle\hat{o}_j\rangle$, where $\{\hat{o}_i, \hat{o}_j\}$ is the anticommutator. Since the entanglement of the system is invariant with respect to displacement, its entanglement properties are completely described by this covariance matrix. By performing partial transposition on the state under scrutiny (which corresponds to leaving the position unchanged while inverting the momentum in the covariance matrix), a criterion for separability can be derived [7]. A slightly different strategy was applied by Duan et al. [6], who found that entanglement between two modes implies that

$$\frac{V(\hat{X}_1 + \hat{X}_2)}{V_{\text{coh}}(\hat{X}_1 + \hat{X}_2)} + \frac{V(\hat{Y}_1 - \hat{Y}_2)}{V_{\text{coh}}(\hat{Y}_1 - \hat{Y}_2)} < 2. \tag{32.2}$$

Here $\hat{X}_1 + \hat{X}_2$ and $\hat{Y}_1 - \hat{Y}_2$ are joint variables, which are ultimately the operators of the co-eigenstates of the two-mode system. $V(\ldots)$ are the variances of the joint variables and $V_{\text{coh}}(\ldots)$ are the corresponding coherent state variances. The derivation of this result is based on the assumption that the covariance matrix is in its standard form ($C_{11} = C_{22} = C_{33} = C_{44}$, $C_{31} = C_{13} = -C_{24} = -C_{42}$, otherwise $C_{ij} = 0$). Any two-mode covariance matrix can be transformed into this form by local unitary operations, which are known not to change the entanglement of the system [6, 7]. In the lab the transformation of the two-mode state into the standard form is, however, non-trivial. Therefore it might be interesting to transform the criterion into a form that complies with a generalized covariance matrix instead of transforming the actual state. We could then apply local unitary operations that transform the covariance matrix into a generalized form and correspondingly the criterion into a generalized form. For example, if an asymmetry is encountered between the $\hat{X}$ and $\hat{Y}$ quadratures in the individual modes, local squeezing operations are needed to symmetrize the state into the standard form. Instead of transforming the covariance matrix, we transform the criterion and end up with the product form [10]

$$\sqrt{\frac{V(\hat{X}_1 + \hat{X}_2)}{V_{\text{coh}}(\hat{X}_1 + \hat{X}_2)} \frac{V(\hat{Y}_1 - \hat{Y}_2)}{V_{\text{coh}}(\hat{Y}_1 - \hat{Y}_2)}} < 1. \tag{32.3}$$

This criterion is more general, it implies Eq. (32.2), and includes a larger range of covariance matrices.

Although the non-separability of a state is the most fundamental entanglement criterion, historically, continuous variable entanglement was first discussed in the spirit of the demonstration of the EPR paradox for continuous variables. The notion of continuous EPR-like correlations of the amplitude and phase quadratures was first introduced by Reid and Drummond in 1988 [11]. Along these lines EPR entanglement is present if the correlations between the two subsystems allow inference better than the shot-noise limit of conjugate quadrature fluctuations in one system based on the knowledge of those in the other system. The optimum inference errors are given by the conditional variances $V_{X_1|X_2}$ and $V_{Y_1|Y_2}$ for the position and momentum inferences, respectively. According to Reid and Drummond the domain of EPR entanglement is given by

$$V_{X_1|X_2} V_{Y_1|Y_2} < 1. \tag{32.4}$$

This is a very stringent criterion for entanglement, and it should be stressed that this criterion is not necessary as opposed to the non-separability criterion defined above. The measure of Reid is still very interesting since it proves the usefulness of the entanglement source for some quantum communication protocols. For example, for successful entanglement swapping, the entanglement involved has to satisfy the criterion (32.4). For experimental demonstration of entangled light beams, see Refs. [12–18].

32.3 Multimode Quantum Correlations

In the quantum domain, solitons were found to possess a complex spectral structure, which may be exploited to engineer fiber solitons with well-defined non-classical properties [19–22]. The multimode quantum correlations are present in higher-order solitons as well, and can even be enhanced there [23]. In an experiment, spectrally resolved quantum measurements of ultrashort pulses have stimulated new developments in multimode quantum optics and complemented the single-mode quantum picture of earlier years. The two effects so far most exploited experimentally are quantum intra-pulse and quantum inter-pulse photon-number spectral correlations arising due to the Kerr nonlinear interactions in an optical fiber. Interestingly, such quantum correlations within the pulse are also predicted to lead to intra-pulse entanglement, which might turn into a useful resource for quantum communication [24–26]. Another important point to be discussed is the appropriate quantum description of an optical pulse taking account of the whole complexity of its multimode quantum noise structure.

For a pulse propagating down a nonlinear fiber, the self-phase modulation process due to the Kerr nonlinearity generates sub-shot-noise intensity correlations between different spectral intervals of the pulse. By filtering out the spectral components with the correlated quantum noise and retaining those with the anticorrelated noise — as far as it can be done technically — one can reduce the overall photon-number noise in the pulse substantially below the coherent level, an effect known as squeezing by spectral filtering [19, 27–29]. The best photon-number squeezing via spectral filtering was measured to be -3.8 ± 0.2 dB (70 %) [30] and was predicted to reach up to ≈ -7 dB when optimizing the filter function [31]. Further improvement is expected from shaping the input pulse [26, 32]. Numerical studies of the quantum nonlinear Schrödinger equation [33, 34] have provided a qualitative explanation of soliton amplitude squeezing experiments with spectral filtering [19, 27, 29, 30] by calculating the intensity variances of the individual components in the pulse spectrum. In the experiments it was verified that the main squeezing mechanism is related to intra-pulse multimode photon-number correlation [19]. The correlation coefficients between the various spectral components can be determined experimentally and an intensity variance matrix, a full portrait of the intensity correlation distribution within the pulse, was recorded as in Fig. 32.1 [19]. The correlation coefficient $C(\hat{n}_i, \hat{n}_j)$ is defined as

$$C(\hat{n}_i, \hat{n}_j) = \mathrm{cov}(\hat{n}_i, \hat{n}_j) \Big/ \sqrt{V(\hat{n}_i)V(\hat{n}_j)},$$

where the normalization ensures $|C(\hat{n}_i, \hat{n}_j)| \le 1$. The variance matrix (Fig. 32.1a) has a characteristic butterfly pattern reflecting the internal quantum noise structure of the pulse [19]. The asymmetry of the correlation pattern with respect to the center wavelength is due to the intra-pulse Raman effect. The intensity correlations provide a clear insight into the mechanism of squeezing by spectral filtering: the noise reduction is obtained when the filter function is adjusted so that the negatively correlated parts in the central part of the spectrum are preserved while the positively correlated ones are cut off as much as possible.

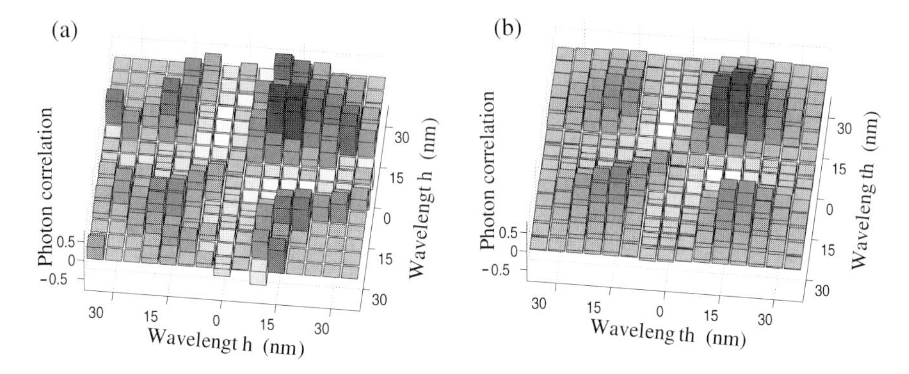

Figure 32.1: Map of intra-pulse quantum correlations, (a) measured and (b) reconstructed in a simple model. Each square represents the correlation coefficient between the wavelength components as indicated on the axes of the diagram. Elements with very large fluctuations (>0.8) are omitted. (a) The measured data points are displayed on one side of the diagonal and copied to the other side [$C(i,j) = C(j,i)$] for clarity. (From Spälter *et al.* [19].) Note that the diagonal elements are $V(\hat{n}_i)/V_{\mathrm{coh}}(\hat{n}_i)$. (b) The best fit to the results measured in [19] using a reconstructed three-mode variance matrix $V_{\hat{X}k,\hat{X}k'}$. (From Opatrny *et al.* [40].)

The Kerr interaction in the fiber also provides a nonlinear interaction between two solitons giving rise to cross-phase-modulation as two solitons of different speeds pass through each other. This permits delicate quantum measurements on one soliton, where one measures its photon number introducing only little disturbance to this variable, a technique referred to as quantum non-demolition (QND) measurement or back-action-evading (BAE) measurement [35, 36]. Multimode inter-pulse photon-number correlations that arise between spectral domains during a two-color soliton collision and disappear after the collision is complete have been predicted theoretically and exploited in a recent successful QND experiment [21, 37]. The dominant effect is an enhancement of the spectral separation of the two solitons. The resulting frequency shift of one pulse depends on the photon number and relative velocity of the collision partner. Using spectral filtering techniques, the spectral changes can be transformed into a direct measurement of the photon number. For the first time in fiber QND detection, the QND criteria were clearly fulfilled in this experiment based on the transient inter-pulse spectral intensity correlations [37]. Note that QND interactions potentially represent an important requisite for quantum information processing. For example, entanglement generated in the QND cross-coupling can be used for quantum teleportation [38]. Recently, the idea of entanglement generation in soliton collisions was extended to the continuous interaction of two solitons [39]: photon-number entangled soliton pairs are predicted when two time-separated single-component temporal solitons experience continuous nonlinear interaction in a fiber.

For a complete quantum description of an optical pulse, one thus has to use a multimode density matrix. The question is this: How many modes are necessary for such a description? If one works with *monochromatic* modes, then an infinite number of modes would be needed, which is impracticable. On the other hand, one can construct different sets of modes as linear combinations of the monochromatic modes (see Ref. [40] and references therein). These modes (which are however *non-monochromatic*) may be more suitable for the quantum description of pulses. In particular, it would be useful to find such a modal structure so that the quantum state of the pulse can be described by the density operator $\hat{\varrho} = \hat{\varrho}_{\mathrm{exc}} \otimes \hat{\varrho}_{\mathrm{vac}}$, where $\hat{\varrho}_{\mathrm{exc}}$ is some non-trivial density operator of a few modes and $\hat{\varrho}_{\mathrm{vac}}$ is the vacuum state density operator of all the remaining modes. Thus, we could work with a concise description of the pulse by means of $\hat{\varrho}_{\mathrm{exc}}$ [40]. Knowing the answer to the question of what is the complete quantum description of a pulse is very helpful in quantum information processing: one would be able to fully utilize the squeezing and entanglement properties of the quantum sources. After forming a soliton pulse in a fiber, one can determine its multimode quantum state in the most comprehensive way. One can then optimize the medium properties to achieve maximum squeezing, or maximum purity of an entangled state. Working with pairs of correlated pulses, one could apply a proper measurement scheme and use entanglement criteria for multimode bipartite Gaussian states (see Refs. [6–8] and Sec. 32.5) to check whether the pulse pair is entangled or separable. It would also be possible to better understand the influence of the medium on the propagating pulse: provided that the output pulse is deformed, what happens with the quantum information carried by the pulse? Is it washed out by decoherence processes or by an eavesdropper, or is it just unitarily transformed into other modes of the same pulse? A simple and correct measurement and description of the multimode state is highly desirable.

Important properties of multimode quantum states are given by the vectors of the quadrature mean values, $\overline{X}_j \equiv \mathrm{Tr}[\hat{\varrho}\hat{X}_j]$ and by the variance matrices V, given as

$$V_{\hat{X}k,\hat{X}'k'} = \tfrac{1}{2}\langle\{(\hat{X}_k - \overline{X}_k),\ (\hat{X}'_{k'} - \overline{X'}_{k'})\}\rangle, \tag{32.5}$$

where index k denotes a mode k in a new set of non-monochromatic modes. Figure 32.1b shows the reconstruction of the measured intra-pulse correlation structure using the non-monochromatic mode decomposition to calculate the $V_{Xk,X'k'}$ and the corresponding photon-number variances and covariances [40]. For a complete description of the experimental results measured on optical solitons, it appears that a very small number of non-monochromatic modes is enough (Fig. 32.1). The results cannot be described by means of a single-mode field, but already three non-monochromatic modes were sufficient to reproduce the experimental data of [19]. The "butterfly" pattern of the measured photon covariances is easy to interpret in this approach: three different non-monochromatic modes overlap and contribute with their fluctuating in-phase quadratures to the photon statistics [40].

Provided that the pulses are of Gaussian statistics, a complete quantum description of the pulse can be done by means of the multimode variance matrix, and the elements of the matrix can be determined using homodyne detection with specially shaped local oscillator pulses. In principle, one can also separate individual modes to match the requirements of the pulse user, e.g., to prepare a single mode maximally squeezed field, or to extract a maximally entangled two-mode field, etc. However, currently this is still an experimental challenge.

32.4 Generation of Bright Entangled Beams

The quantum intra-pulse correlation within fiber solitons and their rich internal noise structure discussed in a previous section give rise to a whole range of quantum effects, which are worth further studies for quantum communication and other applications. However, such studies are still in their infancy and so far no efficient experimental methods are known to exploit the internal entanglement of soliton pulses for quantum information purposes. Even the impact on soliton squeezing of a variety of multimode correlations within the pulse is not fully understood. All existing experiments do not seem to take full advantage of the internal quantum structure of soliton pulses.

Currently, the most efficient quantum communication schemes with fiber solitons are based on a single-mode model of the involved optical fields. The use of optical solitons here is justified by two main practical reasons. First, fiber solitons are intense ultrashort pulses that take full advantage of the optical nonlinearity in a fiber to generate non-classical light. Second, the stability of a soliton during the propagation allows one to achieve high visibility levels while using quantum interferometers, which are important building blocks in quantum information systems with continuous variables. For the generation of bright entangled beams, the Kerr nonlinearity in an optical fiber is employed to produce two independently amplitude squeezed beams. To create continuous variable entanglement, these squeezed light fields are made to interfere at a balanced 50/50 beam splitter [41]. The quality of the entanglement generated by this interference depends on the initial squeezing of the two input beams.

In our experiment we use pulses of about 130 fs at a center wavelength of 1530 nm and at a repetition rate of 82 MHz produced by an optical parametric oscillator (Spectra Physics, OPAL) pumped by a Ti:sapphire laser (Spectra Physics, Tsunami). These pulses are injected into an asymmetric fiber Sagnac interferometer (Fig. 32.2). This is operated simultaneously at two orthogonal polarizations to obtain the two squeezed beams with equal optical power for the entanglement generation. The Sagnac loop consists of an 8 m long polarization-maintaining fiber (3M, FS-PM-7811) and an asymmetric 90/10 beam splitter. At the output of the Sagnac interferometer, up to 4 dB of amplitude squeezing can be observed [15], depending on the detection process and the losses. To generate quadrature entanglement, the two amplitude squeezed outputs of the Sagnac interferometer are superimposed at a 50/50 beam splitter. The interference phase at the 50/50 beam splitter is such that the output beams have equal optical power. The entanglement of the output beams is then verified by the detection of quantum correlations for two conjugate variables, the amplitude and phase quadratures.

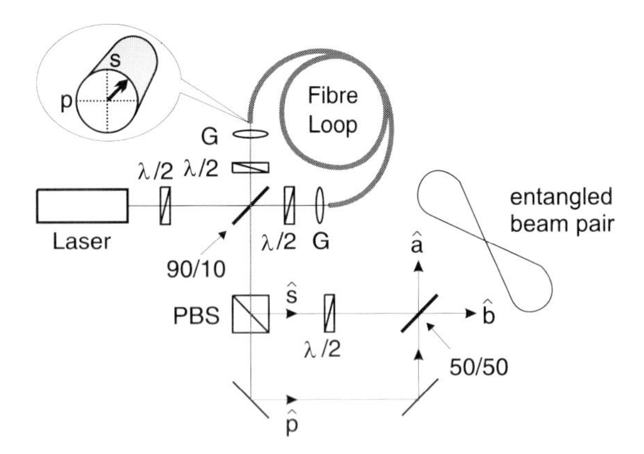

Figure 32.2: Schematic of the experimental setup for intense beam entanglement generation.

32.5 Detection of Entanglement of Bright Beams

32.5.1 Sub-shot-noise Phase Quadrature Measurements on Intense Beams

While the detection of the amplitude anticorrelations can be performed in direct detection with a pair of balanced detectors, the detection of the phase correlations is more involved. The fluctuations of a light field can be interpreted as the result of the beating of a carrier wave with sideband modes [42,43]. The relative phase between the carrier and the sideband modes determines the quadrature that is measured. A homodyne detector [44] is usually used to perform such phase-sensitive measurements by interference of the signal beam and hence probing the sidebands with a local oscillator, which has to be much brighter than the signal beam. For intense signal beams, this gives rise to technical difficulties, as the high intensities may saturate the detectors. Alternatively, phase measurements of bright beams can be achieved via the reflection of light from a cavity [45]. A frequency-dependent phase shift is introduced due to multiple beam interference to rotate the bright carrier with respect to the sidebands. This technique was used in some early quantum optical experiments using fibers [46,47].

For ultrashort light pulses, however, the requirements for the dispersion properties of the resonator are quite demanding. We use another approach to measure the fluctuations of the phase quadrature at a certain sideband frequency without an external local oscillator or a cavity. The scheme is thus suitable for intense beams. A Mach–Zehnder interferometer with an unbalanced arm length reminiscent of that used by Inoue and Yamamoto [48] to measure the longitudinal mode partition noise is used to perform quantum optical measurements of the phase quadrature below the shot-noise level [49]. In this interferometer, a phase shift between the carrier and the sidebands is introduced due to two-beam interference. The setup allows easy switching between measurements of the phase quadrature and the amplitude quadrature and is used to fully characterize the quadrature entanglement of the intense, pulsed beams described above.

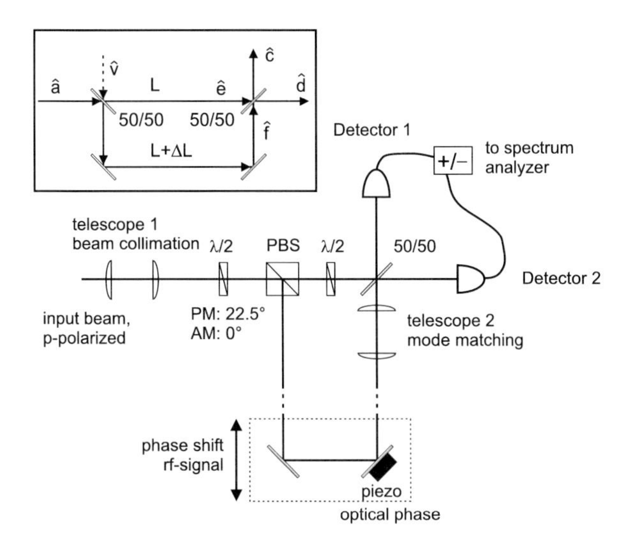

Figure 32.3: Interferometric setup for experimental characterization of bright entangled beams (AM: amplitude measurement; PM: phase measurement).

The setup of the interferometer is depicted in Fig. 32.3. The input mode of the interferometer (labeled a) is split into two parts (e and f) and, after delaying one beam with respect to the second beam by ΔL, the beams interfere on a second beam splitter. The delay is such that a phase shift of π is introduced at the measurement sideband frequency Ω, in our case 20.5 MHz, corresponding to a delay ΔL of 7.3 m. The optical phase is adjusted such that the two output ports (c and d) of the interferometer are equally intense. The fluctuations of the photon number of the two output modes are then recorded in direct detection. The difference signal of the photocurrents corresponds to the fluctuations of the phase quadrature of the input state at the frequency Ω [49]. The first beam splitter consists of a $\lambda/2$-plate and a polarizing beam splitter; thus switching between amplitude measurements and phase measurements is possible. In the former case, all light propagates through one arm of the interferometer (corresponding to a balanced detection scheme); while in the latter case, equal intensity in both arms of the interferometer is set up for phase measurements.

We directed the outputs of our entanglement source into two such phase-measuring interferometers to look for quantum correlations. These were checked by recording the noise level of the difference (sum) signal of each of the photocurrents of the two entangled beams for the phase (amplitude) measurement. The phase correlations were measured to be -1.2 dB below the shot-noise level, corresponding to a squeezing variance $V(\hat{Y}_1 - \hat{Y}_2)/2 = 0.76 \pm 0.02$, whereas the amplitude correlations were at -2.0 dB, corresponding to a squeezing variance $V(\hat{X}_1 + \hat{X}_2)/2 = 0.63 \pm 0.02$ (indices 1 and 2 refer to the two entangled modes). Plugging the measured numbers in the non-separability criterion of Eq. (32.3), we have

$$\Delta = [V(\hat{X}_1 + \hat{X}_2)/2 \times V(\hat{Y}_1 - \hat{Y}_2)/2]^{1/2} = (0.76 \times 0.63)^{1/2} = 0.69 < 1,$$

thus indicating the state as being non-separable. These correlations have been observed, although the individual beams are very noisy: the amplitude noise of the individual beams was measured to be 21.1 dB and the phase noise 20.8 dB above the respective shot-noise levels.

Using the values measured above, we can also fully characterize the state in terms of the covariance matrix of our system (see Eq. 32.1), assuming that there are no cross-quadrature-correlations present either within the two individual beams ($C_{12} = C_{21} = C_{34} = C_{43} = 0$) or between the two beams ($C_{32} = C_{41} = C_{14} = C_{23} = 0$). In units of shot noise, the matrix is given by

$$\gamma = \begin{pmatrix} 128.8 & 0 & -128.2 & 0 \\ 0 & 120.2 & 0 & 119.4 \\ -128.2 & 0 & 128.8 & 0 \\ 0 & 119.4 & 0 & 120.2 \end{pmatrix} . \qquad (32.6)$$

The relatively low degree of squeezing and correlations is due to optical losses in the interferometers (several optical components and a finite visibility of the interferences due to imperfect mode matching) and the simultaneous use of four imperfect detectors with non-optimum balancing for the measurements. When performing the amplitude measurements with a pair of balanced detectors, up to 4 dB of noise reduction can be observed, while having no access to phase quadrature measurements. The amplitude anticorrelations have also been measured earlier [15] in direct detection; however, the phase quadrature information was not directly accessible.

Having this interferometer as a tool to perform phase measurements, several quantum information and communication protocols can be performed using bright beams where phase measurements are required. The full verification of the entanglement swapping experiment with intense, pulsed light beams [50] seems to be possible (see Sec. 32.6) as well as the implementation of a quantum key distribution scheme relying on quadrature entanglement [51]. The experimental demonstration of a continuous variable analog of quantum erasing has recently been performed successfully [52] employing this interferometer to access the phase quadrature information.

32.5.2 Direct Experimental Test of Non-Separability

The criterion of Duan et al. [6] and Simon [7] introduced in Sec. 32.2 has a distinct advantage for experimental quantum communication as it can be expressed in terms of a single directly observable quantity [53]. The non-separability criterion in the form (32.2) can be readily observed in an experiment. In Fig. 32.4 the entanglement source is treated as a black box with two output beams. It could be any source of continuous variable entanglement, an OPO setup, two interfering squeezed beams, or any other generation process. The particular setup of Fig. 32.4 is designed for the intense output beams 1 and 2 of the same coherent amplitude α. The two relevant output modes are superimposed at a beam splitter. The field operators of the beams after this interference are denoted $\hat{c}$ and $\hat{d}$. To adjust the relative interference phase, a variable phase shift θ can be introduced in one of the arms. The balanced detection is performed in one of the output arms, $\hat{c}$ or $\hat{d}$. Difference and sum photocurrents are recorded and the relative phase θ is scanned. The difference photocurrent provides the shot-noise level reference. The sum photocurrent delivers the amplitude noise variance of the signal in the output $\hat{c}$ or $\hat{d}$. For $\theta = \pi/2$ this measured normalized noise variance is directly proportional to the non-separability criterion (Eq. 32.2) for continuous variables:

$$V(X_{\text{c,d}}) = \frac{1}{2}\left[\frac{V(\hat{X}_1 + \hat{X}_2)}{V_{\text{coh}}(\hat{X}_1 + \hat{X}_2)} + \frac{V(\hat{Y}_1 - \hat{Y}_2)}{V_{\text{coh}}(\hat{Y}_1 - \hat{Y}_2)}\right]. \tag{32.7}$$

If the interfering amplitude squeezed beams are of equal squeezing, the output entangled beams have symmetric circular uncertainty regions, thus $V(\hat{X}_1 + \hat{X}_2) = V(\hat{Y}_1 - \hat{Y}_2) = V_{\text{sqz}}$. Then Eq. (32.7) reduces to $V(X_{c,d}) = V_{\text{sqz}}$ and the non-separability condition reads $V_{\text{sqz}} < 1$ independent of the relative interference phase θ. Using (32.7), the non-separability of the state can be verified experimentally in a single measurement [15, 53].

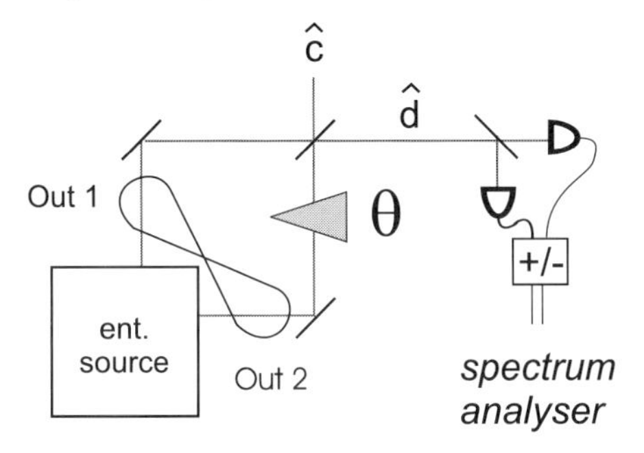

Figure 32.4: Direct detection of the non-separability criterion.

The experimental setup for the direct test of the non-separability criterion can be extended to a multi-purpose quantum interferometer for measurements below the quantum limit and for quantum communication [22]. Quantum interferometry with bright soliton pulses covers a wide range of applications from high-precision measurements of small phase modulation and quantum dense coding [53–55] to quantum teleportation and entanglement swapping [22, 41, 53, 56–60], which involves more complex forms of entanglement.

32.6 Entanglement Swapping

In this section, we first consider the generation of more complex forms of entanglement, namely multipartite entanglement [61–63]. In analogy to discrete variable entanglement, where one can have quantum correlations between more than two photons (e.g., the GHZ states [64–66]), one can construct more than two continuous modes, which show mutual correlations. A possible setup capable of producing four-partite quantum correlated states is shown in Fig. 32.5a.

Alice and Bob both have a source of two-mode entangled continuous variable beams. They send one of their beams to Claire while keeping the other. Claire interferes the two beams — which have never interacted before and stem from independent sources — at a 50/50 beam splitter with an interference phase such that the two output modes are equally intense.

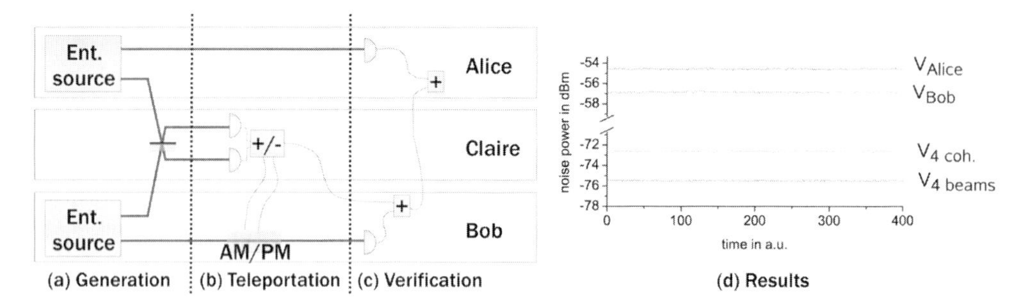

Figure 32.5: Schematic setups for (a) multipartite correlated state, (b) teleportation of entanglement, and (c) verification of teleportation of amplitude correlations. (d) Experimental results for the amplitude correlations observed.

The two resulting modes, together with Alice's and Bob's remaining modes, show four-partite correlations. For the experimental realization [50] of the multipartite correlated state, two entanglement sources were used as described in Sec. 32.4. An interference visibility of 85 % was achieved at Claire's beam splitter, although the two modes originated from different entanglement setups. All detected photocurrents were recorded by spectrum analyzers (resolution bandwidth 300 kHz, video bandwidth 30 Hz) at a frequency of 17.5 MHz. We found that any combinations of up to three photocurrents give a sum variance higher than the shot-noise level of four coherent states. Only if all four photocurrents are added does the variance of the photocurrents I,

$$V_{4\,\text{beams}} = V(I_{\text{Alice}} + I_{\text{Claire1}} + I_{\text{Claire2}} + I_{\text{Bob}}) < 4V_{\text{coh}}, \tag{32.8}$$

drop 3 dB below the shot-noise limit. This is a clear hint that the state is indeed four-fold correlated, and not only pair- or triple-wise. A complete characterization of the state requires the measurement of the correlation matrix, which is technically more challenging. A simple criterion for continuous variable multipartite entanglement that can be applied to our experiment is still not available, though their development advances steadily [67].

Figure 32.5b shows how the setup for multipartite correlations can be transformed into a continuous variable quantum teleportation experiment. Claire has to perform a continuous variable Bell measurement to transfer the quantum properties of the beam she received from Alice onto Bob's beam at his site [13, 59, 68]. A case of special interest arises if Alice's beam itself is entangled with another beam, here with Alice's second beam, which she keeps. If Bob performs a unitary operation depending on Claire's measurement outcome, he ends up with a beam that is entangled with Alice's remaining beam, although these two beams have never interacted directly. This "entanglement swapping" or "teleportation of entanglement" is especially useful in quantum communication protocols to establish entanglement between parties that initially did not share quantum correlations [58, 60, 69]. The unitary transformation consists of an amplitude modulation on Bob's beam, which is proportional to Claire's added photocurrents, and a phase modulation proportional to Claire's subtracted photocurrents. When checking for amplitude correlations between Alice and Bob, one can leave out the phase modulation. As the photocurrents of Alice's and Bob's detectors are proportional to their beams' amplitudes, it is sufficient to apply the amplitude modulation not to Bob's beam directly, but to Bob's photocurrent (Fig. 32.5c). From the expected correlations between Alice and Bob

$$V(I_{\mathrm{Alice}} + I_{\mathrm{Bob\,with\,correction}}) < 2V_{\mathrm{coh}}, \tag{32.9}$$

one can infer the correlations between their photocurrents, which are given by Eq. (32.8). This correlation was shown experimentally. The variances of the added photocurrents, as well as for Alice's and Bob's own beams, are shown in Fig. 32.5d. The high individual noise levels of Alice's and Bob's photocurrents are labelled V_{Alice} and V_{Bob}. Due to the non-minimum uncertainty character of the squeezed beams used in the experiment, an extra penalty of 3 dB variance appears in the final correlations. Thus, the correlations $V_{4\,\mathrm{beams}}$ are exactly at the shot-noise level of two coherent beams, when entanglement sources with 3 dB of initial correlations are used. Although no direct sub-shot-noise correlations are recorded, one has to compare the result with the classical expectation. In the classical case, two extra units of vacuum noise have to be added, one coming from Claire's measurement and one from the modulation of the beam [13]. Thus, the correlations measured in the experiment cannot be explained by classical means. They can be increased by improving the initial squeezing of the entanglement sources, increasing the visibility of Claire's interference, and improving the balance of the detection system. The last point is most crucial, as four detectors are involved, each of which has to detect a highly noisy state (more than -58 dBm), while the correlations sum up to a very low noise level (lower than -75 dBm). With more advanced detection tools, like the phase-sensitive interferometer (see Sec. 32.5.1), it will be possible to completely characterize the entanglement of Alice's and Bob's beams.

32.7 Polarization Variables

The continuing development of the field of quantum communication has sparked a search for new and novel resources. Prominent among these are the quantum Stokes operators, which describe the polarization state of light. These parameters can be measured in direct detection, in particular without cumbersome phase measurements such as interferometers or local oscillators as is the case with the quadrature variables [70]. In quantum cryptography and other quantum communication protocols requiring the measurement of conjugate variables, this is of great value, as an extra phase reference must not be included in the communication. Research into non-classical continuous variable polarization states began in 1993 with the theoretical proposal [71], although polarization squeezing had already been unwittingly produced in 1987 [72]. In more recent years many advances in polarization squeezing [73–75] and polarization entanglement [76, 77] have greatly increased our knowledge of non-classical polarization states and have paved the way for applications. The polarization state of light also shows great promise for quantum memory [78], as the atomic spin operators and the polarization operators both obey the SU(2) Lie algebra and thus quantum states can be transferred directly between these systems [79, 80].

The classical Stokes parameters are a well-known description of the polarization state of light [81, 82]. In direct analogy to these classical parameters, the quantum Stokes operators are [70, 83]

$$\hat{S}_0 = \hat{a}_x^\dagger \hat{a}_x + \hat{a}_y^\dagger \hat{a}_y, \qquad \hat{S}_1 = \hat{a}_x^\dagger \hat{a}_x - \hat{a}_y^\dagger \hat{a}_y,$$
$$\hat{S}_2 = \hat{a}_x^\dagger \hat{a}_y + \hat{a}_y^\dagger \hat{a}_x, \qquad \hat{S}_3 = i(\hat{a}_y^\dagger \hat{a}_x - \hat{a}_x^\dagger \hat{a}_y), \tag{32.10}$$

where $\hat{a}_{x/y}$ and $\hat{a}_{x/y}^\dagger$ refer to the photon annihilation and creation operators of two orthogonal polarization modes x and y, $\hat{S}_0$ corresponds to the beam intensity, and $\hat{S}_1$, $\hat{S}_2$, and $\hat{S}_3$ describe the polarization state. These operators commute as follows:

$$[\hat{S}_0, \hat{S}_i] = 0, \qquad [\hat{S}_i, \hat{S}_j] = 2i\hat{S}_k, \qquad i, j, k = 1, 2, 3. \tag{32.11}$$

Further commutators are found by cyclic permutation of the indices. These relations are similar to those for angular momentum operators. Resulting from these commutators are a set of state-dependent uncertainty relations and thus conjugate variable pairs. Polarization squeezing is generated by decreasing the noise in one of the conjugate parameters below the minimum uncertainty state (MUS) limit given by an equal sign in the uncertainty relations. A polarization squeezed state can be easily visualized if we consider a state with only one non-zero Stokes parameter, say S_2. The conjugate or squeezing variables are then $\hat{S}_1$ and $\hat{S}_3$, which define a plane orthogonal to $\hat{S}_2$. As the beam is S_2 polarized, this plane is "dark", as all Stokes measurements in this plane have a zero mean value [84]. This is mathematically similar to vacuum squeezing as in [85] where the quadrature variables have been measured.

In our experiment we used the efficient amplitude squeezing source afforded by the Kerr effect ($\chi^{(3)}$ nonlinearity) of a glass fiber. Ultrashort pulses (130 fs) at 1495 nm from a Cr:YAG^{4+} laser were used to simultaneously produce two orthogonally polarized, independent amplitude squeezed pulses in a double asymmetric (93:7 splitting ratio) fiber Sagnac interferometer with 14.2 m of polarization-maintaining fiber (FS-PM-7811 from 3M). This setup is similar to that in Fig. 32.2; however, there the fiber birefringence was compensated after the fiber. In this experiment the birefringence was pre-compensated to minimize squeezing losses. Polarization squeezed pulses were generated by simply overlapping the two amplitude squeezed pulses after the Sagnac loop with a constant relative phase. This was actively controlled to give a stable polarization squeezing source.

The mean values of the Stokes operators were $\langle \hat{S}_2 \rangle \neq 0$ and $\langle \hat{S}_1 \rangle = \langle \hat{S}_3 \rangle = 0$, which determined the form of the commutators. We saw polarization squeezing of -3.4 dB in $\hat{S}_1$. Thus $\hat{S}_3$, the conjugate variable of $\hat{S}_1$, is anti-squeezed but also carries the extra noise inherent to fiber propagation, with a value of $+23.5$ dB. Traces below the shot-noise limit were also seen in $\hat{S}_0$ and $\hat{S}_2$, but these correspond to amplitude squeezing below the coherent limit and not polarization squeezing below a polarization MUS limit.

This polarization squeezing source can be used to generate polarization entanglement [70]. Polarization entanglement refers to a quantum non-separable state of two light beams and implies correlation of the quantum uncertainties between one or more pairs of Stokes operators of two (spatially separated) beams. It has the nature of two-mode squeezing. This entanglement has also been observed in experiments with OPOs [76]. Our scheme was designed to use minimal resources, and thus only a 50/50 beam splitter needed to be placed in the polarization squeezed beam [77]. The variances of the Stokes parameters of the two resulting beams (A and B) were measured separately and correlations were seen in the measurements of the conjugate $\hat{S}_1$ and $\hat{S}_3$ variables.

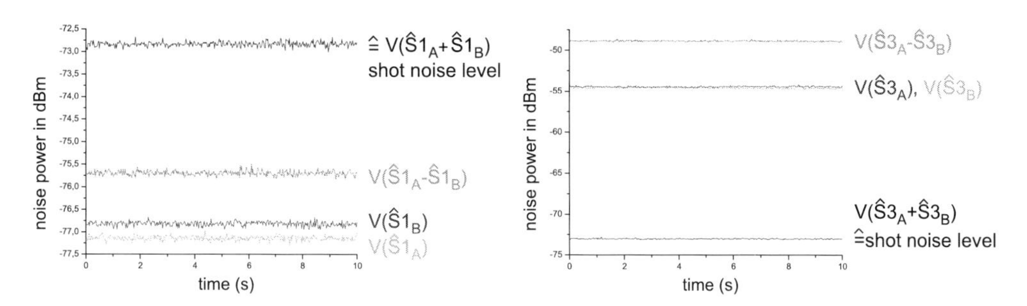

Figure 32.6: Noise traces at 17.5 MHz over 10 s of a polarization squeezed beam mixed with a coherent vacuum. Traces of $\hat{S}_1$ (left) and $\hat{S}_3$ (right) for subsystems A and B, and the sum and difference of these subsystems. The difference of $\hat{S}_1$ and the sum of $\hat{S}_3$ exhibit quantum correlations. Note that the shot-noise level shown is for two beams; shot noise for single beams lies 3 dB lower.

Our experimental results, carried out with the polarization squeezing source described above, are seen in Fig. 32.6. The first graph shows traces of $\hat{S}_1$ for the individual subsystems A and B as well as their sum and difference. If the two beams A and B were uncorrelated, the difference noise signal, $V(\hat{S}_{1A} - \hat{S}_{1B})$, would lie 3 dB above the individual squeezed traces. It was seen that the difference variance was only 1.3 dB above these variances, indicating a strong correlation; this trace is itself squeezed by -2.9 dB below the shot-noise limit. The traces for $\hat{S}_3$ are seen in the second graph; here the individual traces are very noisy, measured to be greater than -55 dBm. However, the sum variance, $V(\hat{S}_{3A} + \hat{S}_{3B})$, lies far below, at the two-beam shot-noise level, showing the strong correlation of the signals. Despite the $\hat{S}_3$ sum variance only being at the shot-noise level, we find that this state violates the non-separability criterion (Eq. 32.2) extended for the Stokes operators

$$\frac{V(\hat{S}_{1A} - \hat{S}_{1B})}{V_{\text{coh}}(\hat{S}_{1A} - \hat{S}_{1B})} + \frac{V(\hat{S}_{3A} + \hat{S}_{3B})}{V_{\text{coh}}(\hat{S}_{3A} + \hat{S}_{3B})} = 0.52 + 1 < 2. \tag{32.12}$$

Polarization squeezed and entangled states can now be stably generated, and thus are a viable, directly detectable alternative to the amplitude and phase quadratures. They can be used advantageously in many quantum protocols which require the measurement of conjugate variables, such as quantum cryptography. The transfer of quantum polarization states to matter also holds promise for quantum memory.

References

[1] S. L. Braunstein, and A. K. Pati (Eds.), *Quantum Information Theory with Continuous Variables*, Kluwer, Dordrecht (2002).

[2] J. Wenger, R. Tualle-Brouri, and P. Grangier, *Phys. Rev. Lett.* **92**, 153601 (2004).

[3] A. I. Lvovsky, H. Hansen, T. Aichele, O. Benson, J. Mlynek, and S. Schiller, *Phys. Rev. Lett.* **87**, 050402 (2001).

[4] R. F. Werner, *Phys. Rev. A* **40**, 4277 (1989).

[5] A. Peres, *Phys. Rev. Lett.* **77**, 1413 (1996);
M. Horodecki et al., *Phys. Lett. A* **223**, 1 (1996);
P. Horodecki, *Phys. Lett. A* **232**, 333 (1997).

[6] L.-M. Duan, G. Giedke, J. I. Cirac, and P. Zoller, *Phys. Rev. Lett.* **84**, 2722 (2000).

[7] R. Simon, *Phys. Rev. Lett.* **84**, 2726 (2000).

[8] G. Giedke, B. Kraus, M. Lewenstein, and J. I. Cirac, *Phys. Rev. Lett.* **87**, 167904 (2001);
G. Giedke, L.-M. Duan, J. I. Cirac, and P. Zoller, *Quantum Inf. Comput.* **1**, 79 (2001);
G. Giedke, and J. I. Cirac, *Phys. Rev. A* **66**, 032316 (2002).

[9] R. F. Werner, and M. M. Wolf, *Phys. Rev. Lett.* **86**, 3658 (2001).

[10] V. Giovannetti, S. Mancini, D. Vitali, and P. Tombesi, *Phys. Rev. A* **67**, 022320 (2003).

[11] M. D. Reid, and P. D. Drummond, *Phys. Rev. Lett.* **60**, 2731 (1988);
M. D. Reid, *Phys. Rev. A* **40**, 913 (1989);
M. D. Reid, quant-ph/0112038 (2001).

[12] Z. Y. Ou, S. F. Pereira, H. J. Kimble, and K. C. Peng, *Phys. Rev. Lett.* **68**, 3663, (1992).

[13] A. Furusawa, J. Sørensen, S. Braunstein, C. Fuchs, H. Kimble, and E. Polzik, *Science* **282**, 706 (1998).

[14] Y. Zhang, H. Wang, X. Y. Li, J. T. Jing, C. D. Xie, and K. C. Peng, *Phys. Rev. A* **62**, 023813 (2000).

[15] Ch. Silberhorn, P. K. Lam, O. Weiß, F. König, N. Korolkova, and G. Leuchs, *Phys. Rev. Lett.* **86**, 4267 (2001).

[16] C. Schori, J. Sørensen, and E. Polzik, *Phys. Rev. A* **66**, 033802 (2002).

[17] W. P. Bowen, R. Schnabel, and P. K. Lam, *Phys. Rev. Lett.* **90**, 043601 (2003).

[18] J. Laurat, T. Coudreau, G. Keller, N. Treps, and C. Fabre, *Phys. Rev. A* **70**, 042315 (2004).

[19] S. Spälter, N. Korolkova, F. König, A. Sizmann, and G. Leuchs, *Phys. Rev. Lett.* **81**, 786 (1998).

[20] I. Abram, *Phys. World* **12**(February), 21 (1999).

[21] F. König, M. A. Zielonka, and A. Sizmann, *Phys. Rev. A* **66**, 013812 (2002).

[22] G. Leuchs, and N. Korolkova, *Optics & Photonic News*, February, 64–69 (2002).

[23] E. Schmidt, L. Knöll, D.-G. Welsch, M. Zielonka, F. König, and A. Sizmann, *Phys. Rev. Lett.* **85**, 3801 (2000).

[24] E. Schmidt, L. Knöll, and D.-G. Welsch, *Opt. Commun.* **194**, 393 (2001).

[25] V. V. Kozlov, and M. Freyberger, Colloquium of the DFG Focused Research Program QIV, Bad Honnef, January 2002, Poster presentation.

[26] D. Fujishima, F. Kannari, M. Takeoka, and M. Sasaki, *Opt. Lett.* **28**, 275 (2003).

[27] S. R. Friberg, S. Machida, M. J. Werner, A. Levenson, and T. Mukai, *Phys. Rev. Lett.* **77**, 3775 (1996).

[28] F. König, S. Spälter, I. Shumay, A. Sizmann, T. Fauster, and G. Leuchs, *J. Mod. Opt.* **45**, 2425 (1998).

[29] S. Lorenz, Ch. Silberhorn, N. Korolkova, R. S. Windeler, and G. Leuchs, *Appl. Phys. B* **73**, 855 (2001).

[30] S. Spälter, M. Burk, U. Strößner, A. Sizmann, and G. Leuchs, *Opt. Express* **2**, 77 (1998).

[31] D. Levandovsky, M. Vasilyev, and P. Kumar, *Opt. Lett.* **24**, 43 (1999).
[32] M. Takeoka, D. Fujishima, and F. Kannari, *Opt. Lett.* **26**, 1592 (2001).
[33] M. Werner, *Phys. Rev. A* **54**, R2567 (1996).
[34] M. Werner, and S. Friberg, *Phys. Rev. Lett.* **79**, 4143 (1997).
[35] A. Sizmann, and G. Leuchs, *Progr. Opt.* **34**, 373 (1999).
[36] G. Leuchs, Ch. Silberhorn, F. König, P. K. Lam, A. Sizmann, and N. Korolkova, in: *Quantum Information Theory with Continuous Variables*, S. L. Braunstein, and A. K. Pati (Eds.), Kluwer, Dordrecht (2002).
[37] F. König, B. Buchler, T. Rechtenwald, G. Leuchs, and A. Sizmann, *Phys. Rev. A* **66**, 043810 (2002).
[38] D. B. Horoshko, and S. Ya. Kilin, *Phys. Rev. A* **61**, 032304 (2000).
[39] R.-K. Li, Y. Lai, and B. A. Malomed, Generation of photon-number entangled soliton pairs through interactions, quant-ph/0405138 (2004).
[40] T. Opatrný, N. Korolkova, and G. Leuchs, *Phys. Rev. A* **66**, 053813 (2002).
[41] G. Leuchs, T. C. Ralph, C. Silberhorn, and N. Korolkova, *J. Mod. Opt.* **46**, 1927 (1999).
[42] C. M. Caves, and B. M. Schumaker, *Phys. Rev. A* **31**, 3068 (1985).
[43] M. D. Levenson, R. M. Shelby, A. Aspect, M. D. Reid, and D. F. Walls, *Phys. Rev. A* **32**, 1550 (1985).
[44] H. P. Yuen, and V. Chan, *Opt. Lett.* **8**, 177 (1983).
[45] P. Galatola, L. A. Lugiato, M. G. Porreca, P. Tombesi, and G. Leuchs, *Opt. Commun.* **85**, 95 (1991).
[46] R. M. Shelby, M. D. Levenson, S. H. Perlmutter, R. G. DeVoe, and D. F. Walls, *Phys. Rev. Lett.* **57**, 691 (1986).
[47] H.-A. Bachor, M. D. Levenson, D. F. Walls, S. H. Perlmutter, and R. M. Shelby, *Phys. Rev. A* **38**, 180 (1988).
[48] S. Inoue, and Y. Yamamoto, *Opt. Lett.* **22**, 328 (1997).
[49] O. Glöckl, U. L. Andersen, S. Lorenz, C. Silberhorn, N. Korolkova, and G. Leuchs, *Opt. Lett.* **29**, 1936 (2004).
[50] O. Glöckl, S. Lorenz, C. Marquardt, J. Heersink, M. Brownnutt, C. Silberhorn, Q. Pan, P. van Loock, N. Korolkova, and G. Leuchs, *Phys. Rev. A* **68**, 012319 (2003).
[51] C. Silberhorn, N. Korolkova, and G. Leuchs, *Phys. Rev. Lett.* **88**, 167902 (2002).
[52] U. L. Andersen, O. Glöckl, S. Lorenz, R. Filip, and G. Leuchs, *Phys. Rev. Lett.*, **93**, 100403 (2004).
[53] N. Korolkova, Ch. Silberhorn, O. Glöckl, S. Lorenz, Ch. Marquardt, and G. Leuchs, *Eur. Phys. J. D* **18**, 229 (2002).
[54] S. L. Braunstein, and H. J. Kimble, *Phys. Rev. A*, **61**, 042302 (2000).
[55] X. Li, Q. Pan, J. Jing, J. Zhang, C. Xie, and K. Peng, *Phys. Rev. Lett.* **88**, 047904 (2002).
[56] S. L. Braunstein, and H. J. Kimble, *Phys. Rev. Lett.* **80**, 869 (1998).
[57] P. van Loock, and S. L. Braunstein, *Phys. Rev. A* **61**, 010302(R) (2002).
[58] X. Jia, X. Su, Q. Pan, J. Gao, C. Xie, and K. Peng, Phys. Rev. Letters 93, 250503(2004).
[59] N. Takei, T. Aoki, S. Koike, K. Yoshino, K. Wakui, H. Yonezawa, T. Hiroka, J. Mizuno, M. Takeoka, M. Ban, and A. Furusawa, arXiv:quant-ph/0311056 (2003).

[60] N. Takei, H. Yonezawa, T. Aoki, and A. Furusawa, in: *Trends in Optics and Photonics Series (TOPS)*, Vol. 97, *International Quantum Electronics Conference (IQEC)*, Technical Digest, Postconference Edition (OSA), IPDA6 (2004).
[61] P. van Loock, and S. L. Braunstein, *Phys. Rev. Lett.* **84**, 3482 (2000).
[62] P. van Loock, and S. L. Braunstein, *Phys. Rev. A* **63**, 022106 (2001).
[63] G. Giedke, B. Kraus, M. Lewenstein, and J. I. Cirac, *Phys. Rev. A* **64**, 052303 (2001).
[64] D. M. Greenberger, M. Horne, A. Shimony, and A. Zeilinger, *Am. J. Phys.* **58**(12), 1131 (1990).
[65] D. M. Greenberger, M. Horne, and A. Zeilinger, *Fortschr. Phys. – Progr. Opt.* **48**(4), 243 (2000).
[66] T. Aoki, N. Takei, H. Yonezawa, K. Wakui, T. Hiroka, A. Furusawa, and P. van Loock, *Phys. Rev. Lett.* **91**(8), 080404 (2003).
[67] P. van Loock, and A. Furusawa, *Phys. Rev. A* **67**, 052315 (2003).
[68] H. Yonezawa, T. Aoki, and A. Furusawa, *Nature* **431**, 430 (2004).
[69] J.-W. Pan, D. Bouwmeester, H. Weinfurter, and A. Zeilinger, *Phys. Rev. Lett.* **80**, 3891 (1998).
[70] N. Korolkova, G. Leuchs, R. Loudon, T. C. Ralph, and Ch. Silberhorn, *Phys. Rev. A* **65**, 052306 (2002).
[71] A. S. Chirkin, A. A. Orlov, and D. Yu. Paraschuk, *Kvant. Electron.* **20**, 999 (1993) [*Quantum Electron.* **23**, 870 (1993)].
[72] P. Grangier, R. E. Slusher, B. Yurke, and A. LaPorta, *Phys. Rev. Lett.* **59**, 2153 (1987).
[73] W. P. Bowen, R. Schnabel, H.-A. Bachor, and P. K. Lam, *Phys. Rev. Lett.* **88**, 093601 (2002).
[74] J. Heersink, T. Gaber, S. Lorenz, O. Glöckl, N. Korolkova, and G. Leuchs, *Phys. Rev. A* **68**, 013815 (2003).
[75] V. Josse, A. Dantan, L. Vernac, A. Bramati, M. Pinard, and E. Giacobino, *Phys. Rev. Lett.* **91**, 103601 (2003).
[76] W. P. Bowen, N. Treps, R. Schnabel, and P. K. Lam, *Phys. Rev. Lett.* **89**, 253601 (2002).
[77] O. Glöckl, J. Heersink, N. Korolkova, G. Leuchs, and S. Lorenz, *J. Opt. B* **5**, 492 (2003).
[78] B. Julsgaard, J. Sherson, J. I. Cirac, J. Fiurasek, and E. S. Polzik, arXiv:quant-ph/0410072 (2004).
[79] J. Hald, J. L. Sørensen, C. Schori, and E. S. Polzik, *J. Mod. Opt.*, **47**, 2599 (2000).
[80] A. Kuzmich, and E. S. Polzik, *Phys. Rev. Lett.* **85**, 5639 (2000).
[81] G. G. Stokes, *Trans. Cambridge Phil. Soc.* **9**, 399 (1852).
[82] M. Born, and E. Wolf, *Principles of Optics*, 7th Edn., Cambridge Univ. Press, Cambridge (1999).
[83] J. M. Jauch, and F. Rohrlich, *The Theory of Photons and Electrons*, Addison-Wesley, Reading, MA (1955).
[84] V. Josse, A. Dantan, A. Bramati, M. Pinard, and E. Giacobino *J. Opt. B* **5**, 513 (2003).
[85] K. Bergman, and H. A. Haus, *Opt. Lett.* **16**, 663 (1991).

Index